An Introduction to SOLID STATE PHYSICS and its Applications

R J Elliott

Reader in Theoretical Physics
University of Oxford

A F Gibson

Professor of Physics
University of Essex

A Title in the Nature–Macmillan
Physics Series

MACMILLAN

First published 1974 by
THE MACMILLAN PRESS LTD
London and Basingstoke
Associated companies in New York Dublin
Melbourne Johannesburg and Madras

SNB 333 11023 4

Printed in Great Britain by
WILLIAM CLOWES & SONS LTD., LONDON, COLCHESTER AND BECCLES

Contents

8 Transport 287

Contents

Preface

Coverage and arrangement of the book

The object of this book is to draw together the basic theory of the solid state and examples of the application of solid state properties in device fabrication. As such it should be of interest to a wide range of students in the general area of 'materials science'. We feel that it is important for the two different aspects of modern solid state physics to be studied by all students. The fundamental theory provides an excellent example of the type of synthesis which underlies all modern science. It is one of the great triumphs of the quantum theory that it describes not only simple atomic systems comprising a few elementary particles but also the complex many-body system which is a solid with its wealth of properties. On the other hand, this understanding has provided the stimulus for the production of a wide range of devices exploiting the manifold properties of solids. We feel it is important for the student to see how this is done, partly because it deepens the understanding of the properties of solids, partly because it is a good example of applied science, and partly because of the actual technological importance of these devices.

However, it is clear that courses which approach the subject from different points of view will emphasise different aspects of the subject. While we would stress the importance of studying both the applied and fundamental sides of solid state physics the book is so constructed that it may be used to emphasise either aspect. Chapters 1–5 cover the basic theory of the fundamental properties of solids emphasising only those experimental results and techniques which give the most complete and direct information and with only occasional digressions into applications. Chapters 6, 7, 8 and 10 describe respectively the optical and microwave properties, transport and magnetic properties of solids, which are of importance in applications. Chapter 9 is concerned with semiconductor junction devices.

To assist the reader we have classified all sections under five heads, marked in the List of Contents as a, A, d, D and AD. These represent the following rough divisions:

- a fundamental theory and the properties of solids;
- d properties, processes and structures of importance in the applications of solid state physics;
- A as a, but either more advanced or more specialised topics;
- D as d, but either more advanced or more specialised topics;
- AD advanced or specialised topics of importance both to the theory and applications of solid state physics.

We would regard:

a + d as a basic solid state physics course suitable for materials scientists;
a + A + AD as suitable for the pure physicist;
a + d + D + AD as a course on applied solid state physics for applied physicists and electrical engineers.

Though some sections designated a or d do refer back to sections in the more advanced categories we would recommend that these sections be studied first and matter discussed in the advanced sections be initially taken on trust.

Units

All equations are written in rationalised (MKSA or SI) units. However, when referring to numerical values in the text or in diagrams we have used whatever units seem most appropriate to the quantity described. Thus lengths are frequently quoted in cm and optical absorption coefficients in cm^{-1}, since this is by far the most widely used convention.

Glossary of symbols

In a subject as extensive as solid state physics there are so many descriptive and measurable quantities that there are not enough ordinary symbols to go round. Moreover, several accepted conventions have grown up in different parts of the subject which use the same symbol for quite different things. We have attempted to maintain most of the conventions, while reducing the duplication of symbols as far as possible. This inevitably leads to some overlap. In the list below, when a symbol is used twice, we have indicated the chapters in brackets. In addition to those listed, a number of symbols appear briefly in the text to simplify the discussion, but these are always defined on the spot. Vector quantities are given in bold type. The lengths of the vectors are denoted by the same symbol in ordinary type.

a	lattice spacing
a	sundry constants
$\boldsymbol{a}_1, \boldsymbol{a}_2, \boldsymbol{a}_3$	basic lattice vectors
a_{at}	atomic radius
a_c	cyclotron orbit radius
a_{ex}	exciton radius
a_I	impurity radius
a_i	state amplitude
a_0	Bohr radius
A	Einstein coefficient
$\mathbf{A}$	vector potential
A'	sundry constants
$\mathcal{A}$	cross-sectional area

$\mathbf{b}$	Burger's vector
b	(3) electric-vibrational coupling
$\boldsymbol{b}_1, \boldsymbol{b}_2, \boldsymbol{b}_3$	basic reciprocal lattice vectors
B	Einstein coefficient
B'	sundry constants
$\mathbf{B}$	magnetic field or induction
$\mathrm{B_i}$	demagnetising field
$\mathbf{B}_0$	applied magnetic field
$\mathbf{B}_c$	critical field
$\mathscr{B}$	Brillouin function
c	velocity of light
c	(1) interatomic spacing (hcp lattice)
c	concentration
C	capacitance
C	(6) Einstein coefficient
C	(10) Curie constant
C_V	specific heat at constant volume
C_{11}, C_{44}	elastic constants
d	various distances
D	diffusion constants
D_e	diffusion constant of electrons
D_h	diffusion constant of holes
D^*	ambipolar diffusion constant
D_2	spin hamiltonian constant
$\mathbf{D}$	electric displacement
$\mathscr{D}$	area of orbit
e	electronic charge
e_p	piezoelectric constant
$\mathrm{E}(\omega)$	energy density
$E(\boldsymbol{k})$	energy of particle or wave
E_A, etc.	various energies depending on suffix
E_F	Fermi energy
E_g	gap energy
E_H	ionisation energy of hydrogen (Rydberg)
$\langle E \rangle$	total energy
f	form factor
f	oscillator strength
$\mathscr{f}$	Fermi–Dirac distribution function
$\mathscr{f}_A, \mathscr{f}_D$	occupied fractions
F	(1) periodic function
F	(5) impurity wave function
$\mathbf{F}$	electric field (this appears with various suffices)
$\mathscr{F}$	free energy
g	level degeneracy
g	spectroscopic splitting factor
$\mathrm{g_e, g_h}$	spectroscopic splitting factor, electrons and holes
$g_n, g(\boldsymbol{Q})$	Fourier coefficients

$g(\mathrm{T})$	Boltzmann distribution factor
G	(8) electron–phonon coupling factor
G	various constants
G_s	shear modulus
h	Planck's constant, $\hbar = h/2\pi$
h_{11} etc.	(9) transistor parameters
$\mathbf{H}$	magnetic field
$\mathscr{H}$	hamiltonian
i	square root of minus one
I	radiation intensity
I	(2 only) current
$\mathscr{I}$	Ising interaction
$\boldsymbol{j}$	total angular momentum of electron
j	electron current
$\mathrm{j_B}$	base current
$\mathrm{j_C}$	collector current
$\mathrm{j_D}$	drain current
$\mathrm{j_E}$	emitter current
$\boldsymbol{J}$	total atomic angular momentum
J	current density
$\mathrm{J_c}$	critical current density
$\mathrm{J_e}$	current density due to electron flow
$\mathrm{J_h}$	current density due to hole flow
$\mathscr{J}$	exchange interaction
$\boldsymbol{k}$	wave vector of excitation, usually electrons
$\boldsymbol{k}$	(10) wave vector of spin waves
k_F	Fermi wave vector
$\mathrm{k_0}$	extinction coefficient
k	Boltzmann's constant
$\boldsymbol{K}, \boldsymbol{K}'$	wave vector of incident and scattered particles
$\boldsymbol{K}_{ex}$	wave vector of excitons
K	absorption coefficient
$\mathscr{K}$	thermal conductivity
$\boldsymbol{l}$	angular momentum
l_f	mean free path
ℓ	section length
ℓ_N, ℓ_P, etc.	space charge lengths in PN junctions
$\boldsymbol{L}$	total angular momentum
L	(8) Lorentz number
L	(2) inductance
$\mathrm{L_e}, \mathrm{L_h}$	diffusion lengths, electrons and holes
$\mathscr{L}$	specimen length
m	electron mass
m^*	effective mass
m_e^*	effective mass of electrons
m_h^*	effective mass of holes
m_r	reduced effective mass

m	integer
$\mathscr{m}$	magnetic dipole moment
M	atomic mass
M_m	mass of melt
M_r	reduced atomic mass
$\mathbf{M}$	magnetisation density
$\mathscr{M}$	total magnetic moment
n	(3) atom number
n	(10) thickness of domain wall
n_i	intrinsic electron density
n_0	equilibrium electron density in P type material
$\mathrm{n_0}$	refractive index
$\mathrm{n, n_i, n_x}$	integers
$\mathscr{n}$	phonon occupation number
N	electron density
N_A	density of acceptors
N_D	density of donors
N_I	density of impurities
N_0	equilibrium electron density in N type material
N_s	density of superelectrons
N	total number of electrons
$\mathscr{N}$	number of unit cells
$\mathscr{N}'$	number of interstitial sites
$\mathscr{N}_0$	number of atoms
$\mathscr{N}_V$	number of vacancies
$\boldsymbol{p}$	momentum
p_0	equilibrium hole density in N type material
p	number of atoms per cell
P	density of holes
P_0	equilibrium density of holes in P type material
$\mathbf{P}$	polarisation density
$\mathscr{P}_T$	tunnelling probability
$\boldsymbol{q}$	phonon wave vector
$\boldsymbol{q}$	(5) fourier transform
q	charge on Cooper pair
Q	(5) displacement co-ordinate
Q	quality factor
$\mathbf{Q}$	reciprocal lattice vector
$\mathscr{Q}$	charge per unit area
$\boldsymbol{r}$	position co-ordinate
r_s	characteristic length
$\mathrm{r_0, r_{0e}, r_{0h}}$	(8) numbers of order unity
R	interatomic distance
$\boldsymbol{R}$	atomic position
R_H	Hall coefficient
R_{ij}	dipole matrix element
$\mathbf{R, R'}$	position of incident and scattered particles

R	resistance
$\mathscr{R}$, $\mathscr{R}_n$, $\mathscr{R}_p$	reflection coefficients
$\mathscr{R}'$	reflected fraction
$\boldsymbol{s}$	spin angular momentum
s	velocity of sound
S	total spin angular momentum
S_f	structure factor
S_0	segregation coefficient
S, S_n, S_s, etc.	entropy
S(E)	density of states
S_v	(9) surface recombination velocity
$\mathscr{S}$	effective spin
t	time variable
t_0	time constant
t_p	deposition time
t_r	transit time, bipolar transistor
T	temperature
T_c	critical or transition temperature
$\boldsymbol{u}$	atomic displacement
U	electronic repulsion energy
U	periodic part of wave function
$\mathscr{U}$	thermal current density
$\boldsymbol{v}$	group velocity of waves or particles
$\boldsymbol{v}_d$	drift velocity
v_e	electron velocity
v_F	Fermi velocity
v_h	hole velocity
V	potential energy
V_{at}	potential energy in atom
V	voltage or potential (appears in (9) with different suffices)
$\mathscr{V}$	crystal volume
$\boldsymbol{w}$	magnetic order vector
w	(4, 8) characteristic length in plasma
w	(9) distance between boundaries
$w(\boldsymbol{r})$	periodic function
W	reduced length
W	incident power per unit area
W_n^i	(10) energy level in magnetic field
x	displacement co-ordinate
X	(4) electron displacement
X	(9) reduced length
X_x	elastic force
y	displacement co-ordinate
Y_x	elastic force
z	displacement co-ordinate
Z	impurity charge
Z	impedance

α	(3) atomic polarisability
α	(8) attenuation coefficient
α	(9) transistor current gain factor
$\alpha_k(\mathbf{q})$	(5) expansion coefficient
α_p	polaron parameter
α_s, α_1, $\alpha(\mathbf{r})$	scattering lengths
$\boldsymbol{\beta}$	photon wave vectors
β	(9) bipolar transistor base transport factor
γ	(3) Gruneison constant
γ	(6) damping constant
γ	(4, 8) coefficient in electronic specific heat
γ	(9) bipolar transistor emitter efficiency
Γ	molecular field constant
$\Gamma(\eta)$	gamma function
δ	phase angle
δ_m	demagnetising factor
$\boldsymbol{\delta}$	nearest neighbour distance
Δ	energy width
Δ	energy splitting
ε	(§ 8.9 only) superconductor energy gap
ε	dielectric constant
ε_0	permittivity of free space
ε_1, ε_2	real and imaginary part of dielectric constant
ε_L	lattice contribution to dielectric constant
ζ	(5, 10) spin-orbit coupling constant
ζ	(6) constant
η	sundry constants
θ	sundry angles
Θ	Curie–Weiss constant
Θ_D	Debye temperature
κ	(8) constant in superconductivity
κ	screening length
κ_B	bulk modulus
$\boldsymbol{\kappa}$	difference of wave vectors
λ	wavelength
λ	(8) penetration depth in superconductors
λ_g	Landé g-factor
λ_s	wavelength of sound wave
λ_1, λ_2, λ_3	(3) expansion parameters
Λ	force constant
Λ', Λ''	anharmonic force constants
μ	(8) charge carrier mobility
μ	(10) magnetic permeability
μ_B	Bohr magneton
μ_e, μ_h	electron, hole mobility
μ_{eff}	effective moment
μ_0	permeability of free space

μ^*	(9) ambipolar mobility
ν	number of electrons per cell
ν_1, ν_2, ν_3	integers
ν_{12}, etc.	transition rates
ξ	(3, 5) vibrational amplitude
ξ, ξ_0	(8) coherence length
$\xi_0\ \xi_T$	(5) mean amplitude
Ξ	stress
ρ	resistivity
$\rho(\boldsymbol{r})$	charge density
σ	electrical conductivity
σ	(5) ionic conductivity
σ_d, σ_{ol}	scattering cross section
τ	mean free time
τ_e	electron lifetime
τ_h	hole lifetime
τ_0	carrier lifetime in junction
τ_c	time between collisions
τ_D	electron distribution relaxation time
τ_E	energy relaxation time
τ_F	electron relaxation time at Fermi energy
ϕ	various angles
ϕ_A, ϕ_S, etc	electron wave functions
Φ	(6) photon flux
Φ	(8) magnetic flux
χ	magnetic susceptibility
χ_1, χ_2	(6) polarisation constants; (8) thermoelectric constants
ω	frequency of waves
ω_c	cyclotron frequency
ω_l	precession frequency
ω_m	maximum frequency
ω_0	optic mode frequency
ω_p	plasma frequency
$\omega(\boldsymbol{q})$	phonon frequency
ω_L	longitudinal optic mode frequency
ω_T	transverse optic mode frequency
Ω	volume of unit cell

Periodic table of the elements

The figure given by the symbol for each element gives the total number of electrons in the neutral atom. The electronic configuration of each element in its lowest energy state is given. The superscript denotes the number of electrons in each single electron state as specified by the principal quantum number n 1–7 and the symbol appropriate to the angular momentum quantum number l (s, p, d, f, for $l = 0, 1, 2, 3$). The closed shell configuration given at the beginning of each row occurs for all elements in that row. The rare earths, in which the 4f shell is filling occur between La (element 57) and Hf (element 72).

PERIODIC TABLE

Closed shells	I Alkalis	II Alkaline earths	Transition elements										III	IV	V	VI	VII Halogens	VIII Rare gases	Closed shells
	H^{1} $1s$																	He^{2} $1s^2$	$1s^2$
$1s^2$	Li^{3} $2s$	Be^{4} $2s^2$											B^{5} $2s^2 2p$	C^{6} $2s^2 2p^2$	N^{7} $2s^2 2p^3$	O^{8} $2s^2 2p^4$	F^{9} $2s^2 2p^5$	Ne^{10} $2s^2 2p^6$	$2s^2 2p^6$
$2s^2 2p^6$	Na^{11} $3s$	Mg^{12} $3s^2$											Al^{13} $3s^2 3p$	Si^{14} $3s^2 3p^2$	P^{15} $3s^2 3p^3$	S^{16} $3s^2 3p^4$	Cl^{17} $3s^2 3p^5$	Ar^{18} $3s^2 3p^6$	$3s^2 3p^6$
$3s^2 3p^6$	K^{19} $4s$	Ca^{20} $4s^2$	Sc^{21} $3d$ $4s^2$	Ti^{22} $3d^2$ $4s^2$	V^{23} $3d^3$ $4s^2$	Cr^{24} $3d^4$ $4s^2$	Mn^{25} $3d^5$ $4s^2$	Fe^{26} $3d^6$ $4s^2$	Co^{27} $3d^7$ $4s^2$	Ni^{28} $3d^8$ $4s^2$	Cu^{29} $3d^{10}$ $4s$	Zn^{30} $3d^{10}$ $4s^2$	Ga^{31} $4s^2 4p$	Ge^{32} $4s^2 4p^2$	As^{33} $4s^2 4p^3$	Se^{34} $4s^2 4p^4$	Br^{35} $4s^2 4p^5$	Kr^{36} $4s^2 4p^6$	$4s^2 3d^{10} 4p^6$
$4s^2 3d^{10} 4p^6$	Rb^{37} $5s$	Sr^{38} $5s^2$	Y^{39} $4d$ $5s^2$	Zr^{40} $4d^2$ $5s^2$	Nb^{41} $4d^4$ $5s$	Mo^{42} $4d^5$ $5s$	Tc^{43} $4d^6$ $5s$	Ru^{44} $4d^7$ $5s$	Rh^{45} $4d^8$ $5s$	Pd^{46} $4d^{10}$	Ag^{47} $4d^{10}$ $5s$	Cd^{48} $4d^{10}$ $5s^2$	In^{49} $5s^2 5p$	Sn^{50} $5s^2 5p^2$	Sb^{51} $5s^2 5p^3$	Te^{52} $5s^2 5p^4$	I^{53} $5s^2 5p^5$	Xe^{54} $5s^2 5p^6$	$5s^2 4d^{10} 5p^6$
$5s^2 4d^{10} 5p^6$	Cs^{55} $6s$	Ba^{56} $6s^2$	La^{57} $5d$ $6s^2$	Hf^{72} $5d^2$ $6s^2$	Ta^{73} $5d^3$ $6s^2$	W^{74} $5d^4$ $6s^2$	Re^{75} $5d^5$ $6s^2$	Os^{76} $5d^6$ $6s^2$	Ir^{77} $5d^9$	Pt^{78} $5d^9$ $6s$	Au^{79} $5d^{10}$ $6s$	Hg^{80} $5d^{10}$ $6s^2$	Tl^{81} $6s^2 6p$	Pb^{82} $6s^2 6p^2$	Bi^{83} $6s^2 6p^3$	Po^{84} $6s^2 6p^4$	At^{85} $6s^2 6p^5$	Rn^{86} $6s^2 6p^6$	$6s^2 4f^{14} 5d^{10} 6p^6$
$6s^2 4f^{14} 5d^{10} 6p^6$	Fr^{87} $7s$	Ra^{88} $7s^2$	Ac^{89} $6d$ $7s^2$																

Ce^{58} $4f^2$ $6s^2$	Pr^{59} $4f^3$ $6s^2$	Nd^{60} $4f^4$ $6s^2$	Pm^{61} $4f^5$ $6s^2$	Sm^{62} $4f^6$ $6s^2$	Eu^{63} $4f^7$ $6s^2$	Gd^{64} $4f^7$ $6s^2 5d$	Tb^{65} $4f^8$ $6s^2 5d$	Dy^{66} $4f^{10}$ $6s^2 5d$	Ho^{67} $4f^{11}$ $6s^2$	Er^{68} $4f^{12}$ $6s^2$	Tm^{69} $4f^{13}$ $6s^2$	Yb^{70} $4f^{14}$ $6s^2$	Lu^{71} $4f^{14}$ $6s^2 5d$	Rare earths
Th^{90} $7s^2 6d^2$	Pa^{91} $5f^2$ $7s^2 6d$	U^{92} $5f^3$ $7s^2 6d$	Np^{93} $5f^4$ $7s^2$	Pu^{94} $5f^6$ $7s^2$	Am^{95} $5f^7$ $7s^2$	Cm^{96} $5f^7$ $7s^2 6d$	Bk^{97}	Cf^{98}	Es^{99}	Fm^{100}	Md^{101}	No^{102}	Lw^{103}	Actinides

1

Crystal structure

1.1 Introduction

Helium is the only material which is not solid under normal conditions at some sufficiently low temperature and it is solidified by pressure. Indeed most materials are solid up to the ambient temperature and beyond. The study and understanding of the solid state is therefore one of the most important tasks of physics. Since the advent of quantum mechanics in the late 1920s, progress towards this end has been extremely rapid and most phenomena in solids are now qualitatively understood even though they may be too complex to be dealt with quantitatively. This understanding of the properties of solids is now sufficient for them to be manipulated in the construction of a large range of practical solid state devices.

A solid is essentially an ordered array of atoms, bound together by electrical forces to form what is effectively a very large molecule. Unlike the small molecules of chemistry, however, the size of a solid may be arbitrarily large, the constituent units being repeated any number of times. The surface of such a system will be different from the interior, but if the array is large enough the most obvious properties of the material will be determined by the atoms in the bulk of the solid and not by the nature of the surface. To be sure, some properties will be determined by the surface states but these require special attention.

The number of constituent particles in a typical solid is very large. In a cubic centimetre there may be as many as 10^{23} nuclei and 10^{24} electrons. At first sight it would seem impossible to study effectively such a large number of interacting particles theoretically. The problem is, however, immensely simplified by the high symmetry of a solid. By contrast, in a liquid where long range order does not exist, the theory is much more complicated; indeed there is no theoretical description of liquids comparable with that of solids. Some materials like glasses are solid-like in that the atoms cannot diffuse easily as in a liquid. On the other hand, they do not have long-range order and hence cannot be described by a simple theory based on symmetry. They are sometimes described as supercooled liquids and are normally only in metastable equilibrium. The true equilibrium state of the system is an ordered solid at low temperatures but the time required to reach this state is enormously long.

Another simplifying feature is that we are mainly concerned with states of the system with low energy. The slow motion of the heavy nuclei can usually be separated from the much more rapid motion of the light electrons and on the energy scale which is usually of interest, the internal motions within the nuclei are irrelevant and the nuclei may be regarded as immutable particles without internal structure.

1.2 Translational symmetry (a)

Even before the microscopic structure of crystals had been investigated it was believed that the atoms of a solid were arranged in a regular array. The way crystals grow and cleave strongly suggests such a possibility and indeed the relationship between the microscopic symmetry and the macroscopic properties of crystals has been exploited over thousands of years to make jewels. The shapes of the gems reflect the planar structure of the atomic arrays.

The symmetry of a crystal is best defined in terms of an ideal crystal composed of atoms arranged in an infinite regular array. Such an array implies *translational symmetry*. This is an ideal, for real crystals have surfaces which disrupt the arrangement, and they will contain imperfections as well. (These imperfections are of great importance in determining certain properties of crystals and will be studied in Chapter 5.) The ideal crystal can be made up of a set of polyhedral blocks which fit exactly together. Each block is identical and may contain several nuclei, each with an associated cloud of electrons. It is called the unit cell and may be conveniently chosen in a number of different ways.

Transitional symmetry is defined in terms of three basic vectors $\boldsymbol{a}_1$, $\boldsymbol{a}_2$ and $\boldsymbol{a}_3$, which must not be coplanar. If a translation is made through any vector which is the sum of integral multiples of these basic vectors, the crystal appears unchanged. The end points of such vectors given by

$$\boldsymbol{R}_l = n_1 \boldsymbol{a}_1 + n_2 \boldsymbol{a}_2 + n_3 \boldsymbol{a}_3 \tag{1.1}$$

form the space lattice. A property such as charge density is the same at all points $\boldsymbol{r} + \boldsymbol{R}_l$ as it is at $\boldsymbol{r}$, i.e. it is periodic in space.

As well as translational symmetry, there is also symmetry associated with various rotations, reflections and similar operations which leave the crystal apparently unchanged. Such operations must of course leave the space lattice (1.1) unchanged and in addition must leave the repeating unit, the unit cell, unchanged. The full symmetry of the lattice is given by the space group, which consists of the very large number of translational symmetry operations as well as the (relatively few) rotational symmetry operations. It is the translational symmetry with its insistence on the equivalence of many points in the crystal which has the most important influence on the properties of crystals.

The unit cell is not, of course, unique. One obvious way to define it is as the parallelepiped bounded by the basic vectors $\boldsymbol{a}_1$, $\boldsymbol{a}_2$ and $\boldsymbol{a}_3$. Even then the midpoint of the cell can be arbitrarily chosen although it is usual to centre it

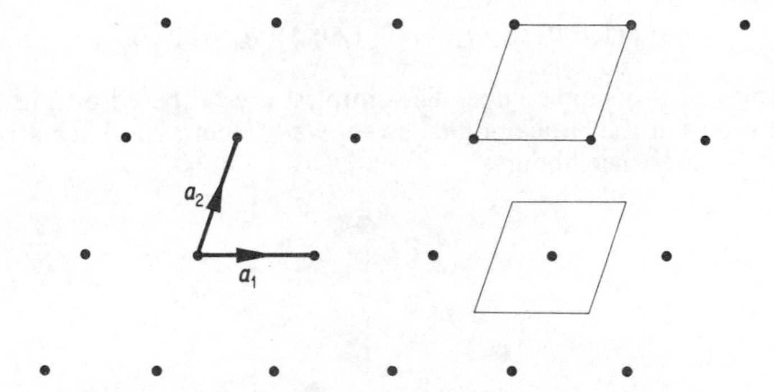

Figure 1.1 General two-dimensional space lattice showing basic vectors, and two forms of the unit cell as a parallelogram with different centres

on one of the atomic positions. Another convenient way of choosing the unit cell is to take the volume bounded by the planes which perpendicularly bisect the nearest $\boldsymbol{R}_l$. Examples of these two methods of choosing the unit cell are shown in figures 1.1 and 1.2 for two-dimensional lattices.

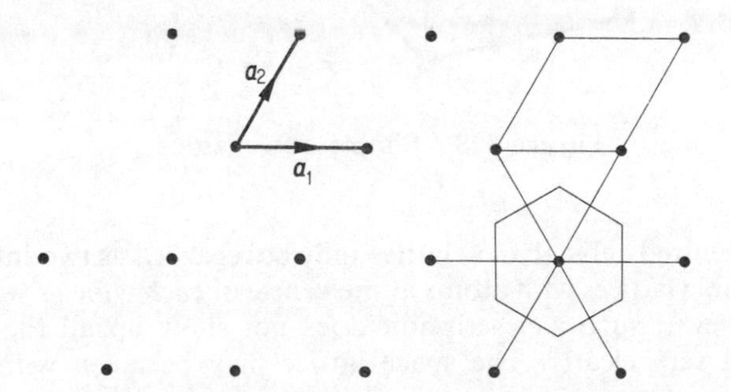

Figure 1.2 Triangular plane lattice. $|\boldsymbol{a}_1| = |\boldsymbol{a}_2|$ and the angle between them is $\pi/3$. Two forms of the unit cell are shown—one as a parallelogram and the other as a hexagon obtained by drawing perpendicular bisectors to each lattice vector from a central site

The full use of the symmetry properties of a crystal is best made with the theory of groups, but this is really outside the scope of this book, and further reference may be made to books such as Heine's *Group Theory in Quantum Mechanics* (1960). Most of the essential consequences of symmetry, however, may be derived from common-sense arguments without the use of the formal theory.

1.3 Simple lattices (a)

The most highly symmetrical lattices which occur naturally are cubic. They are therefore of some practical interest and they also provide useful simple examples which help in visualising the more general case. The simple cubic lattice has basic vectors

$$\boldsymbol{a}_1 = a(1,0,0) \qquad \boldsymbol{a}_2 = a(0,1,0) \qquad \boldsymbol{a}_3 = a(0,0,1) \tag{1.2a}$$

and the unit cell is a simple cube. The simplest crystal based on this lattice has single atoms at the lattice points, as shown in figure 1.3. Each atom has six identical nearest neighbours.

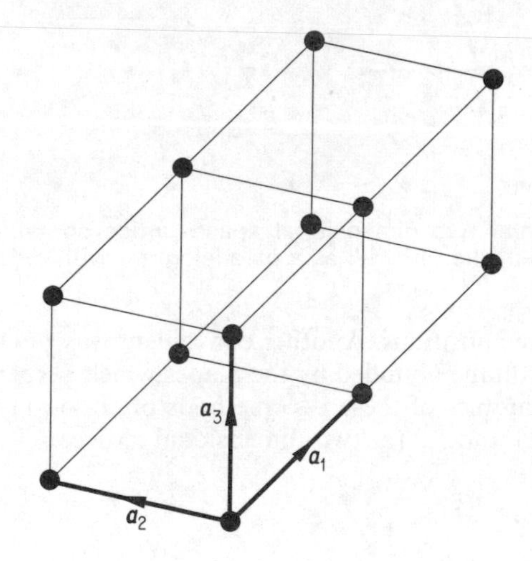

Figure 1.3 Simple cubic lattice

The body-centred cubic (b.c.c.) lattice may be regarded as two interpenetrating simple cubic lattices with atoms at the centre of each cube as well as at the corners. However, such a description does not show up all the symmetry of the crystal very clearly. The space lattice may be taken with the basic vectors

$$\boldsymbol{a}_1 = \tfrac{1}{2}a(-1,1,1) \qquad \boldsymbol{a}_2 = \tfrac{1}{2}a(1,-1,1) \qquad \boldsymbol{a}_3 = \tfrac{1}{2}a(1,1,-1) \tag{1.2b}$$

and a better choice for the unit cell may be the parallelepiped defined by these vectors, or the truncated octahedron constructed as shown in figure 1.4. For the simplest crystal, which has a single atom at each lattice point, these unit cells contain only one atom. If we had continued to regard this crystal as two interpenetrating simple cubic lattices the space lattice would have contained

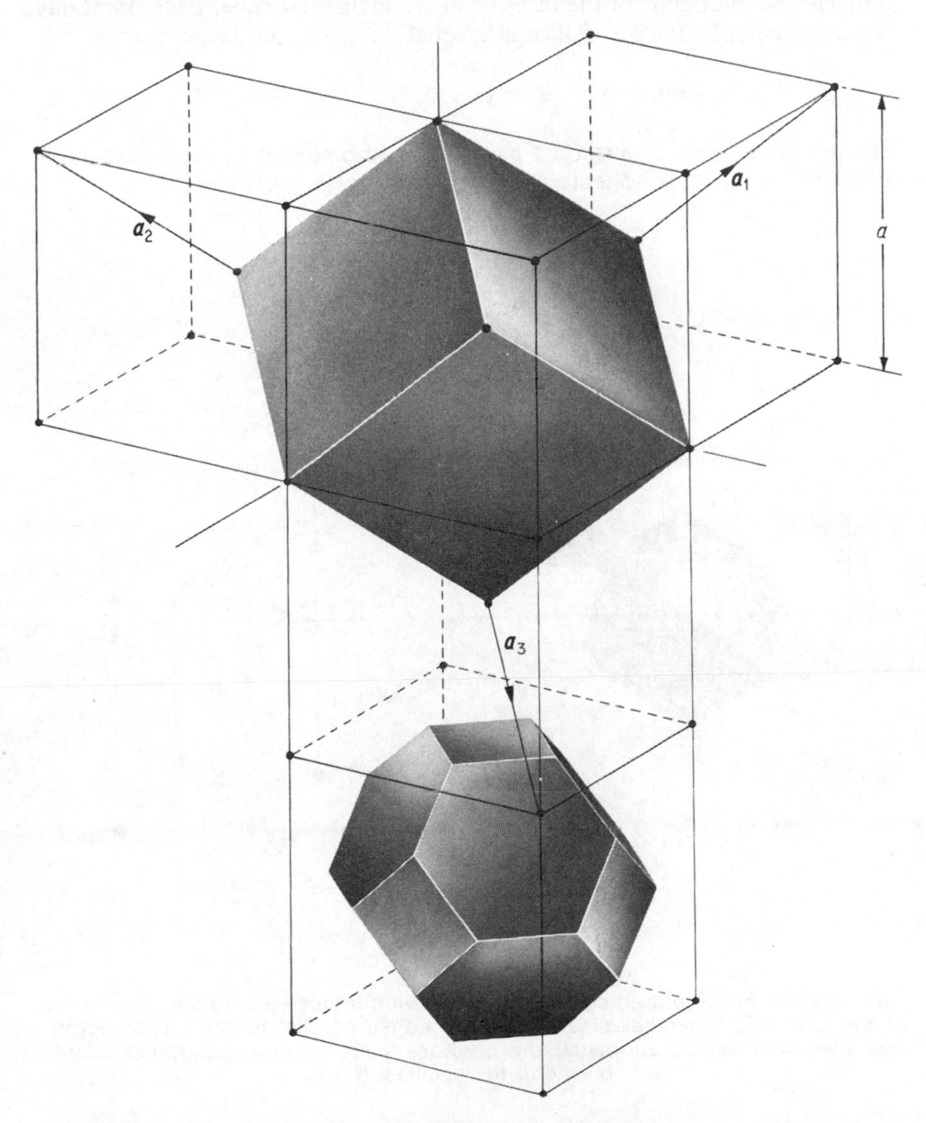

Figure 1.4 Body-centred cubic lattice, showing basic vectors, and two forms of the unit cell, a parallelepiped with the basic vectors as edges and a truncated octahedron where the faces are the planes perpendicularly bisecting the smallest $\boldsymbol{R}_l$

only half the number of translation vectors and the cubic unit cell would have contained two atoms. Thus that description, while sometimes convenient, does not show the full symmetry of the lattice.

The face-centred cubic lattice (f.c.c.) can be considered as four interpenetrating simple cubic lattices giving a cubic unit cell with extra lattice points at the midpoints of the faces of the fundamental cube. Each point has 12 nearest neighbours. The full translational symmetry has basic vectors.

$$\boldsymbol{a}_1 = \tfrac{1}{2}a(0, 1, 1) \qquad \boldsymbol{a}_2 = \tfrac{1}{2}a(1, 0, 1) \qquad \boldsymbol{a}_3 = \tfrac{1}{2}a(1, 1, 0) \tag{1.2c}$$

The primitive unit cell is again a parallelepiped bounded by these vectors or alternatively the rhombohedron shown in figure 1.5.

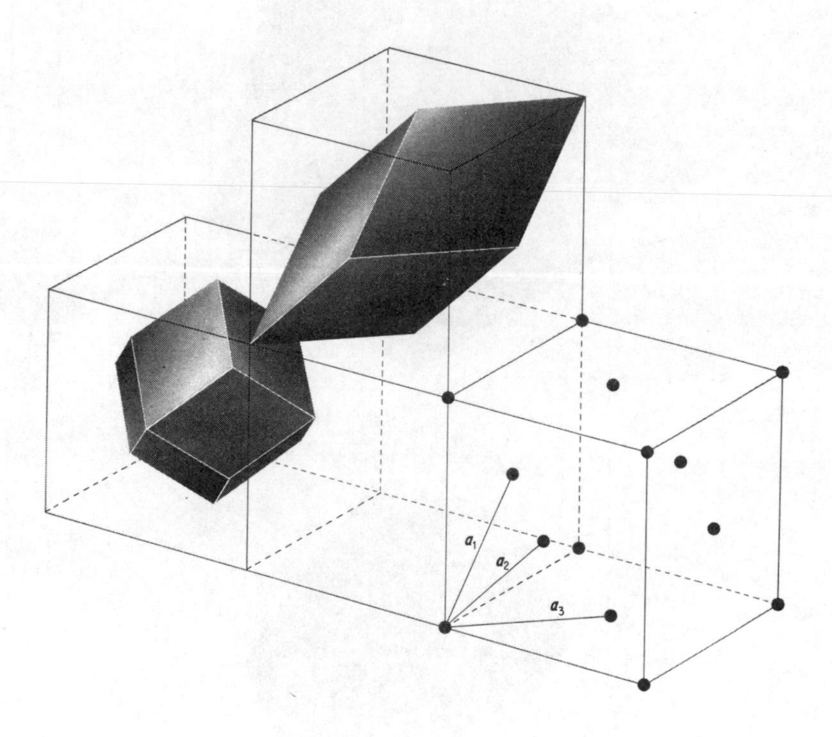

Figure 1.5 Face-centred cubic lattice, showing lattice vectors and two forms of the unit cell, a parallelepiped, with the edges parallel to the basic vectors and a regular rhombic duodecahedron, whose faces are planes perpendicularly bisecting the smallest $\boldsymbol{R}_l$

The crystal structures so far described have a single atom in each unit cell. More complicated crystals can be built with the same space lattices but more complex cells. For example the sodium chloride structure is shown in figure 1.6. It looks like a simple cubic lattice with alternate atoms of different types. On examination, however, we see that the basic vectors which give a repeating structure are $\frac{1}{2}a(0, 1, 1)$, etc., so that the space lattice is again face-centred cubic. There are now two atoms in each primitive unit cell.

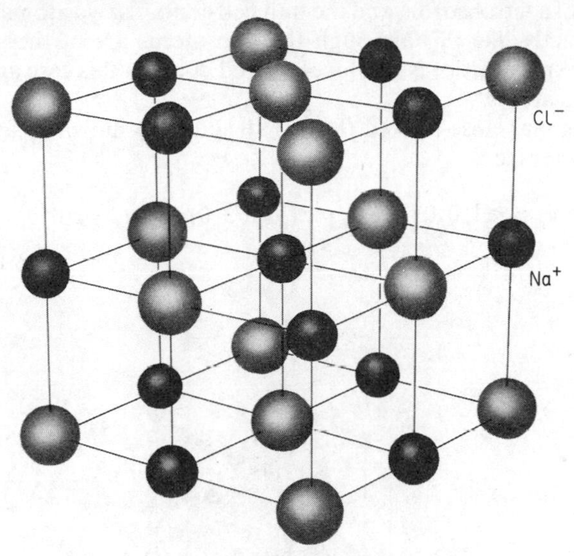

Figure 1.6 The sodium chloride structure—note that the space lattice is face-centred cubic, since each type of ion forms such a lattice

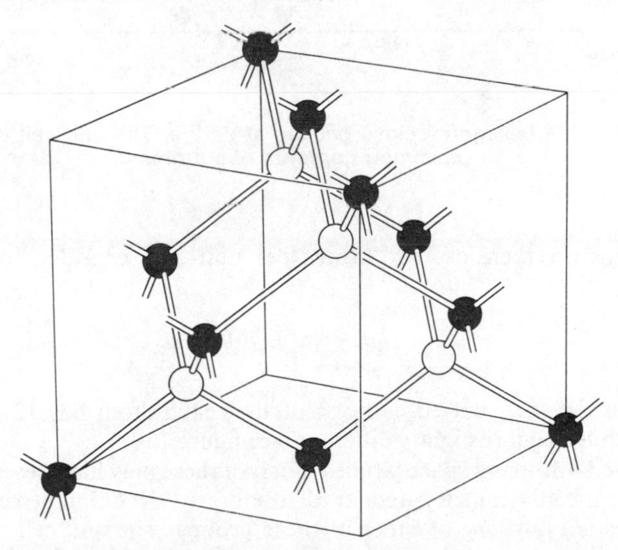

Figure 1.7 The blende structure; again each type of atom forms a face-centred cubic lattice. Diamond is like this but both atoms are the same chemically. Note the tetrahedral arrangement of bonds

Another example is the zinc blend structure of figure 1.7. Here again there are two face-centred cubic lattices but they are arranged rather differently. The space lattice remains face-centred cubic and the unit cell has two atoms in it. Each atom has four neighbours of the other type arranged around it at the corners of a tetrahedron, and the unit cell contains two atoms. The diamond lattice is exactly like this although the two atoms are of the same chemical species. The space lattice is still face-centred cubic in this case and the unit cell contains two atoms.

The hexagonal close-packed (h.c.p.) structure is shown in figure 1.8. Here the basic vectors are

$$\boldsymbol{a}_1 = a(1,0,0) \qquad \boldsymbol{a}_2 = a(\tfrac{1}{2},\tfrac{1}{2}\sqrt{3},0) \qquad \boldsymbol{a}_3 = c(0,0,1) \tag{1.3}$$

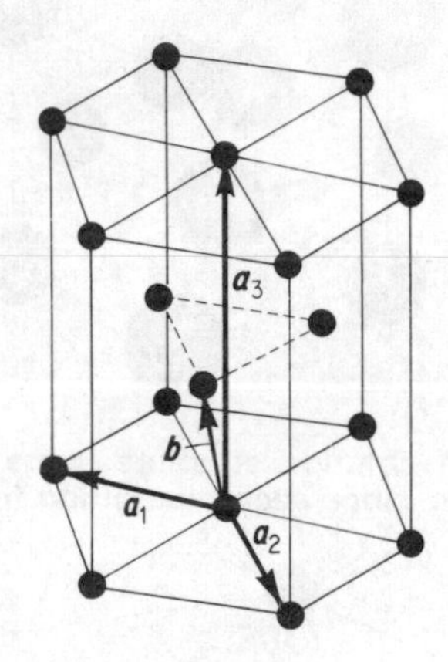

Figure 1.8 The hexagonal close packed structure. The unit cell is a rhombic prism and contains two atoms

In this structure there are two atoms per unit cell separated by the vector

$$(\tfrac{1}{2}a,\ \tfrac{1}{2}a/\sqrt{3},\ \tfrac{1}{2}c) \tag{1.4}$$

Here, as in the face-centred cubic structure, each atom has 12 neighbours, but the arrangement is slightly different (see figure 1.9).

There are 14 different space lattices. Each of these may have several different types of unit-cell symmetry and in all there are 231 different space groups. All crystals fall into one or other of these groups. The unit cell may be very complicated. For example yttrium iron garnet has the chemical formula $Y_3Fe_5O_{12}$ and there are four molecules, that is 80 atoms, in a unit cell although the space lattice is body-centred cubic.

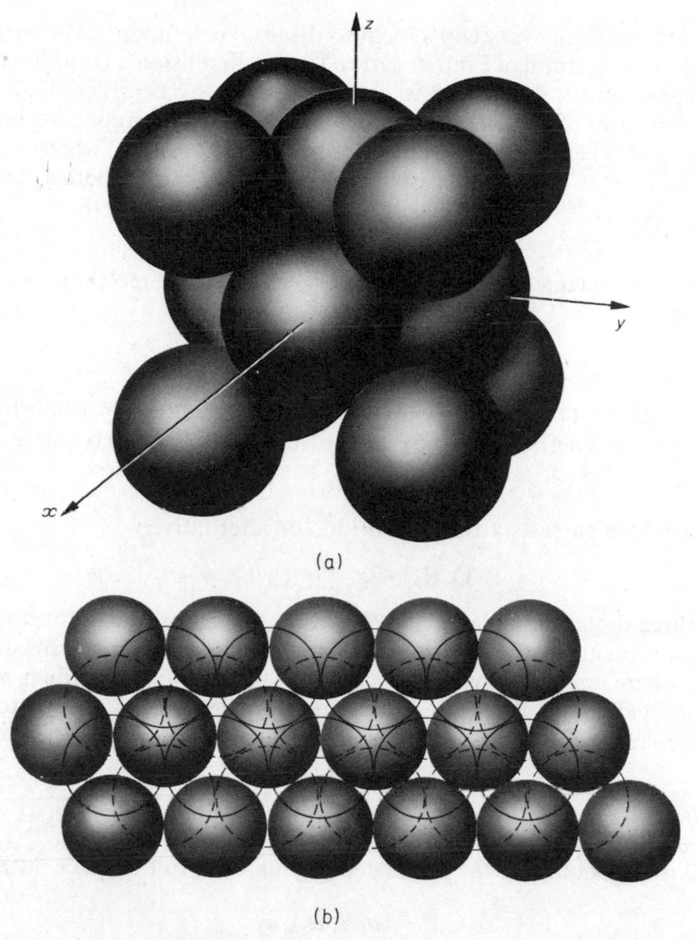

Figure 1.9 (a) Close packed spheres in an f.c.c. arrangement (showing cartesian axes)
(b) Section through (1,1,1) plane of an f.c.c. crystal showing as open and dashed circles the planes above and below respectively. An h.c.p. crystal has this form but with the same type of plane above and below

1.4 The reciprocal lattice (a)

Because of the translational symmetry of a crystal, various properties, for example the electron density or the electrostatic potential, will be the same in each unit cell. These properties are described by a multiply-periodic function which satisfies the condition

$$F(\boldsymbol{r} + \boldsymbol{R}_l) = F(\boldsymbol{r}) \tag{1.5}$$

for all points $\boldsymbol{r}$ in space and for all vectors $\boldsymbol{R}_l$ of the space lattice.

It is mathematically very convenient to discuss such functions in terms of an extension of the method of Fourier series. In one dimension a periodic function which repeats every distance d is defined by $F(x+d) = F(x)$. This can be expanded in terms of a set of functions which have the same repeating property. These are $\sin(\alpha x)$ and $\cos(\alpha x)$ or more conveniently $\exp(i\alpha x)$ where $\alpha = 2n\pi/d$ and n is an integer. Then $F(x)$ can be expanded in a Fourier series,

$$F(x) = \sum_{n} g_n \exp(2n\pi i x/d) \tag{1.6}$$

where n is an integer. A similar property holds in three dimensions, and for F as defined in (1.5) we may write

$$F(\boldsymbol{r}) = \sum_{\mathbf{Q}} g(\mathbf{Q}) \exp(i\mathbf{Q}.\boldsymbol{r}) \tag{1.7}$$

The vectors $\mathbf{Q}$ must be such that $\exp(i\mathbf{Q}.\boldsymbol{r})$ is itself a multiply-periodic function, which repeats when $\boldsymbol{r}$ is changed by any lattice vector $\boldsymbol{R}_l$. This requires

$$\exp(i\mathbf{Q}.\boldsymbol{R}_l) = 1 \tag{1.8a}$$

The vectors which satisfy this condition, or alternatively

$$\mathbf{Q}.\boldsymbol{R}_l = 2\pi \times \text{integer} \tag{1.8b}$$

form a three-dimensional lattice, if $\boldsymbol{R}_l$ is a three-dimensional lattice.

It is convenient for the description of many solid state properties to define a space where vectors have dimensions of inverse length, which we shall call reciprocal space. The lattice of points at the ends of the vectors $\mathbf{Q}$ is called the *reciprocal lattice* and it is easy to show that this lattice has basic vectors

$$\boldsymbol{b}_1 = 2\pi\boldsymbol{a}_2 \times \boldsymbol{a}_3/\Omega \qquad \boldsymbol{b}_2 = 2\pi\boldsymbol{a}_3 \times \boldsymbol{a}_1/\Omega \qquad \boldsymbol{b}_3 = 2\pi\boldsymbol{a}_1 \times \boldsymbol{a}_2/\Omega \tag{1.9a}$$

where Ω is the volume of the parallelepiped unit cell in the space lattice

$$\Omega = (\boldsymbol{a}_1 \times \boldsymbol{a}_2).\boldsymbol{a}_3 \tag{1.9b}$$

If we take the general reciprocal lattice point as

$$\mathbf{Q}_n = \nu_1 \boldsymbol{b}_1 + \nu_2 \boldsymbol{b}_2 + \nu_3 \boldsymbol{b}_3 \tag{1.10}$$

where the ν_i are integers, it can be shown by elementary vector analysis that

$$\boldsymbol{b}_i\ \boldsymbol{a}_i = 2\pi \qquad \text{and} \qquad \boldsymbol{b}_i\ \boldsymbol{a}_j = 0 \qquad (i \neq j) \tag{1.11}$$

and hence

$$\begin{aligned}\mathbf{Q}_n.\boldsymbol{R}_l &= (\nu_1 \boldsymbol{b}_1 + \nu_2 \boldsymbol{b}_2 + \nu_3 \boldsymbol{b}_3)(n_1 \boldsymbol{a}_1 + n_2 \boldsymbol{a}_2 + n_3 \boldsymbol{a}_3) \\ &= 2\pi(\nu_1 n_1 + \nu_2 n_2 + \nu_3 n_3) \\ &= 2\pi \times \text{integer}\end{aligned}$$

as required.

The unit cell commonly used in reciprocal space is constructed with the planes which perpendicularly bisect the smallest reciprocal lattice vectors. This is called the *Brillouin zone* (or occasionally the first zone or the reduced zone). Its volume is the same as that of the parallelepiped constructed from the basic reciprocal lattice vectors $(\boldsymbol{b}_1 . \boldsymbol{b}_2 \times \boldsymbol{b}_3)$ which by manipulation of the vectors can be shown to be

$$\frac{(2\pi)^3}{\Omega^3}(\boldsymbol{a}_1 . \boldsymbol{a}_2 \times \boldsymbol{a}_3)^2 = \frac{(2\pi)^3}{\Omega} \tag{1.12}$$

i.e. it is inversely proportional to the volume of the unit cell in real space.

It is straightforward to demonstrate that the reciprocal lattice of a simple cubic lattice is itself a simple cubic; the reciprocal of a body-centred cubic lattice is face-centred and the reciprocal of a face-centred cubic lattice is body-centred. The Brillouin zone of the body-centred cubic lattice is therefore the rhombohedral unit cell shown in figure 1.5 while that of the face-centred cubic lattice is the truncated octahedron in figure 1.4.

The basic vectors of the reciprocal lattice are perpendicular to the planes defined by pairs of basic vectors in the real lattice. For example the vector

$$\boldsymbol{b}_1 = 2\pi(\boldsymbol{a}_2 \times \boldsymbol{a}_3)/\Omega$$

is perpendicular to the planes defined by $\boldsymbol{a}_2$ and $\boldsymbol{a}_3$, i.e. the planes containing $\boldsymbol{R}_l$ with n_3 fixed but any values of n_1 and n_2. Moreover the lengths of the basic reciprocal lattice vectors are inversely proportional to the spacing between adjacent planes of atoms. For example $|\boldsymbol{a}_2 \times \boldsymbol{a}_3|$ is equal to the area of the base of the unit cell parallelepiped. Hence $\Omega/|\boldsymbol{a}_2 \times \boldsymbol{a}_3)|$ or $2\pi/b_1$ is the perpendicular height of the unit cell, that is, the distance between adjacent planes.

General vectors of the reciprocal lattice have the same properties. All lattice points $\boldsymbol{R}_l$ which satisfy (1.8b) for the same integer form a plane which is perpendicular to $\mathbf{Q}_n$. There will be a set of parallel planes for different values of the integer. There are also several parallel vectors $\mathbf{Q}_n$ which give the same set of perpendicular planes. The indices ν_1, ν_2 and ν_3 corresponding to the smallest vector in a particular direction are called the Miller indices of this set of planes. The magnitude of this smallest vector is again 2π divided by the interplanar spacing.

1.5 Crystal binding (a)

One of the basic problems of solid state physics is to understand why, at low temperatures, atoms bind together to form a regular array. From our knowledge of atoms in other branches of science it is clear that electrostatic forces among the constituent electrons and nuclei must be responsible for this binding and that the attractive Coulomb forces between electrons and nuclei must somehow overcome the repulsive Coulomb forces between electrons and between nuclei so as to hold the structure together. It is also possible to see some graduation of crystalline properties so that familiar solid materials appear to fall into distinct groups. We can gain some insight into the differences between these types of solids by thinking in a qualitative way about the electrostatic forces.

To a first approximation, the electrons in an atom of a solid are bound in an electrostatic potential comprising the Coulomb field of the nucleus and the average repulsion of all the other electrons in the atom. In this approximation the electrons move independently in well-defined orbits. The quantum mechanical states describing the electronic motion are often called 'orbitals'. When the atom is in its lowest energy state the orbitals with the lowest energy, which are those closest to the nucleus, are completely filled with electrons, one to each possible state in accordance with the Pauli exclusion principle. In all but the lightest atoms their binding energy is very large indeed on the energy scale of typical solid state processes. It seems unlikely that these tightly bound orbitals will be modified at all when the atoms come together to form the solid. By contrast, the orbitals of higher energy (which are filled last in accordance with the Pauli principle) lie towards the outside of the atoms and it is relatively easy for their paths and hence their energies to be modified by neighbouring atoms. It is these so-called valence electrons in the high-energy orbitals which are most important in holding solids together. If there are just enough electrons to fill all the atomic orbitals of a particular type to form a closed shell it takes rather more energy to break this shell and modify the electrons' orbits. We would therefore expect that atoms with closed shells of electrons would be relatively little changed, even in a solid. Atoms which can be changed to a closed shell configuration by the addition or removal of a few electrons (thus forming ions) are also expected to be particularly stable. Elements in the columns at the beginning and the end of the Periodic Table are of this type. Elements in the columns near the middle of the Periodic Table are harder to ionise to form closed shells (see page xxii).

Chemists are by tradition particularly concerned with the way in which small numbers of atoms bind together to form molecules. Since a solid is in a sense a very large molecule, we would expect to gain an insight into the behaviour of solids by studying the binding of molecules. Chemical binding can be crudely divided into two types. If two atoms readily form ions and can do so by transferring electrons from one to the other, then they are likely to donate charges in this way and bind together by the mutual Coulomb attraction of the two ions so formed. This type of binding is known as ionic. The other method of binding is for atoms to share one or more electrons between them. This 'covalent' bonding can in principle be understood by reference to the simplest molecules.

The simplest of all is the hydrogen molecule ion. This consists of two protons and a single electron moving in their electric field. While the electron could remain in an orbit around one proton and leave the other as a hydrogen ion, the energy of the system is lower if the electron spends most of its time in the region between the two protons. In this region the electrostatic binding energy of the electron is largest and the electron partially screens the Coulomb repulsion between the two protons. Detailed calculations show how the binding energy of the ion varies as a function of the distance R between the protons.

A curve is shown in figure 1.10 based on the best available calculations. At small R when the repulsive forces between the protons are greater than the attractive forces, the system has a large positive energy; at a larger value of R the energy goes through a minimum where the binding is strongest (about 2.6 eV) and then the energy of the system rises as the protons are moved still farther apart and the binding becomes gradually weaker. The electron spends an equal amount of time around each proton and is shared equally between them.

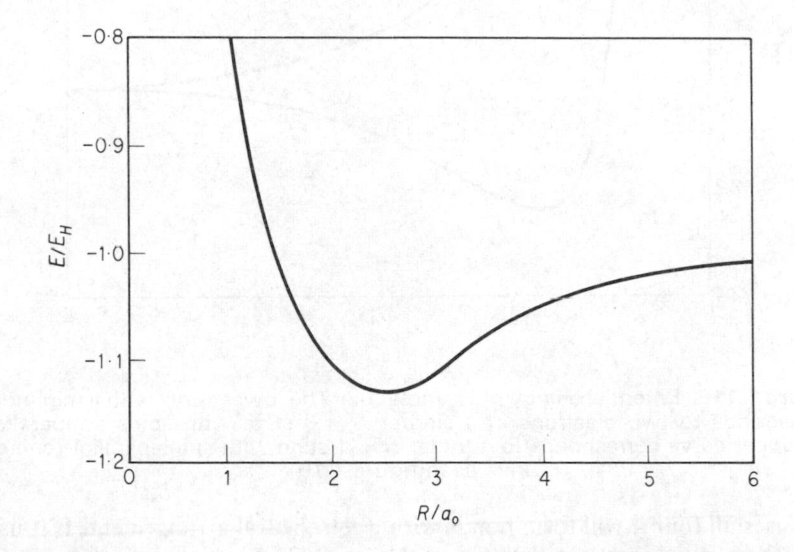

Figure 1.10 Potential energy of H_2^+ molecule, as a function of internuclear distance. The energy is given in units of the ionisation energy of a hydrogen atom (the Rydberg $E_H = 13.6$ eV)

The hydrogen molecule has two electrons moving in the same way as the single electron in the ion. Both of them lower the potential energy of the system by moving between the protons, but this decrease is partly offset by the repulsive interaction between the electrons. Because of the exclusion principle the binding is larger when the two electrons have antiparallel spins and such an arrangement is called a covalent bond. Theoretical curves for the hydrogen molecule are given in figure 1.11. When the two electrons have antiparallel spins there is a maximum binding energy of 4.7 eV. In fact the state with parallel spins does not bind at all in this system.

Molecules more complicated than H_2 bind together well by sharing pairs of electrons in this way. If an atom has several electrons it can form such bonds to several atoms. It is well known from chemistry that such bonds favour particular geometrical arrangements. (For a more detailed discussion see L. Pauling's *The Nature of the Chemical Bond*.) For example a carbon atom in its ground state has four electrons in 2s and 2p orbitals, available for bonding.

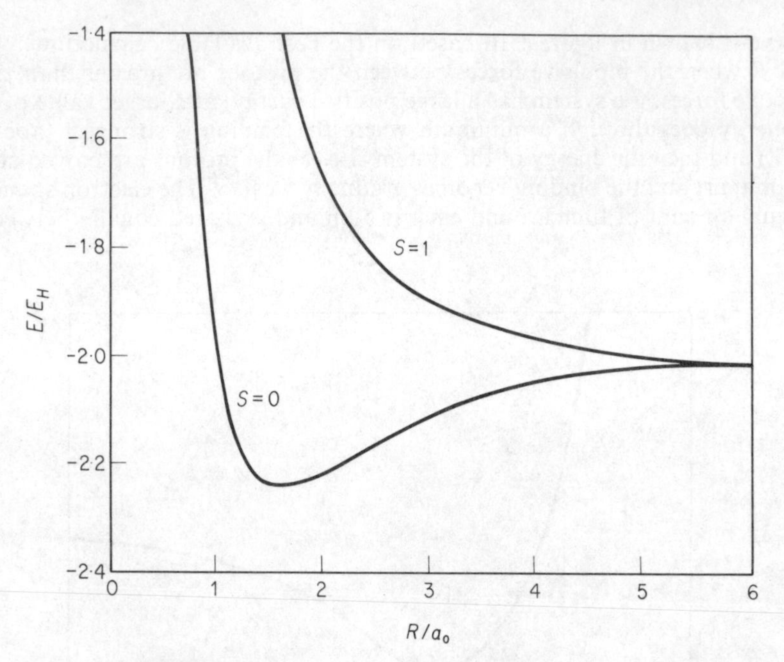

Figure 1.11 Potential energy of H_2 molecule. The lower curve with a minimum corresponds to two electrons in a singlet $S = 0$ state with spins antiparallel. The upper curve corresponds to a triplet $S = 1$ state with spins parallel (energy units as in figure 1.10)

If it uses all four it will form bonds with a tetrahedral arrangement. If it uses only three it will arrange them at angles of $2\pi/3$ in a plane. d electrons are usually fairly tightly bound in atoms and do not form good bonds. Atoms containing such electrons tend to make four equally spaced coplanar bonds or six bonds to atoms at the vertices of an octahedron.

These ideas may be extended to the large arrays of atoms we call solids. They lead us to a classification of different solid types. This is an idealised and not an exact classification, since the concepts of ionic and covalent bonding are themselves only qualitative. In any actual system both methods of bonding will play a role. Nevertheless there are some systems where one or the other type clearly predominates, and these are the ideal examples of their type. Many other systems cannot be so easily classified and fall somewhere in between. There are five broad divisions within this classification.

1.5.1 Van der Waals crystals (a)

A crystal formed entirely of atoms with closed electronic shells would at first sight exhibit neither ionic nor covalent binding. The so-called rare gases in the final column of the Periodic Table form such systems. But in fact Van der Waals showed that there would be attraction between such atoms because of the induced dipole moments arising from a slight distortion of the electronic charge clouds when the atoms are brought into close proximity. These attractive Van der Waals forces give rise to a potential energy of the form $e^2 a_{at}^5/R^6$ where a_{at} is the approximate atomic radius. There is also a repulsive force between the atoms which becomes strong when the electron charge clouds of each atom begin to overlap. Since the charge densities fall off exponentially with distance one might expect this repulsion to increase exponentially at small distances, although in practice it can be reasonably represented by a large inverse power of R, say as $e^2 a_{at}^{11}/R^{12}$ as in the empirical potential proposed by Lennard-Jones. The attractive and repulsive forces together give rise to a total potential energy which varies with distance in much the same way as the potential energy curves shown in figures 1.10 and 1.11. However, the maximum binding energy in this case is very much smaller. One would therefore expect such solids to be weakly held together, to be soft and have low melting points. They will also be bad conductors since there are no electrons available to carry charge through the crystals. As well as the rare gases, many crystals formed of molecular species which are themselves gases at room temperature bind in this way. For example hydrogen seldom occurs as atoms but normally occurs in nature as H_2 molecules. If hydrogen gas of this form is cooled it forms a solid in which the units are essentially unchanged H_2 molecules. Methane (CH_4) and other similar gases form solids in this way.

The energy of such crystals is minimised when as many atoms as possible are near the bottom of the potential energy curve. Since the attractive potential falls off so rapidly (like R^{-6}) only the atoms which are close together contribute much to the binding energy. This suggests that the crystals will be as dense as possible with the maximum number of neighbours to each atom. If spheres are packed as closely as possible, each will have approximately 12 neighbours (see figure 1.9) and it is not surprising that crystals of this type tend to favour the face-centred cubic arrangement in which each atom has twelve nearest neighbours.

1.5.2 Ionic crystals (a)

Compounds of those elements which readily form ions may be thought of as being built up of constituent ions which are little changed when made into a solid. The crystal now binds together because of the electrostatic forces between the ions of opposite charge. These forces are much stronger than the Van der Waals forces and ionic crystals therefore have higher melting points and are harder. Like Van der Waals crystals they are bad conductors of electricity since there are no weakly bound electrons in these materials.

The crystal structures are largely determined by the requirement that unlike charges are arranged together while like charges are kept apart. The Coulomb interaction energy falls off slowly (like R^{-1}) and so many neighbours contribute to the binding energy. Close packing is not now an overriding consideration and many ionic structures are relatively open. The sodium chloride structure (figure 1.6), for example, has a face-centred cubic space lattice and each atom has as nearest neighbours six atoms of the opposite kind. This is the type of structure favoured by alkali halides where the constituent atoms are rather different in size. For compounds with atoms of similar size, the caesium chloride structure is favoured. This is rather like the body-centred cubic lattice of figure 1.4 with a different type of atom at the cube centre. The space lattice in this case is simple cubic. In many ionic crystals the constituents are themselves small molecules, e.g. $[NO_3]^-$, $[SO_4]^{--}$, etc.

1.5.3 Metals (a)

Atoms which do not have a closed shell electronic configuration have loosely bound electrons in the outer shells. For example sodium has an electronic configuration $1s^2 2s^2 2p^6 3s$ which is a closed shell (like Ne) with an extra 'valence' electron on the outside. These 3s electrons could be shared between pairs of Na atoms to form a covalent bond but it is known from chemistry that this is very weak. Instead, in the solid the weakly bound electrons tend to be shared by many atoms forming what are sometimes called metallic bonds. They can in fact be thought of as providing a more or less uniform sea of negative charge in which the positive ions are embedded. A relatively small binding energy is obtained by this arrangement. As one moves to the right in the Periodic Table the larger number of valence electrons improves the binding energy, and in transition metals, where there are also d electrons, the binding can become strong. The metal bonds do not have well-defined directional characteristics, and most metals therefore occur in the closely packed face-centred or hexagonal structures. However, as will be seen in Chapter 4, high density increases the kinetic energy of the electrons and this will favour large lattice spacings and sometimes the less closely packed body-centred cubic structure.

The main characteristic of metals with their large number of relatively free electrons is their high electrical conductivity.

1.5.4 Covalent crystals (a)

Crystals in which two electrons are available for each nearest neighbour pair can form covalent bonds of the type described above for the H_2 molecule. Perhaps the best example of this is carbon, whose bonding characteristics are basically responsible for the properties of all the molecules studied in organic chemistry. In the solid state, carbon exists in two forms. In graphite each carbon atom uses three of its four valence electrons to bind strongly with three neighbours thus forming a two-dimensional hexagonal net. The fourth electron is relatively free and helps the weak binding between the planes. Diamond, on the other hand, uses all four valence electrons in tetrahedral bonds. Because of the directional properties of the bonds these structures are very open. An atom in the diamond lattice (see figure 1.7) has only four nearest neighbours. As one goes down the Periodic Table electrons in higher shells form relatively weaker bonds. Nevertheless Si and Ge, which are in the same column as C in the Periodic Table, form in the diamond structure. Sn on the other hand has two possible allotropic forms and changes phase to the diamond structure at relatively low temperatures.

Relatively few structures have simple covalent bonding, but in many more it plays a part, assisted to varying degrees by ionic binding. An interesting example of this is given by the Group III-V and II-VI compounds where the constituents lie in the Periodic Table on either side of the covalent elements of Group IV just mentioned. GaAs and ZnSe for example have the blend structure of figure 1.7 which retains the tetrahedral co-ordination characteristic of covalent bonding. They are isoelectronic with Ge, but because of the different constituents they clearly have some ionic character. If one moves one step farther to I-VII compounds like CuBr a quite different crystal structure is found, which is more characteristic of ionic binding (see page xxii).

Since the valence electrons are used in bonding, this type of crystal is not as good a conductor as is a metal. Nevertheless it is easier to find free electrons in covalent crystals than in Van der Waals or ionic crystals and many of them are semiconductors of great commercial importance.

1.5.5 Hydrogen bonded crystals (a)

A slightly different type of bonding, known to chemists as hydrogen bonding, can be of great importance in biological systems with large organic molecules although it is of limited interest in traditional solid state physics. It may be thought of as arising because a positive hydrogen ion is a bare proton which has a negligible radius on the atomic scale. By sitting between two negative ions this proton can overcome their repulsive interaction and bind them together. The binding energy is quite small, being of the order 0.1 eV. This type of bonding plays an important role in determining the properties of HF, water and similar materials, as well as ferroelectric crystals such as potassium di-hydrogen phosphate (cf. Chapter 3).

1.6 Crystal diffraction (a)

The actual structures of crystals are determined by studying the diffraction of waves from them. To reveal structure on an atomic scale, wavelengths of the order of the interatomic spacings are required. For electromagnetic radiation, waves of length about 10^{-8} cm are in the X-ray region with an energy quantum $\hbar\omega$ of order 10^4 eV. Bragg and von Laue first used X-rays in 1913 and X-ray crystallography has now developed into a very sophisticated technique. More recently the wave nature of particles such as neutrons and electrons has been exploited in similar experiments. Thermal neutrons with wavelengths of the right order are obtained in large numbers from atomic piles. At a temperature of 10^3 K neutrons have energies of about 0.08 eV corresponding to wavelengths of roughly 10^{-8} cm. Electrons with similar wavelengths have energies of about 10^2 eV. However, because of their charge, electrons interact much more strongly with the constituents of a crystal than do X-rays and neutrons. They therefore penetrate only a small distance into the crystal before losing all their energy and are consequently most useful for studying the surface properties of crystals.

The regular array of atoms in the crystal gives rise to what is effectively a three-dimensional diffraction grating. The more familiar ideas of two-dimensional diffraction gratings as used in optics may be readily extended to this three-dimensional array.

1.6.1 Scattering of waves by a lattice (A)

To understand how an incident wave is scattered by a crystal we first consider what happens when a wave interacts with a single atom which we regard as fixed in space.

An incident plane wave with wave vector $\boldsymbol{K}$ whose amplitude varies in space as

$$\exp \mathrm{i}(\boldsymbol{K}.\mathbf{R} - \omega t) \tag{1.13}$$

stimulates the atom to oscillate in phase, thus generating a second, scattered, wave. If the position vector of the atom is $\boldsymbol{r}$, the scattered wave has the form

$$\alpha_s \exp(\mathrm{i}K|\mathbf{R}' - \boldsymbol{r}|)\exp(-\mathrm{i}\omega t)/|\mathbf{R}' - \boldsymbol{r}| \tag{1.14}$$

i.e. it has the same wavelength and frequency as the incident wave. α_s is a measure of the magnitude of the scattered wave. The scattered wave is spherical, depending only on the distance of an observation at point $\mathbf{R}'$ from the scattering centre at $\boldsymbol{r}$, and not on the direction. The wave observed at $\mathbf{R}'$ depends on the phase of the incident wave and is given, from (1.13) and (1.14) as

$$\alpha_s \exp(\mathrm{i}\boldsymbol{K}.\boldsymbol{r})\exp(\mathrm{i}K|\mathbf{R}' - \boldsymbol{r}|)/|\mathbf{R}' - \boldsymbol{r}| \tag{1.15}$$

This is illustrated in figure 1.12. Since the time dependence is the same throughout, the factor $\exp(-\mathrm{i}\omega t)$ is omitted for simplicity.

The distance $|\mathbf{R}' - \boldsymbol{r}|$ of the observation from the scatterer is given by the cosine rule

$$|\mathbf{R}' - \boldsymbol{r}|^2 = \mathrm{R}'^2 + r^2 - 2\mathbf{R}'.\boldsymbol{r} \tag{1.16}$$

If R' is very large we can expand this in powers of r/R' to give

$$|\mathbf{R}' - \boldsymbol{r}| \simeq \mathrm{R}' - (\mathbf{R}'.\boldsymbol{r})/\mathrm{R}' \tag{1.17}$$

substituting this into (1.15) gives

$$\alpha_s \exp(\mathrm{i}\boldsymbol{K}.\boldsymbol{r})\exp(\mathrm{i}\boldsymbol{K}.\mathbf{R}')\exp[-\mathrm{i}K(\mathbf{R}'.\boldsymbol{r})/\mathrm{R}']/\mathrm{R}' \tag{1.18}$$

If we define $\boldsymbol{K}'$ as a vector of length K in the direction of $\mathbf{R}'$, i.e.

$$K(\mathbf{R}'.\boldsymbol{r})/\mathrm{R}' = \boldsymbol{K}'.\boldsymbol{r}$$

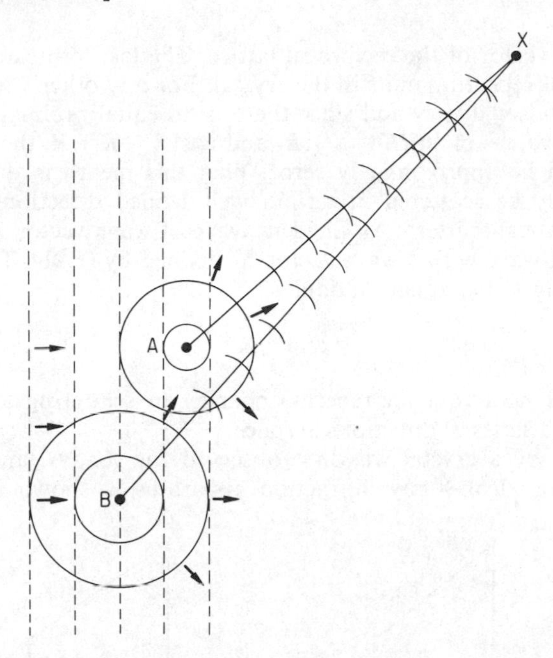

Figure 1.12 Diagram of a plane wave scattered by two centres. The dashed lines are the incident plane wave fronts; the full lines the spherical scattered wave fronts. With the geometry chosen the scattered waves at X interfere constructively

(1.18) takes the simpler form

$$\alpha_s[\exp(\mathrm{i}\boldsymbol{K}.\mathbf{R}')/\mathrm{R}']\exp[\mathrm{i}(\boldsymbol{K}-\boldsymbol{K}').\boldsymbol{r}] \tag{1.19}$$

$\boldsymbol{K}'$ may thus be regarded as the wave vector of the outgoing wave since at large distances near point R′ the spherical wave looks like a plane wave with this wave vector. Then $\boldsymbol{K}-\boldsymbol{K}'$ is the change in wave vector on scattering. On the de Broglie hypothesis $\hbar\boldsymbol{K}$ is the momentum of the particle described by the wave, and $\hbar(\boldsymbol{K}-\boldsymbol{K}')$ is the change in momentum on scattering.

If there is a lattice of atoms of the same type at points $(\boldsymbol{R}_l+\boldsymbol{r})$ which all scatter with the same intensity, the scattered wave at R′ is

$$\alpha_s[\exp\mathrm{i}(\boldsymbol{K}.\mathbf{R}')/\mathrm{R}']\exp[\mathrm{i}(\boldsymbol{K}-\boldsymbol{K}').\boldsymbol{r}]\sum\exp[\mathrm{i}(\boldsymbol{K}-\boldsymbol{K}').\boldsymbol{R}_l] \tag{1.20}$$

The last term describes the interference between the scattered waves. A simple example is shown in figure 1.12. The interference is said to be constructive if each term in the sum is unity and since the $\boldsymbol{R}_l$ are lattice vectors this is just the condition (1.8) that

$$(\boldsymbol{K}-\boldsymbol{K}')=\mathbf{Q} \tag{1.21}$$

where **Q** is a vector of the reciprocal lattice. The last term then sums to $\mathcal{N}$, the number of repeating units in the crystal. For any other value of $(\boldsymbol{K}-\boldsymbol{K}')$ the exponentials will vary and since there is an equal probability of positive and negative values of $\sin(\boldsymbol{K}-\boldsymbol{K}').\boldsymbol{R}_l$ and $\cos(\boldsymbol{K}-\boldsymbol{K}').\boldsymbol{R}_l$ the sum over the $\mathcal{N}$ terms will be approximately zero. What this means is that an incident wave can only be scattered in certain well-defined directions by a perfect lattice of identical scatters. An incident wave of wave vector $\boldsymbol{K}$ will give rise to a scattered wave with a wave vector $\boldsymbol{K}'$ defined by (1.21). The intensity of such scattering is the square modulus of (1.20)

$$\alpha_s^2 \mathcal{N}^2/\mathrm{R}'^2 \tag{1.22}$$

Since there is no frequency (energy) change on scattering and $|\boldsymbol{K}| = |\boldsymbol{K}'|$, (1.21) defines a series of directions in space.

Diffraction by a crystal was first observed for X-rays, and a schematic diagram of an ideal X-ray diffraction apparatus is shown in figure 1.13.

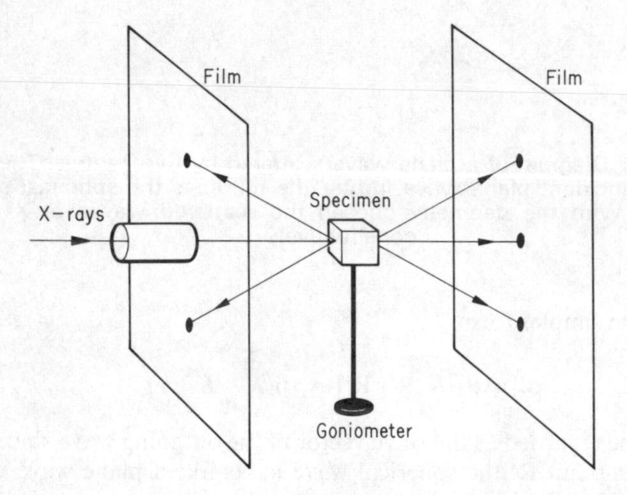

Figure 1.13 Diagram of X-ray scattering apparatus. The goniometer allows the specimen orientation to be varied

Photographic film exposed to the scattered radiation develops a pattern of spots as shown in figure 1.14 and from these the directions of the scattered waves and hence **Q** can be found. This leads to the determination of the reciprocal lattice and from it the space lattice of the crystal.

It is not necessary to use monochromatic radiation for the incident wave—Laue himself used a spectrum of wavelengths. Spots are formed on the film for the special values of $\boldsymbol{K}$ and $\boldsymbol{K}'$ that satisfy the condition (1.21). However, unless the value of $|\boldsymbol{K}'|$ is measured, the information on the film will not be sufficient to determine the length of **Q**, and hence the lattice spacing remains unknown. However, the method can be used to orient crystals and to measure

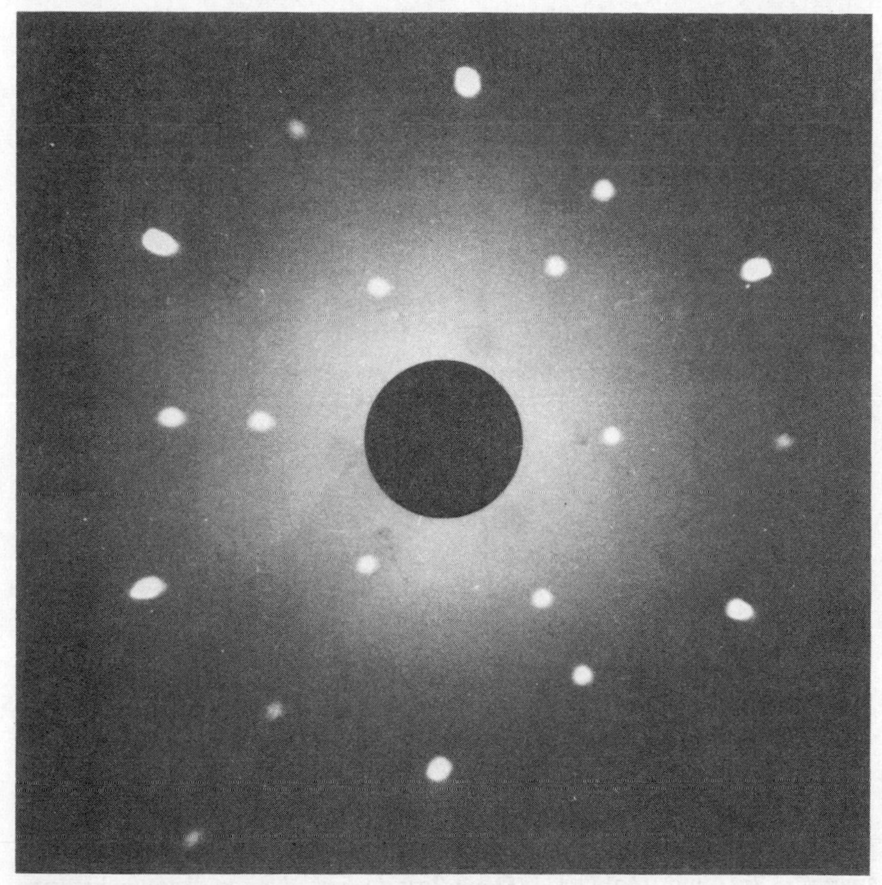

Figure 1.14 Actual pattern of spots obtained by X-ray scattering from a silicon crystal

their rotational symmetry. Since waves are scattered only for definite values of $\boldsymbol{K}'$ this property of crystals is sometimes used to obtain monochromatic waves. For example thermal neutrons from an atomic pile have a wide range of energies and if a collimated beam of these neutrons is scattered from a single crystal, mono-energetic beams are produced.

Some information on crystal structure can also be obtained from powdered samples of the material. In this case $\mathbf{Q}_n$ varies in direction from crystallite to crystallite. If $\boldsymbol{K}$ is fixed in length and direction and $|\boldsymbol{K}| = |\boldsymbol{K}'|$ this means that $\boldsymbol{K}'$ can vary in direction as long as

$$|\boldsymbol{K} - \boldsymbol{K}'| = 2K \sin\theta = |\mathbf{Q}| \tag{1.23}$$

where 2θ is the angle between K and K' (see figure 1.15). The scattered radiation now appears in cones at well-defined angles 2θ to the direction of the incident

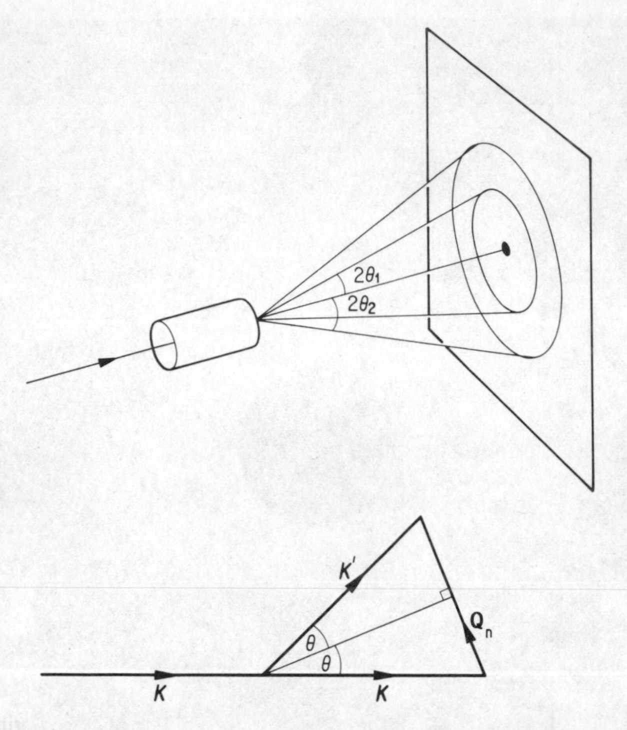

Figure 1.15 Bragg scattering in cones from a powdered sample. The vector diagram illustrates equations (1.21) and (1.23)

beam. This form of the scattering law is due to Bragg. He pointed out that since $K = 2\pi/\lambda$ and from (1.11) $|\mathbf{Q}_n| = 2\pi n/d$ where d is the distance between the nearest lattice planes in the direction $\mathbf{Q}_n$ (1.23) may be written

$$2d \sin\theta = n\lambda \tag{1.24}$$

This gives us some physical feeling for the origin of the original diffraction condition. When the wavelength of the radiation matches the interatomic spacing in this way the scattered wave builds up constructively to give a scattered wave of large amplitude.

The most detailed information is obtained using (1.21) directly in an experiment with mono-energetic radiation on a single crystal.

1.6.2 Incoherent scattering (A)

In the discussion of the previous section we assumed that all atoms of the same species scatter in the same way, i.e. that the α_s in (1.20) are all the same. This is true for X-rays which scatter from the electrons but neutrons are able to distinguish between different types of nuclei in the same chemical species.

For example different isotopes of the same element will scatter neutrons differently, and if a nucleus has a spin the scattering will be different for different spin orientations. In a random arrangement of such isotopes and spins α_s will have a non-zero average and coherent scattering will be proportional to this average. The fluctuations in α_s will give rise to scattering at all $\boldsymbol{K}'$ as if from $\mathcal{N}$ independent single nuclei and will contain no information about the structure. This is known as incoherent scattering. The proton is a very strong incoherent scatterer and neutrons cannot be used to find the position of hydrogen atoms in solids unless deuterium is substituted. It is also difficult to distinguish light atoms like H using X-rays because of the small number of electrons they possess.

1.6.3 Structure factors (A)

So far we have demonstrated only that a diffraction experiment can determine the space lattice of a crystal. Further information is needed about the arrangement of atoms in a unit cell in order to get a fuller description of the lattice. Some information can be obtained by measuring the intensity of the scattering.

In general there will be scattering from all parts of the unit cell with an intensity which depends on the position $\boldsymbol{r}$. To include this the middle term in (1.20) may be generalised to

$$\frac{1}{\Omega}\int \alpha(\boldsymbol{r}) \exp \mathrm{i}(\mathbf{Q}.\boldsymbol{r}).\mathrm{d}\boldsymbol{r} \tag{1.25}$$

and this is called the *structure factor*. For neutron scattering this is relatively simple since the neutron interacts mainly with the nuclei which in an ideal crystal are static and, on the scale of unit cell dimensions (10^{-8} cm), can be regarded as points. If the scattering strength of each nucleus at $\boldsymbol{r}_i$ is α_i the factor becomes

$$\sum \alpha_i \exp \mathrm{i}(\mathbf{Q}.\boldsymbol{r}_i) \tag{1.26}$$

However, it is only in an ideal lattice that the nuclei are stationary at the lattice points. The atomic vibrations cause the nuclei to move so that on the average they are found in space with a Gaussian probability distribution so that

$$\alpha(\boldsymbol{r}) \propto \exp(|\boldsymbol{r}-\boldsymbol{r}_i|^2/\langle u_i^2\rangle) \tag{1.27}$$

where $\langle u_i^2\rangle$ is the mean square displacement. On performing the integral in (1.25) this gives for the structure factor

$$\sum \alpha_i \exp(\mathrm{i}\mathbf{Q}.\boldsymbol{r}_i)\exp(-\tfrac{1}{4}\mathbf{Q}^2\langle u_i^2\rangle) \tag{1.28}$$

Because of the finite size of the scatterer the structure factor falls off at large $\mathbf{Q}$. This exponential factor is called the *Debye–Waller factor*. It decreases as the temperature, T, increases since classically we would expect the thermal agitation to increase $\langle u^2\rangle \sim kT$. But $\langle u^2\rangle$ does not vanish at $T=0$ because in quantum mechanics the atomic oscillators have a zero-point motion.

X-rays are scattered by the electrons in the solid and in this case $\alpha(\boldsymbol{r})$ in (1.25) is proportional to the charge density, $\rho(\boldsymbol{r})$, say. For a single atom this will be spherically symmetric to a good approximation and will fall off with $\boldsymbol{r}$ in some exponential way. Then (1.25) takes the form

$$f_i(\mathbf{Q}) = \frac{1}{\Omega}\int \rho(\boldsymbol{r}-\boldsymbol{r}_i)\exp\left[i\mathbf{Q}.(\boldsymbol{r}-\boldsymbol{r}_i)\right]d\boldsymbol{r} \tag{1.29}$$

where f is called the *form factor* and this will also fall off with increasing Q. For several atoms in a cell the total structure factor for X-ray scattering will be

$$\sum_i f_i(\mathbf{Q})\exp(-\tfrac{1}{4}Q^2\langle u_i^2\rangle)\exp(i\mathbf{Q}.\boldsymbol{r}_i) \tag{1.30}$$

since the thermal agitation will also affect the electron clouds as it does the nuclei.

In principle if intensity measurements are made at all $\mathbf{Q}_n$ the Fourier transform (1.25) can be inverted to derive the actual charge density. In practice

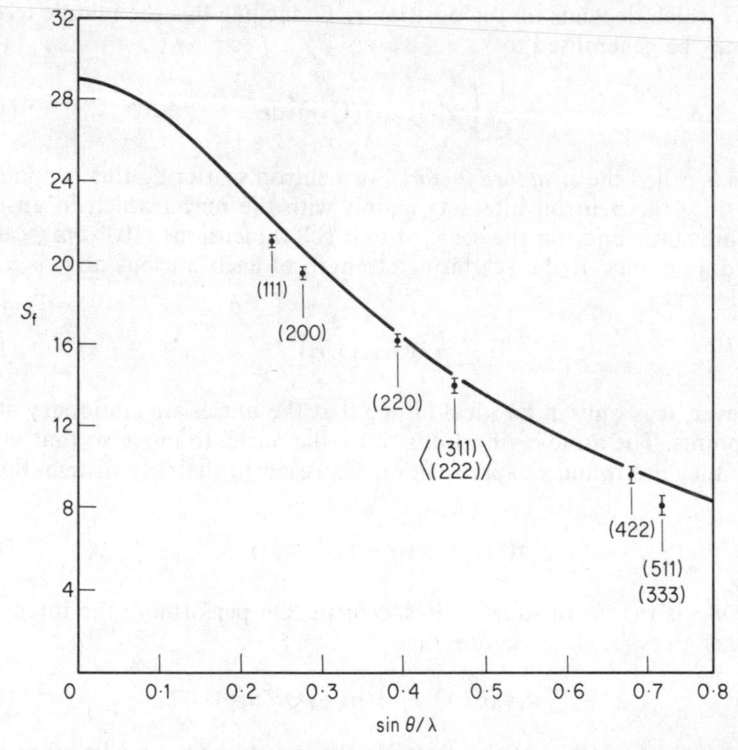

Figure 1.16 The form factor of Cu (after B. W. Batterman *et al., Phys. Rev.* (1961) **122**, 68). Experimental points correspond to particular reflections. The curve is based on a calculation of charge density

this is impossible, if only because the low intensity at large **Q** makes measurement difficult. But in fact $\rho(\boldsymbol{r})$ in solids is essentially the same as for the atom or ion in free space since most of the charge density comes from inner shells of electrons which are little changed in the solid. These can be computed or measured in simple systems (see figure 1.16 for the form factor in Cu). Hence the f_i (**Q**) of the constituent atoms is usually known, as is α_i in the neutron case. It is therefore a simpler problem to determine the $\boldsymbol{r}_i$ from (1.30) if only a few intensities are known. Crystallographers have brought great skill and ingenuity to this problem so that with the use of computers even the structures of huge biological molecules have been unravelled.

In some crystals attempts have been made to determine the charge density induced in the outer electrons by solid state effects. For example in an alkali halide with the rock salt structure the space lattice is face-centered cubic. The two atoms in the cell give a structure factor

$$S_f = f_A + f_C \exp i(\mathrm{Q}.\boldsymbol{r}_i) \tag{1.31}$$

taking $\boldsymbol{r}_i = \frac{1}{2}a(0,0,1)$ between the anion A and the cation C. The reciprocal lattice of **Q** is body-centred cubic with basic vectors $\boldsymbol{b}_1 = (2\pi/a)(1,1,1)$, etc. (cf. §1.4). Thus if $\mathbf{Q}_n$ contains an odd total number $\sum v_i$ (as defined in (1.10))

$$S_f = f_A - f_C \tag{1.32}$$

For an even number of $\sum v_i$

$$S_f = f_A + f_C \tag{1.33}$$

If the anion and cation have similar charge densities the odd reflections have zero intensity—in fact if they were really identical the lattice could be simple cubic and these odd **Q** would be absent. K^+ and Cl^- are isolectronic so odd reflections should be very weak indeed in KCl. A detailed study of such reflections should yield detailed information about the difference of ionic radius arising from the different nuclear charge.

A similar situation is of interest in diamond. In the blende lattice (figure 1.7) the two atoms give reflections at all reciprocal lattice points $\mathbf{Q}_n$. If the atoms are identical some of these vanish. In diamond the two atoms are the same chemically but differ if the covalent bonding distorts the charge clouds. The weak reflections may therefore be used to study the form of the bond in diamond, or the ionicity in similar atoms as in GaAs.

PROBLEMS

(Some answers, where appropriate, are given on page 480 *et sequi.*)

1.1 Show that, for the closest packing of spheres, the densities of the face-centred cubic, body-centred cubic, simple cubic and diamond lattices are approximately in the ratio 1.4:1.3:1.0:0.65.

1.2 Show that the ideal hexagonal close packed lattice has a c/a ratio of 1.633 $(=2\sqrt{2}/\sqrt{3})$ and a density equal to that of the face-centred cubic lattice.

1.3 A two-dimensional direct lattice is formed by repetition of a parallelogram ABCD where AB = 4 units, AC = 3 units and the angle BAC equals $\pi/3$. Find the reciprocal lattice vectors.

1.4 Graphite has layers of atoms, each layer consisting of hexagonal rings of atoms like a honeycomb so that each atom has three neighbours at a distance a. Show that there are two atoms in the smallest unit cell, and find the direct lattice and reciprocal lattice.

1.5 The body-centred cubic lattice can be considered as simple cubic with a unit cell in the form of a cube and two atoms per unit cell. Obtain an expression for the structure factors of the Bragg reflections and show that the only ones that are not zero correspond to the reciprocal lattice vectors of the b.c.c. lattice.

1.6 Cuprous oxide has a cubic unit cell with oxygen atoms at the centre $(0,0,0)$ and at the corners $a(\pm1,\pm1,\pm1)$. The copper atoms are arranged in a tetrahedron around the central oxygen, at $\frac{1}{2}a(1,1,1)$; $\frac{1}{2}a(1,-1,-1)$; $\frac{1}{2}a(-1,1,-1)$ and $\frac{1}{2}a(-1,-1,1)$. Calculate the structure factors and so show that some reflections are determined only by the copper, others only by the oxygen atoms.

1.7 A crystal of cubic symmetry and lattice spacing 50 nm is mounted with its $(0,0,1)$ axis perpendicular to an incident beam of X-rays of wavelength 10 nm. Initially the crystal is set so as to produce a diffracted beam whose direction relative to the crystal lattice vectors is (h_1,k_1,l_1). Calculate the angle through which the crystal should be turned to produce a beam in the direction (h_2,k_2,l_2) where

(i) $(h_1,k_1,l_1)=(0,0,0)$ and $(h_2,k_2,l_2)=(0,4,0)$,
and
(ii) $(h_1,k_1,l_1)=(0,2,0)$ and $(h_2,k_2,l_2)=(2,0,0)$.

1.8 Neutrons from a pile are collimated and filtered so that the beam contains only those of energy less than 15 meV. The beam strikes a single crystal of f.c.c. aluminium (nearest neighbour interatomic distance 28 nm) along a $(1,0,0)$ direction. What scattering pattern will be observed?

1.9 If the neutrons in the beam of 1.8 above are distributed in energy with the maximum number having an energy of 12 meV, how should the crystal be orientated to obtain the most intense scattered beam?

1.10 The potential energy of an ionic crystal with a rock-salt structure and containing $\mathcal{N}$ ions of each type is

$$V(R)=-\mathcal{N}\left(\frac{a\,e^2}{4\pi\varepsilon_0 R}-\frac{B}{R^n}\right)$$

where a is a number (Madelung's constant), B/R^n has the dimensions of energy and $n > 1$.

(i) Derive an expression for the bulk modulus of compressibility, $V(dP/dV)$, and

(ii) show that in extension there is a critical stress beyond which the crystal is unstable.

2

Excitations

2.1 Basic approximations (a)

In addition to the static properties of a crystal, discussed in the last chapter, the dynamic properties of the constituent particles of a crystal are also of great interest. These determine the changes that occur in the crystal because of thermal agitation at a finite temperature, T, and also the response of the crystal to external stimuli. For a quantum mechanical description we need to know the low-lying energy states of the crystal which will become populated in thermal equilibrium when $T > 0$ or to which transitions can take place if external stimuli are applied. This is in general a very complex problem but fortunately the very high symmetry of the system simplifies the discussion of these states. At low temperatures the deviations from the perfectly symmetrical ground state are relatively small, and this fact can also be used to simplify the problem.

Because of the very large number of particles involved a detailed description of the system is still a formidable task. There are, however, a number of good approximations which can be employed. The first and probably the most important of these is that in which the motion of the atoms, which are heavy, is treated separately from that of the light electrons. This is called the *adiabatic approximation*. The same approximation can be made for molecules, for which it is usually called the Born–Oppenheimer approximation. There are two basic consequences:

(a) The nuclei move so slowly that the motion of the electrons can be determined on the assumption that the nuclei are fixed.
(b) The nuclei can be thought of as moving in a potential determined by the average motion of the electrons.

Figure 1.10 shows the way in which the energy of a simple molecule varies with the separation of the nuclei. This curve is obtained by determining the electron orbits with the nuclei fixed at a separation R, in accordance with (a). On assumption (b) the nuclear motion is determined by this curve. Near the minimum the energy of the molecule varies quadratically with the distance as it would if the nuclei were bound together with a perfectly elastic spring. At low energies, therefore, the nuclei vibrate with harmonic motion. At higher energies this is no longer the case, and the motion is said to be anharmonic. At high enough energies the force which tends to keep the nuclei bound together, given by $\partial V/\partial R$, vanishes at large distances although at small distances there is a strong repulsion. The interaction between all the nuclei of the crystal may normally be expected to have potential energies in this form, so that at low temperatures and energies the motion of the nuclei will be largely vibrational. This will be discussed in detail in Chapter 3.

Small molecules also have rotational degrees of freedom, but in crystals these are usually suppressed by the interaction between the large number of particles involved. It is true, of course, that the whole crystal can rotate and its centre of mass can move freely but such motions are normally constrained

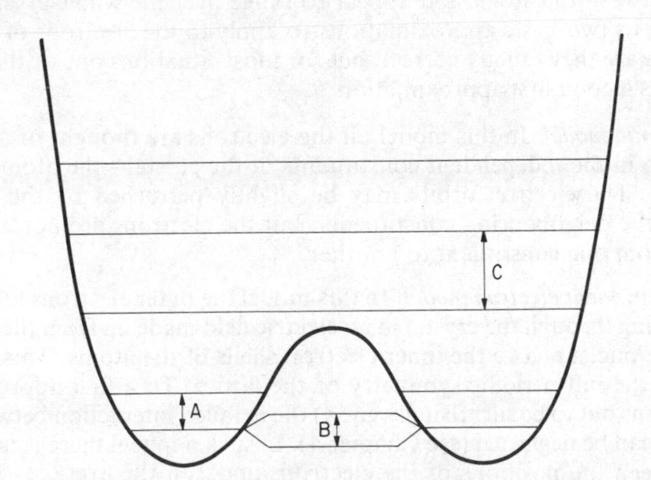

Figure 2.1 Diagram of a double potential well, with energy levels. A, vibrational states in separate wells. B, tunnelling frequency between A states. C, vibrational states of whole well

by external influences and are of no interest. Some crystals whose basic units are small molecules can show degrees of freedom in which the molecules themselves rotate. However, interaction with the surroundings is usually strong enough to hinder this rotation at low temperatures and turn it into an harmonic rocking motion, which is sometimes called a libration. Only solid hydrogen appears to show nearly free rotation of its constituents at low temperatures.

There are some other special situations in solids, although they are rare, in which the form of the energy variation is very different from the potential well of figure 1.10. A double well of the type shown in figure 2.1 can occur, for instance, when a small atom is substituted for a big one in the crystal. An example of this is Li in KCl. The hydrogen atoms in some crystals with hydrogen bonds also appear to move in such double wells between adjacent ions. At low temperatures the hydrogen atoms may tunnel through the small potential barrier in the middle and at higher temperatures and energies they move fairly freely between the steep sides of the well. Motion in such wells is anharmonic even at low energies.

2.2 Approximations to the electron motion (a)

In the discussion of crystal binding in § 1.5 we noted that many of the electrons in a solid could be expected to continue to move in their atomic orbits. In Van der Waals crystals and in ionic crystals this is approximately true of all the electrons. For covalent and metallic crystals, however, while the inner electrons of each atom still move in this way the outer valence electrons are shared between the atoms and appear to range over the whole crystal. These ideas give us two basic approximations to apply to the electrons of a crystal. In no case are they exactly correct, but for most situations one or the other of them gives a good first approximation.

(a) *Atomic model.* In this model all the electrons are thought of as moving as they do in the independent constituents of the crystal—the atoms, ions or molecules. The electron orbits may be slightly perturbed by the electrical forces of the neighbouring constituents, but the electrons are not allowed to wander from one constituent to another.

(b) *Independent electron model.* In this model the outer electrons are thought of as moving through the crystal in an electric field made up from the potential due to the nuclei and to the inner electron shells of the atoms. This potential will have the full periodic symmetry of the lattice. To a first approximation (which turns out to be surprisingly good) the detailed interaction between these electrons can be neglected (see Chapter 4). In such a model there is no correlation between the positions of the electrons and, on the average, situations where two or more electrons are found on the same atom will be possible. This is in sharp contrast to the atomic model where each atom always has its correct number of electrons.

2.3 Elementary excitations (a)

For electrons in metallic and covalent bonds which are shared between all the atoms the description of the states of a crystal in terms of independent particles ranging over the whole crystal is a natural one. In the terminology of quantum mechanics such electrons will exist in stationary states and be described by wave functions defined over the whole crystal. *Excitations* of the crystal will occur when electrons are promoted from one state to another of higher energy.

Other motions in a crystal are at first sight rather different in that they involve several particles. One example is the motion of heavy nuclei at low energies which, as has been indicated, is vibrational. If one nucleus is set in motion the coupling to the other atoms will ensure that the vibrational energy does not remain fixed at this site but moves across the crystal from nucleus to nucleus. It is therefore convenient to think in terms of a vibrational excitation which has a finite amplitude at all of the crystal atoms. In a large solid body such as an elastic string or an elastic continuum we know that the mechanical vibrations take the form of waves. It is reasonable to expect, as we shall show in the next section, that the vibrations of the atoms in a crystal also have wave form. Quantum mechanics tells us that the energy in such waves must be quantised. Using de Broglie's ideas of wave-particle duality it is convenient to associate a particle with a quantum of excitation in such a wave. For vibrations in a crystal, such a quantum is called a phonon by analogy with the photon which represents a quantum in the electromagnetic field. Phonons and vibrational waves will be discussed in detail in Chapter 3. For a general thermal agitation of a crystal one would expect vibrations to be generated from many points in a crystal. As long as the total energy is low, however, this can be described in terms of non-interacting excitations of the phonon type.

The wave-particle duality is well displayed by the examples of an electron and a phonon. Classically the electron is thought of as a particle but in quantum mechanics it is described in terms of wave functions which take the form of plane waves in free space. By contrast, the vibrational motion of a crystal is a collective oscillation involving all the atoms which, in classical physics, is regarded as a wave. In quantum mechanics the motion of these waves is quantised and it is convenient to associate a particle with each quantum. It is generally true that all excitations in crystals will have some wave-like character although it is modified from the simplest form as discussed below.

Other types of excitation can occur in crystals and will be discussed in later chapters of this book. By analogy we would expect to be able to describe the excitations in terms of waves and to associate a particle with each energy quantum. Apart from electrons, phonons and photons we shall discuss excitations called magnons, excitons, plasmons, polarons, polaritons, helicons, etc. All these excitations have one thing in common: they are excitations of the whole crystal and not of particular atoms. As a consequence we can deduce certain general properties of excitations which will be of value in our subsequent discussion. The remainder of this chapter is primarily devoted to the treatment of these general properties.

2.4 The symmetry of excitations (a)

In a uniform body such as an elastic continuum, where all points are equivalent, the excitations discussed in the last section take the form of waves. In free space the wave functions of particles such as electrons can also be taken as plane waves which vary as $\exp(i\boldsymbol{k}.\boldsymbol{r})$. The vibrational energy density in the first case and the electron density in the second case are given by the square modulus $|\exp(i\boldsymbol{k}.\boldsymbol{r})|^2$. This is simply a constant independent of position, as we would expect since all positions in such a homogeneous medium are equivalent. The fact that the excitations take the form of waves may be said to arise from the complete translational symmetry of these systems.

In a crystal not every point in space is equivalent but there is appreciable translational symmetry given by the space lattice discussed in Chapter 1. We therefore expect the plane wave form to be modified in a crystal. We know that if a simple wave of wave vector $\boldsymbol{k}$ is confined to a lattice it will be scattered as shown in (1.21) from $\boldsymbol{k}$ to $\boldsymbol{k}'$ where

$$\boldsymbol{k} \to \boldsymbol{k}' = \boldsymbol{k} + \boldsymbol{Q} \tag{2.1}$$

that is through a reciprocal lattice vector. Continual Bragg scattering of this kind will lead to the combination of such scattered waves—eventually giving a stable wave of the form

$$\exp i(\boldsymbol{k}.\boldsymbol{r}) \to \psi(\boldsymbol{r}) = \sum_{\mathrm{n}} \mathrm{C_n} \exp[\mathrm{i}(\boldsymbol{k} + \boldsymbol{Q}_{\mathrm{n}}).\boldsymbol{r}] \tag{2.2}$$

But, as shown in §1.4, $\exp(\mathrm{i}\boldsymbol{Q}_{\mathrm{n}}.\boldsymbol{r})$ has the periodic symmetry of the lattice and so

$$\sum_{\mathrm{n}} \mathrm{C_n} \exp(\mathrm{i}\,\boldsymbol{Q}_{\mathrm{n}}.\boldsymbol{r}) = \mathrm{U}(\boldsymbol{r}) \tag{2.3}$$

where $\mathrm{U}(\boldsymbol{r})$ has the full translational symmetry of the lattice, that is

$$\mathrm{U}(\boldsymbol{r} + \boldsymbol{R}_l) = \mathrm{U}(\boldsymbol{r}) \tag{2.4}$$

$\mathrm{U}(\boldsymbol{r})$ defines the form of the excitation within a unit cell and is the same in every cell. The total wave function on the other hand, contains an exponential factor

$$\psi(\boldsymbol{r}) = \mathrm{U}(\boldsymbol{r}) \exp(\mathrm{i}\boldsymbol{k}.\boldsymbol{r}) \tag{2.5}$$

so that it changes in phase from one unit cell to the next, exactly as a wave does. This property of the wave function is defined by the relation

$$\psi(\boldsymbol{r} + \boldsymbol{R}_l) = \exp(\mathrm{i}\boldsymbol{k}.\boldsymbol{R}_l)\psi(\boldsymbol{r}) \tag{2.6}$$

An alternative way of writing (2.5) is

$$\psi(\boldsymbol{r}) = \sum_{l} \phi(\boldsymbol{r} - \boldsymbol{R}_l) \exp(\mathrm{i}\boldsymbol{k}.\boldsymbol{R}_l) \tag{2.7}$$

where $\phi(\boldsymbol{r})$ is a function centred at the origin. This also satisfies the relation (2.6).

As before the square modulus of the wave function represents the energy or the particle density. For the wave function (2.5) this is $|U(\boldsymbol{r})|^2$ and has full periodic symmetry. Thus the wave function $\psi(\boldsymbol{r})$ has the essential symmetry required by the problem and the physically measurable properties like the electron charge density have the same value (as they must) at all equivalent sites in each unit cell. The wave function itself can change from cell to cell, but only in the way in which a wave changes. The relation (2.6) for excitations in a crystal is known as Bloch's theorem. A proper proof would require the formalism of group theory but the essential aspects are brought out by the argument above.

In an infinite perfect lattice we could have begun with a wave with any value of $\boldsymbol{k}$. There is, however, no essential difference between $\boldsymbol{k}$ and $(\boldsymbol{k}+\boldsymbol{Q})$ as may be seen from (2.5)—a change of origin by any $\boldsymbol{Q}$ would leave the form unchanged;

$$U(\boldsymbol{r})\exp[i(\boldsymbol{k}+\boldsymbol{Q}).\boldsymbol{r}] = w(\boldsymbol{r})\exp(i\boldsymbol{k}.\boldsymbol{r}) \tag{2.8}$$

where

$$w(\boldsymbol{r}) = U(\boldsymbol{r})\exp(i\boldsymbol{Q}.\boldsymbol{r})$$

is still, like $U(\boldsymbol{r})$, periodic. Thus $\boldsymbol{k}$ can be conveniently defined so that its end point lies in a unit cell of reciprocal space which is conventionally taken to be that centred at the origin. This is the Brillouin zone defined in §1.4.

Any excitation in a lattice therefore has two important components as described by (2.5) and (2.7). First there is the actual nature of the excitation in a cell. Secondly there is the phase variation as one goes from cell to cell. This is specified by a wave vector $\boldsymbol{k}$ in the relation (2.6). There is nothing special to quantum mechanics or to solid state physics in these properties. They hold for any periodic system. In §2.10 we examine the analogous situation of electromagnetic waves in periodic structures in some detail and see that they have similar properties.

For an infinite crystal $\boldsymbol{k}$ can take any value but, as has been shown by (2.8), it is unnecessary to define it outside the first Brillouin zone. In the next section we show that in a finite crystal the allowed values of $\boldsymbol{k}$ are not completely continuous but are very nearly so. We expect excitations having adjacent $\boldsymbol{k}$ values to be very similar, and their properties to vary continuously as $\boldsymbol{k}$ varies over the zone. In particular the energy or frequency of the excitation $E(\boldsymbol{k})$ will vary continuously with $\boldsymbol{k}$. This again is a property of periodic systems and is discussed for electromagnetic waves in §2.10. Simple examples of lattice waves and electron waves in solids are given in Chapters 3 and 4.

The range of energy covered by the variation of $E(\boldsymbol{k})$ when $\boldsymbol{k}$ goes across the zone is called an *allowed energy band* and such a set of excitations is called a band or branch. There can be many bands or branches in a system, and these may be separated by energy ranges within which no excitations occur. Such energy ranges are called *forbidden gaps*. We shall see many examples in the rest of the book.

2.5 The number of waves in a lattice (AD)

For a finite lattice there can only be a finite number of excitations and hence the number of allowed values of $\boldsymbol{k}$ must be restricted to some finite number and it is necessary to enumerate them if extensive physical quantities which are proportional to the size of the crystal are to be evaluated. This will involve defining the surface of the crystal in some way and using the boundary conditions. We know, however, that the bulk properties of the crystal cannot depend on the nature of the surface, so that our result should be independent of the actual boundary conditions used. Physically these would be changed by making the surface rough, oxidising it, or painting it red. We wish, however, to choose some convenient mathematical device to ease the computation.

One simple way is to end the crystal abruptly so that the wave functions fall to zero there. We consider a cube of sides $\mathscr{L}_x\mathscr{L}_y\mathscr{L}_z$ and one corner at the origin. It is then necessary to take a linear combination of plane waves in the form

$$\exp(\pm ik_x x)\exp(\pm ik_y y)\exp(\pm ik_z z) \tag{2.9}$$

To meet the boundary conditions for motion in the x-direction we must have $\psi = 0$ at the faces $x = 0$ and $x = \mathscr{L}_x$. We can do this by combining the waves into $\sin k_x x$ and imposing the restriction

$$k_x = \mathrm{n}_x \pi/\mathscr{L}_x \tag{2.10}$$

for all integers n_x. Similarly the boundary conditions in the other directions are given by

$$k_y = \mathrm{n}_y \pi/\mathscr{L}_y \qquad k_z = \mathrm{n}_z \pi/\mathscr{L}_z \tag{2.11}$$

and the whole wave function takes the form

$$\sin k_x x \sin k_y y \sin k_z z \tag{2.12}$$

Since changing the sign of any component of $\boldsymbol{k}$ only changes the sign of the whole function, each term may be taken as positive and $\boldsymbol{k}$ confined to the positive octant of space. Within this octant the ends of the allowed $\boldsymbol{k}$ vectors form a lattice of closely spaced points—their separation in the x-direction is $\pi/\mathscr{L}_x$, where $\mathscr{L}_x$ might be about 1 cm. Reciprocal lattice points, on the other hand, are separated by about π/a, which is $\sim 10^8$ times greater. This 'spottiness' of $\boldsymbol{k}$ space has no physical consequence except for very small crystals or films where the discrete allowed wavelengths of standing waves may be detected. An example of this effect is shown in figure 2.2. The excitations there are actually magnons (see §10.10.2) in a thin ferromagnetic film and they assist the transmission of electromagnetic energy if the wavelength fits into the film thickness. For larger crystals the possible values of $\boldsymbol{k}$ are so numerous that they may be regarded as uniformly distributed with a density

$$(\pi/\mathscr{L}_x)^{-1}(\pi/\mathscr{L}_y)^{-1}(\pi/\mathscr{L}_z)^{-1} = \mathscr{V}/\pi^3 \tag{2.13}$$

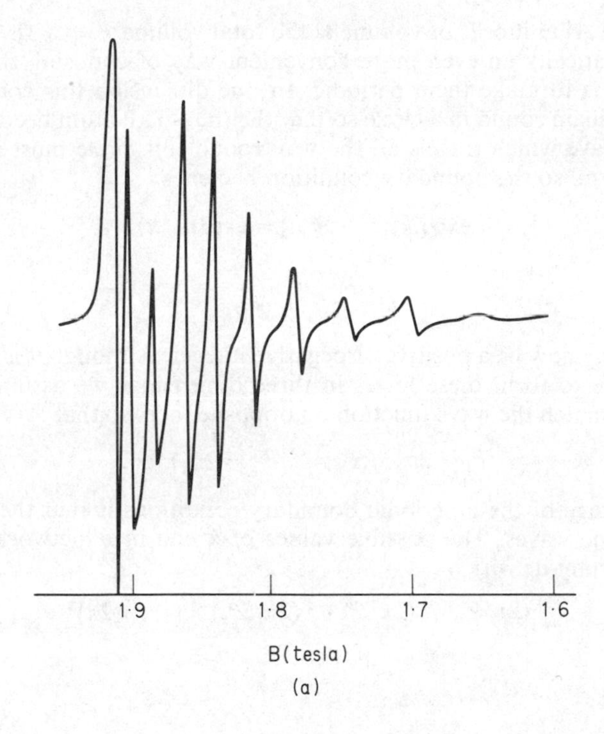

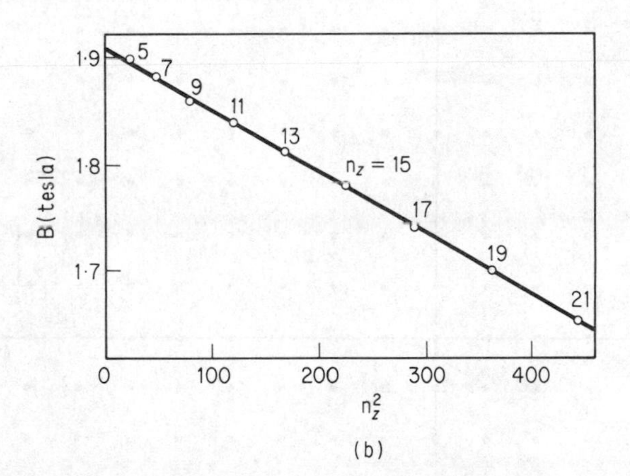

Figure 2.2 (a) Differential of the absorption in a thin ferromagnetic film at fixed frequency as a function of applied field B, showing discrete standing spin waves allowed in the film (after H. M. Rosenberg and T. G. Phillips, *Phys. Lett.* **8**, 298). (b) The energy of the spin waves and hence B varies as n_z^2 (see § 10.10.2) where n_z is defined by (2.11). Because of the boundary conditions only odd values of n_z are seen

If there are $\mathcal{N}$ unit cells of volume Ω the total volume $\mathcal{V} = \mathcal{N}\Omega$.

Mathematically an even more convenient way of choosing the boundary conditions is to make them periodic. In one dimension this corresponds to bending a chain round in a circle so that the $(n_x + 1)$th atom becomes the first atom. A wave which travels all the way round this circle must return to its original form, so the boundary condition becomes

$$\exp[ik_x(x + \mathcal{L}_x)] = \exp(ik_x x) \tag{2.14}$$

or

$$k_x = 2n_x \pi/\mathcal{L}_x$$

where n_x may now be a positive or negative integer. Although it is not physically possible to form these loops in three dimensions we assume that it is possible to match the wave function on opposite faces so that

$$k_y = 2n_y \pi/\mathcal{L}_y; \qquad k_z = 2n_z \pi/\mathcal{L}_z \tag{2.15}$$

The advantage of these periodic boundary conditions is that the excitations remain plane waves. The possible values of $\boldsymbol{k}$ end in a network (cf. figure 2.3) of constant density

$$(2\pi/\mathcal{L}_x)^{-1}(2\pi/\mathcal{L}_y)^{-1}(2\pi/\mathcal{L}_z)^{-1} = \mathcal{V}/(2\pi)^3 \tag{2.16}$$

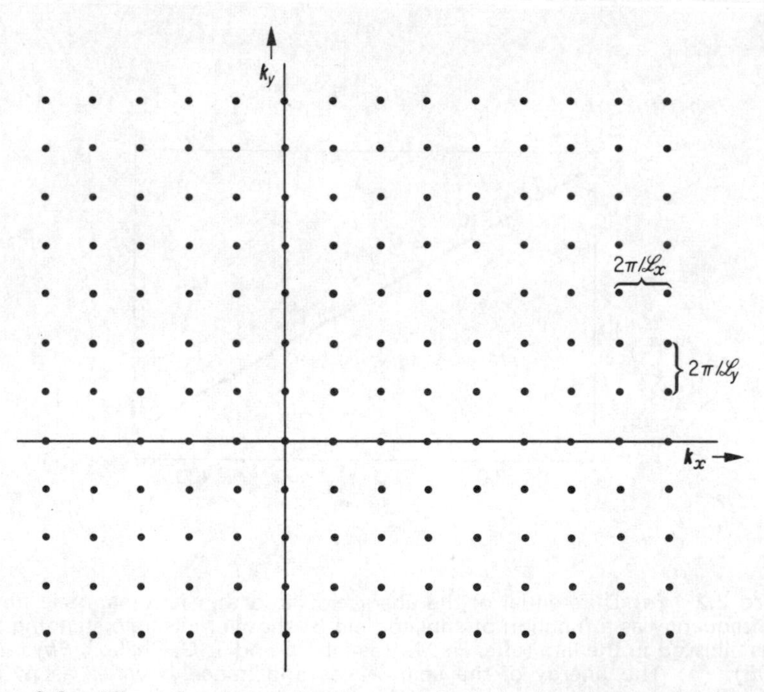

Figure 2.3 Allowed values of $\boldsymbol{k}$ in x–y plane (end points marked) using periodic boundary conditions showing 'spottiness' of $\boldsymbol{k}$ space

This is one of the most useful formulae in theoretical solid state physics in that it allows us to relate macroscopic properties to elementary microscopic excitations. If we wish to sum over all excitations we sum over all allowed $\boldsymbol{k}$ values in the Brillouin zone (B.Z.) and all bands i, then

$$\sum_i \sum_k = \sum_i \frac{\mathcal{N}\Omega}{(2\pi)^3} \int_{\text{B.Z.}} \mathrm{d}k_x \mathrm{d}k_y \mathrm{d}k_z \tag{2.17}$$

The periodic boundary conditions and this form of the summation will be used throughout the book unless it is explicitly stated that other boundary conditions are being used. This may occur, for example, because the physics of the surface is important.

At first sight (2.13) and (2.16) look different, but in fact they are the same (as they must be since the nature of the boundary is immaterial): in (2.13) the density is eight times greater than in (2.16) but on the other hand $\boldsymbol{k}$ is restricted to the positive octant—one-eighth of space.

In the previous chapter it was shown (1.12) that the volume of a Brillouin zone is $(2\pi)^3/\Omega$. Thus performing the integral in (2.17) the number of different $\boldsymbol{k}$ values allowed is just equal to $\mathcal{N}$, the number of unit cells,

$$\sum_k 1 = \mathcal{N} \tag{2.18}$$

2.6 Conservation rules in a lattice (a)

In the preceding sections we have discussed the consequences of wave-particle duality in a solid. Entities such as electrons or neutrons which are classically regarded as particles have wave functions and wave-like properties. Elastic and electromagnetic waves on the other hand are quantised and have 'particles' associated with these quanta. In each case the particle or quantum of excitation has a wave vector $\boldsymbol{k}$ defined in the above sense. On the de Broglie hypothesis $\hbar\boldsymbol{k}$ is a sort of momentum sometimes called the crystal momentum. The energy is a continuous function of $\boldsymbol{k}$, $E(\boldsymbol{k})$ say, and may have a complicated dependence on $\boldsymbol{k}$ across the zone. The group velocity

$$\boldsymbol{v} = \frac{1}{\hbar} \nabla_{\boldsymbol{k}} E(\boldsymbol{k}) \tag{2.19}$$

is also in general complicated. Sometimes electrons behave almost as if they are free. Then

$$E(\boldsymbol{k}) = \hbar^2 k^2/2m \tag{2.20}$$

where m is the mass and the velocity

$$\boldsymbol{v} = \hbar\boldsymbol{k}/m \tag{2.21}$$

A simple elastic wave in a continuum (e.g. a sound wave), on the other hand, has

$$E(\boldsymbol{k}) = \hbar s k \tag{2.22}$$

and velocity s.

The particles may interact with one another. The interactions between the various particles in a crystal and between these and other particles impinging on the crystal from outside are the basic mechanisms which determine many properties of solids. The interactions or collisions take various forms depending on the particles involved. At the low energies, which are usually of interest in solid state physics, electrons and neutrons cannot be created or destroyed and their numbers must be conserved. However, other quanta (phonons and photons for instance) can be created or destroyed.

The simplest interaction is a collision involving two quanta, one of which is created by destroying the other. A good example of this is a photon–phonon interaction (cf. Chapter 6). The essential requirement for this interaction to occur is that both excitations have the same waveform at each site; that is if $\boldsymbol{q}$ and $\boldsymbol{\beta}$ are the wave vectors of the phonon and photon respectively

$$\exp(\mathrm{i}\boldsymbol{q}\boldsymbol{R}_l) = \exp(\mathrm{i}\boldsymbol{\beta}\boldsymbol{R}_l) \tag{2.23}$$

or

$$\boldsymbol{q} = \boldsymbol{\beta}$$

Energy must also be conserved in the collision so that

$$E(\boldsymbol{q}) = E(\boldsymbol{\beta}) \tag{2.24}$$

In fact this is a simple coupling of two waves which is strongest at a resonance, when wave vector and frequency match. If the coupling is strong, frequent collisions will occur and the excitation will continually change from one type to the other. It can be said to spend part of its time in each form and to be a mixed mode: in this case partly vibrational and partly electromagnetic (cf. Chapter 3).

A collision between a particle like an electron and a vibrational excitation (phonon) will result in the scattering of the electron, with a change in electron wave vector from, say, $\boldsymbol{k}$ to $\boldsymbol{k}'$, and the creation or destruction of a vibrational excitation with wave vector $\boldsymbol{q}$. For strong interaction we must have completely constructive interference in each cell. The scattered wave has a form similar to (1.20)

$$(1/\mathrm{R}')\exp(\mathrm{i}\boldsymbol{k}.\mathbf{R}')\exp[\mathrm{i}(\boldsymbol{k}-\boldsymbol{k}').\boldsymbol{r}]\sum_l \alpha_s \exp[\mathrm{i}(\boldsymbol{k}-\boldsymbol{k}').\boldsymbol{R}_l]$$

except that the excitation now varies the scattering amplitude at each site so that

$$\alpha_l = \exp(\mathrm{i}\boldsymbol{q}.\boldsymbol{R}_l)$$

The last term now takes the form

$$\sum_l \exp[\mathrm{i}(\boldsymbol{k} - \boldsymbol{k}' + \boldsymbol{q}) . \boldsymbol{R}_l] \tag{2.25}$$

which is equal to $\mathcal{N}$ if the conservation rule

$$\boldsymbol{k} - \boldsymbol{k}' + \boldsymbol{q} = \boldsymbol{Q} \tag{2.26}$$

is satisfied, where $\boldsymbol{Q}$ is a vector of the reciprocal lattice. Energy conservation will also require

$$\mathrm{E}(\boldsymbol{k}) - \mathrm{E}(\boldsymbol{k}') = \pm \mathrm{E}(\boldsymbol{q}) \tag{2.27}$$

The conservation rule (2.26) is similar to (2.23) and to the requirement of momentum conservation for particles in free space, except for the appearance of $\boldsymbol{Q}$ on the right hand of the equation. This extra term is hardly surprising since the crystal momentum is not defined to within an addition of $\boldsymbol{Q}$. It may be said that this corresponds to momentum being absorbed by the lattice as a whole, but since the lattice is so massive the centre of mass motion induced in this way is beyond detection (and experimentally would in any case normally be suppressed by outside constraints). Processes in which $\boldsymbol{Q} \neq 0$ are sometimes called 'Umklapp' processes and those in which $\boldsymbol{Q} = 0$ 'Normal' processes. The conservation rule (2.26) is illustrated in figure 2.4.

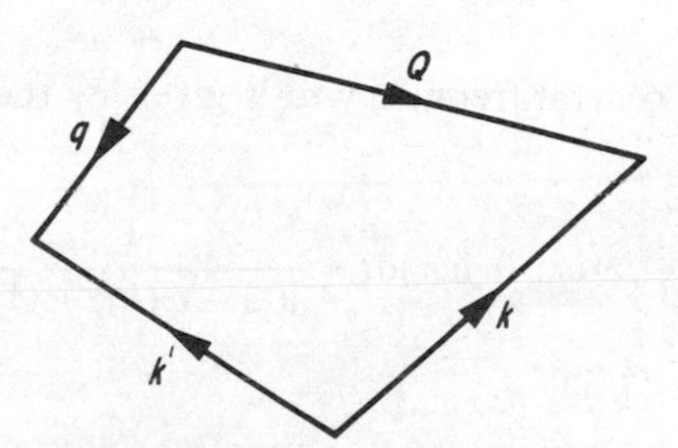

Figure 2.4 Quadrilateral formed by vectors satisfying equation (2.26). In addition the lengths of $\boldsymbol{k}$, $\boldsymbol{k}'$ and $\boldsymbol{q}$ are related by (2.27)

A more direct analogy with momentum conservation is seen in particle collisions, e.g. neutrons scattering from electrons. Then we find the requirement

$$\boldsymbol{K} + \boldsymbol{k} = \boldsymbol{K}' + \boldsymbol{k}' + \boldsymbol{Q} \tag{2.28}$$

which, when multiplied by $\hbar$, is the usual form if $\boldsymbol{Q} = 0$. Here the neutron is scattered from $\boldsymbol{K} \to \boldsymbol{K}'$ and the electron from $\boldsymbol{k} \to \boldsymbol{k}'$. Energy conservation requires

$$E_{\mathrm{n}}(\boldsymbol{K}) + E(\boldsymbol{k}) = E_{\mathrm{n}}(\boldsymbol{K}') + E(\boldsymbol{k}') \tag{2.29}$$

where $E_{\mathrm{n}}(\boldsymbol{K})$ is the energy of a neutron with wave vector $\boldsymbol{K}$.

As we shall see in subsequent chapters the simultaneous operation of energy and momentum conservation is often sufficient to determine the general physical behaviour of many processes.

2.7 Mean free time and mean free path (a)

The interactions between excitations described in the last section allows them to exist in a given state for only a finite time before they are scattered to a new state or annihilated. The average time is called the mean free time τ. The uncertainty principle states that precise measurements of frequency and energy can only be made if an infinitely long time is available so that a state with a finite lifetime must have an uncertainty in energy

$$\Delta E \sim \hbar/\tau \tag{2.30}$$

Such a broadening may be observed experimentally in, for example, the optical properties which measure the electric response at varying frequences. The existence of a mean free time means that the wave associated with an excitation is damped. It has a time dependence given by

$$\exp(\mathrm{i}\omega(\boldsymbol{k})t)\exp(-t/\tau) \tag{2.31}$$

The response at a general frequency ω is given by the Fourier transform of this, viz.

$$\int_0^\infty \exp(\mathrm{i}\omega(\boldsymbol{k})t)\exp(-t/\tau)\exp(-\mathrm{i}\omega t)\mathrm{d}t = \frac{1}{\mathrm{i}(\omega-\omega(\boldsymbol{k}))+(1/\tau)} = \frac{\mathrm{i}(\omega-\omega(\boldsymbol{k}))\tau^2-\tau}{(\omega-\omega(\boldsymbol{k}))^2\tau^2+1} \tag{2.32}$$

The real and imaginary parts of this function, plotted in figure 2.5, show a characteristic width in frequency of $1/\tau$.

The finite lifetime also ensures that a wave packet built from such excitations will have a finite free path length. If the wave packet travels with velocity v this is

$$l_\mathrm{f} = v\tau \tag{2.33}$$

Such a finite free path leads to an uncertainty in the momentum or wave vector by the same uncertainty principle. As can be seen in figure 2.6 a broad $E(\boldsymbol{k})$ curve has a vertical width ΔE related to a horizontal width Δk by

$$\hbar/\tau = \Delta E = \left(\frac{\partial E}{\partial k}\right)\Delta k = \left(\frac{\partial E}{\partial k}\right)\cdot\frac{1}{l_\mathrm{f}} \tag{2.34}$$

which agrees with (2.33) using expression (2.19) for the group velocity.

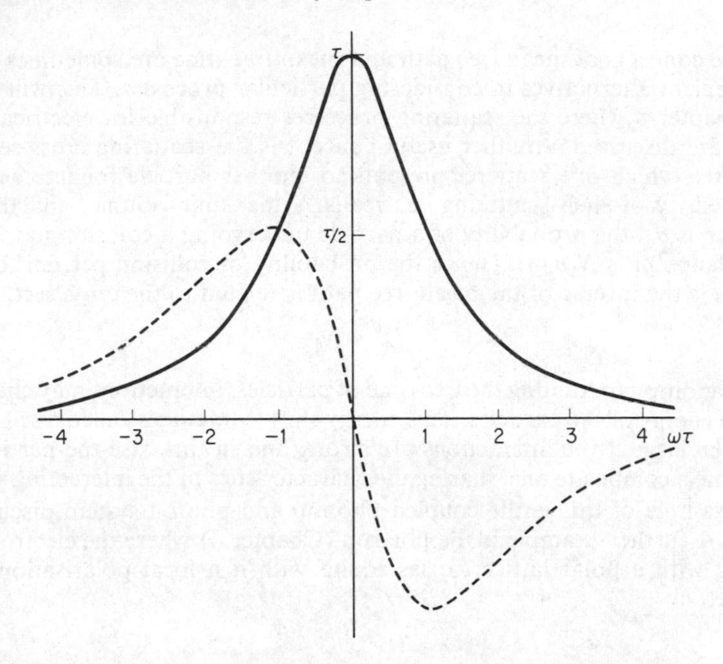

Figure 2.5 The real and imaginary parts of the frequency response of a damped oscillator. The 'bell-shaped' curve is sometimes called a Lorentzian

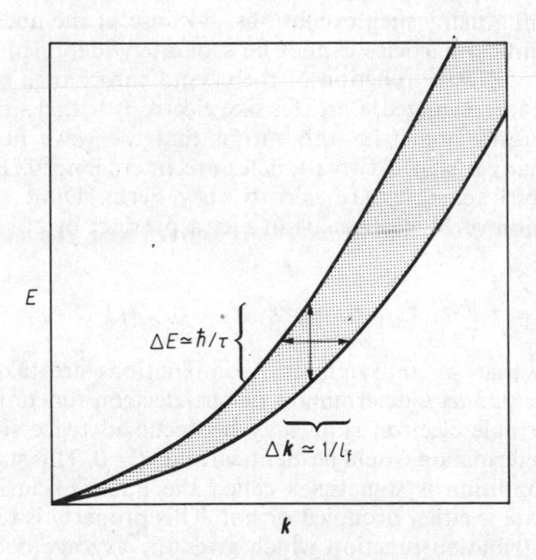

Figure 2.6 Relation of widths ΔE, $\Delta \boldsymbol{k}$ in a broadened $E(\boldsymbol{k})$ curve. ΔE is inversely proportional to the mean free time τ and $\Delta \boldsymbol{k}$ to the mean free path l_f

The concepts of mean free path and mean free time are sometimes used as convenient alternatives in considering particular processes. This will be seen in Chapter 8 where the scattering processes responsible for electrical resistivity are discussed. Another useful concept is the scattering cross section—the area which one scatterer presents to another particle for interaction. If the density of such scattering centres is N_I per unit volume, and the cross section is σ_d, the probability of a particle undergoing a collision in travelling a distance Δl is $N_I \sigma_d \Delta l$. Hence the probability of collision per unit distance which is the inverse of the mean free path is related to the cross section by

$$l_f^{-1} = N_I \sigma_d \tag{2.35}$$

In addition to limiting the free time of particles, interactions may change the mean energy of the particles. This energy shift (sometimes called a self energy) may be large if the interactions are strong and in this case the particle may become a composite one, sharing the characteristics of the interacting systems. An example of this is the coupled phonon and photon system discussed in §2.6. A further example is the polaron (Chapter 4) where the electron interacting with a polar lattice carries round with it a local polarisation of the lattice.

2.8 Statistics of excitations (a)

Even in the simple approximation in which the excitations are regarded as non-interacting, quantum mechanics places some restrictions on the states of a crystal containing many such excitations. Because of the uncertainty principle a set of identical particles cannot be separately identified. This in turn means that the total wave function of the crystal can change only in sign if two particles are interchanged. Particles like electrons with a spin of one-half are found to be antisymmetric—this means that the wave function of the whole system changes sign if two particles are interchanged. Electrons and other antisymmetric particles are said to obey Fermi–Dirac statistics. The total wave function of N electrons is in fact a product of N single electron wave functions

$$\Psi(\boldsymbol{r}_1, \boldsymbol{r}_2 \ldots \boldsymbol{r}_N) = \mathscr{P}\psi_1(\boldsymbol{r}_1) \ldots \psi_N(\boldsymbol{r}_N) \tag{2.36}$$

where $\mathscr{P}$ means that all antisymmetric combinations are taken. Equation (2.36) may be written as a determinant of one-electron functions. From this we see that no single electron state may be occupied twice since then two columns of the determinant would be identical and $\Psi = 0$. This statement of the antisymmetry condition is sometimes called the Pauli Exclusion Principle. Each electron state is either occupied or not. This property is reflected in the Fermi–Dirac distribution function which gives the average occupation of a state with energy $E_i(\boldsymbol{k})$ in thermal equilibrium at temperature T as

$$f_i(\boldsymbol{k}) = \{\exp[(E_i(\boldsymbol{k}) - E_F)/kT] + 1\}^{-1} \tag{2.37}$$

where E_F is called the Fermi energy. At $T=0$ K, $f_i(\boldsymbol{k})=1$ if $E_i(\boldsymbol{k})<E_F$ and zero if $E_i(\boldsymbol{k})>E_F$. In other words at zero temperature all the low-lying energy states are fully occupied up to the Fermi energy. As T is increased the sudden decrease in $f(\boldsymbol{k})$ at E_F is smoothed out over an energy region about kT wide as excitations of energy kT are allowed. The total number of electrons N in a solid is fixed and this fixes the parameter E_F by the equation

$$\sum_i \sum_k f_i(\boldsymbol{k}) = \mathrm{N} \tag{2.38}$$

E_F in general depends on T but at low T it is nearly constant if N is large (see Chapter 4).

The other types of excitation we have mentioned are particles with no spin. In this case the total wave function must be unchanged if two particles are interchanged. It is said to be symmetric and is in the form of (2.36) except that a symmetric combination of product functions is taken. There is now no restriction on the number of particles occupying any state and no restriction on the total number of such particles which may be created and destroyed. This is reflected in the Bose–Einstein distribution function which gives the average number of such particles in thermal equilibrium at temperature T as

$$f_i(\boldsymbol{k}) = \{\exp(E_i(\boldsymbol{k})/kT) - 1\}^{-1} \tag{2.39}$$

At $T=0$ there are no excitations and $f_i(\boldsymbol{k})=0$. At finite T there can be large numbers of excitations with low energies ($E_i(\boldsymbol{k}) \ll kT$) and for these

$$f_i(\boldsymbol{k}) \sim kT/E_i(\boldsymbol{k}) \tag{2.40}$$

but at high energies ($E_i(\boldsymbol{k}) \gg kT$) the number falls off exponentially

$$f_i(\boldsymbol{k}) \sim \exp(-E_i(\boldsymbol{k})/kT) \tag{2.41}$$

Phonons and magnons are particles of this type which have low energies and appreciable thermal densities at normal temperatures. Excitons and plasmons on the other hand generally have high energies and their thermal density is usually very small. They are mainly observed when they are created with energy supplied from outside the crystal.

2.9 Density of states (a)

The thermal distribution of particles as reflected in the functions (2.37) and (2.39) naturally depends on the energies of the single-particle states. The total thermodynamic function in a crystal will contain the sum over all such states, as for example in (2.38). This sum is a simple uniform sum over $\boldsymbol{k}$ values but it will be convenient to change it to an integral over energies. From (2.17) a typical expression of this form is

$$\sum_i \sum_k f_i(\boldsymbol{k}) = \frac{\mathcal{N}\Omega}{(2\pi)^3} \sum_i \int_{\text{B.Z.}} f_i(\boldsymbol{k})\mathrm{d}\boldsymbol{k} = \int f(E)\mathrm{S}(E)\mathrm{d}E \tag{2.42}$$

Here S(E) is termed the *density of states* per unit energy range. S(E)dE gives the number of states of various $\boldsymbol{k}$ whose energies lie between E and $(E+\mathrm{d}E)$. It is proportional to the volume of the crystal and plays an essential role in the determination of thermodynamic functions. In general the relation between E and $\boldsymbol{k}$ is so complicated that no analytic form for S(E) can be obtained but in simple cases it can be. For free electrons which obey the quadratic energy law (2.20) $E(\boldsymbol{k})$ is independent of direction. The $\boldsymbol{k}$ values for a fixed E lie on a sphere of area $4\pi k^2$ and those with energies between E and $E+\mathrm{d}E$ in a shell of volume $4\pi k^2\,\mathrm{d}k$. Then

$$\frac{\mathcal{N}\Omega}{(2\pi)^3}\int \mathrm{d}\boldsymbol{k} = \frac{\mathcal{N}\Omega}{(2\pi)^3}\int 4\pi k^2\,\mathrm{d}k = \frac{\mathcal{N}\Omega}{(2\pi)^2}\left(\frac{2m}{\hbar^2}\right)^{3/2}\int E^{1/2}\,\mathrm{d}E = \int \mathrm{S}(E)\,\mathrm{d}E \tag{2.43}$$

and S(E) is proportional to $E^{1/2}$ and the total volume $\mathcal{N}\Omega$. If the same particles are confined in two dimensions to an area $\mathcal{A}$ the density of modes in $\boldsymbol{k}$ space, following the argument of §2.5, is $\mathcal{A}/(2\pi)^2$. In this case the area of $\boldsymbol{k}$ space corresponding to energies between E and $E+\mathrm{d}E$ is an annulus of area $2\pi k\mathrm{d}k$. Thus

$$\frac{\mathcal{A}}{(2\pi)^2}\int \mathrm{d}\boldsymbol{k} = \frac{\mathcal{A}}{(2\pi)^2}\int 2\pi k\mathrm{d}k = \frac{\mathcal{A}}{4\pi}\left(\frac{2m}{\hbar^2}\right)\int \mathrm{d}E = \int \mathrm{S}(E)\,\mathrm{d}E \tag{2.44}$$

and S(E) is constant independent of E and proportional to $\mathcal{A}$. In one dimension a similar treatment gives

$$\frac{\mathcal{L}}{2\pi}\int \mathrm{d}k = \frac{\mathcal{L}}{2\pi}\left(\frac{2m}{\hbar^2}\right)^{1/2}\int E^{-1/2}\,\mathrm{d}E = \int \mathrm{S}(E)\,\mathrm{d}E \tag{2.45}$$

These formulae will prove useful in Chapter 4 when we consider the behaviour of electrons in magnetic fields.

For excitations with a linear energy variation $E(\boldsymbol{k}) = \hbar sk$ which is independent of the direction of $\boldsymbol{k}$

$$\frac{\mathcal{N}\Omega}{(2\pi)^3}\int 4\pi k^2\,\mathrm{d}k = \frac{\mathcal{N}\Omega}{2\pi^2(\hbar s)^3}\int E^2\,\mathrm{d}E = \int \mathrm{S}(E)\,\mathrm{d}E \tag{2.46}$$

so that $\mathrm{S}(E) \propto E^2$. Such excitations have S(E) proportional to $\mathcal{A}E$ in two dimensions and S(E) independent of E and proportional to $\mathcal{L}$ in one dimension.

Even though the actual form of S(E) is complicated for an allowed energy band with a general $E(\boldsymbol{k})$, it will contain several well-defined features called *critical points*. These occur at extrema where

$$\nabla_{\boldsymbol{k}} E(\boldsymbol{k}) = 0 \tag{2.47}$$

The simplest examples are the true maximum and minimum of the band where $E(\boldsymbol{k})$ will be a quadratic function of $\boldsymbol{\kappa} = \boldsymbol{k} - \boldsymbol{k}_0$, $\boldsymbol{k}_0$ being the extremum value. Choosing simple axes this will have the form

$$E(\boldsymbol{k}) = E_m + \gamma_x \kappa_x^2 + \gamma_y \kappa_y^2 + \gamma_z \kappa_z^2 \tag{2.48}$$

where E_m is the minimum or maximum value. γ_x, γ_y and γ_z are all positive for a minimum; for a maximum they are all negative. The quadratic dependence

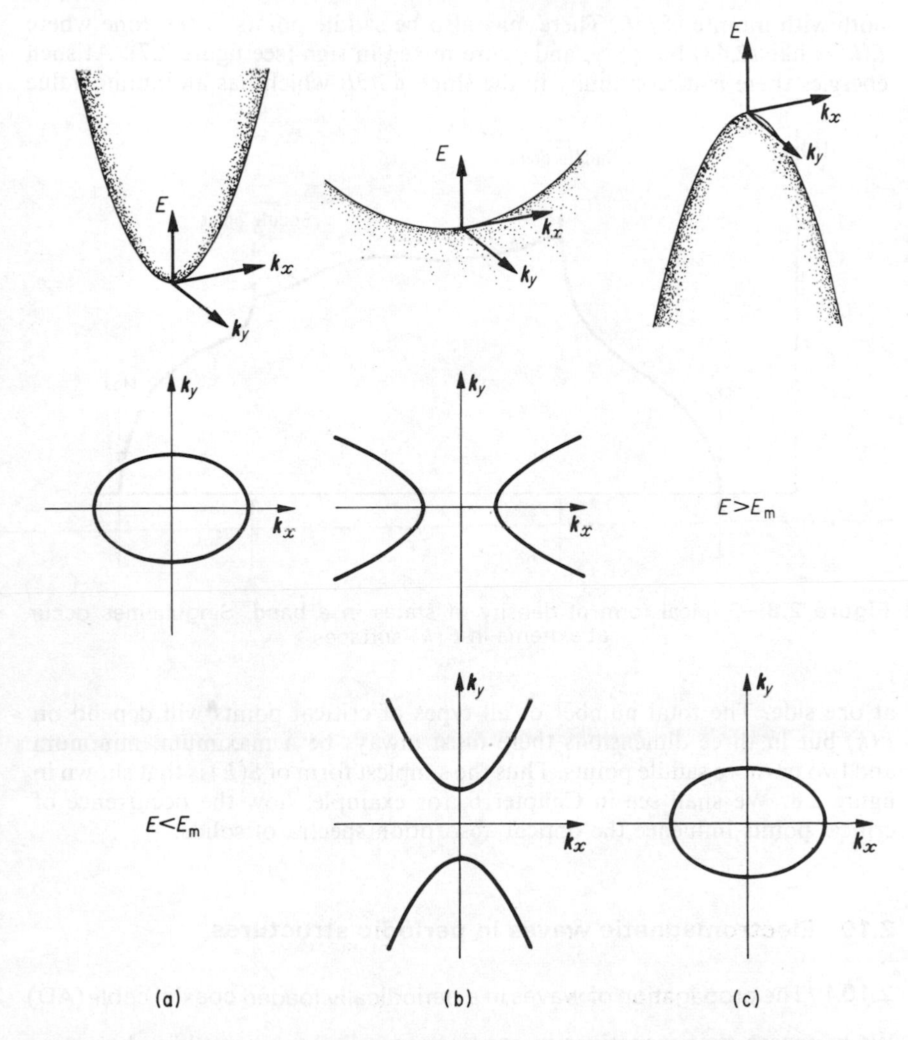

Figure 2.7 Energy surfaces of $E(\boldsymbol{k})$ as a function of $\boldsymbol{k}_x$ and $\boldsymbol{k}_y$ for fixed $\boldsymbol{k}_z$ showing (a) minimum, (b) saddle points and (c) maximum. Also curves of constant energy which are sections of these surfaces at $E \gtrless E_m$. Near the extremes these curves are ellipses or hyperbolae

gives S(E) the same energy dependence as (2.43) with a different multiplying constant depending on γ_x, γ_y and γ_z. Near the minimum

$$S(E) \propto (E - E_m)^{1/2} \tag{2.49}$$

and near a maximum

$$S(E) \propto (E_m - E)^{1/2} \tag{2.50}$$

both with infinite $\partial S/\partial E$. There may also be saddle points in the zone where $E(\boldsymbol{k})$ is like (2.48) but γ_x, γ_y and γ_z are mixed in sign (see figure 2.7). At such energies there is discontinuity in the slope $\partial S/\partial E$ which has an infinite value

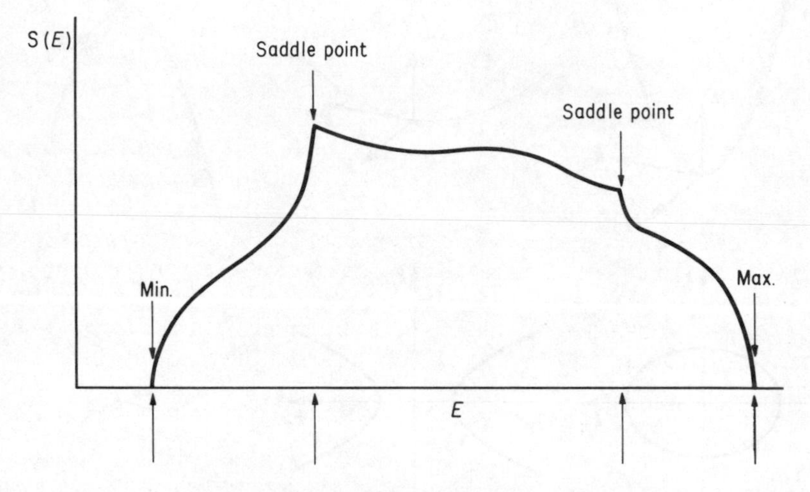

Figure 2.8 Typical form of density of states in a band. Singularities occur at extrema in $E(\boldsymbol{k})$ surfaces

at one side. The total number of all types of critical points will depend on $E(\boldsymbol{k})$ but in three dimensions there must always be a maximum, minimum and two or more saddle points. Thus the simplest form of S(E) is that shown in figure 2.8. We shall see in Chapter 6, for example, how the occurrence of critical points influence the optical absorption spectra of solids.

2.10 Electromagnetic waves in periodic structures

2.10.1 The propagation of waves in a periodically loaded coaxial cable (AD)

We have seen that excitations in crystals generally have wave-like characteristics. The properties of excitations and the operation of the conservation rules which limit their interaction (§2.6) are largely determined by the way in which the energy of the excitations varies with wave vector. This, in turn, is affected by the crystal structure since the excitation is a wave travelling through a periodic structure—the crystal lattice. In the next chapter we shall show how the energy of phonons varies with wave vector and in Chapter 4 we discuss the same relationship with respect to electrons. We have, however, already intimated that we may expect the energy, E, to be a nearly continuous function of wave vector while $\boldsymbol{k}$ lies within a unit cell of reciprocal space. The range of energies so covered constitutes an allowed energy band for the excitation in question (§2.4). There may be one or more such bands separated by energy regions within which the wave vector of the excitation is imaginary so that the wave is attenuated.

The existence of allowed and forbidden energy (frequency) bands is characteristic of waves propagating through a periodic structure and examples occur in many branches of physics and engineering. In the belief that it will be of value to electrical engineers, and not without value to others, we now describe a well-known situation in microwave engineering, namely the propagation of electromagnetic waves through a periodically loaded transmission line. A practical example is a coaxial cable into which small inductances are inserted at periodic (e.g. $\frac{1}{4}$ mile) intervals. Such periodically loaded lines are used in long distance telephone circuits. For consistency with later chapters we use β for the wave vector of an electromagnetic wave (photon) and deduce, by the standard methods of circuit theory, the relationship between frequency, ω, $(=E/\hbar)$ and β.

The relationships between voltage, V, and current, I, at the input (1) and output (2) of a four terminal linear network (figure 2.9) can always be written:

$$\begin{aligned} V_1 &= AV_2 + BI_2 \\ I_1 &= CV_2 + DI_2 \end{aligned} \tag{2.51}$$

or in matrix form

$$\begin{bmatrix} V_1 \\ I_1 \end{bmatrix} = \begin{bmatrix} A & B \\ C & D \end{bmatrix} \cdot \begin{bmatrix} V_2 \\ I_2 \end{bmatrix} \tag{2.52}$$

where A, B, C and D are constants (cf. § 9.4.2). If the network is symmetric so that the input and output are interchangeable $A = D$ and $A^2 - BC = 1$. A section of length ℓ of a very long coaxial cable can be represented by a symmetric four terminal network, the conductors having finite inductance and

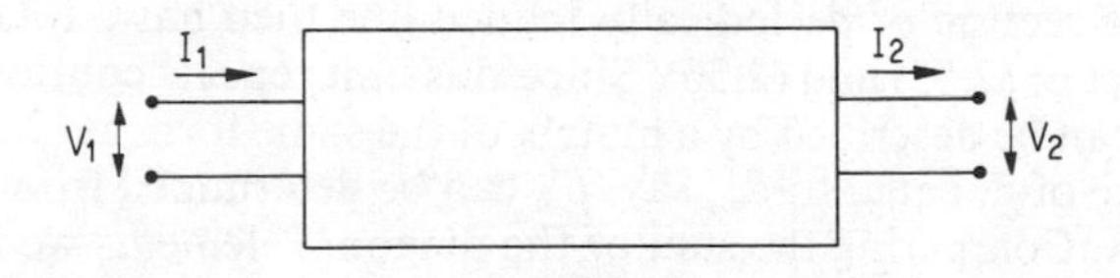

Figure 2.9 Definition of a four terminal network

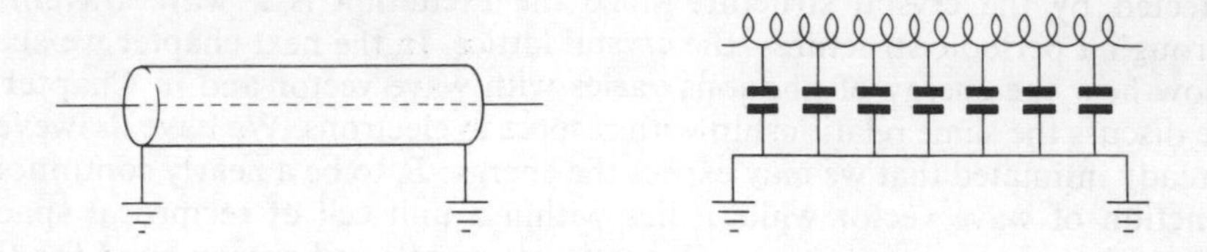

Figure 2.10 A coaxial transmission line and its equivalent circuit

a capacitance between them (figure 2.10). In such a lossless line, the electrical disturbance takes the form of a wave. The amplitude of V and I stays constant but the phase may vary so that

$$\frac{V_2}{V_1} = \frac{I_2}{I_1} = \exp(-i\beta\ell) \tag{2.53}$$

and since adding a section of length ℓ to a very long line cannot change the input impedance, Z_0, we have

$$\frac{V_1}{I_1} = \frac{V_2}{I_2} = Z_0 \tag{2.54}$$

The matrix in (2.52) then follows by substitution of these results into (2.51) and is

$$\begin{bmatrix} \cos(\beta\ell) & iZ_0 \sin(\beta\ell) \\ \dfrac{i \sin(\beta\ell)}{Z_0} & \cos(\beta\ell) \end{bmatrix} \tag{2.55}$$

where $\beta\ell = \omega\ell/c$ and c is the velocity of a wave travelling along the line.

The matrix describing a periodically loaded line, i.e. a line with lumped impedances of value Z_1, inserted at points d apart where $d \ll \ell$, can now be derived by matrix multiplication. The matrix of a series impedance which is physically much smaller than the wavelength is

$$\begin{bmatrix} 1 & Z_1 \\ 0 & 1 \end{bmatrix} \tag{2.56}$$

since (2.51) then takes the form $V_1 - V_2 = Z_1 I_1$ and $I_1 = I_2$.

A complete section of periodically loaded line then has a total effect given by the product of (2.55) and (2.56). Since this unit repeats continuously after a distance d it can be described by a matrix of the same form as (2.55) but with a different value of β, equal to β_m, say. β_m can be determined from the elements of this matrix. Comparing the sum of the diagonal elements we find

$$\cos(\beta_m d) = \cos(\omega d/c) + (iZ_1/2Z_0) \sin(\omega d/c) \tag{2.57}$$

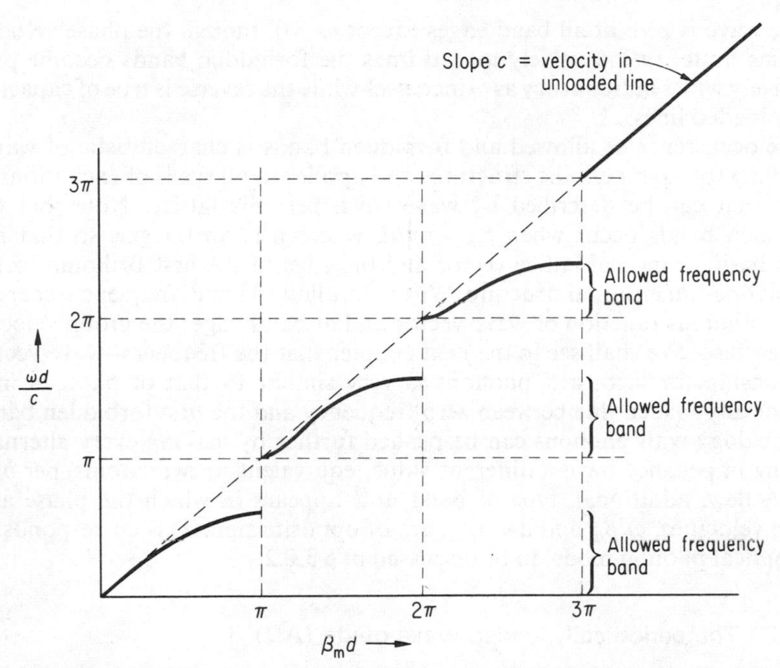

Figure 2.11 Dispersion characteristic of a long coaxial line into which inductances have been inserted at periodic intervals d

If the inserted loads are inductances, as in telephone cables, $Z_1 = i\omega L$ and the second term on the right-hand side is real and negative. Equation (2.57), with $Z_1 = i\omega L$, is illustrated in figure 2.11. It will be clear from (2.57) that real values of β_m are possible if

$$-1 < \cos(\beta_m d) < +1 \tag{2.58}$$

and these conditions correspond to the allowed frequency bands of figure 2.11. Use of (2.58) in conjunction with (2.57) shows that an electromagnetic wave may propagate in the inductively loaded line if its frequency conforms to

$$0 < \frac{\omega d}{c} < \pi - \sin^{-1}\left[\frac{\omega L/Z_0}{1 + (\omega L/2Z_0)^2}\right]$$

but between the latter upper limit and $\omega d/c = \pi$ there is a band within which β_m is imaginary. Equation (2.53) then shows that the wave is strongly attenuated.* As ω is further increased transmission is again possible at $\omega d/c = \pi$ and successive bands appear within which propagation is allowed and forbidden. It is easy to show by differentiation of (2.57) that the group velocity

* Transmission is, of course, still possible if ℓ is small enough, i.e. of the order of the wavelength. Electron 'tunnelling' through thin insulating films is a solid state equivalent of this process (see § 9.2.5).

of the wave is zero at all band edges except $\omega = 0$, though the phase velocity remains finite. In inductively loaded lines the forbidden bands become progressively wider in frequency as ω increases while the reverse is true of capacitatively loaded lines.

The occurrence of allowed and forbidden bands is characteristic of waves travelling through periodic structures and applies to all types of excitations in solids that can be described by waves in a periodic lattice. Note that the forbidden bands occur when $\beta_m = n\pi/d$, where n is an integer, so that π/d is the basic reciprocal lattice vector and the edge of the first Brillouin 'zone' for this one-dimensional structure. Within an allowed band, frequency (energy) is a continuous function of wave vector and at band edges the group velocity is often zero. We shall see in the next chapter that the frequency–wave vector relationship for 'acoustic' phonons is very similar to that of photons in a periodically loaded line between zero frequency and the first forbidden band. The analogy with phonons can be pushed further by making every alternate loading impedance have a different value, equivalent to two 'atoms' per unit cell. A new, additional, type of band now appears in which the phase and group velocities, ω/β_m and $d\omega/d\beta_m$, are of opposite sign. This corresponds to the 'optical phonon mode' to be discussed in § 3.2.2.

2.10.2 The periodically loaded wave guide (AD)

Wave guides are quantitatively more complicated than coaxial lines because Z_0 and ω/β are frequency dependent even when the guide is unloaded. Qualitatively, however, some instructive analogues of solid state behaviour can be found.

Many ingenious ways of periodically loading a wave guide have been devised. One method is to insert metal plates, a central hole having been cut in each plate, normal to the propagation direction and at regular intervals along the guide. The cross section of such a guide is illustrated in figure 2.12. This,

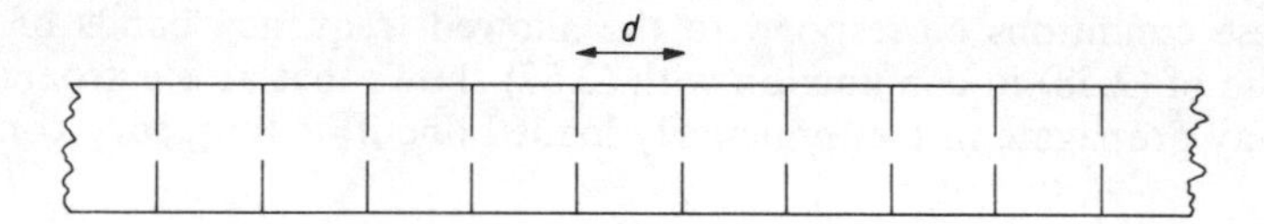

Figure 2.12 The cross section of a waveguide into which metal plates, with a hole in the centre of each plate, have been inserted at periodic intervals

incidentally, is the structure adopted in linear accelerators where it is necessary to reduce the phase velocity of the electromagnetic wave in the structure to equal the group velocity of electrons fired as a beam through the holes. That periodic loading can reduce the wave velocity is evident from figure 2.11, the main features of which apply in this case. The most striking difference between wave guide and coaxial line is that in the former propagation is not possible until ω exceeds a critical value determined by the cross-sectional dimensions.

Hence there is a forbidden band at zero frequency and alternate allowed and forbidden bands thereafter. The photon group velocity is zero at all band edges.

When analysing a structure of the type shown in figure 2.12 two extreme approximations are possible. First, if the holes are very small the structure resembles an assembly of resonant cavities weakly coupled together through the holes. An equivalent extreme approximation, called 'tight binding', is adopted in the analysis of the band structure of electrons in solids (§ 4.2.2). The resonant frequency of each cavity is only slightly affected by its neighbours and only narrow allowed frequency bands, separated by wide forbidden gaps, appear. The frequencies of the allowed bands are near the resonant frequencies of the cavities.

A second extreme approximation is possible when the holes are very large, almost equal to the cross-sectional area of the guide. The frequency–wave vector relationship is then similar to that of an unloaded guide and the loads are only a small perturbation. This is the microwave equivalent of the 'nearly free electron' approximation of electron band theory (§ 4.2.1). The effect of the loads is to produce narrow forbidden frequency bands within an otherwise continuous variation of ω with β. These forbidden bands occur whenever the reflections from the loads add up in phase, i.e. whenever

$$d = \mathrm{n}\lambda_{\mathrm{m}}/2 \quad \text{or} \quad \beta_{\mathrm{m}} = \mathrm{n}\pi/d$$

as in the periodically loaded coaxial line.

Throughout this discussion we have assumed that the coaxial line or wave guide was essentially infinitely long. We could, of course, have used periodic

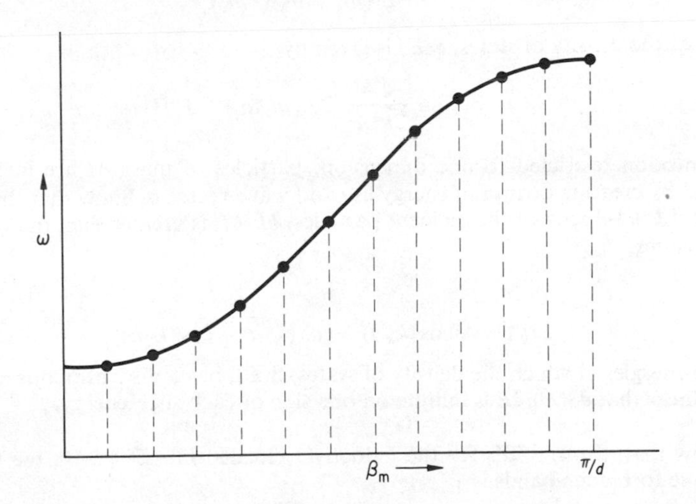

Figure 2.13 Dispersion characteristic of a periodically loaded waveguide of finite length, indicating resonances

boundary conditions, as in § 2.5. To permit a final analogy between periodically loaded wave guide and excitations in solids consider a finite length of guide containing $\mathcal{N}$ coupled cavities each of length d (figure 2.12) and terminated by a reflector at one end. Resonances will occur whenever the length of the guide, $\mathcal{N}d$, is an integral number of half wavelengths or

$$\beta_m = n\pi/\mathcal{N}d$$

and in the range $0 < \beta_m < \pi/d$ there will be $\mathcal{N}$ such resonances equally spaced along the β_m axis, as shown in figure 2.13. Due to dispersion, they will not be equally spaced along the ω-axis but crowded towards the band edges where $d\omega/d\beta_m$ tends to zero. These resonances correspond with the available states that excitations may occupy (§ 2.9) in a one-dimensional structure. Since they are equally spaced in one dimension in wave vector space their density increases as $\beta_m^2 \, d\beta_m$ in a three-dimensional microwave system or crystal, as in § 2.9.

PROBLEMS

(Some answers, where appropriate, are given on page 481.)

2.1 Prove that the density of modes in a one-dimensional crystal of length $\mathcal{L}$ in the x-direction is $(\mathcal{L}/2\pi)\,dk_x$ and that in a two-dimensional crystal of area $\mathcal{A}$ it is $[\mathcal{A}/(2\pi)^2]\,dk_x\,dk_y$. Find the density of states per unit energy range in each case if $E = ck^n$.

2.2 If the energy of a crystal excitation varies with wave vector $\boldsymbol{k}$ like

$$E(k) = \frac{\hbar^2}{2}\left(\frac{k_x^2}{m_x} + \frac{k_y^2}{m_y} + \frac{k_z^2}{m_z}\right)$$

show that the density of states, $S(E)$ is given by

$$S(E) = \frac{\mathcal{N}\Omega}{2\pi^2 \hbar^2}(2m_x m_y m_z)^{1/2} E^{1/2}$$

2.3 In motion restricted to one dimension, particles of mass M are inelastically scattered by creating quanta of energy $\hbar sq$ and wave vector $\boldsymbol{q}$. Show that this is only possible if the velocity of the incident particles, $\hbar k/M$, is greater than the velocity s of the quanta.

2.4 If

$$E(k) = A[\cos(k_x a) + \cos(k_y a) + \cos(k_z a)]$$

find the energies at which the density of states, $S(E)$, has a discontinuous change of slope. Show that $\partial S(E)/\partial E$ is infinite on one side of each such energy.

2·5 Show that, if $\omega L \ll 2Z_0$ for the inductively loaded line of § 2.10, the width of successive forbidden bands is

$$\Delta\omega = n\pi L c^2/d^2 Z_0$$

where n is an integer and c the velocity in the unloaded line.

3

Lattice vibrations

3.1 Introduction (a)

Using the adiabatic approximation described in § 2.1 every pair of heavy atomic nuclei has an effective interaction V between them such that the potential energy curve forms a well which varies with distance as in figure 1.10. For low energies this well may be regarded as parabolic and the force, given by $-\partial V/\partial R$ then varies linearly with distance. The system behaves like an array of particles joined by harmonic springs. The classical motion of such a system is the well-known one of small oscillations and it may be described in terms of *normal modes*, i.e. independent motions of characteristic frequency. If one nucleus of such a network is set in motion this motion is soon transmitted by the springs to all the others. In a normal mode all the atoms will participate in the motion with well-defined amplitude and phase and if the motion is begun in this way it will persist in the same form at a fixed frequency. For example, if an elastic string is plucked it will move with components of motion at all harmonic frequencies. However, if it is started in motion as a simple sine wave it will keep the same form and oscillate at a fixed frequency.

The problem of finding the normal modes and characteristic frequencies of a crystal is a classical one. The high degree of symmetry is very helpful in determining these modes. As was shown in Chapter 2 the modes must have the same amplitude in each cell, but will vary from cell to cell across the crystal like a wave with a particular wave vector $\boldsymbol{q}$. The crystal will have independent oscillations in these modes. The quantum mechanical description of the system is quite straightforward. Each oscillator will have its energy quantised with energy levels equally spaced at $(\mathrm{n}+\frac{1}{2})\hbar\omega$. A full description of the state of the crystal will have the occupation number n specified as a positive integer or zero for each type of oscillator. Even in the lowest energy state allowed by quantum mechanics ($\mathrm{n}=0$) the energy of each oscillator is not zero but $\frac{1}{2}\hbar\omega$. This is usually referred to as the *zero-point motion*.

It is convenient to associate the concept of a particle which we call a phonon with the quanta of vibrational energy. The situation is entirely analogous with electromagnetic waves where the quanta are called photons. If one type of oscillator is excited into its nth state we can say that n phonons of this type are present. If the oscillator passes from its nth to its $(\mathrm{n}+1)$th state in a transition we say a phonon has been created in the process; in passing from the nth to the $(\mathrm{n}-1)$th state a phonon has been destroyed.

3.2 Vibrations of a one-dimensional atomic chain

3.2.1 Monatomic linear chain (a)

The simple problem of a linear chain of atoms of mass M bound to each neighbour by a spring obeying Hooke's law will serve to show the essential features of the problem while keeping the mathematics simple. We assume there are $\mathscr{N}$ atoms and periodic boundary conditions so that atom $\mathscr{N}$ is attached to atom 1 as if it were atom $\mathscr{N}+1$ (i.e. the chain is bent into a ring). In equilibrium they are each separated by a distance a, and the displacement of atom n from this position is called u_{n} (see figure 3.1). Motion is assumed

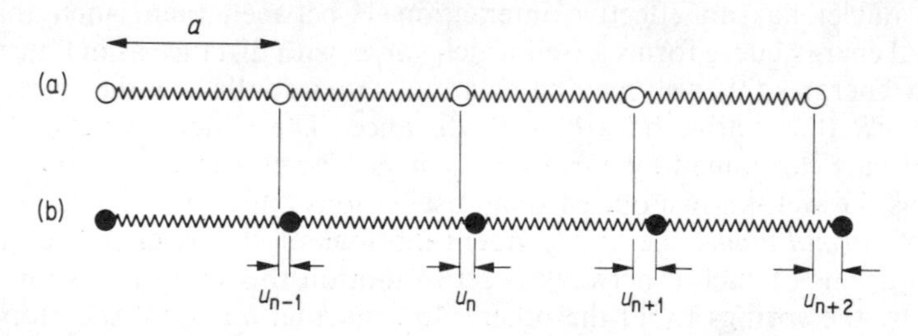

Figure 3.1 Displacement of a linear chain of identical atoms joined by springs

to be confined to the x-direction, that of the chain. The potential energy of the system is

$$V=\sum_{\mathrm{n}=1}^{\mathscr{N}}\tfrac{1}{2}\Lambda(u_{\mathrm{n}+1}-u_{\mathrm{n}})^2 \tag{3.1}$$

and differentiation gives the force on the nth atom. Λ is the constant defined in Hooke's law. Newton's equation of motion for the nth atom is

$$M\frac{\mathrm{d}^2 u_\mathrm{n}}{\mathrm{d}t^2} = \Lambda[(u_{\mathrm{n}+1} - u_\mathrm{n}) + (u_{\mathrm{n}-1} - u_\mathrm{n})] \tag{3.2}$$

The two terms on the right represent the forces to the right from the stretched and compressed springs to right and left respectively. There is an equation like this for each atom. Now, as discussed in Chapter 2, we expect from symmetry a normal mode solution in the form

$$u_\mathrm{n} = \xi \exp \mathrm{i}(qna - \omega t) \tag{3.3}$$

i.e. a wave varying like $\exp(iqx)$ along the line with arbitrary amplitude ξ. Equation (3.3) is clearly a solution of (3.2) since on substituting in that equation

$$-M\xi\omega^2 \exp \mathrm{i}(qna - \omega t) = \Lambda\xi[\exp(iqa) + \exp(-iqa) - 2] \exp \mathrm{i}(qna - \omega t) \tag{3.4}$$

which holds for any value of ξ provided

$$\omega^2 = (2\Lambda/M)(1 - \cos qa) \tag{3.5}$$

or

$$\omega(q) = 2(\Lambda/M)^{1/2} \sin \tfrac{1}{2}qa \tag{3.6}$$

if we choose the positive root.

This ω is plotted as a function of q as figure 3.2. It is symmetric between q and $-q$, i.e. waves to the left and the right are identical. It also repeats after each $2\pi/a$. It was shown in § 2.4 that q need not be defined outside the first Brillouin zone. For a linear chain the reciprocal lattice is also a chain

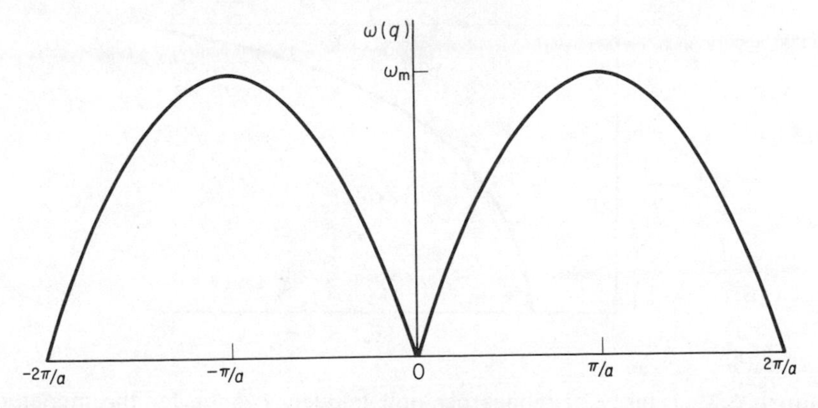

Figure 3.2 $\omega(q)$ curves for a monatomic linear chain. The pattern repeats every $2\pi/a$ and need only be defined in the Brillouin zone: $-\pi/a < q < \pi/a$

with $Q_n = 2\nu\pi/a$. The Brillouin zone stretches from $-\pi/a$ to $+\pi/a$, which is just the distance over which the curve in figure 3.2 repeats. With $\mathcal{N}$ atoms and periodic boundary conditions there are just $\mathcal{N}$ allowed values of q in this range. For $\mathcal{N}$ atoms there are $\mathcal{N}$ degrees of freedom so all the modes have been found. There is one branch to the vibrational spectrum given by (3.6).

One or two features of the $\omega(q)$ curve are of interest. When q is small, i.e. long wavelengths, $\omega = (\Lambda/M)^{1/2}qa$ and so is linear in q. This is the form of sound waves in a continuum and since at long wavelengths many atoms are vibrating together the discrete nature of the lattice does not show itself. The displacement u may be regarded as a continuous function of position x and $u(x)$ expanded in a Taylor series. Then

$$u_{n+1} = u(x+a) = u(x) + a\frac{\partial u(x)}{\partial x} + \tfrac{1}{2}a^2\frac{\partial^2 u(x)}{\partial x^2} \tag{3.7}$$

Equation (3.2) then becomes the simple wave equation

$$\frac{1}{s^2}\frac{\partial^2 u}{\partial t^2} = \frac{\partial^2 u}{\partial x^2} \tag{3.8}$$

with the velocity of sound, s, being given by $s^2 = \Lambda a^2/M$.

As q increases and the wavelength shortens to something approaching the interatomic spacing the curve bends away from the linear law as the discrete structure of the chain becomes important. At the zone edge $q = \pi/a$ (wavelength $\lambda = 2a$), the group velocity is zero and no waves of higher frequency can propagate.

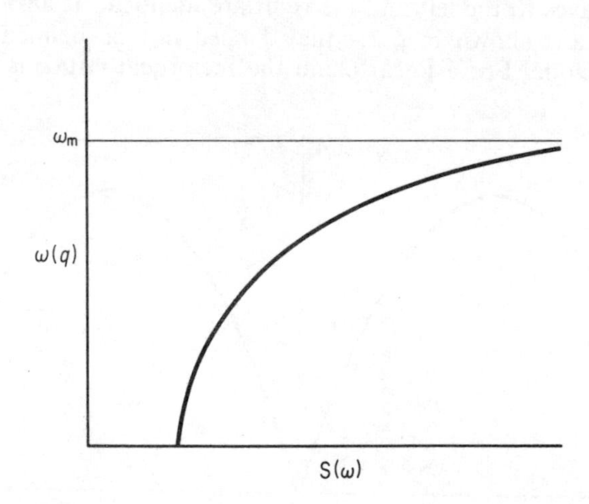

Figure 3.3 Density of modes per unit frequency range for the monatomic linear chain [equation (3.9)]. It diverges at ω_m where the $\omega(q)$ curve of figure 3.2 is flat and tends to a constant as $q \to 0$ (cf. § 2.9)

The density of modes as defined in (2.43) is given by

$$\frac{\mathscr{L}}{2\pi}\int_{-\pi/a}^{\pi/a} dq = \frac{2\mathscr{L}}{2\pi}\int_0^{\omega_m}\left(\frac{\partial\omega}{\partial q}\right)^{-1} d\omega = \frac{\mathscr{L}}{\pi}\int_0^{\omega_m}\frac{d\omega}{\frac{1}{2}a\omega_m\cos\frac{1}{2}qa}$$

$$= \frac{2\mathscr{L}}{a\pi}\int_0^{\omega_m}(\omega_m^2-\omega^2)^{-1/2}\,d\omega$$

where the density of modes

$$S(\omega) = (2\mathscr{L}/\pi a)(\omega_m^2-\omega^2)^{-1/2} \qquad (0<\omega<\omega_m) \tag{3.9}$$

is shown in figure 3.3. The maximum frequency $\omega_m = 2(\Lambda/M)^{1/2}$.

3.2.2 Diatomic linear chain (a)

A linear chain with two types of atoms with masses M and M' alternating regularly also shows instructive features. We take a chain with $2\mathscr{N}$ atoms—$\mathscr{N}$ unit cells each of length $2a$—and periodic boundary conditions. Each type of atom is coupled by identical springs to its neighbours. The potential energy is

$$V = \sum_{n=1}^{\mathscr{N}} \Lambda(u_n - u_n')^2 + \Lambda(u_n' - u_{n+1})^2 \tag{3.10}$$

The equation of motion is now slightly different for each atomic type

$$\left.\begin{aligned} M\frac{\partial^2 u_n}{\partial t^2} &= \Lambda(u_n' + u_{n-1}' - 2u_n) \\ M'\frac{\partial^2 u_n'}{\partial t^2} &= \Lambda(u_{n+1} + u_n - 2u_n') \end{aligned}\right\} \tag{3.11}$$

Again we expect each like atom to have the same amplitude but the different atom types to have different relative amplitude. We assume a form which varies from cell to cell like $\exp(iqx)$, namely:

$$u_n = \xi\exp i(2nqa - \omega t), \qquad u_n' = \xi'\exp i[(2n+1)qa - \omega t] \tag{3.12}$$

This is in fact a solution of (3.11) if

$$\text{and} \quad \left.\begin{aligned} -M\omega^2\xi &= \Lambda\{\xi'[\exp(iqa)+\exp(-iqa)] - 2\xi\} \\ -M'\omega^2\xi' &= \Lambda\{\xi[\exp(-iqa)+\exp(iqa)] - 2\xi'\} \end{aligned}\right\} \tag{3.13}$$

i.e. if

$$\omega^2(q) = \Lambda\left(\frac{1}{M}+\frac{1}{M'}\right) \pm \Lambda\left[\left(\frac{1}{M}+\frac{1}{M'}\right)^2 - \frac{4\sin^2 qa}{MM'}\right]^{1/2} \tag{3.14}$$

and

$$\frac{\xi}{\xi'} = \frac{2\Lambda\cos qa}{2\Lambda - M\omega^2} = \frac{2\Lambda - M'\omega^2}{2\Lambda\cos qa} \tag{3.15}$$

The form of $\omega(q)$ is plotted in figure 3.4. There are now two separate solutions which repeat every π/a. This is to be expected since the unit cell here has size $2a$ and the first zone runs from $-\pi/2a$ to $\pi/2a$. There are $\mathcal{N}$ values of q in this range and so two complete branches are required to provide the $2\mathcal{N}$ normal modes corresponding to the $2\mathcal{N}$ degrees of freedom of $2\mathcal{N}$ atoms confined to a line.

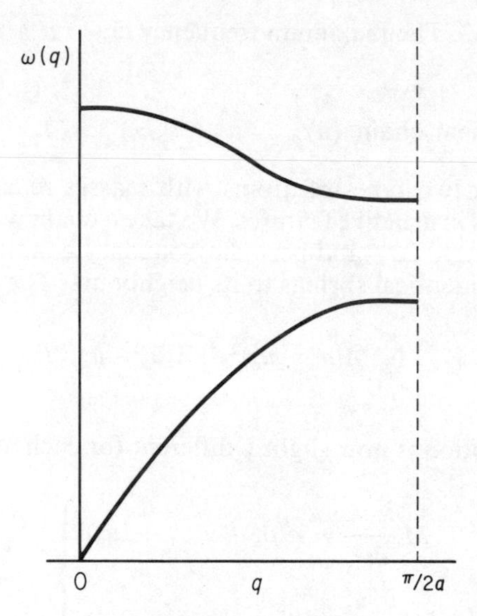

Figure 3.4. $\omega(q)$ curve for diatomic linear chain showing two branches. The upper branch is called optic and the lower acoustic. Only half the Brillouin zone $0 < q < \pi/2a$ is shown

The two modes of the same q have different relative motions of the two atoms in the unit cell. For small q the negative sign in (3.14) gives

$$\omega(q) = \left(\frac{2\Lambda}{M+M'}\right)^{1/2} qa; \qquad \frac{\xi}{\xi'} = 1 \tag{3.16}$$

and the positive sign gives

$$\omega(q) = \left[\frac{2\Lambda(M+M')}{MM'}\right]^{1/2}\left(1 - \frac{MM'q^2a^2}{2(M+M')^2}\right); \qquad \frac{\xi}{\xi'} = -\frac{M'}{M} \tag{3.17}$$

The lower frequency branch has the typical characteristics of a sound wave, and for small q the detailed structure of the chain has no effect on the mode. For this reason the lower branch is called the *acoustic branch.* In the higher mode the two atoms in the cell move in opposite directions (ξ/ξ' is negative in (3.17); see also figure 3.5). If the two atoms had opposite charges as in a

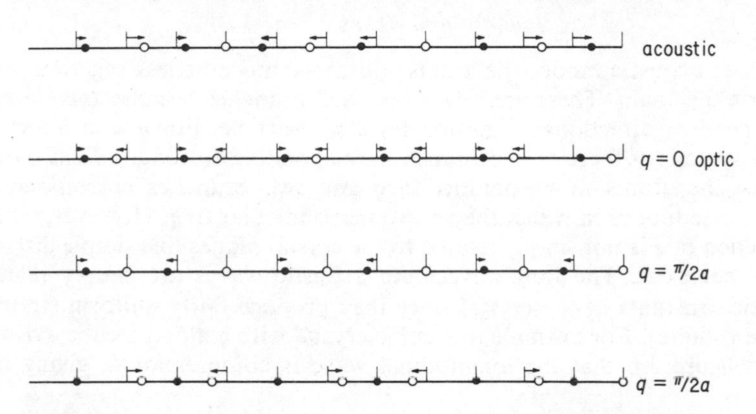

Figure 3.5 Displacement of atoms in various modes in a diatomic chain. (a) acoustic, (b) $q = 0$ optic, (c) and (d) $q = \pi/2a$ at zone boundary

simple ionic crystal such a mode would create a set of oscillating dipoles. These would be strongly coupled to an electric field and would be set in motion by a field of the right frequency. For most solids this frequency is similar to that of infrared light and for this reason the upper branch is called an *optic branch* (see Chapter 6). Away from $q = 0$ no such simple distinction exists. For example at the zone boundary, $q = \pi/2a$, the two solutions are

$$\left.\begin{aligned} &\omega(q) = (2\Lambda/M)^{1/2}; \qquad \xi' = 0 \\ \text{and} \qquad &\omega(q) = (2\Lambda/M')^{1/2}; \qquad \xi = 0 \end{aligned}\right\} \tag{3.18}$$

with one type of atom stationary in each case.

As $M \to M'$ this spectrum becomes like that of the monatomic chain. The gap at $q = \pi/2a$ disappears and the top branch should be folded out to cover the region $\pi/2a$ to π/a.

3.3 Vibrational modes of crystals

3.3.1 Modes of a general lattice (a)

Using the symmetry properties discussed in Chapter 2 and analogies from the detailed treatment of linear chains in § 3.2 it is now possible to see the essential form of the mode spectrum in a general three-dimensional lattice. Suppose we have a crystal with p atoms in each cell and $\mathcal{N}$ cells. There will be $3p\mathcal{N}$ degrees of freedom and hence $3p\mathcal{N}$ modes. These modes will be described by wave vectors $\boldsymbol{q}$ which are uniformly distributed over the Brillouin zone. By (2.18) there are $\mathcal{N}$ such allowed $\boldsymbol{q}$ values together. Thus the frequencies $\omega(\boldsymbol{q})$ will form continuous functions over the zone, and to accommodate all the modes there must be 3p branches of the spectrum.

Of these branches three will be acoustic such that

$$\omega(q) \to sq \qquad \text{as} \qquad q \to 0$$

In these acoustic modes the atoms will move more or less together, as they did on the chain. There are now three such branches because there are three independent directions of motion for each particle. For $\boldsymbol{q}$ in a direction of high symmetry there is one branch corresponding to longitudinal modes in which the atoms move parallel to $\boldsymbol{q}$ and two branches corresponding to transverse modes in which they move perpendicular to $\boldsymbol{q}$. However, when the direction of $\boldsymbol{q}$ is not simply related to the crystal planes this simple distinction does not exist. The long-wavelength acoustic waves are simply related to elastic constants of the crystal since they produce fairly uniform strain over large volumes. For example in a cubic crystal with $\boldsymbol{q}$ along a cube axis we see from figure 3.6 that the longitudinal wave is compressional, giving rise to

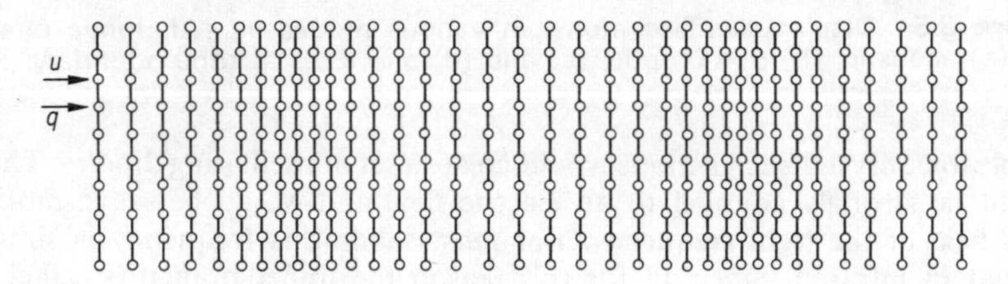

Figure 3.6 Compressional (longitudinal) acoustic wave in a cubic crystal. The displacement of the atoms is parallel to $\boldsymbol{q}$ and gives alternate regions at compressed and extended crystal

alternate dense and rarefied regions. No lateral strain takes place and the motion is governed by the axial modulus for this direction. It is normally called C_{11} and is given by (longitudinal load per unit area of cross section)/(increase in length per unit length). On a microscopic scale it relates the x-component of the force to displacement in the x-direction u_x by

$$X_x = C_{11} \frac{\partial u_x}{\partial x}$$

The total force on a small element of width δx is

$$X_x(x + \delta x) - X_x(x) = \delta x \frac{\partial X_x}{\partial x} \tag{3.19}$$

giving an equation of motion

$$\rho \frac{\partial^2 u_x}{\partial t^2} = C_{11} \frac{\partial^2 u_x}{\partial x^2} \tag{3.20}$$

where ρ is the density. The velocity of these longitudinal waves is then

$$s_l = (C_{11}/\rho)^{1/2} \tag{3.21}$$

The transverse waves on the other hand give rise to a shear strain (see figure 3.7). This is governed by the appropriate rigidity modulus C_{44} given by

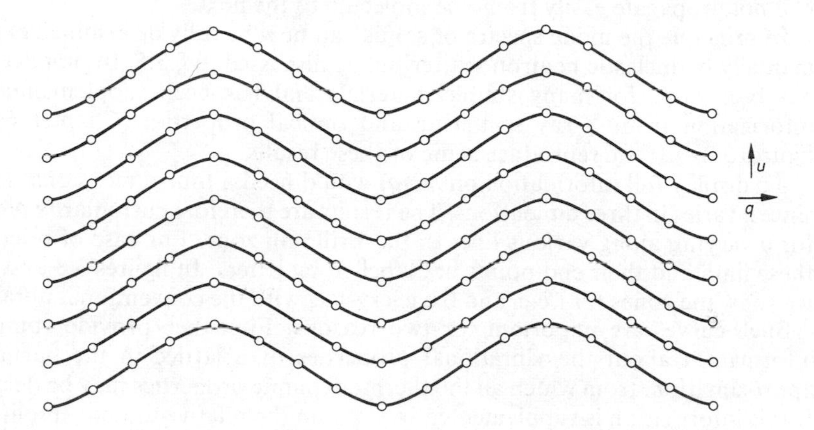

Figure 3.7 Shear (transverse) acoustic wave in a cubic crystal. The atomic displacements are perpendicular to $\boldsymbol{q}$

(tangential force per unit area)/(angular deformation). The force in the y-direction is related to displacement u_y in the y-direction by

$$Y_x = C_{44}\frac{\partial u_y}{\partial x} \tag{3.22}$$

The total force on a small slab is

$$Y_x(x + \delta x) - Y_x(x)$$

and again the equation of motion takes the form

$$\rho\frac{\partial^2 u_y}{\partial t^2} = C_{44}\frac{\partial^2 u_y}{\partial x^2} \tag{3.23}$$

and the wave velocity is $s_t = (C_{44}/\rho)^{1/2}$. The velocities of the waves in general directions are related to the stiffness constants in a much more complicated way.

The other 3(p − 1) branches are optic branches in that $\omega(\boldsymbol{q}) \to$ constant as $\boldsymbol{q} \to 0$ although not all of them will produce dipole moments which interact strongly with an electric field. The rotational and other types of symmetry in the unit cell will affect the details of the spectrum. For example, in a cubic crystal some optic modes at $\boldsymbol{q} = 0$ will be three-fold degenerate because of the equivalence of motion along the direction of the three crystal axes. All the optic modes at $\boldsymbol{q}=0$ are characterised by different relative motion of the atoms in the cell.

If the crystal has well-defined molecular constituents we expect branches corresponding to the internal motion of the molecule. Such branches will tend to be flat in $\boldsymbol{q}$ space, particularly when they are at high frequencies, since they will not propagate easily from one molecule to the next.

In principle the mode spectra of solids can now be fully determined experimentally by inelastic neutron scattering, as discussed in § 3.5. In practice this has been done for many simple materials and has been supplemented by information from X-ray scattering and optical properties (Chapter 6). In figures 3.8–3.16 we reproduce some of these results.

To display full information on $\hbar\omega(\boldsymbol{q})$ would need a four-dimensional figure since $\boldsymbol{q}$ varies in three dimensions. The results are therefore customarily plotted for $\boldsymbol{q}$ varying along various lines in the Brillouin zone. For ease of reference these lines and their end points are labelled by letters. In figures 3.8 and 3.11 we show the zones for f.c.c. and b.c.c. crystals with the conventional notation.

Such curves are important for two reasons. First they provide complete information about the vibrational properties of a lattice in the harmonic approximation, from which all the thermodynamic properties may be derived. If this information is supplemented by data on the relative atomic amplitudes of the modes, which may also be determined from the intensity of neutron scattering, many other properties may be derived. In the second place the results may in principle be analysed to find the detailed force constants between the atoms and hence to learn about atomic forces in solids. It is convenient to use a general potential energy function which, in the harmonic approximation, is a quadratic function of the atomic displacements as in (3.1) and (3.10). If we specify the cell centred at position $\boldsymbol{R}_l$ by l we must define three perpendicular displacements for each of the p atoms in this cell by $u_\alpha(l)$ where α has 3p values. The potential energy, built from expressions like those in (3.1) will now be of the form

$$\sum_{\alpha,\beta}\sum_{l,l'}\Lambda_{\alpha\beta}(l,l')\,u_\alpha(l)\,u_\beta(l') \tag{3.24}$$

The Λ are called the force constants. Because of the translational symmetry they depend only on the relative position

$$\boldsymbol{R}=\boldsymbol{R}_l-\boldsymbol{R}_{l'}$$

and not the actual position $\boldsymbol{R}_l$ in the crystal. The range and the form of Λ are of particular interest in determining the origin of interatomic forces. The characteristic frequencies $\omega_\mathrm{i}^2(\boldsymbol{q})$ as in (3.14) depend on the Fourier transform

$$\Lambda_{\alpha\beta}(\boldsymbol{q})=\sum_{\boldsymbol{R}}\Lambda_{\alpha\beta}(\boldsymbol{R})\exp(\mathrm{i}\boldsymbol{q}\,.\,\boldsymbol{R}) \tag{3.25}$$

3.3.2 Rare gas crystals (A)

Figure 3.8 shows the frequency spectrum for the rare gas solid neon which is face-centred cubic (see figure 1.5). The reciprocal lattice is body-centred cubic and the Brillouin zone is a truncated octahedron as shown in figure 1.4. The curves are shown for three directions of $\boldsymbol{q}$ in the zone. In the (1,0,0) direction (q points labelled Δ) and the (1,1,1) direction (labelled Λ) there are two degenerate transverse modes (i.e. two modes of the same frequency for

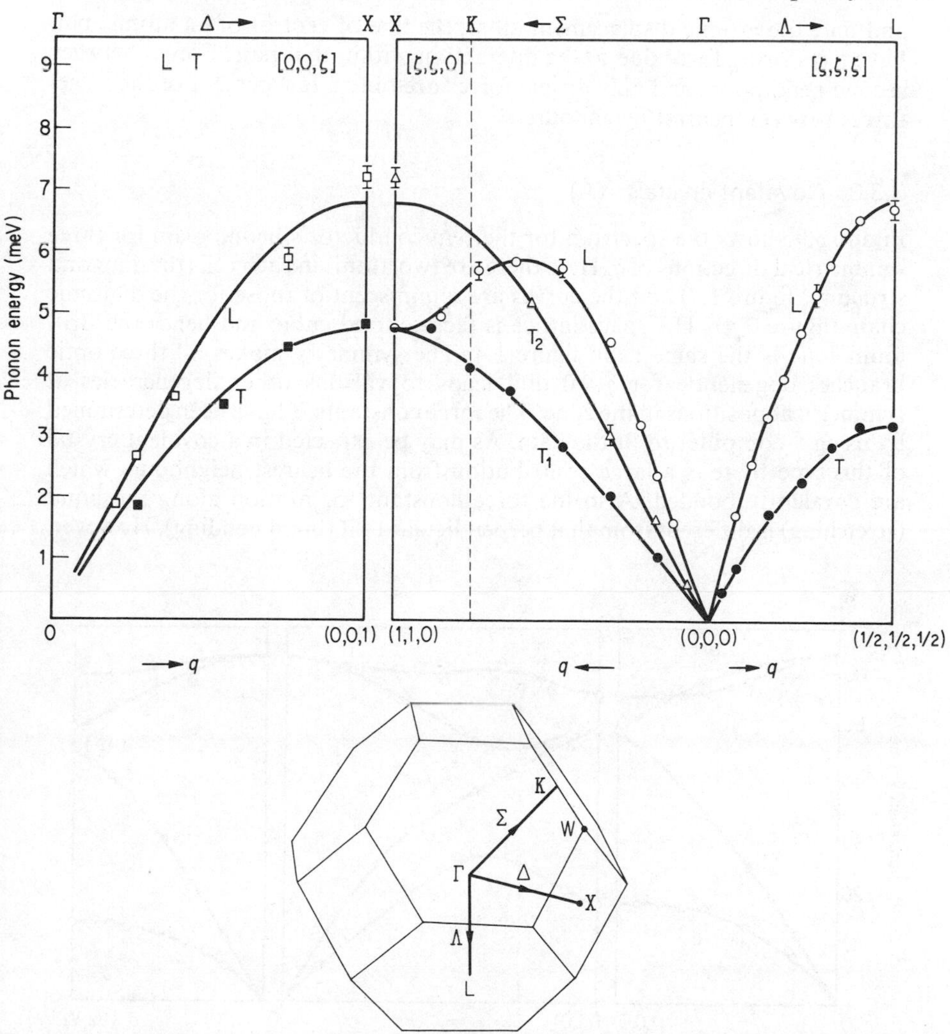

Figure 3.8 Measured vibrational spectrum $\hbar\omega(\boldsymbol{q})$ for various directions in the zone of f.c.c. Ne at 4.7 K (after J. A. Leake *et al.*, *Phys. Rev.* **181**, 125 (1969)). The lines of $\boldsymbol{q}$ depicted are shown in a Brillouin zone, with conventional lettering. $\boldsymbol{q}$ is given in units of $2\pi/a$

each $\boldsymbol{q}$) so that only two curves are seen. All the curves have a simple shape similar to figure 3.2 for a linear chain. This shows that Λ is here of short-range—as is to be expected in a Van der Waals crystal—so that only the nearest neighbours are important. The forces are also found to be what is called central, i.e. $V(\boldsymbol{R})$ depends only on the length of $\boldsymbol{R}$. In this case the harmonic part of the potential energy is

$$(\partial^2 V/\partial R^2)\,[(\boldsymbol{R}\,.\,\boldsymbol{u})/R]^2$$

and only the atomic displacement along the line of centres of an atomic pair have a restoring force due to the interaction within that pair. Forces between second neighbours and anisotropic forces are only a few percent of the larger forces between nearest neighbours.

3.3.3 Covalent crystals (A)

Figure 3.9 shows the spectrum for the semiconductor silicon, again for three symmetrical directions of $\boldsymbol{q}$. Here there are two atoms in each cell (the diamond structure, figure 1.7) and the curves are reminiscent of those for the diatomic chain (figure 3.4). The space lattice is face-centred cubic and hence the Brillouin zone is the same as in figure 1.4. The symmetry makes all three optic branches degenerate for $\boldsymbol{q} = 0$ and leads to various other degeneracies at symmetrical positions in the zone. The force constants Λ have been determined by using a computer to fit the data. As may be expected in a covalent crystal of this type there is a large contribution from the nearest neighbours which are covalently bonded. Also the force constant for motion along the bond (stretching) is different from that perpendicular to it (bond bending). However,

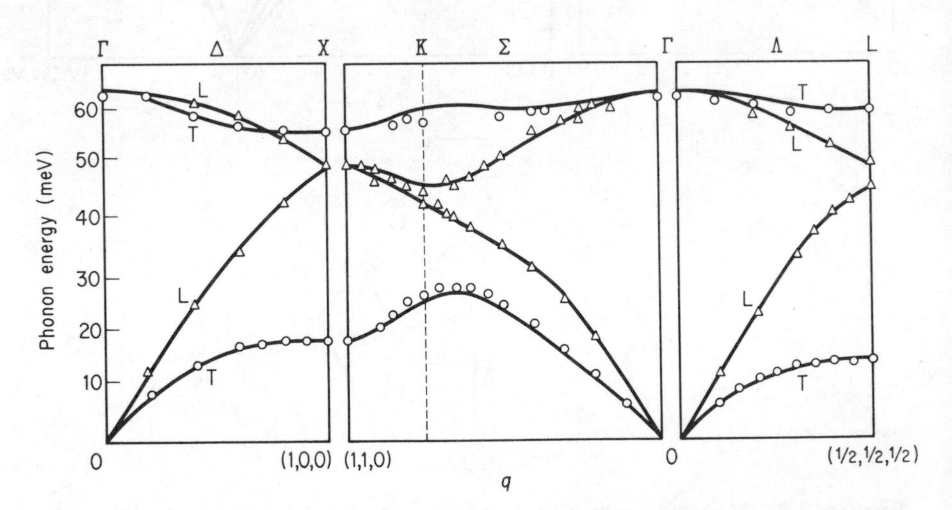

Figure 3.9 Measured vibrational spectrum $\hbar\omega(\boldsymbol{q})$ for Si (after G. Dolling *Inelastic Scattering of Neutrons*. Chalk River Conf. p. 41 I.A.E.A. (1963)). The Brillouin zone has the same shape as in figure 3.8. $\boldsymbol{q}$ is given in units of π/a

the actual shape of the curves (particularly the flatness of the transverse branches) indicates the influence of longer range forces. These can be described in terms of the 'shell model'. This notes that it is inadequate to regard the constituent atoms as rigid balls: the nuclear motion and the electronic motion may not follow each other exactly and the atoms may distort during the vibration. This situation may be formalised by regarding the electrons around each atom as a shell of light mass, separate but coupled to the nucleus. This electron shell will also be coupled by a spring to the shells of neighbouring atoms (see figure 3.10), and the nuclei, too, will be coupled. Of course this is only a very approximate way of representing the interaction between the electrons and the nuclei. Still, the shell model allows the relative motion of the electrons and the nuclei to be represented in a simple way as shown in the figure 3.10 and shows that dipoles may interact over a long distance by electric forces, thus yielding an effective long-range vibrational coupling. In fact the spectrum of silicon can be explained to quite a good approximation by nearest neighbour harmonic forces and a single-parameter shell model of this kind. The shell model also explains certain optical effects discussed in Chapter 6, since the induced dipoles can couple to the electric fields of light waves to give two-phonon effects.

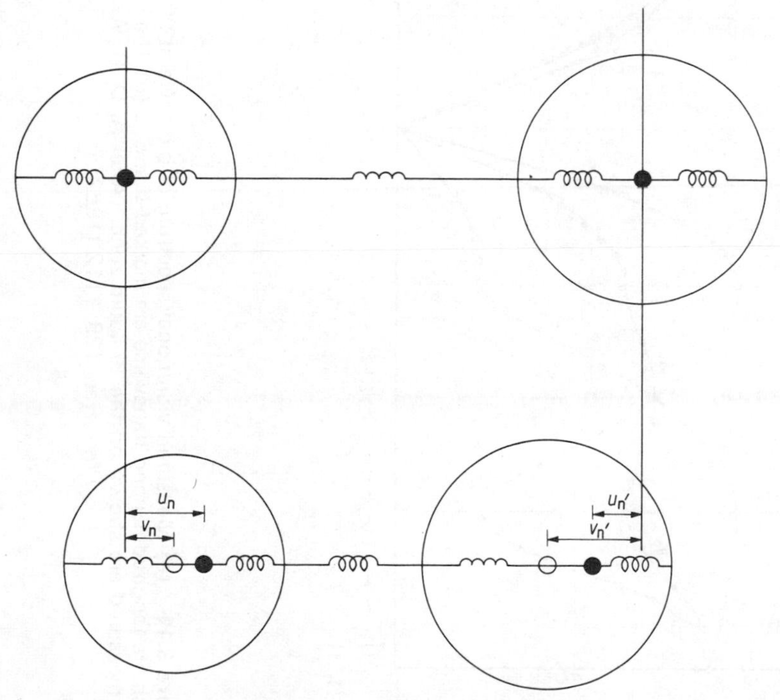

Figure 3.10 Representation of shell model. The cores (full circles) may move a distance u_n while the centres of the shells (open circles) move v_n. This gives a dipole moment on each atom proportional to $(u_n - v_n)$. Shells and cores are shown connected by springs

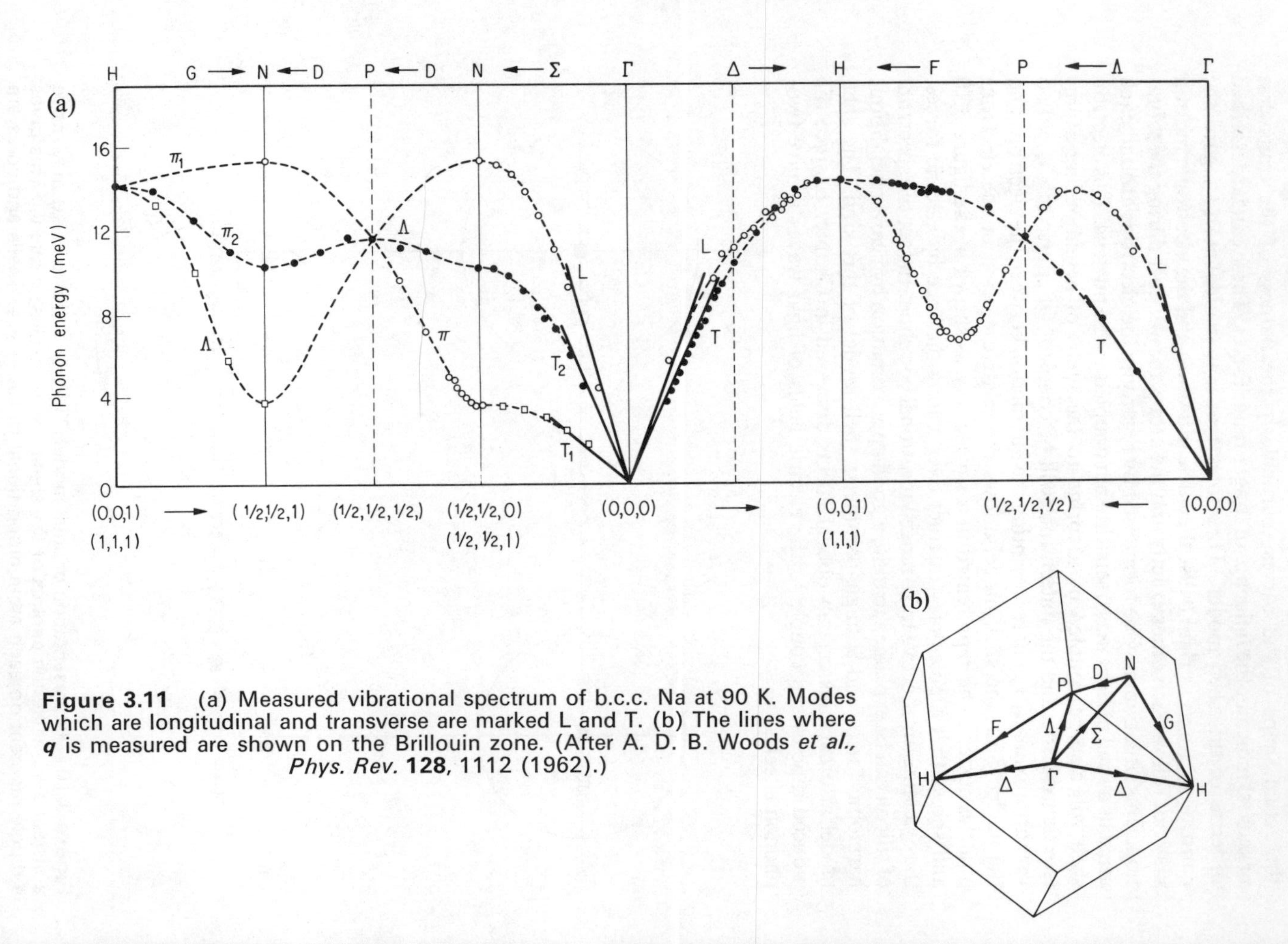

Figure 3.11 (a) Measured vibrational spectrum of b.c.c. Na at 90 K. Modes which are longitudinal and transverse are marked L and T. (b) The lines where $\boldsymbol{q}$ is measured are shown on the Brillouin zone. (After A. D. B. Woods *et al.*, *Phys. Rev.* **128**, 1112 (1962).)

3.3.4 Metals (A)

A typical vibrational spectrum of a simple metal (body-centred cubic sodium) is shown in figure 3.11. The Brillouin zone for Na is the duodecahedron shown in figure 1.5. The density of states computed from a detailed fit of the measured curves is shown in figure 3.12 and critical points are clearly seen where there are saddle points in the $\omega(\boldsymbol{q})$ curves. In this case the interatomic forces may be of much longer range since they can be transmitted by distorting the charge distribution of conduction electrons but we shall not discuss these

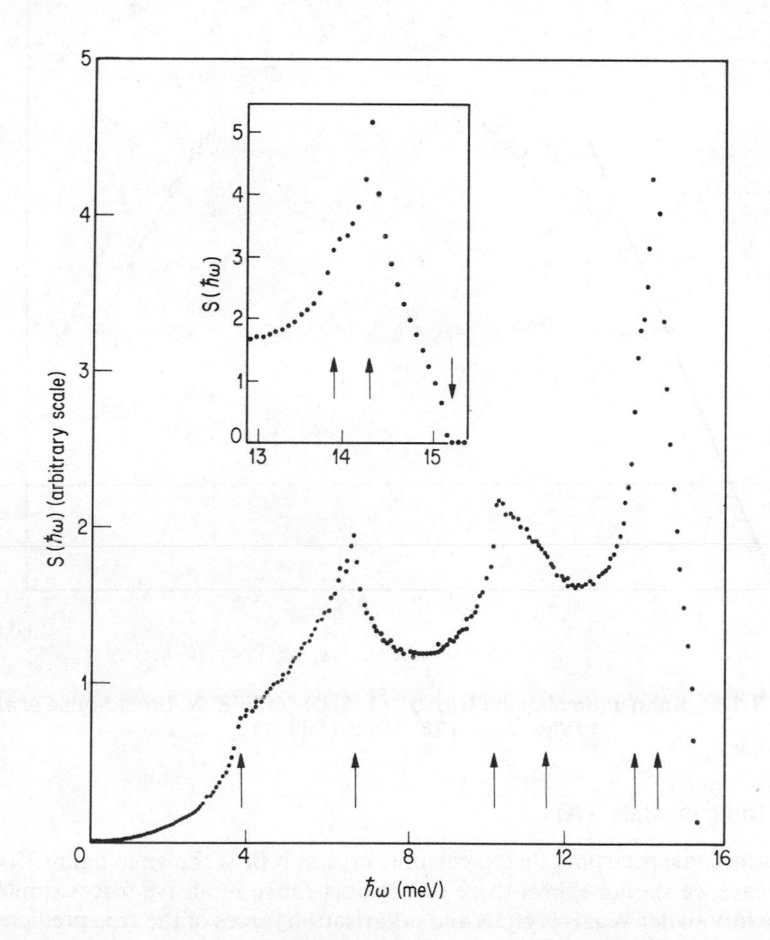

Figure 3.12 Density of modes in Na (after A. E. Dixon *et al., Proc. Phys. Soc.* **81**, 973 (1963)). The arrows indicate critical points—which correspond to points of zero slope in figure 3.11. (For example, the lowest critical point at 4 eV corresponds to the lowest branch at N in figure 3.11. The maximum energy 15.5 meV is also due to a mode at N, the high peak in S just below this due to the flat branch on the zone face N → G → H. H is also a critical point for this upper branch.)

forces in detail here. Some special features arise from the details of the distribution of the conduction electrons. As will be shown in Chapter 8 only phonons in a restricted range of $\boldsymbol{q}$ can interact with the electrons. There is therefore a fairly abrupt change in $\omega(\boldsymbol{q})$ as $\boldsymbol{q}$ passes out of this region—the resulting kink in the $\omega(\boldsymbol{q})$ curve is called a Kohn anomaly. These anomalies are very small in light metals like Na and Al but they are clearly visible in Pb (figure 3.13).

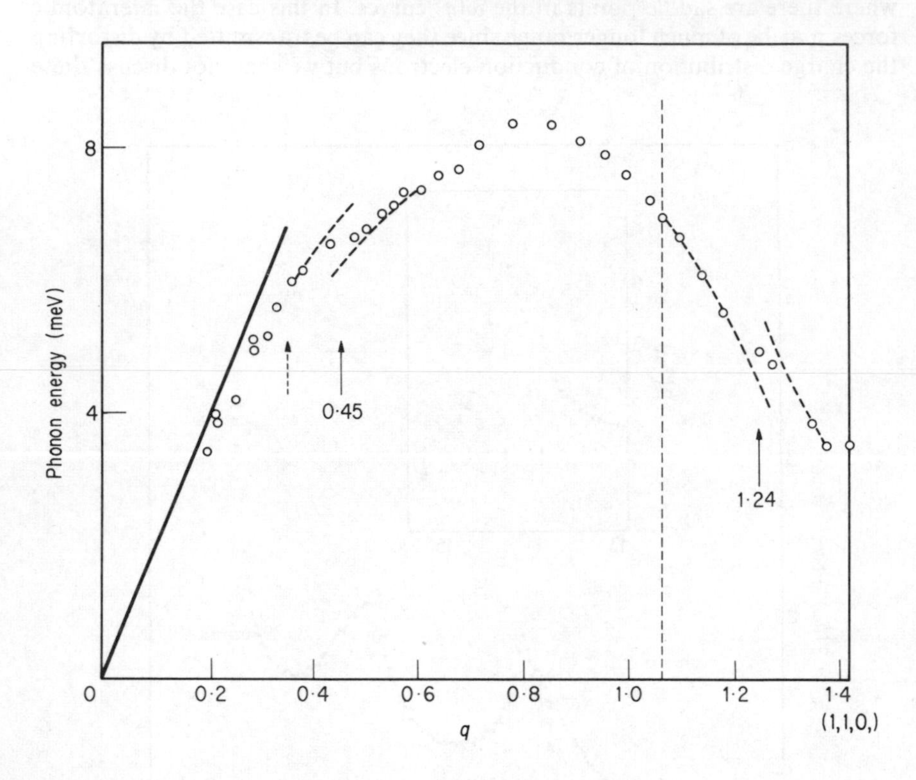

Figure 3.13 Kohn anomalies in Pb for $\boldsymbol{q} \parallel (1, 1, 0)$ (after B. N. Brockhouse *et al.*, *Phys. Rev.* **128**, 1099 (1962))

3.3.5 Ionic crystals (A)

The vibrational spectrum of a typical ionic crystal KBr is shown in figure 3.14. In this case we should expect there to be short-range repulsive forces similar to those in Van der Waals crystals and polarisation forces of the type predicted by the shell model. In addition we should expect long-range forces to arise from the Coulomb interaction between the ions. The most striking feature of the spectrum in figure 3.14 compared to those of previous sections is the large splitting between transverse and longitudinal optic branches. This apparently persists up to $\boldsymbol{q} = 0$ where we might expect degeneracy on symmetry grounds. The splitting is due to the Coulomb forces.

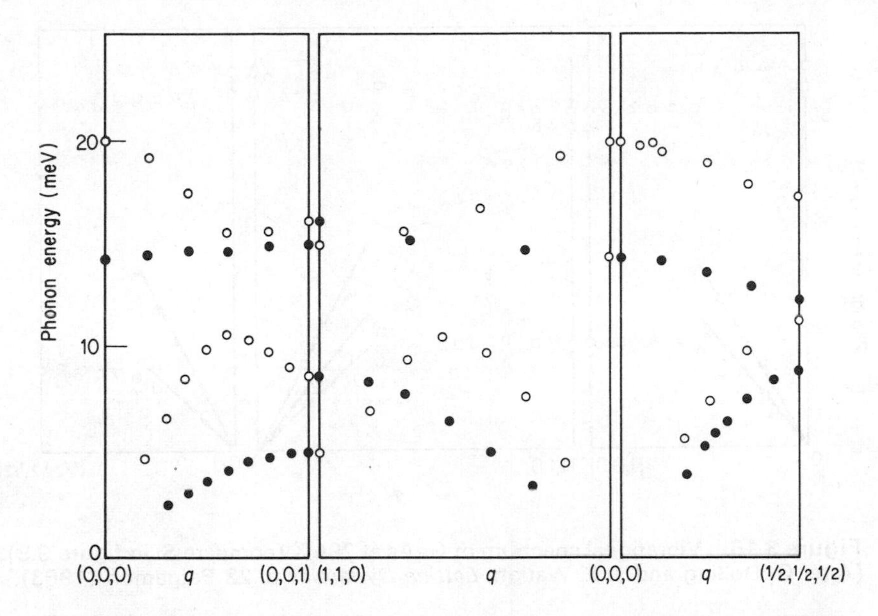

Figure 3.14 Vibrational spectrum in KBr at 90 K (after B. N. Brockhouse *et al.*, *Phys. Rev.* **131**, 1025 (1963)). The open circles are longitudinal modes and the full circles are transverse (and doubly degenerate). Note the large splitting between these for the optic branch at $\boldsymbol{q} = 0$. The Brillouin zone is as in figure 3.8

If an ion with charge e is moved a distance $\boldsymbol{u}(l)$ from its equilibrium position it creates a dipole of magnitude $e\boldsymbol{u}(l)$ relative to the equilibrium state. The change in potential energy due to displacements of a pair of ions has the usual dipole–dipole form

$$V(\boldsymbol{R}) = \frac{e^2}{R^5}\{[\boldsymbol{u}(l).\boldsymbol{u}(l')]\,R^2 - 3[\boldsymbol{u}(l).\boldsymbol{R}]\,[\boldsymbol{u}(l').\boldsymbol{R}]\} \qquad (3.26)$$

The long-range nature of this force gives anomalous properties in the Fourier transform $\Lambda(\boldsymbol{q})$ (3.25) which is related to $\omega^2(\boldsymbol{q})$. For $\boldsymbol{q} = 0$

$$\Lambda(\boldsymbol{q}) = \sum_{\boldsymbol{R}} \Lambda(\boldsymbol{R}) \qquad (3.27)$$

and this depends on the shape of the crystal. These shape effects disappear as q increases so that the wavelength becomes much less than the crystal size (which is typically one-millionth of the way to the zone boundary). Nevertheless the Coulomb forces cause a splitting between longitudinal and transverse optic branches, as will be seen from the next section.

Partially ionic crystals like the III-V compound GaAs have a similar but smaller splitting of the optic branches (figure 3.15). The spectrum is like that of a covalent crystal Si (figure 3.10) apart from the splitting.

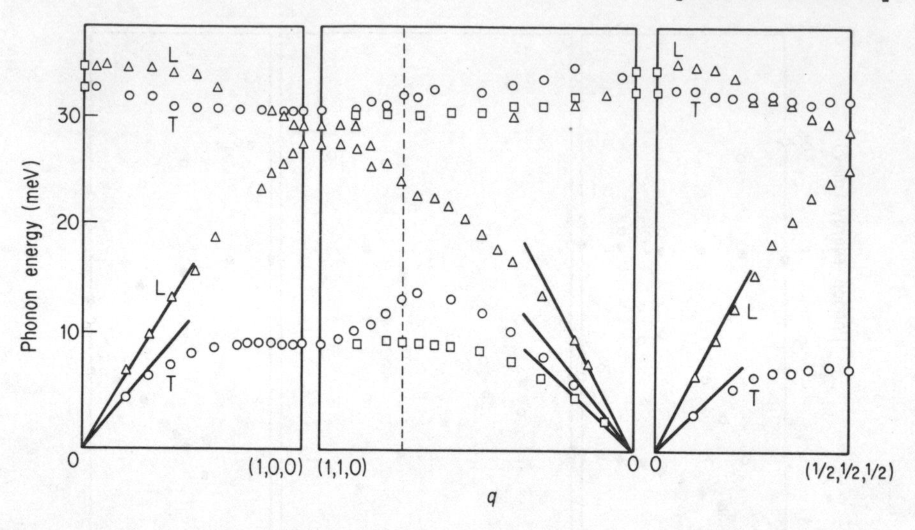

Figure 3.15 Vibrational spectrum of GaAs at 296 K (compare Si in figure 3.9). (After G. Dolling and J. L. Waugh, *Lattice Dynamics* p. 23 Pergamon (1963).)

3.4 Optic modes in ionic crystals (A)

Because electric fields are set up by the oscillating charges when the constituents of an ionic crystal vibrate, the electrical and optical properties of such crystals are strongly affected by the vibrations. Some of these effects are discussed in later chapters but we shall now show that the splitting of the longitudinal and transverse optic branches is related to the dielectric constant.

Let $\boldsymbol{u}$ be the amplitude of displacement in the optic mode and let the electric field in the crystal be $\mathbf{F}$. The equation of motion in this mode can be written

$$\frac{\partial^2 \boldsymbol{u}}{\partial t^2} = -\omega_0^2 \boldsymbol{u} + b\mathbf{F} \tag{3.28}$$

since $\mathbf{F}$ forces the oscillations through the coupling b. The displacement will also create a polarisation density $\mathbf{P}$ and this will be enhanced by the polarisation induced by the field. Then

$$\mathbf{P} = (b'\boldsymbol{u} + \alpha\mathbf{F})\,\varepsilon_0 \tag{3.29}$$

where ε_0 is the permittivity of free space ($\sim 10^{-9}/36\pi$ farad metre^{-1}), α is the polarisability of the constituent atoms and b' gives the polarisation due to displacement. In these modes the displacement varies with position in the crystal like a wave: $\exp(\mathrm{i}\boldsymbol{q}.\boldsymbol{r})$. Then

$$\begin{aligned} \nabla.\boldsymbol{u} = \mathrm{i}\boldsymbol{q}.\boldsymbol{u} &= 0 \qquad \text{for transverse waves } \boldsymbol{u}_\mathrm{T} \perp \boldsymbol{q} \\ &\neq 0 \qquad \text{for longitudinal waves } \boldsymbol{u}_\mathrm{L} \parallel \boldsymbol{q} \end{aligned} \tag{3.30}$$

and

$$\nabla \times \boldsymbol{u} = i\boldsymbol{q} \times \boldsymbol{u} \neq 0 \quad \text{for transverse waves}$$
$$= 0 \quad \text{for longitudinal waves} \tag{3.31}$$

In addition the electric field obeys the electrostatic equations

$$\nabla . \mathbf{D} = \nabla . (\varepsilon_0 \mathbf{F} + \mathbf{P}) = 0; \qquad \nabla \times \mathbf{F} = 0 \tag{3.32}$$

By taking the curl of (3.28) and using (3.31) and (3.32) we get

$$\frac{\partial^2 \boldsymbol{u}_\mathrm{T}}{\partial t^2} = -\omega_0^2 \boldsymbol{u}_\mathrm{T} \qquad \text{or} \qquad \omega_\mathrm{T}^2 = \omega_0^2$$

i.e. the frequency of the transverse mode, ω_T, is unaffected by the electric fields. On the other hand, taking the divergence of (3.28) and using (3.29), (3.30) and (3.32), we find

$$\frac{\partial^2 \boldsymbol{u}_\mathrm{L}}{\partial t^2} = -\left(\omega_0^2 + \frac{bb'}{1+\alpha}\right) \boldsymbol{u}_\mathrm{L} \tag{3.33}$$

This expression may be written more simply in terms of a dielectric constant. From (3.28) and (3.29)

$$\varepsilon \mathbf{F} = \mathbf{D} = \varepsilon_0 \mathbf{F} + \mathbf{P} = \varepsilon_0 \left(1 + \alpha - \frac{bb'}{\omega^2 - \omega_0^2}\right) \mathbf{F} \tag{3.34}$$

The expression in brackets is the frequency-dependent dielectric constant of the system. As $\omega \to \infty$, ε tends to its high frequency value, $\varepsilon(\infty)$, where

$$\varepsilon(\infty) = 1 + \alpha \tag{3.35}$$

while as $\omega \to 0$, ε tends to its low frequency value

$$\varepsilon(0) = \varepsilon(\infty) + bb'/\omega_0^2 \tag{3.36}$$

With these definitions we can write (3.34) as

$$\varepsilon = \varepsilon(\infty) + \frac{\varepsilon(0) - \varepsilon(\infty)}{1 - (\omega/\omega_0)^2} \tag{3.37}$$

and using them in (3.33) we find the longitudinal mode frequency, ω_L, is given by

$$\omega_\mathrm{L}^2 = \frac{\varepsilon(0)}{\varepsilon(\infty)} \omega_\mathrm{T}^2 \tag{3.38}$$

This remarkable result is called the Lyddane, Sachs, Teller relation. The dielectric constant (3.37) becomes infinite at $\omega = \omega_T$ and is negative when $\omega_T < \omega < \omega_L$. This has striking effects on the optical properties as will be discussed in § 6.6.*

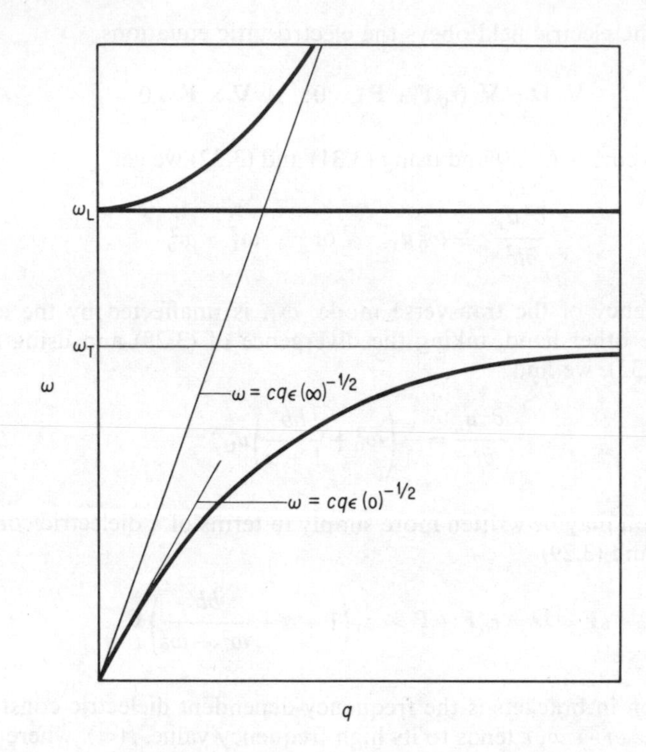

Figure 3.16 Dispersion in mixed electromagnetic and vibrational modes. If the interaction between them is neglected the faint lines are obtained. The crossing between transverse vibrational and the electromagnetic modes is removed by the interaction

* It should be noted that the equations (3.32) define **F** as the true electric field, not the local field. This latter will contain a contribution from the dipole distribution in the crystal which depends on $\boldsymbol{u}$ and thus may be regarded as included in the $-\omega_0^2 \boldsymbol{u}$ term of (3.28). For this reason ω_0^2 does not simply depend on the short-range forces. A wave of long wavelength effectively sets up a slab of polarisation as in figure 3.17. The longitudinal mode where **P** is along the axis perpendicular to the slab sets up surface charges which contribute a field $-\mathbf{P}$ inside. In the transverse mode **P** lies in the plane of the slab and the surface charges give nothing. In each case a contribution $\mathbf{P}/3$ arises from the Lorentz field. If there is no electronic polarisability $\mathbf{P} = \mathcal{N} e\boldsymbol{u}/\mathcal{V}$ and

$$M_r \omega_T^2 = \Lambda - \eta/3; \quad M_r \omega_L^2 = \Lambda + 2\eta/3$$

where $\eta = \mathcal{N}^2/\varepsilon_0 \mathcal{V}$, Λ is the appropriate short-range force constant and M_r is the reduced mass of the ions.

Further insight into this coupling can be obtained if it is remembered that the electric field itself oscillates and further that its motion is quantised. From Maxwell's equations, the electrostatic relations (3.32) must be replaced by

$$\nabla \times \mathbf{F} = -\frac{\partial \mathbf{B}}{\partial t}; \qquad \nabla \times \mathbf{B} = \mu_0 \frac{\partial \mathbf{D}}{\partial t} \tag{3.39}$$

where μ_0 is the magnetic permeability of free space ($=4\pi \times 10^{-7}$ Henry metre^{-1}) and it is assumed that the relative magnetic permeability of the crystal is unity. Hence

$$\nabla \times (\nabla \times \mathbf{F}) = -\frac{1}{c^2}\varepsilon \frac{\partial \mathbf{F}}{\partial t} \tag{3.40}$$

where c is the velocity of light *in vacuo* and $c^{-2} = \mu_0 \varepsilon_0$. If **F** and **D** vary as waves, i.e. as $\exp i(\boldsymbol{\beta}.\boldsymbol{r} - \omega t)$, **F** and **B** are both transverse (perpendicular to $\boldsymbol{\beta}$) and perpendicular to each other, as in a free space electromagnetic wave. Substituting into (3.40) gives a relationship between ω and the wave vector β for these transverse waves

$$c^2 \beta^2 = \omega^2 \varepsilon$$

or

$$(\omega_0^2 - \omega^2)\, c^2 \beta^2 = \omega^2[\omega_0^2 \varepsilon(0) - \omega^2 \varepsilon(\infty)] \tag{3.41}$$

from (3.37). There are two solutions—in the limit of no coulombic interaction $[\varepsilon(0) = \varepsilon(\infty)]$ they correspond to an ordinary electromagnetic wave with $\omega = c\beta$ and a vibrational mode with $\omega = \omega_0$; but with interaction they mix, and the resulting dispersion is shown in figure 3.16. Near $\beta = 0$ the roots are

$$\omega^2 = c^2 \beta^2/\varepsilon(0) \tag{3.42}$$

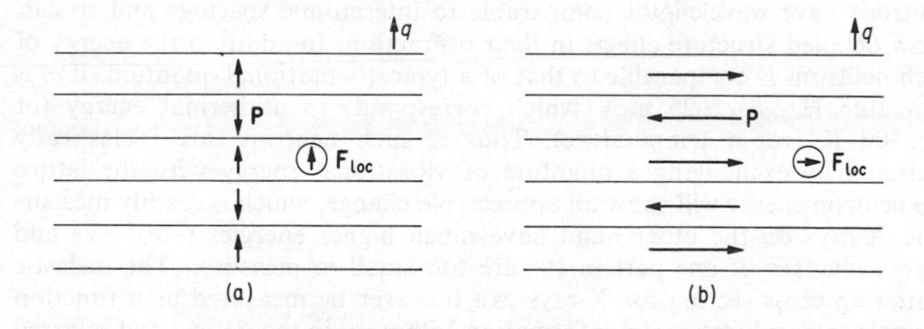

Figure 3.17 Slab of polarisation in a dielectric corresponding to (a) longitudinal and (b) transverse waves. The lines represent planes of nodes in the wave where the polarisation reverses. These are a wavelength $2\pi/q$ apart and this is chosen so that the atomic spacing $a \ll 2\pi/q \ll \mathscr{L}$, the specimen size. The polarisation may be regarded as giving free poles along these nodes. In (a) the slab containing the point where **F** is measured contributes a field **P** and the contributions from all the others cancel. In (b) the poles are the ends of the slab and give nothing. In each case there is a Lorentz field **P**/3 (cf. figure 10.13)

appropriate to electromagnetic waves in a medium with dielectric constant $\varepsilon(0)$ and

$$\omega^2 = \omega_0^2[\varepsilon(0)/\varepsilon(\infty)] \tag{3.43}$$

a transverse mechanical vibration at the longitudinal mode frequency (3.38). At large β, well beyond the crossing point the roots are

$$\omega^2 = c^2\beta^2/\varepsilon(\infty) \tag{3.44}$$

and

$$\omega^2 = \omega_0^2, \tag{3.45}$$

the transverse mode frequency.

The coupled modes described by these equations have been observed directly. Such quanta are usually called polaritons. They provide one of the best examples of mixed excitations discussed in Chapter 2. In the intermediate frequency region the modes have a partly electromagnetic and partly mechanical character.

3.5 The determination of vibrational spectra by inelastic neutron scattering (a)

The detailed and valuable information on the vibrational spectra of many simple solids discussed in § 3.3 has been derived in the main from inelastic neutron scattering experiments. The advantage of slow neutrons for this measurement lies in their favourable energy–momentum relation. Thermal neutrons have wavelengths comparable to interatomic spacings and so can show detailed structure effects in their diffraction. In addition the energy of such neutrons is comparable to that of a typical vibrational quantum. If ω is $2\pi \times 10^{12}$ Hz, $\hbar\omega \simeq 25$ meV which corresponds to a thermal energy for $T \simeq 300$ K (room temperature). Thus if such neutrons are inelastically scattered by exchanging a quantum of vibrational energy with the lattice the neutron energy will show an appreciable change, which is readily measurable. X-rays on the other hand have much higher energies ($\sim 10^4$ eV) and energy changes of one part in 10^6 are too small to measure. (The inelastic scattering cross section for X-rays can however be measured as a function of angle to give some useful information.) Photons in the optical and infrared regions have the same energies as vibrational quanta but their wave vectors are so small that diffraction effects cannot be seen.

The essential conditions for scattering are given in § 1.6. A beam of neutrons of fixed energy is incident on the sample, their energy and direction of motion being defined if their wave vector $\boldsymbol{K}$ is specified. The neutrons which are scattered from the sample have their energy and direction measured so that their wave vector $\boldsymbol{K}'$ is also specified.

If a phonon of wavevector $\boldsymbol{q}$ is created or destroyed, the conservation of wave vector requires (cf. § 2.3)

$$\boldsymbol{K} - \boldsymbol{K}' = \pm\boldsymbol{q} + \boldsymbol{Q} \tag{3.46}$$

and energy conservation requires

$$\frac{\hbar^2}{2m_n}(\boldsymbol{K}^2 - \boldsymbol{K}'^2) = \pm\hbar\omega(\boldsymbol{q}) \tag{3.47}$$

Both energy loss experiments ($|\boldsymbol{K}| > |\boldsymbol{K}'|$) for relatively low-frequency excitations and energy gain experiments ($|\boldsymbol{K}| < |\boldsymbol{K}'|$) with low-energy incident neutrons have been performed in appropriate circumstances. As may be seen from (3.46) and (3.47) if $\boldsymbol{K}$ is known, a direct measure of $\boldsymbol{K}'$ gives $\boldsymbol{q}$ and $\hbar\omega$ and therefore $\omega(\boldsymbol{q})$ directly. The intensity of scattering depends on the amplitudes in the mode. The probability of a phonon being destroyed is simply proportional to the number $n(\boldsymbol{q})$ present in thermal equilibrium—given by the Bose–Einstein distribution (2.39). The probability of the phonon being created is somewhat larger—it can occur even at $T = 0$ K when $n(q) = 0$. If the neutrons are in thermal equilibrium with the sample the numbers created and destroyed must just balance. But the number of neutrons available with the higher energy is smaller by the Boltzmann factor $\exp(-E/kT)$. Thus the probability of creation must be given by

$$n(q)\exp(E(q)/kT) = \frac{\exp[E(q)/kT]}{\exp[E(q)/kT] - 1} = n(q) + 1 \tag{3.48}$$

This argument is sometimes called the *principle of detailed balance.*

Neutrons are emitted with a range of energies by an atomic pile, and in numbers roughly given by the Maxwell–Boltzmann distribution appropriate to the pile temperature, though with some extra high-energy particles which have been incompletely moderated. The most numerous have an energy of about kT as expected from kinetic theory. The number of low-energy neutrons can be increased by putting a cold body (such as liquid hydrogen) in the pile. Monochromatic neutrons are obtained by selecting only a narrow range from this spectrum. One way to do this is to Bragg scatter from a single crystal as described in Chapter 1. The energy of these neutrons after they have been inelastically scattered may also be determined by Bragg scattering from a further known single crystal. The apparatus which uses this technique is called a triple axis spectrometer and is shown schematically in figure 3.18. It is also possible to use mechanical methods to form a monochromatic beam and to determine the energy of the scattered particles. The velocity of a thermal neutron is about 10^5 cm/s, similar say to that of a rifle bullet. A chopper consisting of a rapidly spinning disc with a curved path cut in it will only allow neutrons of a fixed energy to pass along this path. The energy of the scattered particles can be determined by measuring electronically their time of flight along the path to the counters.

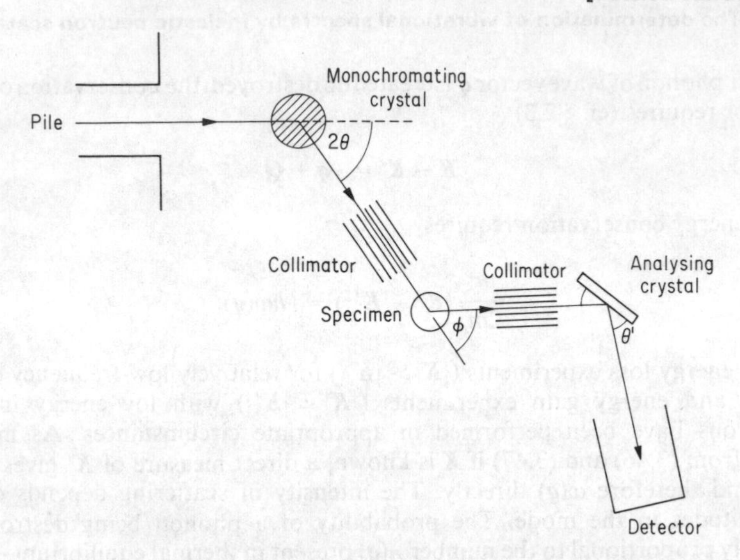

Figure 3.18 Diagram of triple axis neutron spectrometer used to measure phonon spectra

3.6 The thermal properties of solids (a)

The oscillations of the atoms of a crystal are directly related to the heat content of the system. The thermodynamic properties of a crystal are readily written down in the harmonic approximation for a known spectrum of $\omega_i(\boldsymbol{q})$. Each mode behaves as an independent oscillator whose mean thermal excitation is given by the Bose–Einstein distribution function (2.39). The total energy in the system, $\langle E \rangle$, is then

$$\langle E \rangle = \sum_{i,\boldsymbol{q}} [n_i(\boldsymbol{q}) + \tfrac{1}{2}]\, \hbar\omega_i(\boldsymbol{q}) = \sum_{i,\boldsymbol{q}} \tfrac{1}{2}\hbar\omega_i(\boldsymbol{q}) \coth\left(\frac{\hbar\omega_i(\boldsymbol{q})}{2kT}\right) \tag{3.49}$$

after manipulation of the expression. Classically we expect an average energy kT to be associated with each mode of motion and at high temperatures (3.49) reduces to the classical result

$$\langle E \rangle = kT \sum_{i,\boldsymbol{q}} 1 = 3\mathcal{N}kT \tag{3.50}$$

since the total number of modes is $3\mathcal{N}$. The specific heat is constant in this region and given by

$$C_V = \partial\langle E \rangle/\partial T = 3\mathcal{N}k \tag{3.51}$$

At other temperatures it is necessary to replace the sum over modes by an integral over the density of states, $S(\hbar\omega)$, as in (2.42) and

$$\langle E \rangle = \int_0^{\omega_m} \tfrac{1}{2}\hbar\omega \coth\left(\frac{\hbar\omega}{2kT}\right) S(\hbar\omega)\, \hbar\, d\omega \tag{3.52}$$

so that

$$C_V = k \int_0^{\omega_m} \left(\frac{\hbar\omega}{2kT}\right)^2 \frac{S(\hbar\omega)\,\hbar\,d\omega}{\sinh^2(\hbar\omega/2kT)} \tag{3.53}$$

where ω_m is the maximum phonon frequency.

The density of states is in general a complicated function (cf. figure 3.12). However, at low T the integral (3.53) is dominated by the contribution from low ω. At low frequencies, because of the linear law $\omega = sq$ for acoustic waves, $S(E) \propto E^2$ as in (2.46). Now, however, the actual value of s depends on the direction of $\boldsymbol{q}$ and there are three branches. Defining some suitable average of the sound velocity, s, we can write

$$S(\hbar\omega) = \frac{3\mathcal{N}\Omega\omega^2}{2\pi^2\hbar s^3} \tag{3.54}$$

Since only the low ω contribution is important in (3.53) at low temperatures we use this value at all ω so that

$$C_V = 3\mathcal{N}k\left[\frac{(2kT)^3\,\Omega}{\pi^2\hbar^3 s^3}\right]\int_0^{x_m} x^4 \sinh^{-2} x\,dx \tag{3.55}$$

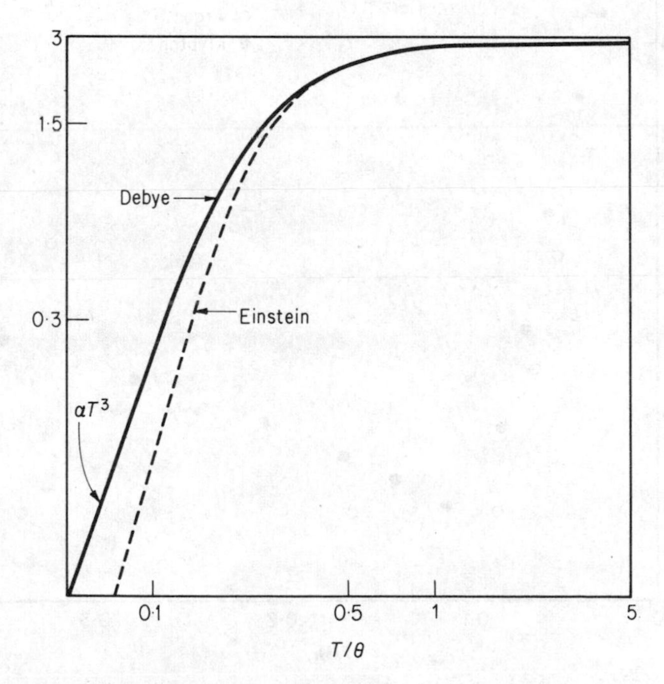

Figure 3.19 Theoretical form of specific heat curve (log plot) for Debye and Einstein models. Both give the same classical result at high T but differ at low T

where $x_m = \hbar\omega_m/2kT$. This may be regarded as infinite when kT is small and the integral is then equal to $\pi^4/60$. The important point is that $C_V \propto T^3$ at low T. The general form of the C_V curve is shown in figure 3.19. The intermediate region around $kT \sim \frac{1}{2}\hbar\omega_m$ depends on the details of $S(\hbar\omega)$.

Historically the vibrational specific heat of solids was important in the development of quantum theory. It is not, however, a very sensitive tool for studying the detailed vibrational spectrum since it contains an average over all the modes. Approximate forms for the density of states have been used to give a very good description of the specific heat. The earliest and simplest was due to Einstein who assumed all modes had the same frequency ω_2 so that

$$C_V = 3\mathcal{N}k\left[\frac{(\hbar\omega_2/2kT)}{\sinh(\hbar\omega_2/2kT)}\right]^2 \tag{3.56}$$

which gives the correct high T classical limit (3.51) but an exponential dependence on T at low T which is wrong. Debye improved the result by assuming that the low ω form (3.54) holds at all ω as it would in an elastic continuum.

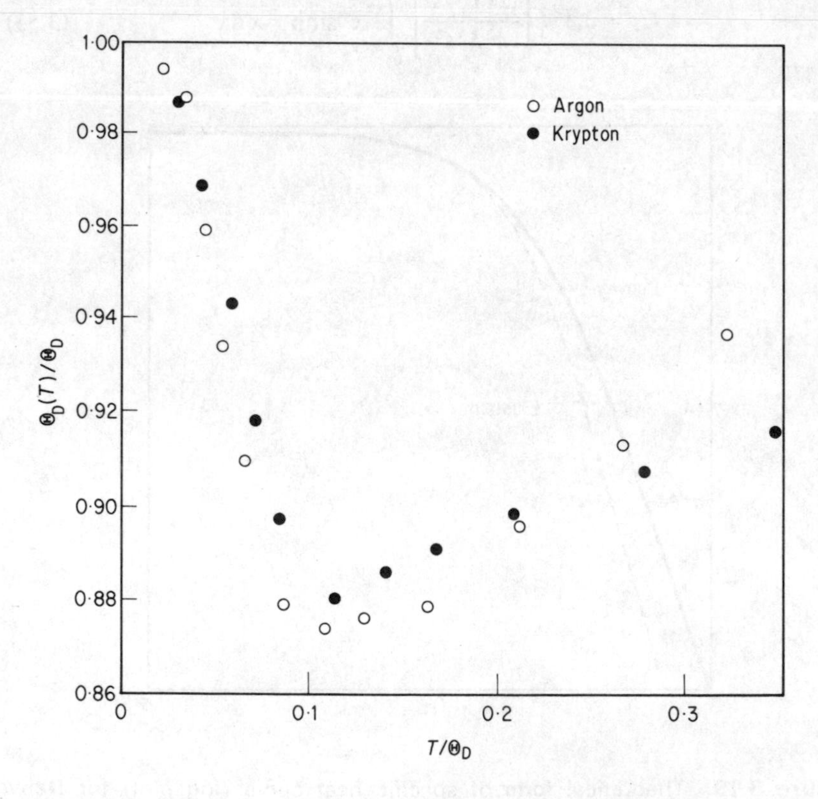

Figure 3.20 Variation of effective Debye Θ with T in A and Kr. (After J. Grindley and R. Howard, *Lattice Dynamics*, p. 131 Pergamon (1963).)

Assuming three branches only

$$S(\hbar\omega) = 9\mathcal{N}(\omega^2/\hbar\omega_m^3) \tag{3.57}$$

and

$$C_V = 18\mathcal{N}k\left(\frac{2kT}{\hbar\omega_m}\right)^3 \int_0^{x_m} x^4 \sinh^{-2} x \, dx \tag{3.58}$$

(This result is plotted in figure 3.19.)

The effect of the actual $S(\hbar\omega)$ can be demonstrated by comparing (3.58) with measured values of C_V. In order to obtain agreement it is found necessary to allow $\hbar\omega_m$ to change with T. A plot of this effective *Debye temperature*, Θ_D, defined by $k\Theta_D = \hbar\omega_m$ is given in figure 3.20. The dip is caused by the peaking of $S(\hbar\omega)$ due to the lowest transverse acoustic branches.

3.7 Anharmonic effects (AD)

So far is has been assumed that the potential between atoms is such as to give completely harmonic motion. In fact the potential well of a molecule is known to deviate from this form as the energy increases (cf. figure 1.10) and indeed in special cases it may do so at low energies (cf. figure 2.1). These deviations give rise to anharmonic effects, which normally increase with E and hence with temperature. Some very crude idea of these effects can be obtained by examining the molecular potential well of figure 1.10. Expanding in terms of u, the deviation from the equilibrium position R_0, gives the potential in the form

$$V = \Lambda u^2 - \Lambda' u^3 - \Lambda'' u^4 \tag{3.59}$$

where all the coefficients are positive. At low energies the characteristic frequency is $(2\Lambda/M)^{1/2}$. The classical value of the mean square displacement is

$$\langle u^2 \rangle = \tfrac{1}{2}(kT/\Lambda) \tag{3.60}$$

The mean displacement $\langle u \rangle$ is non-zero because of the cubic term. It is given very crudely by taking $\partial V/\partial u = 0$ and neglecting the term in Λ''. Then

$$\langle u \rangle = 3\Lambda' \langle u^2 \rangle / 2\Lambda = 3\Lambda' kT/4\Lambda^2 \tag{3.61}$$

From this we expect an increase in interatomic spacings and hence a thermal expansion of the crystal. This is proportional to T at high T, but at low T it increases like T^4, in the same way that $\langle u^2 \rangle$ does in a lattice.

Another consequence of a non-zero value of Λ' is the phenomenon of piezoelectricity, namely the appearance of an electric field when the crystal dimensions are altered by mechanical stress. The converse effect also occurs. The piezoelectric effect is used for the conversion of acoustic vibrations to electrical signals and *vice versa* (see § 8.4). Furthermore, extremely sharp mechanical resonances can occur in suitably cut crystals. Electronic oscillators using such crystals have well-defined oscillation frequencies which can be made relatively insensitive to temperature and form the basis of electronic clocks and watches. The most widely used material in this application is quartz which, though it has a relatively small piezoelectric coefficient, can be used at frequencies as high as 1 GHz. If mounted in a constant temperature enclosure and isolated from extraneous vibrations, quartz crystal controlled oscillators can achieve stabilities as high as 1 in 10^9, quite sufficient to detect the irregularities of the Earth's motion around the Sun.

The effective frequency of oscillations also varies as the energy increases. In fact in the molecular well the spacing between successive vibrational levels decreases as may be seen from optical spectra. The mean curvature of (3.59) may be regarded as approximately

$$(\Lambda - \Lambda'\langle u\rangle - \Lambda''\langle u^2\rangle)\, u^2 = \Lambda'\, u^2 \tag{3.62}$$

where using (3.60) and (3.61)

$$\Lambda' = \Lambda - \left(\frac{3\Lambda'^2}{4\Lambda^2} + \frac{\Lambda''}{2\Lambda}\right) kT \tag{3.63}$$

so that the characteristic frequencies decrease with T.

An alternative way to think of these effects is to assume that the characteristic frequencies in a crystal change with volume. The simplest assumption is that they all suffer the same relative change

$$\frac{\Delta\omega}{\omega} = \gamma \frac{\Delta\Omega}{\Omega} \tag{3.64}$$

where γ is called the Gruneisen constant. The total free energy of the crystal may be written as a function of volume

$$\mathscr{F} = \frac{1}{2\kappa}\left(\frac{\Delta\Omega}{\Omega}\right)^2 + kT \sum_{i,q} \log\left(2\sinh\frac{\hbar w_i(q)}{2kT}\right) \tag{3.65}$$

where the first term is the potential energy of a compressed elastic continuum of bulk modulus κ. Differentiating with respect to volume gives a minimum of $\mathscr{F}$ when

$$\frac{1}{\kappa}\frac{\Delta\Omega}{\Omega} = \sum_{i,q} \gamma\hbar\omega_i(q) \coth\frac{\hbar\omega_i(q)}{2kT} = 2\gamma\langle E\rangle \tag{3.66}$$

using (3.49).

Thus the volume change is proportional to $\langle E\rangle$, which is proportional to T at high T, and like T^4 at low T. Experimentally γ is usually of order unity. The thermal variation of $\omega(q)$ is seen to be like $\gamma^2\langle E\rangle$.

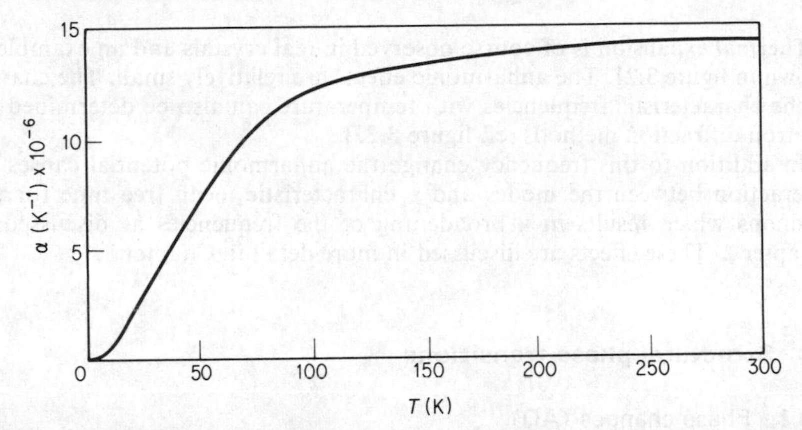

Figure 3.21 The coefficient of linear expansion in Au. By (3.66) $\alpha = \frac{1}{3}(1/\Omega)/(\partial\Omega/\partial T)$ is proportional to $\partial\langle E\rangle/\partial T = C_V$ and the curve does have a form like that of C_V (figure 3.19)

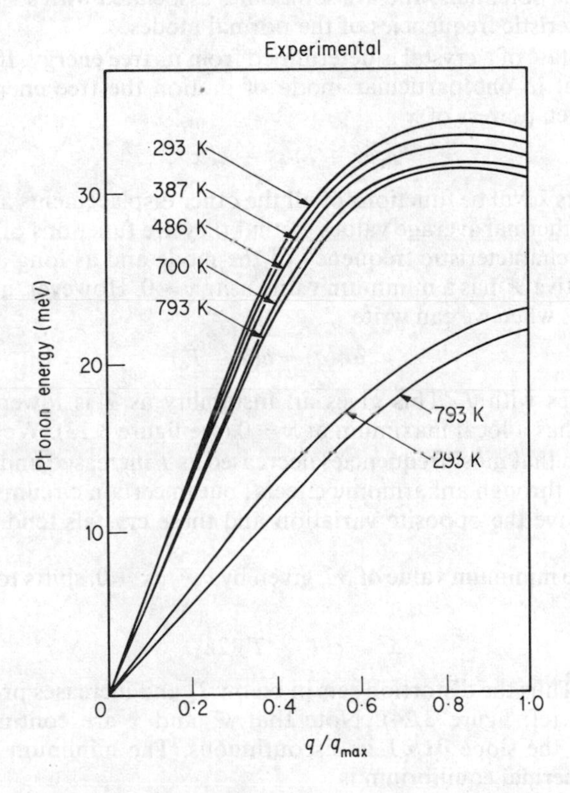

Figure 3.22 Variation of $\hbar\omega(\boldsymbol{q})$ with T in Al (after K. E. Larssen *et al.*, *Inelastic scattering of Neutrons*, p. 593 *I.A.E.A.* Vienna (1961)) for one direction of $\boldsymbol{q}$

Thermal expansion is of course observed in real crystals and an example is shown in figure 3.21. The anharmonic effects are relatively small. The change in the characteristic frequencies with temperature can also be determined by neutron diffraction methods (cf. figure 3.22).

In addition to this frequency change the anharmonic potential causes an interaction between the modes and a characteristic mean free time for the phonons which results in a broadening of the frequencies as discussed in Chapter 2. These effects are discussed in more detail in Chapter 8.

3.8 Structural phase transitions

3.8.1 Phase changes (AD)

Many crystals show changes from one structure to another as the temperature is raised, or under external forces. These changes must reflect properties of the interatomic potentials, and are sometimes associated with striking changes in the characteristic frequencies of the normal modes.

The stable state of a crystal is determined from its free energy. If x represents a displacement in one particular mode of motion the free energy will be a function of even powers of x

$$\mathscr{F} = \lambda_1 x^2 + \lambda_2 x^4 + \lambda_3 x^6 \ldots \tag{3.67}$$

The coefficients λ will be functions of all the other displacements and these can be given their thermal average values so that they are functions of T. λ_1 will be related to the characteristic frequency of the mode and as long as it remains large and positive $\mathscr{F}$ has a minimum value near $x = 0$. However, an interesting situation arises when we can write

$$\lambda_1 \propto \omega_{\mathrm{i}}^2(q) = c(T - T_{\mathrm{c}}) \tag{3.68}$$

i.e. λ_1 increases with T. This gives an instability as T is lowered below T_{c} since $\mathscr{F}$ then has a local maximum at $x = 0$ (see figure 3.23). We remarked in the last section that most frequencies decreased as T increased and the 'springs' became softer through anharmonic effects; but in certain circumstances some modes may have the opposite variation and these crystals tend to distort at low T.

If $\lambda_2 > 0$ the minimum value of $\mathscr{F}$, given by $\partial \mathscr{F}/\partial x = 0$, shifts to the position $\bar{x}$ given by

$$\bar{x}^2 = c(T_{\mathrm{c}} - T)/2\lambda_2 \tag{3.69}$$

when $T < T_{\mathrm{c}}$. Thus the distortion sets in below T_{c} and increases proportionally to $(T_{\mathrm{c}} - T)^{1/2}$ (cf. figure 3.24). Note that $\mathscr{F}$ and $\bar{x}$ are continuous at the transition but the slope $\partial \bar{x}/\partial T$ is discontinuous. The minimum of $\mathscr{F}$ corresponding to thermal equilibrium is

$$\begin{aligned} \mathscr{F} &= -c^2(T_{\mathrm{c}} - T)^2/4\lambda_2 \quad && T < T_{\mathrm{c}} \\ &= 0 && T > T_{\mathrm{c}} \end{aligned} \tag{3.70}$$

The specific heat is given by

$$C_V = \frac{\partial}{\partial T} T^2 \frac{\partial}{\partial T} \frac{\mathscr{F}}{T} = \frac{c^2 T}{2\lambda_2} \qquad T < T_c \tag{3.71}$$

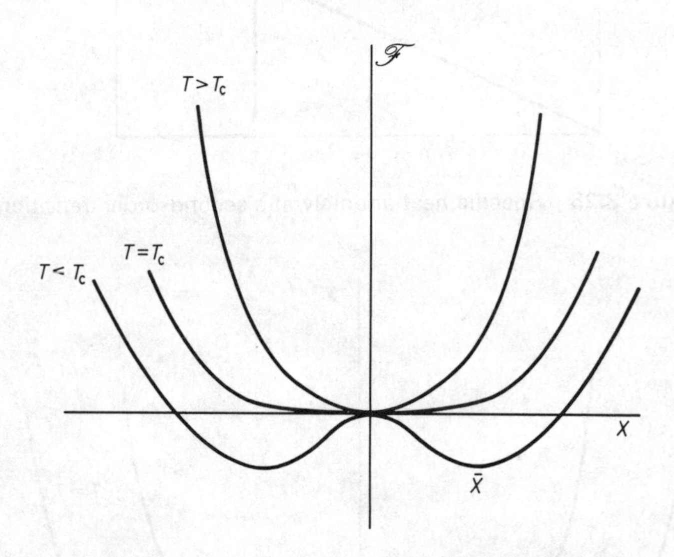

Figure 3.23 Free energy as a function of displacement at various T showing properties of a second-order phase transition (equations (3.67) and (3.68))

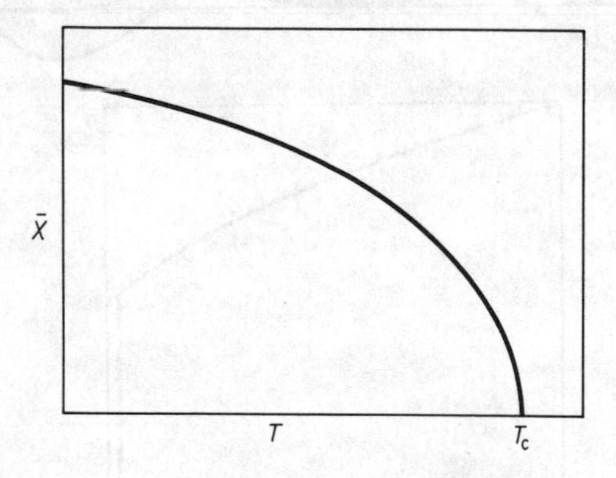

Figure 3.24 Order parameter below transition temperature T_c in a second-order phase transition

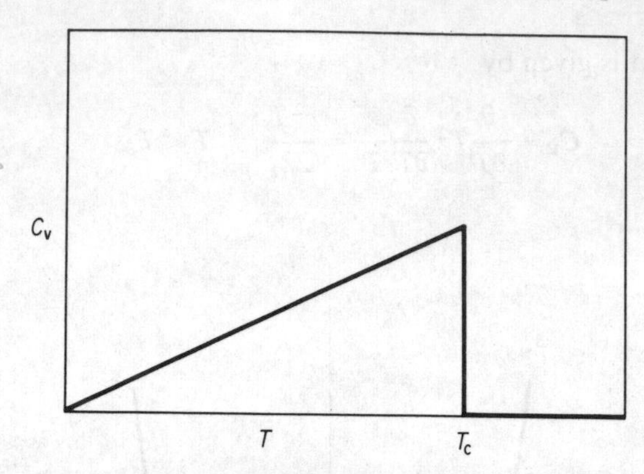

Figure 3.25 Specific heat anomaly at a second-order transition

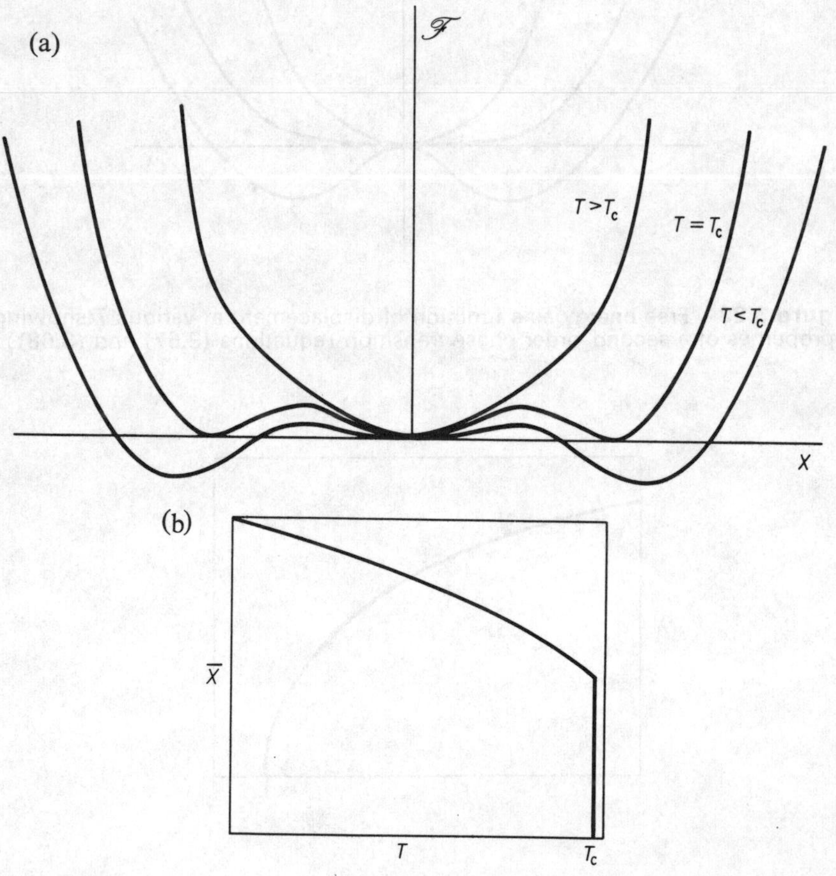

Figure 3.26 (a) Free energy and (b) order parameter for a first-order phase transition (compare figures 3.23 and 3.24)

and falls discontinuously to zero at $T = T_c$ (see figure 3.25). This is described as a second-order phase transition since the discontinuities are in the second derivatives of $\mathscr{F}$. There is no latent heat at the transition.

As is seen from (3.68) the transition takes place when $\omega_i^2(\boldsymbol{q}) \to 0$. This decrease in the mode frequency, which is called 'softening', means that the harmonic restoring forces are becoming very weak and eventually they allow a large displacement which is only curtailed by the anharmonic forces. An alternative way of viewing this is to note that as $\omega_i(\boldsymbol{q}) \to 0$ excitations in the mode cost no energy and a large number may be spontaneously created to produce the distortion. As $\omega_i(\boldsymbol{q})$ becomes imaginary the motion becomes exponential in time (like $\exp \gamma t$) rather than harmonic (like $\exp i\omega t$).

Not all transitions are second order. For example if $\lambda_2 < 0$ while eventual stability is maintained by $\lambda_3 > 0$, there can be a second minimum in $\mathscr{F}$ away from $x = 0$. As T decreases this minimum falls below $\mathscr{F} = 0$ and becomes the absolute minimum in the system. When this occurs $\bar{x}$ changes discontinuously (see figure 3.26). This may be well away from the condition $\lambda_1 = 0$ and the softening of the mode will not be seen in this case. This corresponds to a first-order transition with a latent heat.

3.8.2 Ferroelectric crystals (AD)

One of the most interesting forms of phase transitions accompanied by a lattice distortion occurs in ferroelectric crystals. In this case the instability occurs in a $q = 0$ optic mode which has an electric dipole moment associated with it. In the low T, distorted phase the crystal has a macroscopic dipole moment. The electric polarisability of such crystals increases dramatically as T_c is approached.

A good simple example of this type of crystal is barium titanate ($BaTiO_3$) which has the perovskite crystal structure shown in figure 3.27. In the distorted phase the positive ions move relative to the negative ions. The Ti ion moves away from the centre of the tetrahedron of oxygens. This distortion corresponds to a transverse optic mode. As discussed in the footnote in § 3.4 this can be at a frequency lower than that expected from short-range forces because of a negative contribution from the dipole–dipole forces. If anharmonic effects play a role, ω_T may show a temperature dependence of the form

$$\omega_T^2 = C(T - T_c)$$

In fact such a dependence is observed experimentally in $BaTiO_3$ by neutron scattering (figures 3.28 and 3.29). We note from (3.38) that

$$\varepsilon(0) = \varepsilon(\infty)\, \omega_L^2/\omega_T^2 \sim A/(T - T_c) \tag{3.72}$$

since only ω_T^2 is expected to have a large dependence on T.

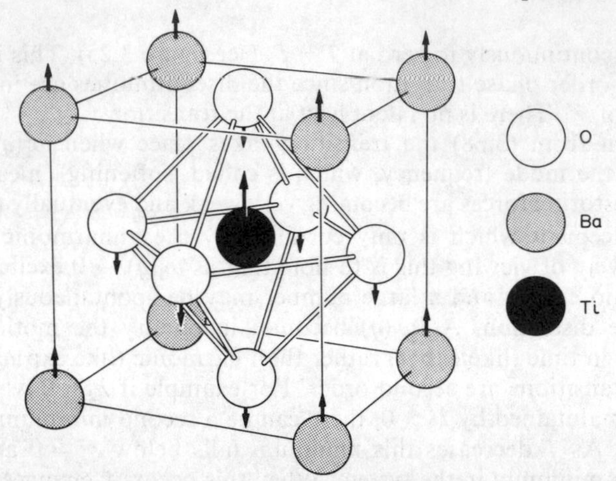

Figure 3.27 Unit cell of a perovskite ferroelectric such as $BaTiO_3$ showing main distortion which gives rise to ferroelectricity

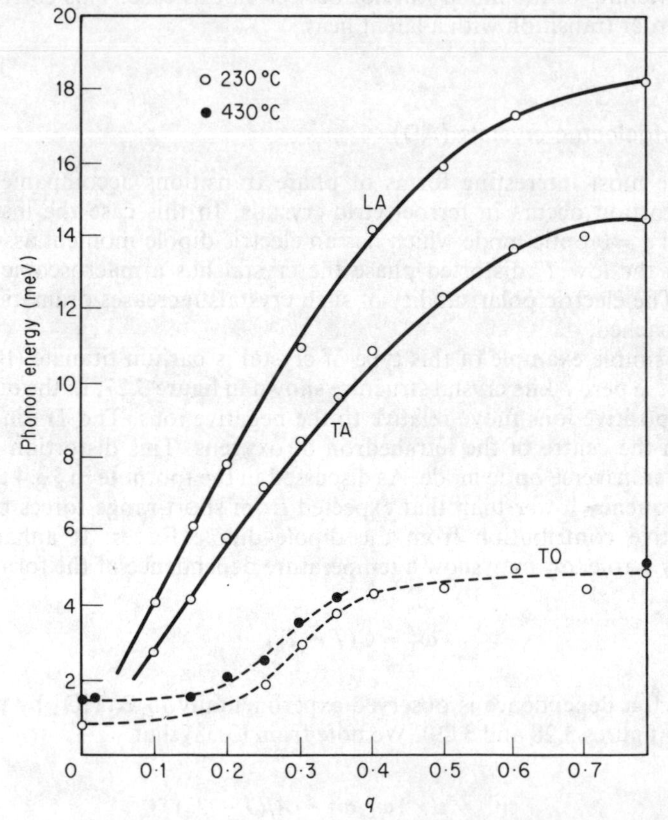

Figure 3.28 Variation of $\hbar\omega(q)$ with T in $BaTiO_3$ showing transverse optic (T.O.) soft mode at two temperatures. The acoustic modes change hardly at all. (After G. Shirane *et al., Phys. Rev. Lett.* **19**, 234 (1969).)

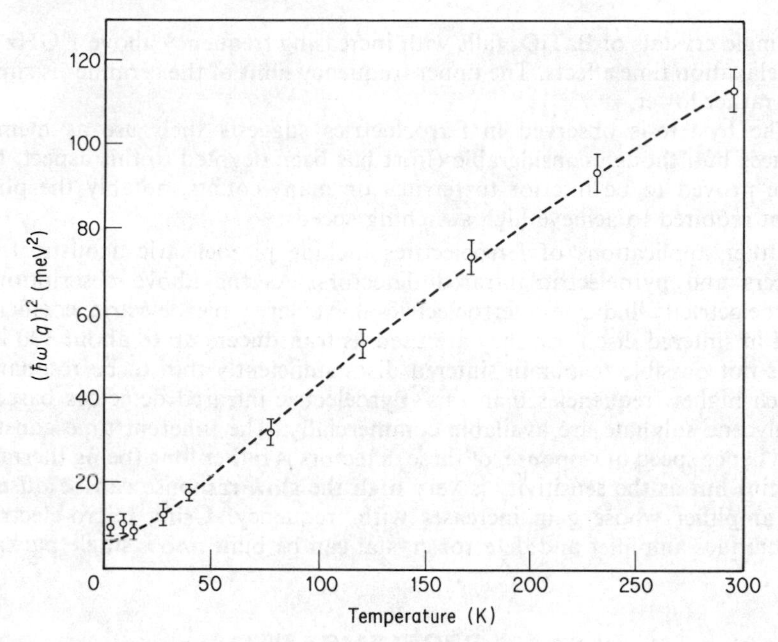

Figure 3.29 Variation of $(\hbar\omega(q))^2$ for the transverse optic mode in $KTaO_3$ (points) compared with the inverse dielectric constant (dashed line) (after G. Shirane *et al., Phys. Rev.* **157**, 396 (1967))

This dependence of $\varepsilon(0)$ on temperature is typical of all ferroelectrics. Below T_c they will have a spontaneous polarisation. Actually most transitions are first order although $\varepsilon(0)$ reaches a very high value and the discontinuity in the polarisation is small, as is the latent heat. However, not all such transitions can be immediately discussed in terms of soft phonon modes. In crystals like KH_2PO_4 the ferroelectricity is associated with the distribution of hydrogens between anharmonic potential wells with double minima (as shown in figure 2.1). In other crystals with organic constituents like Rochelle salt $NaK(C_4H_4O_6).4H_2O$ and triglycine sulphate where the crystal structure is complex the detailed mechanism is unknown.

Ferroelectric crystals show hysteresis effects and structure like ferromagnetic crystals. We defer discussion of this type of behaviour until Chapter 10. By analogy with magnetic systems, antiferroelectric order also occurs.

The high dielectric constant of ferroelectrics is exploited commercially in the manufacture of high-capacitance condensers. $BaTiO_3$ forms the basis of such capacitors since T_c is not much above room temperature. However, the very rapid variation in $\varepsilon(0)$ with temperature around room temperature is undesirable and various mixed compounds, such as $(Ba–Sr–Ca)TiO_3$ are normally used. The mixed materials are prepared as sintered ceramics and it is likely that inhomogeneity also contributes to the reduced temperature dependence of $\varepsilon(0)$. The price of this lower temperature coefficient is a reduction in the value of $\varepsilon(0)$ by an order of magnitude (to $\sim 10^2\ \varepsilon_0$). The dielectric constant of single crystals of $BaTiO_3$ falls with increasing frequency above 1 GHz due to relaxation time effects. The upper frequency limit of the ceramics is similar but rather lower.

The hysteresis observed in ferroelectrics suggests their use as memory devices but, though considerable effort has been devoted to this aspect, they have proved to be inferior to ferrites on many counts, notably the power input required to achieve high switching speeds.

Other applications of ferroelectrics include piezoelectric acoustic transducers and pyroelectric infrared detectors. As the above description of ferroelectricity indicates, ferroelectrics have large piezoelectric coefficients and in sintered disc form they are used as transducers up to about 100 kHz. It is not possible to obtain sintered discs sufficiently thin to be resonant at much higher frequencies than this. Pyroelectric infrared detectors based on triglycene sulphate are available commercially. The inherent time constant, and hence speed of response, of these detectors is rather long (being thermal in origin) but as the sensitivity is very high the slow response can be offset by an amplifier whose gain increases with frequency. Using micro-electronic techniques amplifier and detector crystal can be built into a single package.

PROBLEMS

(Some answers, where appropriate, are given on page 481.)

3.1 Show that, for the diatomic linear chain of § 3.2.2, the density of modes per unit frequency interval diverges at the maximum frequency and at the frequencies on either side of the gap, while it tends to a constant as $\omega \to 0$.

3.2 Show by a symmetry argument that the two transverse modes in a face-centred cubic lattice have the same frequency (i.e. are degenerate) when $\boldsymbol{q}$ is parallel to a (1,0,0) direction or a (1,1,1) direction but not when $\boldsymbol{q}$ is parallel to a (1,1,0) (compare figure 3.8).

3.3 In graphite there are only weak forces between layers of atoms, while the forces between atoms within a layer are strong. Show how this may be used to explain the fact that the specific heat of graphite varies as T^2 at fairly low temperatures, rather than T^3 as expected on a simple Debye theory.

3.4 Show that, because of the changes induced by anharmonicity (§ 3.7) the specific heat at high temperatures might be expected to vary, not as (3.51) but as

$$C_V = 3\mathcal{N}k(1 + akT)$$

where a is proportional to γ^2.

3.5 In a linear chain all the atoms have the same mass but are connected alternately by springs of force constants Γ_1 and Γ_2 respectively. Calculate the frequency, wave vector spectrum. Why are there two branches?

3.6 Estimate the zero-point energy of a solid assuming the Debye model.

3.7 In a face-centred cubic crystal atoms of mass M are joined by nearest neighbour harmonic forces so that the potential energy of displacement is

$$\tfrac{1}{2}a[(\boldsymbol{u}(R) - \boldsymbol{u}(R')).(\boldsymbol{R} - \boldsymbol{R}')]^2$$

Show that $C_{11} = 2C_{44}$, i.e., that the velocity of longitudinal acoustic waves in the (1,0,0) direction is greater than the velocity of transverse acoustic waves by a factor of $\sqrt{2}$. Similarly show that in the (1,1,1) direction this ratio is $2/\sqrt{3}$.

3.8 A crystal of $BaTiO_3$ in its ferroelectric phase has a polarisation of 0.2 C m^{-2} at a certain temperature. What is the electric field strength in a thin slice of such a crystal cut normal to the polarisation direction? Assuming a unit cell of the crystal has a volume 6×10^{-29} m^3 calculate the dipole moment per cell and the displacement this implies if the whole is due to the motion of the Ti ion.

4

Electrons in bands

4.1 Introduction (a)

We have seen in the last chapter that the possible frequencies of lattice vibrations are limited and that the frequencies, or energies, of phonons in a perfect crystal must lie within certain well defined bands. Within a band or branch the discrete phonon states are so close in energy that their energy can be considered to be a continuous function of wave vector. These features are characteristic of excitations in crystals, as described in Chapter 2.

A qualitatively similar situation exists for electrons in a solid for they, too, can be described by waves in a periodic lattice and only certain allowed bands of energy may be occupied by electrons. Within a given band the electron states are so close in energy that the electron energy can be considered (unless the sample is very small, see § 2.5) as a continuous function of electron wavevector, $\boldsymbol{k}$. The electron band structure (i.e. the variation in electron energy, E, with $\boldsymbol{k}$ for all bands) of a solid is, however, much more complicated than the phonon band structure. The energy bands may overlap so that a continuous band of allowed electron energies is obtained. It is found that the width, in energy, of bands tends to increase as energy increases so that overlapping bands are common at high energies while the lower lying bands remain well separated. The existence of a continuum of electron states due to band overlap does not, however, mean that the details of the band structure can be ignored: there will, for example, still be the variation in density of electron states per unit energy interval (§ 2.9) to consider. In this chapter we shall be concerned first with indicating how electron band structures can, in principle, be deduced theoretically and the approximations frequently adopted. We shall then describe, as examples, the band structure of some real solids which have been determined by a judicious mixture of theory and experiment.

The probability of occupation of the band states by electrons is often as important in determining the properties of a solid as the details of the band structure. As electrons obey Fermi statistics (§ 2.8) all states in all the lowest bands are fully occupied while all states in the highest bands are empty. We now note that the electrons in a completely filled band cannot carry an electric current. This important conclusion is a consequence of the exclusion principle. In order to move under the influence of an electric field an electron must be capable of being accelerated by the field. Acceleration implies a continuous change in electron momentum and hence wave vector, but by definition there are no empty states of different $\boldsymbol{k}$ in a completely filled band into which an electron can go. Hence the electrons in a full band cannot contribute to conduction. Since it is obvious that a completely empty band, also, cannot contribute to conduction since there are no electrons in it to move, we conclude that electrical conduction arises from the existence of partially filled electron bands.

Solids are commonly classified as metals, semiconductors and insulators. The observed conductivities cover a range of about $10^{20}:1$. In the band picture this wide range is explained, at least to a rough approximation, in terms of the density of mobile electrons, i.e. those in partially filled bands. It should be noted, however, that band theory is not a good approximation for some solids with insulating properties (see § 4.10).

In metals the Fermi level lies within one or more allowed bands which are consequently partially filled and the material is a good conductor. The energy difference between the bottom of the band and the Fermi energy, E_F, is generally considerably greater than kT so that the density of mobile electrons is substantially independent of temperature. The magnitude of the wave vector, $\boldsymbol{k}_F$, of an electron with energy E_F depends on the details of the band structure. In general E_F is a function of both the magnitude and direction of $\boldsymbol{k}_F$ so the lengths of the vectors $\boldsymbol{k}_F$ define a three-dimensional surface in reciprocal space. Electrons on this surface, which is called the Fermi surface of the metal, all have energy E_F. If the energy of the mobile electrons is simply proportional to $|\boldsymbol{k}^2|$, as it is for free electrons, the Fermi surface is a sphere. So simple a band structure is comparatively rare in real metals but, as we shall show, it is a good approximation in some cases (§ 4.5). The shape of the Fermi surface can be determined experimentally in a variety of ways. The most important of these are described in § 4.7 and in later chapters.

Pure semiconductors are characterised by a rapid, approximately exponential, increase in conductivity with temperature. Here the Fermi level lies in a disallowed energy region between two bands. At zero temperature all bands are either completely filled or completely empty but at finite temperatures the smearing of the Fermi distribution produces a small but finite probability that some states in the lowest empty ('conduction') band will be occupied by electrons which can carry a current. Equally there is a finite probability that some states in the highest filled ('valence') band will not be occupied by electrons. We can consider that the electrons in the conduction band have been promoted from the valence band by gaining sufficient thermal energy. The unfilled states in the valence band permit the electrons remaining in this band to be accelerated by a field so that these, also, contribute to conduction. We shall find it easier to describe the latter contribution by concentrating on the (relatively few) vacant states rather than the electrons in the band and this we do in § 4.6.

4.2 An introduction to electron band theory (a)

In metals and covalent solids the valence electrons are shared between all the atoms of the crystal. Even in ionic and Van der Waals solids some degree of electron sharing takes place. It is therefore often a useful approximation to regard the valence electrons as non-interacting bodies moving through the crystal under a potential which is the average of the potential of all the other valence electrons, the core electrons and the nuclei in the crystal. Although at first sight it might seem that the neglect of the inter-electronic Coulomb repulsion would lead to serious errors, the approximation turns out to be a very good one and the *independent electron model*, as it is called, gives a good description of many aspects of crystals. The detailed justification of the independent electron model is a theoretical problem of some difficulty and we shall consider only some very simple physical arguments later (§ 4.9). For the moment we shall pursue the consequences of the model, secure in the knowledge that the model can be justified *a posteriori* by using it to predict behaviour which is in agreement with experiment.

The wave functions describing the behaviour of the electrons will reflect the crystal structure. On the one hand we can regard the crystal as an assembly of individual atoms. An electron in an isolated atom can exist in a few allowed states which have well-separated energies. If a pair of such atoms come together the electron is shared between them and resonates back and forth with a well-defined frequency. If the wave function for an electron in atom A is written ϕ_A, and that for an electron in atom B is ϕ_B the possible new wave functions for an electron shared between A and B are approximately

$$(\phi_A \pm \phi_B)/\sqrt{2} \tag{4.1}$$

and have energies

$$E_0 \pm E_1 \tag{4.2}$$

where E_0 is the energy of a particular state in the isolated atom and $2E_1$ is the energy difference between the two new states. The splitting is related to the frequency ω with which an electron resonates between the atoms by $\hbar\omega = 2E_1$. If, as in a crystal, an electron is shared between many atoms rather than a pair, each energy level is split into many states which effectively form a continuous band of allowed energies centred around E_0. The width of the band will be related to E_1. Since E_1 arises from the motion of the electron from atom to atom it might be thought of as the kinetic energy of motion across the crystal, while E_0 represents the atomic potential energy.

If, on the other hand, the effect of the atoms is neglected the electron states are those of free particles; they are of simple wave form and their energy $\hbar^2 k^2/2m$ is that of free electrons with a continuous distribution at all positive energies. Both these features may be expected to be reflected in the actual states in crystals. Their form can be derived entirely from the symmetry arguments set out in Chapter 2. The electron moves in a periodic potential

$$V(\boldsymbol{r} + \boldsymbol{R}_l) = V(\boldsymbol{r}) \tag{4.3}$$

which is the same in each unit cell. The wave function satisfies Schroedinger's equation

$$\left(-\frac{\hbar^2}{2m}\nabla^2 + V\right)\psi = E\psi \tag{4.4}$$

and because of the periodicity has the form described in Chapter 2. In this form it is usually called a Bloch function and

$$\psi_{i\boldsymbol{k}} = (\mathcal{N}\Omega)^{1/2}\,\mathrm{U}_{i\boldsymbol{k}}(\boldsymbol{r})\exp(\mathrm{i}\boldsymbol{k}\,.\,\boldsymbol{r}) \tag{4.5}$$

where $\mathrm{U}_{i\boldsymbol{k}}$, like V in (4.3), has the full translational symmetry of the crystal. Since V has deep wells at each atomic site, $\mathrm{U}(\boldsymbol{r})$ will vary rapidly around these positions—in fact we expect $\mathrm{U}(\boldsymbol{r})$ to vary rather like an atomic function in this vicinity. $\mathrm{U}(\boldsymbol{r})$ contains the essential features of the unit cell and reflects the atomic properties. By contrast the plane-wave component of ψ describes the variation on a large scale across the crystal. This component would persist for free electrons when V is independent of position. The wave function (4.5) is normalised and U is normalised in each unit cell, i.e.

$$\int_{\text{Cell}} |\mathrm{U}_{i\boldsymbol{k}}(\boldsymbol{r})|^2\,\mathrm{d}\boldsymbol{r} = 1 \tag{4.6}$$

where $\mathrm{d}\boldsymbol{r} = \mathrm{d}x\,\mathrm{d}y\,\mathrm{d}z$ is the element of volume.

The symmetry arguments of Chapter 2 showed that with a suitable choice of boundary conditions a crystal with $\mathcal{N}$ cells will have $\mathcal{N}$ allowed values of $\boldsymbol{k}$ uniformly spread across the Brillouin zone. The energy of the electron states will be a continuous function of $\boldsymbol{k}$ over this region. Such a set will form a band and there will be many bands specified by the suffix i. The fundamental problem of band theory is to find the energy $E_i(\boldsymbol{k})$ and the functions $\mathrm{U}_{i\boldsymbol{k}}(\boldsymbol{r})$. This has been done for many simple systems by an interrelation of theory and experiment. Some actual band structures will be described below. The main theoretical difficulties are the determination of $V(\boldsymbol{r})$ with sufficient accuracy and an adequate knowledge of the functions $U(\boldsymbol{r})$ for all the atomic states, since the wave functions for the valence electrons must be made orthogonal to those for electrons at lower energies. Some indication of the band structures can, however, be obtained by approximate solutions to the problem.

4.2.1 The nearly free electron model (AD)

The simplest model of all is one in which the valence electrons in a crystal are assumed to be completely free. Then, from (4.5),

$$\psi_{\boldsymbol{k}}(\boldsymbol{r}) = (\mathcal{N}\Omega)^{1/2} \exp(\mathrm{i}\boldsymbol{k}.\boldsymbol{r}) \tag{4.7}$$

$$E(k) = \hbar^2 k^2/2m \tag{4.8}$$

On this model there is no band structure and the existence of a Brillouin zone is irrelevant since the electron cannot detect the crystal structure. If we now add a small atomic potential (4.7) and (4.8) are slightly modified in a way which may be determined by perturbation theory (cf. § 2.10). The potential only 'mixes' states of $\boldsymbol{k}$ and $\boldsymbol{k}+\boldsymbol{Q}$ (cf. (2.1)). This is especially important if $E(\boldsymbol{k}) = E(\boldsymbol{k}+\boldsymbol{Q})$ when the atomic potential couples states of the same energy and causes a splitting, associated with a resonance between the states. For example, in a one-dimensional chain of atoms separated by a this occurs when $k = \pm Q/2$ where $Q = 2\pi \mathrm{n}/a$ (cf. § 3.2.1). The coupled states are similar to (4.1) and have the form

$$(2\mathcal{N}a)^{-1/2} \sin(Qx/2) \quad \text{and} \quad (2\mathcal{N}a)^{-1/2} \cos(Qx/2)$$

The coupling energy analogous to E_1, in (4.2), is given by

$$\int \psi^*_{-k}(x)\, V(x)\, \psi_k(x)\,\mathrm{d}x = \frac{1}{a}\int_0^a V(x) \exp(\mathrm{i}Qx)\,\mathrm{d}x = V(Q) \tag{4.9}$$

so that the energies are $(\hbar^2 Q^2/8m) \pm V(Q)$. Thus a gap of energy $2V(Q)$ is formed at the point which is equivalent to the zone boundary. This is shown in figure 4.1 where the energies are plotted using the actual k values which extend beyond the zone. This is called the extended zone scheme. By translating the region $\pi/a < k < 2\pi/a$ by $-2\pi/a$ and the region $-2\pi/a < k < -\pi/a$ by $2\pi/a$ they revert to the region $-\pi/a < k < \pi/a$ of the Brillouin zone. These regions are sometimes referred to as the second zone. Similar translations through appropriate $Q_{\mathrm{n}} = 2\mathrm{n}\pi/a$ will bring all such regions into the first zone.

Similar effects occur in two and three dimensions where discontinuities appear in the energy all over the zone boundaries. Figure 4.2(a) shows some curves of constant energy in two dimensions as given by the free electron approximation superimposed on a square zone. This figure might also be regarded as a section of the spheres which form the constant energy surfaces in three dimensions (figure 4.2(b)). A small perturbation will cause these

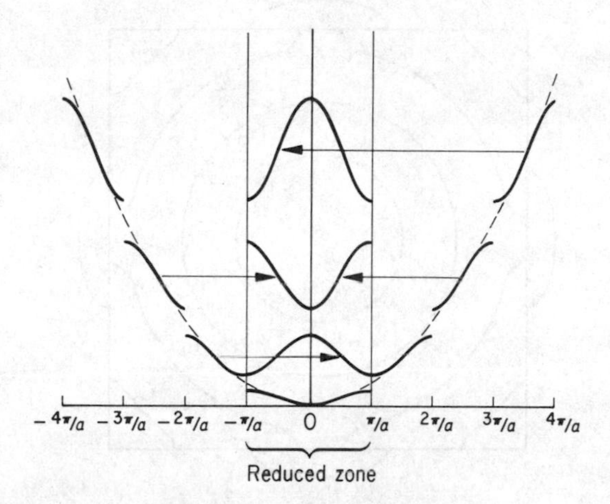

Figure 4.1 Energy curves for a nearly free electron model in one dimension. The dotted curve is the free electron case $E = \hbar^2 k^2/2m$. Gaps appear at the zone boundaries $n\pi/a$. All the curves may be translated into the reduced zone $-(\pi/a) < k < (\pi/a)$ by displacement through a reciprocal lattice vector $\boldsymbol{Q}_n = 2n\pi/a$ as shown

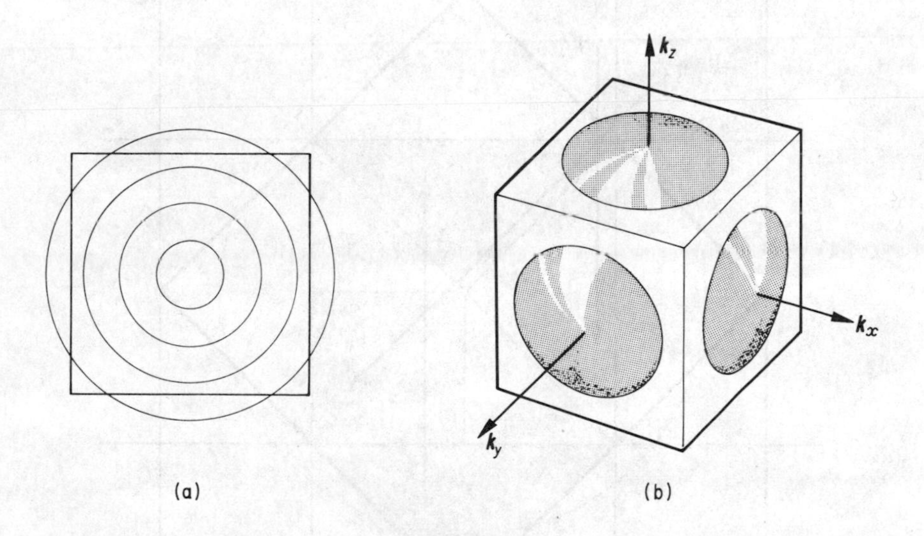

Figure 4.2 Curves of constant energy for the free electron model in two dimensions are circles. At some energies they cut the zone boundary—here represented by a square. (b) Surfaces of constant energy for the free electron model in three dimensions are spheres, shown here with a cubic zone. Figure 4.2(a) may be regarded as a section of this figure with $k_z = 0$

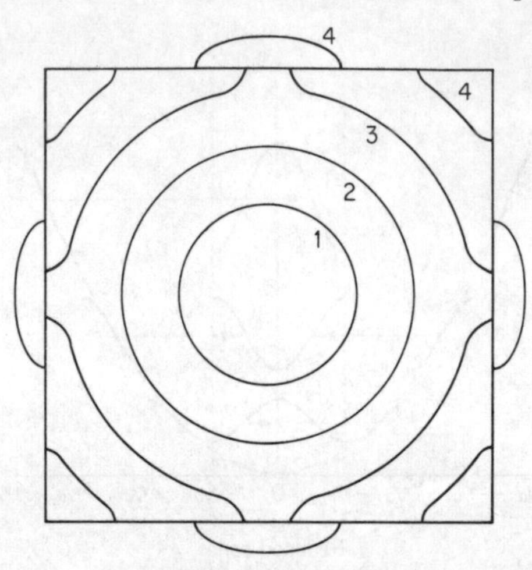

Figure 4.3 Curves of constant energy in the nearly free electron model showing changes caused by the zone boundary. The four curves are perturbed versions of the four circles in figure 4.2(a)

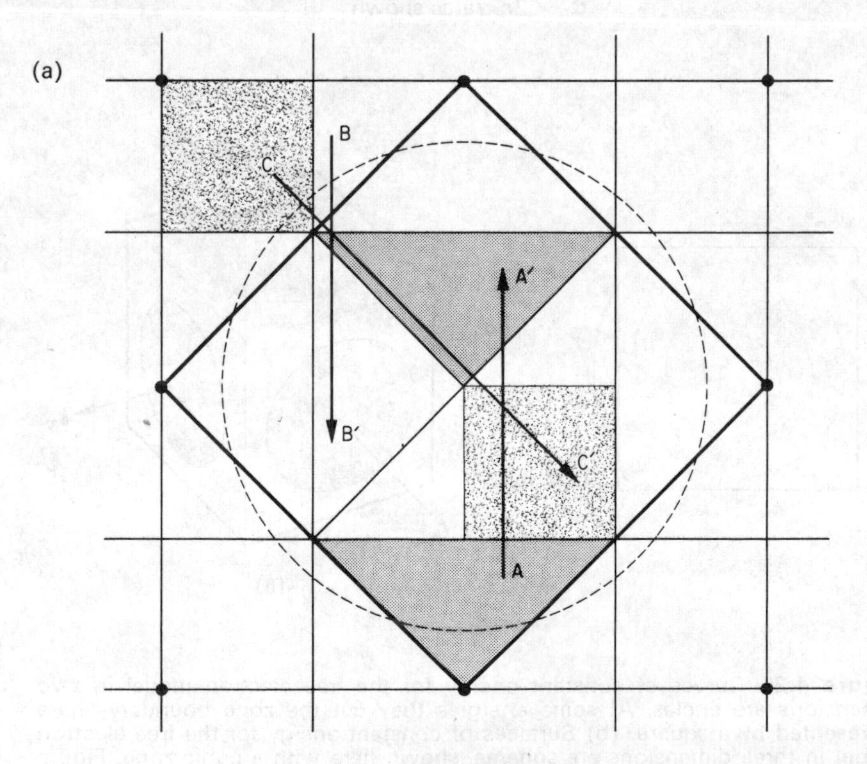

Figure 4.4(a)

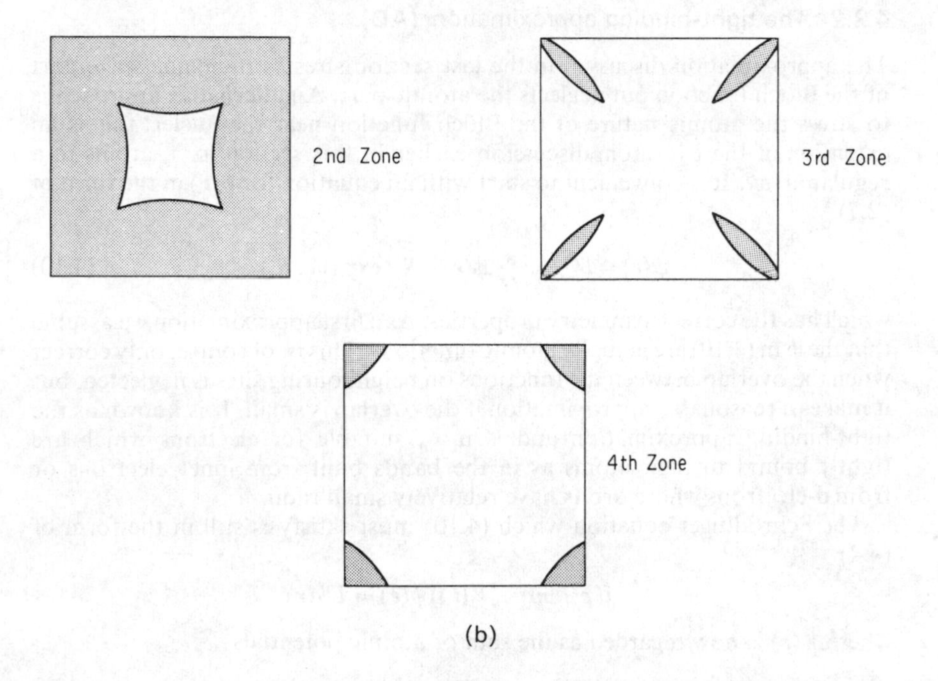

Figure 4.4 Four zones of a square lattice, obtained by drawing the perpendicular bisectors from the origin to Q_n. The region shaded as A may be translated to fill the original square zone as shown. This is called the second zone. Regions like B and C form the second and third zones. The circle is a free electron constant energy curve
(b) The form of the circular constant energy curve of (a) when translated into the first zone. The shaded regions correspond to the areas inside the circle

curves to distort as shown in figure 4.3. They bulge out towards the boundary since the energy in the zone is depressed there. Outside the boundary the sectors are reduced because the energy is increased there (cf. figure 4.1).

If the sections outside the zone are brought back inside by translations of $\boldsymbol{Q}$ they will form different continuous curves (surfaces in three dimensions). An example of this is shown in figure 4.4 for a free-electron curve spreading into four zones. If lines are drawn in space perpendicularly bisecting all $\boldsymbol{Q}$s a pattern shown in figure 4.4(a) is produced. The sections of different shading can, by translations of $\pm Q_x$, $\pm Q_y$ or $\pm(Q_x \pm Q_y)$ be made to cover the whole zone. This leads to continuous energy curves within the zone as shown in figure 4.4(b). In fact $V(\boldsymbol{Q})$ will smooth out the sharp corners which occur, but we would expect there to be three separate curves corresponding to regions 2, 3 and 4 of the extended scheme.

4.2.2 The tight-binding approximation (AD)

The approximation discussed in the last section stresses the plane wave part of the Bloch function but neglects the atomic part. An alternative approach is to stress the atomic nature of the Bloch function near the nuclei: this is an extension of the two-atom discussion earlier in this section to $\mathcal{N}$ atoms in a regular array. It is convenient to start with an equation for $\psi(\boldsymbol{r})$ in the form of (2.7)

$$\psi(\boldsymbol{r}) = \mathcal{N}^{-1/2} \sum_{l} \phi(\boldsymbol{r} - \boldsymbol{R}_l) \exp(\mathrm{i}\boldsymbol{k} . \boldsymbol{R}_l) \tag{4.10}$$

which has the correct symmetry properties. As a first approximation we assume that the ϕ in (4.10) are actually atomic functions. This is, of course, only correct when the overlap between the functions on neighbouring sites is neglected, but it makes a reasonable approximation if the overlap is small. It is known as the tight-binding approximation and is most suitable for electrons which are tightly bound to their atoms as in the bands built from inner electrons or from d-electrons whose orbits have relatively small radii.

The Schrödinger equation which (4.10) must satisfy is still in the form of (4.4)

$$[(\boldsymbol{p}^2/2m) + V(\boldsymbol{r})]\psi(\boldsymbol{r}) = E\psi(\boldsymbol{r})$$

where $V(\boldsymbol{r})$ is now regarded as the sum of atomic potentials

$$V(\boldsymbol{r}) = \sum_{l} V_{\mathrm{at}}(\boldsymbol{r} - \boldsymbol{R}_l) \tag{4.11}$$

As the atoms come close together the individual potentials overlap and the potential is modified, particularly in the regions between the atoms (figure 4.5).

For a single atom the electron has atomic energy E_{at} given by

$$[(\boldsymbol{p}^2/2m) + V_{\mathrm{at.}}(\boldsymbol{r} - \boldsymbol{R}_l)]\,\phi(\boldsymbol{r} - \boldsymbol{R}_l) = E_{\mathrm{at}}\,\phi(\boldsymbol{r} - \boldsymbol{R}_l) \tag{4.12}$$

but if the overlap is included this energy is slightly changed to E_0 where

$$[(\boldsymbol{p}^2/2m) + V(\boldsymbol{r})]\,\phi(\boldsymbol{r} - \boldsymbol{R}_l) = E_0\,\phi(\boldsymbol{r} - \boldsymbol{R}_l) \tag{4.13}$$

In addition the electron can now move from site to site at a frequency determined by an energy analogous to E_1 defined in (4.2). For this problem it has the value

$$E_{\boldsymbol{\delta}} = \int \phi^*(\boldsymbol{r} - \boldsymbol{R}_l)[(\boldsymbol{p}^2/2m) + V(\boldsymbol{r})]\,\phi(\boldsymbol{r} - \boldsymbol{R}_l + \boldsymbol{\delta})\,\mathrm{d}\boldsymbol{r} \tag{4.14}$$

for transfer to a site a distance $\boldsymbol{\delta}$ away. This will be largest on nearest neighbours. Neglecting the fact that the ϕ on different sites are not orthogonal so that (4.10) is incorrectly normalised we see that (4.10) is indeed a solution of (4.4) and

$$E(\boldsymbol{k}) = E_0 + \sum_{\boldsymbol{\delta}} E_{\boldsymbol{\delta}} \exp(\mathrm{i}\boldsymbol{k} . \boldsymbol{\delta}) \tag{4.15}$$

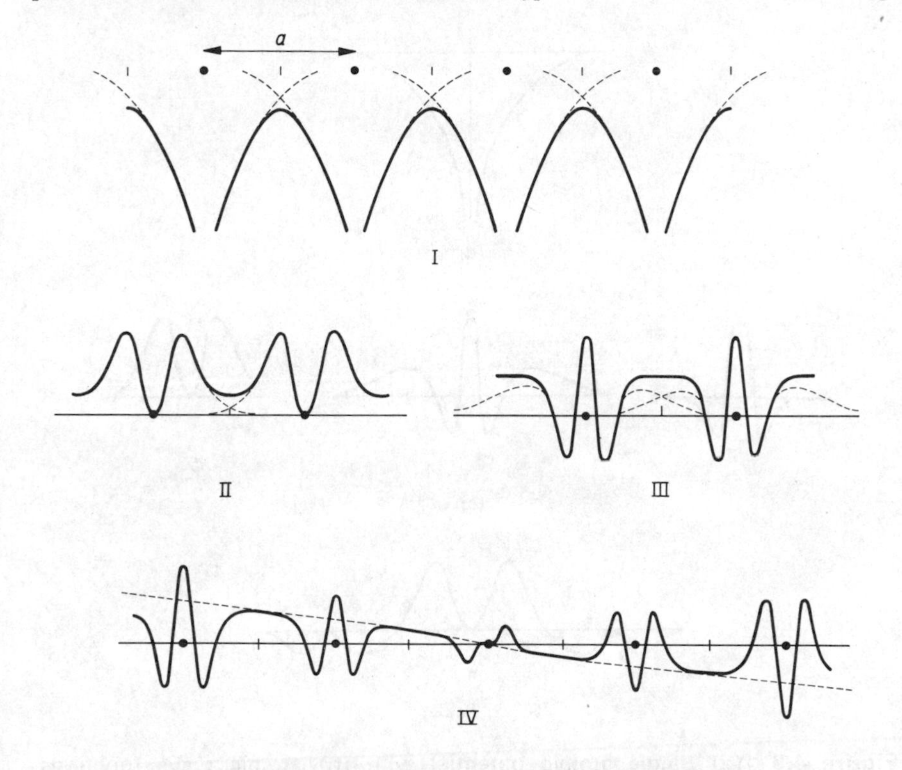

Figure 4.5 I: The potential in a linear array of atoms (full curve) built from overlapping atomic potentials (shown dashed)

II: Bloch functions corresponding to $\boldsymbol{k} = 0$ state built from atomic wave functions (dashed) with small overlap

III: Same type of state built from atomic functions (dashed) of larger overlap

IV: Schematic representation of wave functions corresponding to $\boldsymbol{k} \neq 0$. The real part of ψ is shown. It is approximately in the form (III) multiplied by $\cos k_x$ (shown dashed)

A schematic representation of the potential and the wave functions is given in figure 4.5.

Wannier has shown that for a simple band we can always write the wave function like (4.10) but that the correct form of ϕ is one which spreads over several cells. These ϕs are often called Wannier functions. In effect, the tight-binding approximation replaces these functions (see figure 4.6) by the approximate atomic functions.

For a simple cubic lattice and a band built from s type atomic functions E_δ will be the same for all nearest neighbours. If we neglect all other interactions

$$E(\boldsymbol{k}) = E_0 + 2E_1 (\cos k_x a + \cos k_y a + \cos k_z a) \tag{4.16}$$

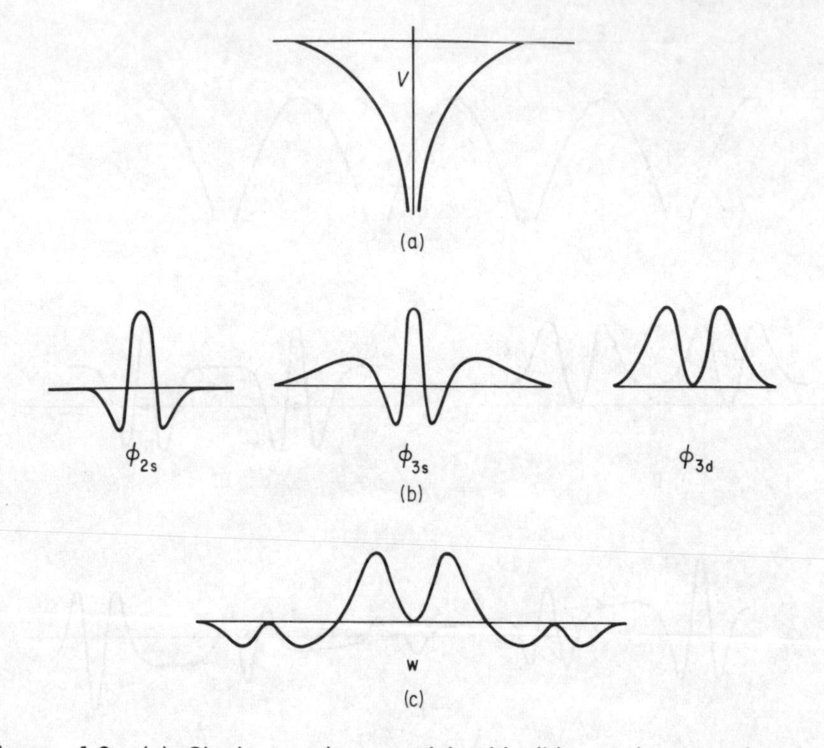

Figure 4.6 (a) Single atomic potential with (b) atomic wave functions corresponding roughly to 2s, 3d and 3s types. The last two were used in forming the tight binding functions in figure 4.5
(c) The Wannier function, W is constructed from 3d functions by negative admixture on neighbouring sites

Curves of constant energy in the plane $k_z = 0$ are shown in figure 4.7. These may be compared with those of figure 4.3, which shows similar but smaller distortions from circles. The density of states in such a band is shown in figure 4.8.

The width of the band in this case is $12E_1$, and it spreads about the atomic energy E_0. As E_1 becomes larger the band will begin to spread into the bands arising from other atomic levels. It may then be necessary to include say both s and p electron states and determine several bands together. Provided the atomic levels are far apart we can refer to bands by labels derived from the atomic orbitals involved, the 3s band and the 3p band for instance, but in fact such states become mixed together so that a band is only predominantly and not wholly of one type.

Band characteristics which arise largely from atomic properties are described qualitatively by the tight-binding approximation. One of these is the effect of spin-orbit coupling, which arises from the relativistic interaction between the spin and orbital motion of an electron. In simple band structures all the band

Figure 4.7 Curves of constant energy for the tight binding approximation for a simple cubic crystal in the plane $k_z = 0$ (compare figure 4.3)

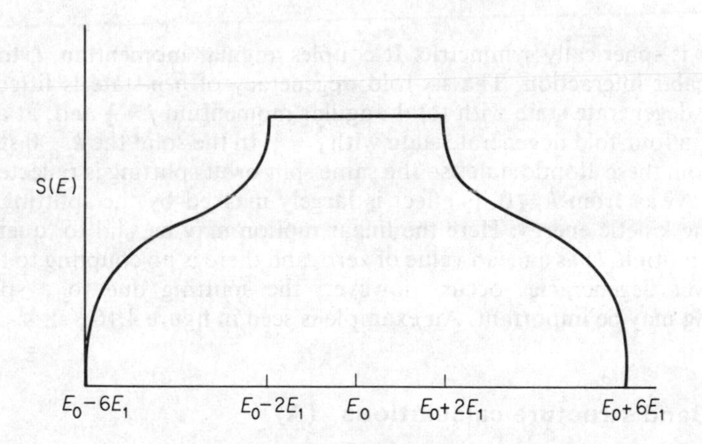

Figure 4.8 The density of states for the tight binding model in the simple cubic crystal. Note the four critical points

states are doubly degenerate because of the two electron spin states. The spin-orbit coupling, even if it is strong, does not remove the degeneracy if the crystal contains a centre of inversion symmetry. The coupling can cause a small splitting if such a centre is lacking, but it is so small as to be negligible for almost all physical properties. However, where several bands are degenerate (i.e. have the same energy) although they are built from different atomic orbitals the change caused by spin-orbit coupling can be appreciable. The spin-orbit coupling has the form

$$\mathscr{H}_{\text{s.o.}} = \frac{\hbar}{2m^2c^2} \nabla \boldsymbol{V} \times \boldsymbol{p} \cdot \boldsymbol{s} \tag{4.17}$$

and depends on the gradient of the potential, V. V is only large near the atomic nuclei where both it and the wave function are most atom-like. We therefore expect to be able to estimate spin-orbit effects from atomic considerations in the tight-binding approximation.

Consider for example a band in a cubic crystal built from atomic p-states. At $\boldsymbol{k} = 0$ the electron states have the full symmetry of the lattice and since the Cartesian directions are all equivalent the three bands built from p_x, p_y, p_z atomic orbitals are degenerate. Away from $\boldsymbol{k} = 0$ the bands split because of the finite kinetic energy of the electrons. For $\boldsymbol{k}$ along the x-direction the p_x bands will have a different overlap and energy from the p_y, p_z bands, in the same way as longitudinal and transverse lattice excitations are of different frequency. In an atom, the spin-orbit coupling is

$$\mathscr{H}_{\text{s.o.}} = \frac{\hbar^2}{2m^2c^2} \frac{1}{r} \left(\frac{\partial V}{\partial r} \right) \boldsymbol{l} \cdot \boldsymbol{s} \tag{4.18}$$

since V is spherically symmetric. It couples angular momentum, $\boldsymbol{l}$, to spin $\boldsymbol{s}$, by a scalar interaction. The six-fold degeneracy of a p-state is lifted into a doubly degenerate state with total angular momentum $j = \frac{1}{2}$ and, at a higher energy, a four-fold degenerate state with $j = \frac{3}{2}$. In the solid the $\boldsymbol{k} = 0$ states are built from these atomic states so the same spin-orbit splitting is reflected in the bands. Away from $\boldsymbol{k} = 0$ its effect is largely masked by the splitting arising from the kinetic energy. Here the linear motion may be said to 'quench' the orbital motion, $\boldsymbol{l}$ has a mean value of zero, and there is no coupling to the spin. Wherever degeneracies occur, however, the splitting due to a spin-orbit coupling may be important. An example is seen in figure 4.16.

4.3 Band structure calculations (A)

The approximations discussed in the last two sections over-emphasise respectively the free electron aspect and the atomic aspect of the band wave functions. However, in so doing they draw attention to the two essential ingredients which must be included in any better theory. Deriving improved wave functions is mathematically tedious but with the extensive use of computers, great progress has been made in recent years and it is now possible to make detailed and fairly accurate predictions about the band structures of simple materials.

It is beyond the scope of this book to give more than a very qualitative account of these methods, the most advanced of which are the augmented plane wave (APW) and the orthogonalised plane wave (OPW) methods. Essentially a band wave function (figure 4.9) must be like a plane wave in the regions between the atoms and like an atomic wave function in the regions where V has a deep potential well.

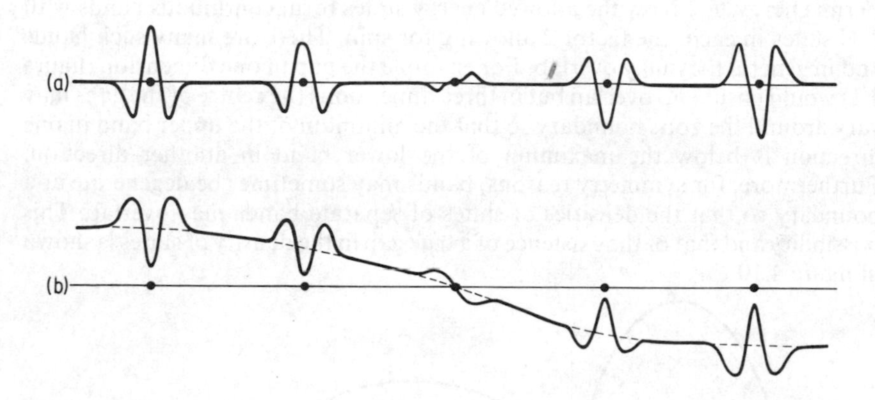

Figure 4.9 Representation of OPW form of Bloch functions (real part shown). Curve (a) is a tight binding core function with very small overlap (compare figure 4.5 IV). A part proportional to this is subtracted from the plane wave $\cos k_x$ to give curve (b)

In the APW method attention is fixed on the plane waves in the outer regions of the cell and these are regarded as scattered by the atomic potentials. The matching of the plane waves on to the atomic functions is a difficult problem. In the OPW method atomic functions are added to the plane wave in order to ensure that the resultant wave function is orthogonal to the core electron wave functions, as it must be.

In fact in many metallic materials the admixture of the core states has remarkably little effect on the energy of the higher states, and deviations from the free-electron energies are rather small. There are, of course, gaps at the zone boundaries, but using the theory of nearly free electrons an effective potential with Fourier components $V(\boldsymbol{Q})$ as defined in (4.9) can be derived to account for these gaps. This is called the pseudopotential, i.e. the potential which gives the correct answer assuming the states to be made up of plane waves. It is fairly smooth and does not have the deep wells of the free potential. It must be realised that this is a device which is only useful because many materials have band energies given by a nearly free electron model; the true wave functions have an atomic component which must be included in order to describe other properties of the electrons.

4.4 The Fermi level and Fermi surface (a)

So far we have considered only single-particle states. Before the properties of a system containing many electrons can be derived, the way in which the individual states are occupied must be known. At $T=0$ the crystal is in its lowest energy state. The Pauli principle (cf. (2.37)) ensures that the available states are filled, with one electron each, from the lowest level upwards until all the electrons are exhausted. The energy of the highest filled state is called the Fermi energy, E_F. Now the allowed energy states form continuous bands with $2\mathcal{N}$ states in each, the factor 2 allowing for spin. There are many such bands and in general they may overlap. For example the gap in one dimension (figure 4.1) would ensure no overlap but in three dimensions the centre of the gaps may vary around the zone boundary so that the minimum of the upper band in one direction is below the maximum of the lower band in another direction. Furthermore, for symmetry reasons, bands may sometimes be degenerate at a boundary so that the densities of states of separate bands may overlap. This possibility and that of the existence of a true gap in the density of states is shown in figure 4.10.

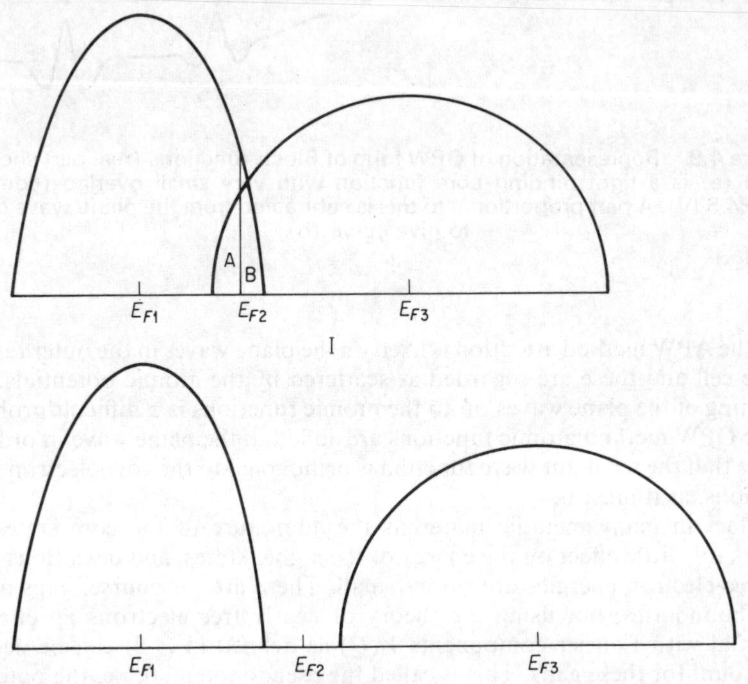

Figure 4.10 (a) Overlapping and (b) non-overlapping bands. The Fermi energies corresponding to one electron per atom and three electrons per atom (E_{F1} and E_{F3} respectively) always lie in the centre of the bands. The energy for two electrons per atom E_{F2} lies in the gap in (b) to give a semiconductor while in (a) it lies in the overlap region. Region A corresponds to electron states occupied in the upper band, region B to unoccupied electron states or hole states in the lower band

If there is an odd number of electrons per cell the Fermi level must lie in the middle of a band (or bands, if they are overlapping). This situation is characteristic of metals. Since the occupied states are separated by infinitesimal energies from empty states, electrons may be promoted to higher energy states by a continuous transfer of energy. In such materials, therefore, electrons can be accelerated by an electric field and the materials are good conductors.

If there is an even number of electrons per cell, the electrons fill an integral number of bands. If, however, the bands overlap, the Fermi level will still lie within a band and the crystal will again be a metal. If the overlap is small so that it occurs at an energy with a low density of states the material is sometimes called a semi-metal.

If the electrons fill an integral number of bands and these are separated by a band gap from further empty bands the solid is a semiconductor. Electrons can then be promoted to new states only by a finite energy gain across the gap. This energy may be obtained thermally so that the system becomes conducting as the temperature increases. Impurities may also change the number of electrons and so move the Fermi level towards one of the bands (see Chapter 5).

In metals there are many states whose energy is practically equal to the Fermi energy, $E_i(\boldsymbol{k}) = E_F$. Because $E_i(\boldsymbol{k})$ is a continuous function of $\boldsymbol{k}$ this relation describes a continuous surface in $\boldsymbol{k}$ space, which separates the region of occupied from the region of unoccupied states. This is called the Fermi surface. On the free electron approximation all such surfaces are spheres. If there are ν conduction electrons per cell the volume of this sphere must be sufficient to contain $\nu\mathcal{N}$ states. Using (2.17)

$$\nu\mathcal{N} = \left(\frac{2\mathcal{N}\Omega}{(2\pi)^3}\right)\left(\frac{4\pi}{3}\right) k_F^3.$$

where k_F is the electron wave vector at the Fermi surface and hence

$$k_F = (3\pi^2\,\nu/\Omega)^{1/3} \tag{4.19}$$

The Fermi energy in the free electron approximation is, of course,

$$E_F = \hbar^2\,k_F^2/2m$$

and the density of states at the Fermi energy is, from (2.43)

$$S(E_F) = \frac{2\mathcal{N}\Omega}{(2\pi)^2}\left(\frac{2m}{\hbar^2}\right)^{3/2} E_F^{1/2} = \frac{3\mathcal{N}}{2\nu E_F} \tag{4.20}$$

including a factor 2 for spin.

If several bands overlap the Fermi surface can have several separate parts. An example of this is the two-dimensional case illustrated in figure 4.4 where the curve of constant energy can be regarded as a Fermi surface. This corresponds to the situation where there are approximately four electrons per cell in the system.

4.5 Actual band structures and Fermi surfaces (a)

We shall now describe some actual band structures. Various aspects of band structures can be obtained by experiment; in semiconductors, for instance, optical properties are the most useful in this regard (see Chapter 6) while in metals the detailed form of the Fermi surface can be deduced from optical and magnetic properties. Indeed in metals it is the states near the Fermi level that dominate most properties since it is only the electrons in these states that can be promoted when small energy quanta are available. Methods of actually determining these properties will be discussed in §§ 4.7 and 4.8. If some experimental information is used in conjunction with band theory calculations the detailed behaviour of the bands can be estimated with greater accuracy.

4.5.1 Simple metals (a)

In an alkali metal like Na the electrons are expected to behave as if they are nearly free. In fact the Fermi surface is found to be very nearly spherical. The partially occupied band has largely 3s atomic character, particularly near $\boldsymbol{k} = 0$. The reciprocal lattice of b.c.c. Na is f.c.c. as shown in figures 1.5 and 3.11, and the nearest point on the zone boundary is half-way to the nearest reciprocal lattice point (at point N as shown in figure 3.11). If the Na–Na distance is $\sqrt{3}a/2$ and the unit cell size in the real lattice is $\frac{1}{2}a^3 = \Omega$, the reciprocal lattice vectors are $(2\pi/a)(1, 1, 0)$, etc. The distance to the boundary is therefore

$$d = \sqrt{2}\pi/a$$

Comparing this with k_F given by (4.19) and putting $v = 1$ we see

$$(k_F/d) = (3/\pi\sqrt{2})^{1/3} = 0.876 \tag{4.21}$$

and hence, while some bulging of the Fermi surface is to be expected in this direction because of the zone boundary effects discussed in § 4.2, it is not very great. The detailed form of the bands is shown in figure 4.11.

In Cu and Ag, which form face-centred cubic lattices, the valence electrons are again s-like and behave like nearly free electrons. For such a lattice (see § 1.3) the reciprocal lattice is b.c.c. (figure 1.4). The nearest point on the zone boundary is half-way to the point $(2\pi/a)(1, 1, 1)$, a distance of $d = \pi\sqrt{3}/a$. Comparing this with k_F given by (4.19) we see

$$(k_F/d) = (4/\pi\sqrt{3})^{1/3} = 0.91 \tag{4.22}$$

so the sphere passes somewhat near the boundary. However, the d-electrons affect the band structure of these materials and increase the effect of the atomic potentials. As a result the distortion of the Fermi surface is sufficient for it to reach the zone boundary (like the curve 3 in figure 4.3). The Fermi surface of these metals is therefore nearly spherical but with small 'necks', the radius of the necks in Cu being about 0.2 of the spherical radius. A picture of this Fermi surface is shown in figure 4.12. Part of the actual band structure of Cu is shown in figure 4.15 and will be discussed later.

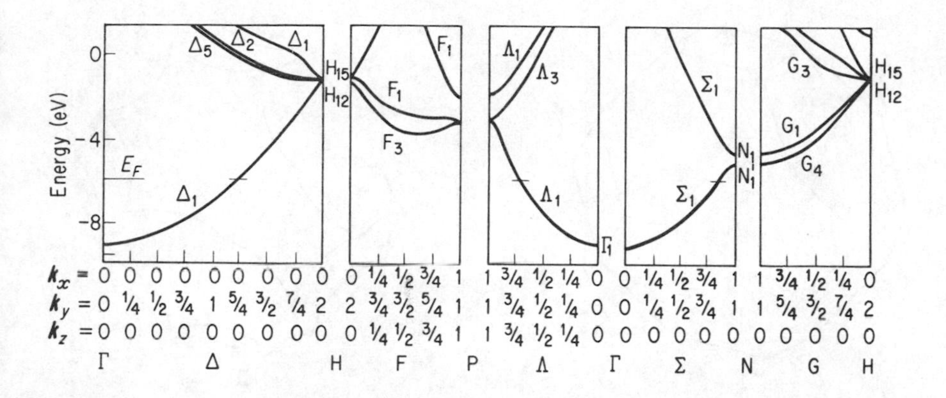

Figure 4.11 Energy bands in Na as calculated by J. F. Kenney (*MIT Prog. Rep.* **53**, 38 (1964)), for various lines in ***k*** space. The lines and points labelled are shown on the Brillouin zone in figure 3.11. The Fermi energy cuts only the three lowest curves which are nearly parabolic (i.e. free electron-like)

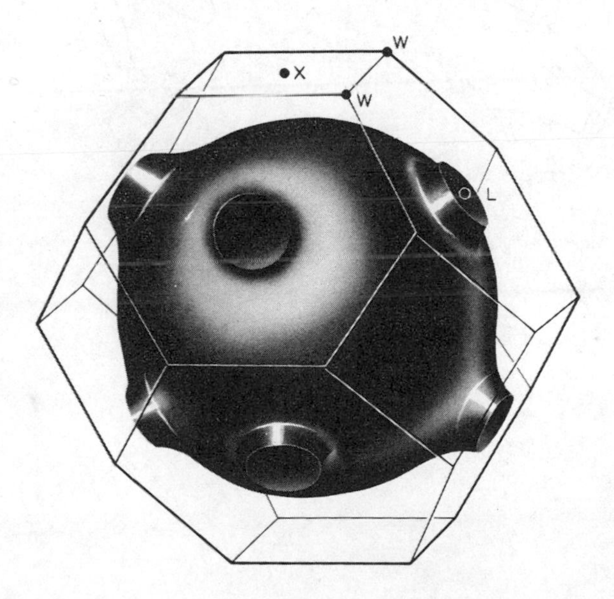

Figure 4.12 The Fermi surface of copper showing 'necks' where it reaches out to the zone boundary. The letters label special points of the zone, X at the midpoint of a square face, L at the midpoint of a hexagonal face and W at the corners

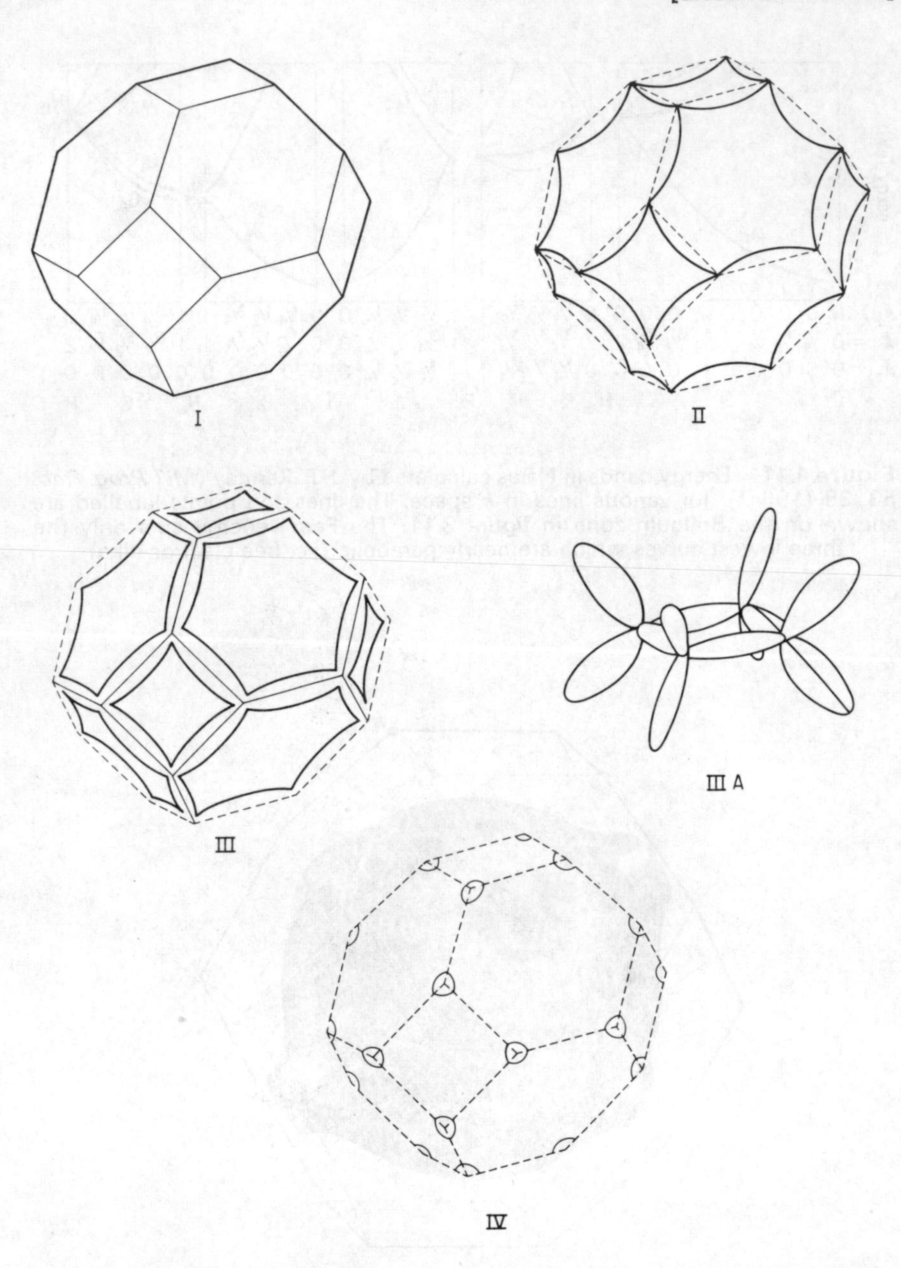

Figure 4.13 The Fermi surface of Al on the free electron model. I, the first band is completely full. II, the region around the faces is full. III, a complex region around the edges with many narrow areas is full. By translation it may be put together to form the 'monster' of III(a). IV, small pockets around W full (after W. A. Harrison, *Phys. Rev.,* **118**, 1182 (1960))

4.5.2 Polyvalent metals (a)

The band structure of Al is in some ways similar to that of Na in figure 4.11. Both bear some similarity to a nearly free electron band reflected back into the Brillouin zone, in this case that appropriate to a f.c.c. crystal. The lower band is predominantly 3s-like while the upper bands have some 3p character. There are three electrons per atom in these bands and as in the two-dimensional example of figure 4.4 the Fermi sphere overlaps four zones in the extended scheme. The lowest band is completely full with $2\mathcal{N}$ electrons. The other $\mathcal{N}$ valence electrons lie partly in a region around the zone surface as shown in figure 4.13 while the rest occupy small pockets in the third and fourth zones. In three dimensions this leads to the rather complicated surface shown in figure 4.13 and sometimes called the 'monster'. Fermi surfaces with complex geometry are not uncommon in polyvalent metals. In fact experiments bear out most of the features predicted by this model although the 'monster' is modified to the form shown in figure 4.14.

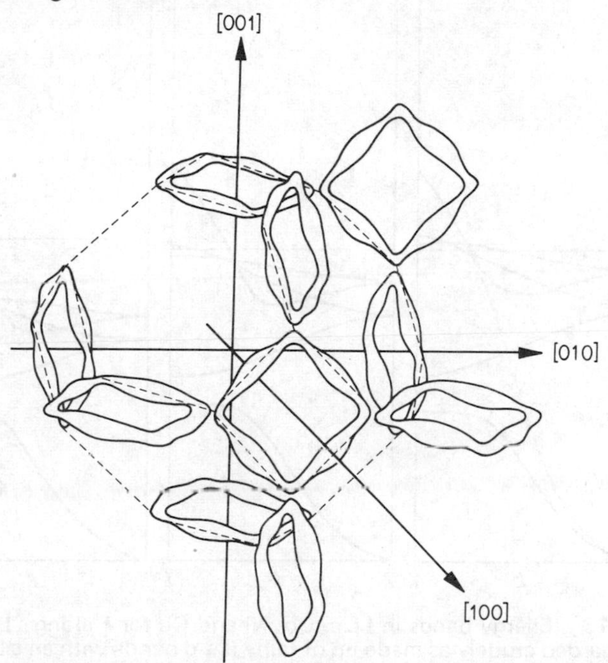

Figure 4.14 Part of actual Fermi surface in Al which corresponds to the monster (after E. P. Volskii, *J.E.T.P.*, **46**, 123 (1963))

4.5.3 Transition metals (a)

The elements of the transition series of the Periodic Table have partly filled shells of d-electrons. We should expect these to give rise to narrow bands fairly well described by the tight-binding approximation. The s and p electrons on the other hand should continue to give nearly-free electron bands. This is borne out by actual calculations. Band structures of face-centred cubic Co, Ni and Cu are shown in figure 4.15, and are closely similar. The band structure consists of an s band which runs steeply from the bottom at Γ to the top at X, and d bands (two doubly degenerate pairs and a single because of the symmetry in this direction) running across the middle. These bands do not in fact cross the s band but bend away from it. The Fermi level for Co and Ni lies in the region of the d bands and the Fermi surface is complex. This is made more complicated by the magnetism of these elements. For Cu the Fermi level lies above the d bands and cuts the single s band leading to the nearly spherical Fermi surface. As discussed in § 4.5.1 it does not cut the s band in the (1, 1, 1) direction giving the necks.

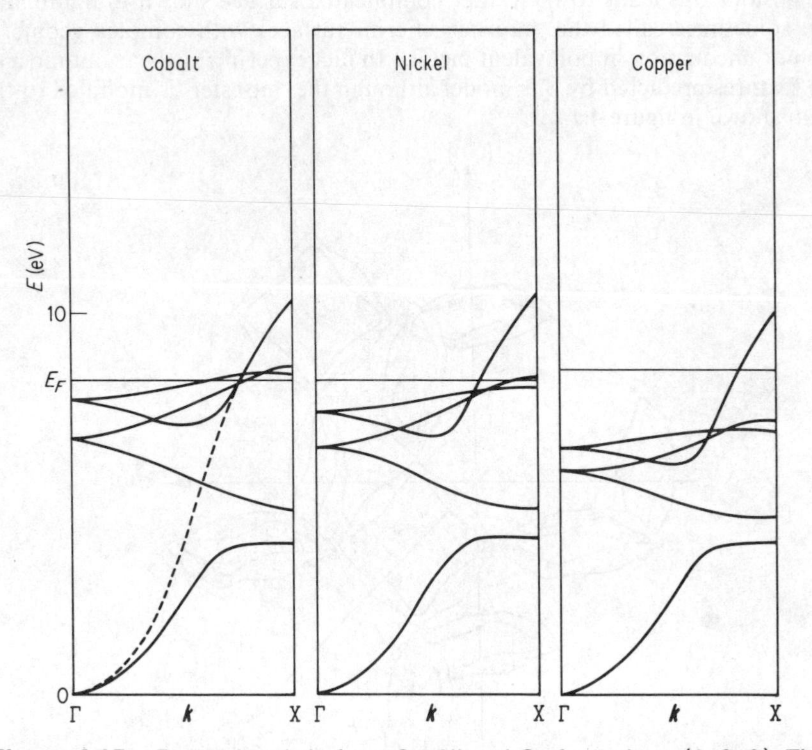

Figure 4.15 Energy bands in f.c.c. Co, Ni and Cu for ***k*** along (1, 0, 0). These may be regarded crudely as made up of fairly flat d bands with an s band crossing them. If the d bands are neglected the s band would be as shown (dashed) in the first figure. The Fermi energy cuts the d bands for Co and Ni giving them a complicated Fermi surface. In Cu it cuts only the s band (after W. M. Lomer and W. A. Gardner, *Prog. Mat. Sci.,* **14**, 117 (1969))

4.5.4 Semiconductors (a)

The band structure of Si is shown in figure 4.16. As is required for a semiconductor there is a gap between the full valence bands and the empty conduction bands. The four valence and four conduction bands shown can be regarded as arising from 3s and 3p electrons. When an electron is shared between two atoms in a covalent bond, the electron can exist in one of two states; one lower and one higher in energy than the corresponding atomic states, as in

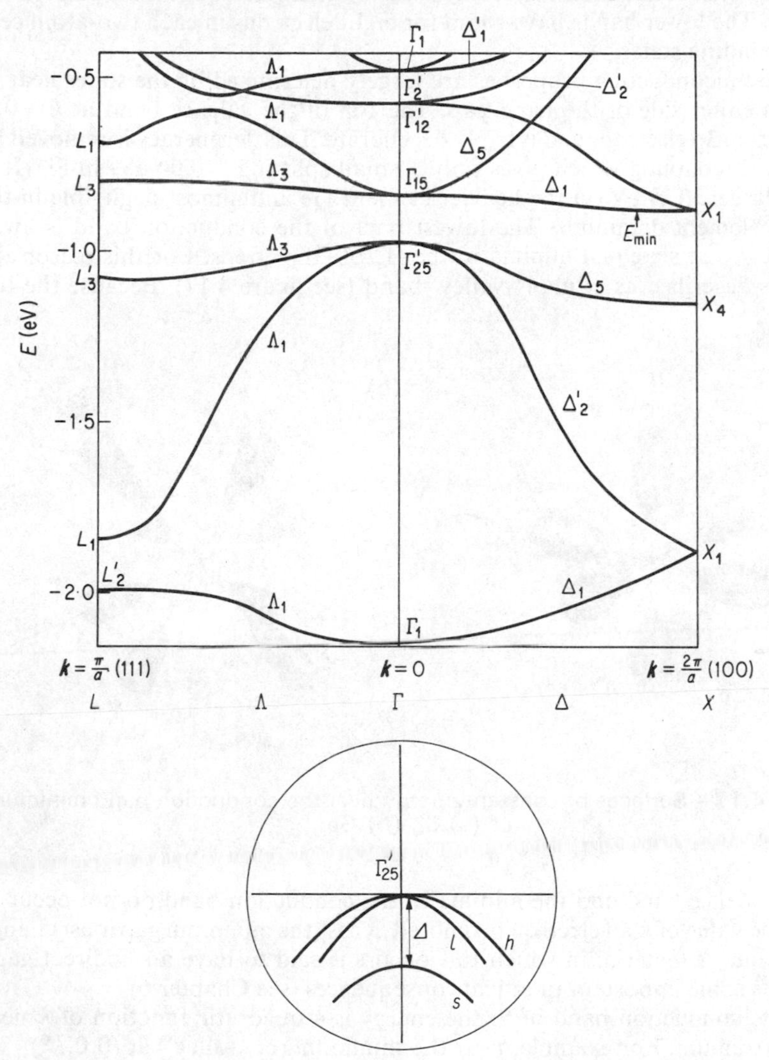

Figure 4.16 The energy band structure of Si (after W. Kleinman and J. C. Phillips, *Phys. Rev.*, **118**, 1164 (1960)). The four valence bands (Λ_3 and Δ_5 are doubly degenerate) are built from 3s and 3p bonding states. The conduction bands show d electron states as well as the 3s and 3p anti-bonding states. The insert shows the spin orbit splitting at the top of the valence bands on an expanded scale. Heavy hole band *h* and light hole band *l* are separated by spin orbit splitting $\Delta = 0.04$ eV from split-off band s

(4.1) and (4.2). The former are referred to as bonding and the latter as anti-bonding states. In the band structure the valence bands are built from bonding states and the conduction bands from anti-bonding states of all the pairs in the crystal. The lower bands have room for eight electrons in each two-atom cell, all in bonding states.

The semiconducting properties are largely determined by the states near to and on either side of the band gap. The top of the valence band at $\boldsymbol{k} = 0$ is made from 3p electrons and is triply degenerate. This degeneracy is removed by spin-orbit coupling which gives only a small splitting, ~0.04 eV, in Si. (It is much larger, 0.29 eV, in the heavier element Ge and almost negligible in the lighter element diamond.) The lowest part of the conduction band is away from $\boldsymbol{k} = 0$ at six equal minima in the (1,0,0) directions. For this reason the band is described as a 'many-valley' band (see figure 4.17). Because the top

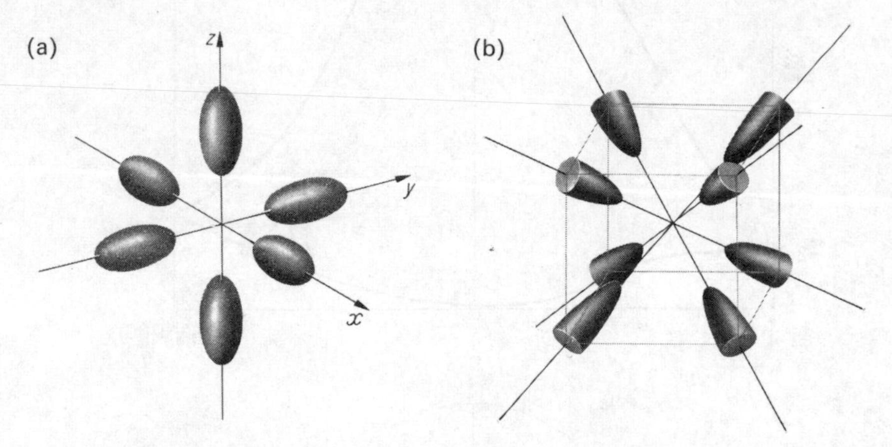

Figure 4.17 Surfaces of constant energy near the conduction band minimum of (a) Si, (b) Ge

of the valence band and the minima in the conduction band do not occur at the same value of $\boldsymbol{k}$ an electron promoted across the minimum gap must change in $\boldsymbol{k}$ value. A material in which this occurs is said to have an 'indirect gap'. This has some important practical consequences (see Chapter 6).

In the conduction band of Si the energy is a quadratic function of $\boldsymbol{k}$ near each extremum. For example, near the minimum (or 'valley') at $(0,0,k_z^0)$

$$E = \hbar^2 \left[\frac{k_x^2 + k_y^2}{2m_t^*} + \frac{(k_z - k_z^0)^2}{2m_l^*} \right] \tag{4.23}$$

m_t^* and m_l^* are parameters with the dimension of mass which are called 'effective masses' (see § 4.6). The other minima have the same shape but different orientations and a constant energy surface consists of six spheroids as shown

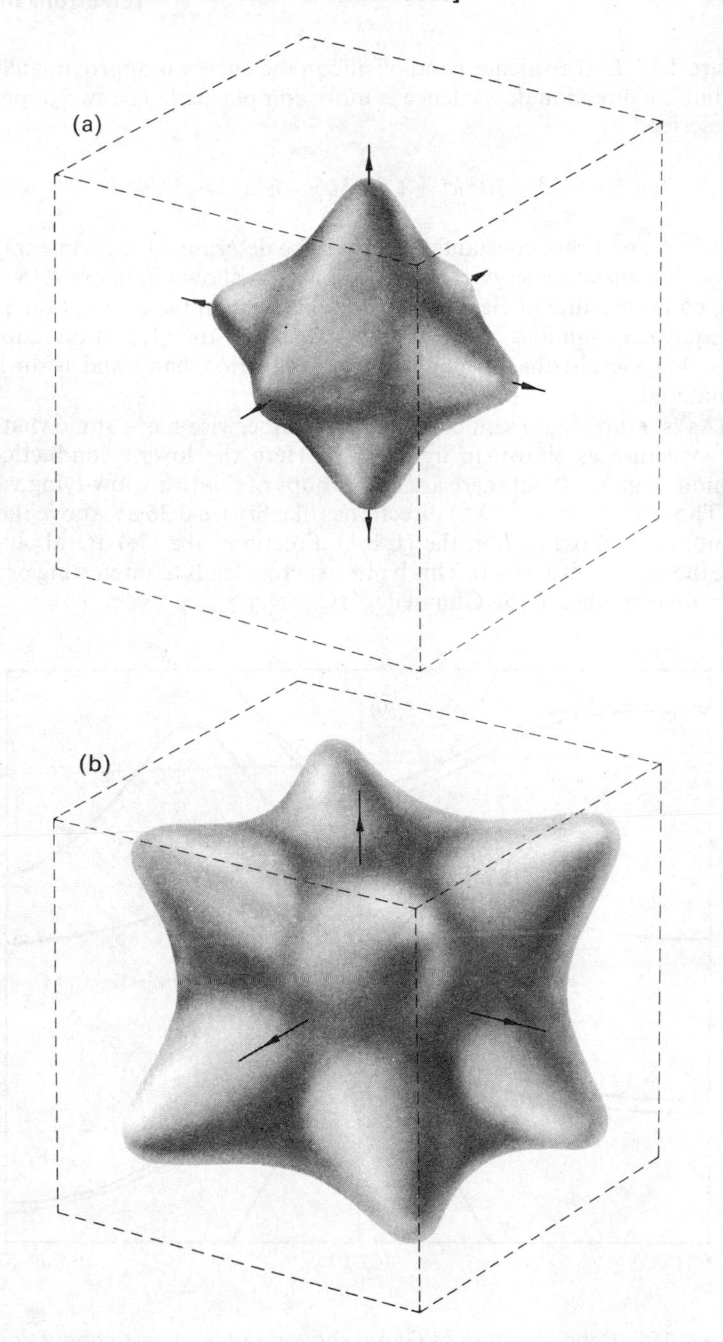

Figure 4.18 Surfaces of constant energy near the top of the valence band of Si (Ge, GaAs, etc. are all similar) ; (a) heavy holes, (b) light holes

in figure 4.17. In the valence band of silicon the energy is approximately quadratic but the direction dependence is more complicated. The two upper bands are described by

$$E = Ak^2 \pm [B^2 k^4 + C^2(k_y^2 k_z^2 + k_z^2 k_x^2 + k_x^2 k_y^2)]^{1/2} \tag{4.24}$$

where A, B and C are constants which can be determined experimentally. The surfaces of constant energy are fluted spheres as shown in figure 4.18.

The band structure of Ge is similar to that of Si but the conduction band has four equivalent minima at the zone boundary in the (1, 1, 1) direction. This means that Ge also has a many-valley conduction band and is an indirect gap material.

GaAs is a direct gap semiconductor but otherwise has a somewhat similar band structure as shown in figure 4.19. Here the lowest conduction band minimum is at $\boldsymbol{k} = 0$ but there are two groups of relatively low-lying valleys as well. Those at X in the (1, 0, 0) directions (like Si) are 0.36 eV above the lowest minimum and those at L in the (1, 1, 1) direction (like Ge) are about 0.5 eV above the minimum at $\boldsymbol{k} = 0$. This band structure leads to interesting properties which are exploited in the Gunn effect (Chapter 8).

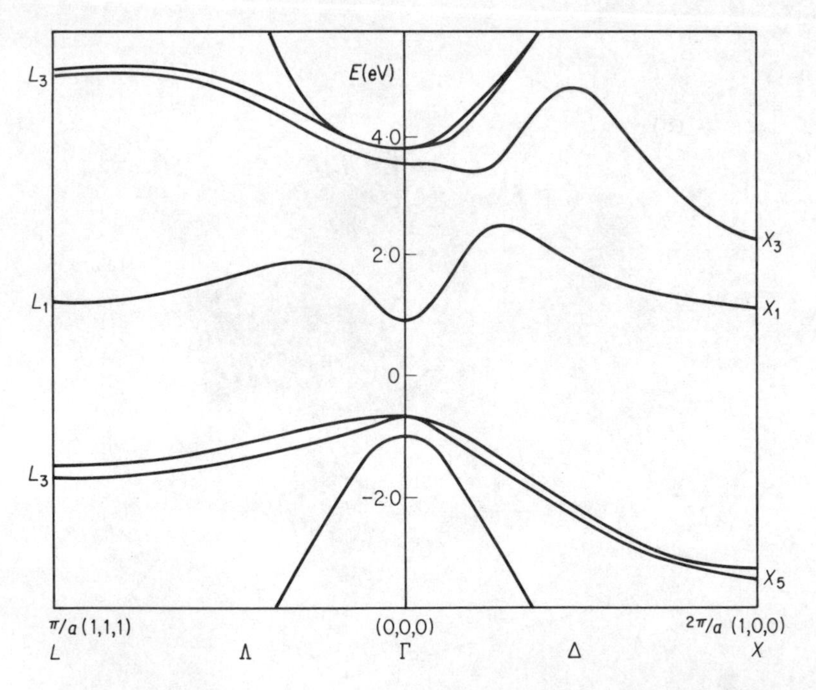

Figure 4.19 Band structure of GaAs, showing only lowest conduction bands and highest valence bands. Note the three close minima in the conduction band at Γ, X and L. The spin orbit splitting of the valence band is larger than in Si. (After M. Pollak *et al., Physics of Semiconductors*, p. 23. Kyoto, (1966).)

Table 4.1 Band Structure Parameters of some Technically Important Semiconductors

Parameter	InSb	InAs	GaAs		Ge	Si	GaP		PbTe
Minimum energy gap at 300 K (eV)	0.18	0.36	1.43		0.67	1.106	2.24		0.30
$\boldsymbol{k}$ at conduction band minimum	0	0	0		at zone edge in (1,1,1)	near zone edge in (1,0,0)	near zone edge in (1,0,0)		at zone edge in (1,1,1)
m_e^*/m	0.012	0.024	0.072	m_l^*/m	1.6	0.97	1.5		0.2
				m_t^*/m	0.082	0.19	0.25		0.03
$\boldsymbol{k}$ at valence band maximum	0	0	0		0	0	0		at zone edge in (1,1,1)
m_h^*/m	0.6	0.41	0.6		0.3	0.5	0.5	m_l^*/m	0.25
	0.015	0.04	0.15		0.04	0.16	0.15	m_t^*/m	0.03

InSb, InAs and GaAs. See figures 4.18, 4.19 and 6.11. Conduction band has single minimum at centre of zone so that effective electron mass, m_e^*, is single valued. Valence band maximum is also at $\boldsymbol{k} = 0$ but doubly degenerate. Hence two values of m_h^* are given.

Ge, Si and GaP. See figures 4.16, 4.17, 4.18 and 6.13(b). Conduction band has several equivalent minima, each being spheroidal in $\boldsymbol{k}$ space and hence specified by effective electron masses m_l^*, m_t^*, reflecting the curvature in $\boldsymbol{k}$ space along the principal axes of the spheroids. Valence band similar to InSb, etc.

PbTe. Conduction band similar to Ge. Valence band has several equivalent maxima.

The band structure of InSb is basically similar with both extrema at the origin ($\boldsymbol{k} = 0$) but with a much smaller gap. The spin-orbit splitting of the valence band is greater than the energy gap and about 1.0 eV. The conduction band has a very large curvature—near the origin $E = \hbar^2 k^2/2m^*$ where $m^* = 0.013\ m$—but is only quadratic over a small region of $\boldsymbol{k}$. The main features of the band structures of these and other technically important semiconductors are summarised in table 4.1.

4.6 The dynamics of electrons and holes (a)

Near the edges of bands the energy is generally a quadratic function of $\boldsymbol{k}$, although the constants involved may depend on the direction of the vector. As we have seen in the previous section, it is convenient to describe the curvature by a parameter of the dimension of mass called the effective mass such that

$$E(\boldsymbol{k}) = \hbar^2 \boldsymbol{k}^2/2m^* \tag{4.25}$$

and m^* can vary with direction as in (4.23) and (4.24). The velocity of a particle in these band states is then

$$\boldsymbol{v} = \hbar\boldsymbol{k}/m^* \tag{4.26}$$

and the momentum

$$\boldsymbol{p} = \hbar\boldsymbol{k} = m^*\boldsymbol{v} \tag{4.27}$$

as for classical particles.

These concepts can be generalised for a general band structure where $E(\boldsymbol{k})$ is not parabolic. The group velocity of the particles with these wave functions is

$$\boldsymbol{v} = \frac{1}{\hbar}\nabla_{\boldsymbol{k}}\, E(\boldsymbol{k}) \tag{4.28}$$

The equations of motion of a wave packet can be determined from the work done by a force $e\mathbf{F}$ due to an electric field $\mathbf{F}$. In time δt this is

$$\delta E = e\mathbf{F}.\boldsymbol{v}\,\delta t \tag{4.29}$$

and since

$$\delta E = \delta\boldsymbol{k}\,.\,\nabla_{\boldsymbol{k}}\, E(\boldsymbol{k}) \tag{4.30}$$

we can combine (4.28) and (4.30) to give

$$e\mathbf{F} = \hbar\, \mathrm{d}\boldsymbol{k}/\mathrm{d}t \tag{4.31}$$

which simply expresses the force as the rate of change of the momentum.

On the other hand the acceleration can be written

$$\frac{\mathrm{d}\boldsymbol{v}}{\mathrm{d}t} = \frac{1}{\hbar}\frac{\mathrm{d}}{\mathrm{d}t}\boldsymbol{\nabla}_k E = \frac{1}{\hbar}\left(\frac{\mathrm{d}\boldsymbol{k}}{\mathrm{d}t}.\boldsymbol{\nabla}_k\right)\boldsymbol{\nabla}_k E = (e\mathbf{F}.\boldsymbol{\nabla}_k)\boldsymbol{\nabla}_k E \tag{4.32}$$

and an effective mass defined by

$$e\mathbf{F} = m^*(\mathrm{d}\boldsymbol{v}/\mathrm{d}t) \tag{4.33}$$

so that m^* can be related to the curvature of the band using (4.32). For an isotropic band

$$\hbar/m^* = \partial^2 E/\partial k^2 \tag{4.34}$$

in agreement with (4.25).

4.6.1 Holes (a)

In some cases a crystal has only a few unoccupied states in an otherwise full band. It is then convenient to consider the properties of the system in terms of entities called holes. A single hole represents a full band with only one state empty. The only transitions possible within the band are those where an electron fills the empty state by vacating its previous state and it is found to be easier to fix attention on the empty state and its properties rather than on the electrons. Similar situations are familiar in everyday experience—it is, for example, easier to discuss the useful properties of a spirit level in terms of the motion of the bubble than in terms of the liquid.

Suppose a band is completely full except for one electron state of wave vector $\boldsymbol{k}_e$ which is empty. Now the total wave vector $\boldsymbol{K} = \sum \boldsymbol{k}$ of a full band is zero so with one electron missing from state $\boldsymbol{k}_e$ the total wave vector is $\boldsymbol{K} = -\boldsymbol{k}_e$. If this is to be described by a single hole its wave vector is

$$\boldsymbol{k}_h = -\boldsymbol{k}_e \tag{4.35}$$

In order to raise the energy of the empty state it is necessary to depress the electron from the higher energy state. In other words the total energy of the system is decreased if the energy of the empty state is increased. (Returning to the analogy with a spirit level we note that when it is placed on an incline the bubble rises in order to lower the potential energy.) If we are to describe the whole effect by the behaviour of a single hole we see that

$$E_h = -E_e \tag{4.36}$$

the hole energy is the negative of the energy of the unoccupied electron state. Using (4.35) and (4.36) in (4.28) we see that

$$\boldsymbol{v}_h = \boldsymbol{v}_e \tag{4.37}$$

Near band edges where the hole concept is most useful the energy is a quadratic function of $\boldsymbol{k}$ as in (4.24). From (4.36) and (4.25) therefore

$$m_h^* = -m_e^* \tag{4.38}$$

At the top of bands m_e^* is negative (4.34) so that the hole mass is positive and more convenient to work with.

Negative effective masses which occur in electron bands arise, as do all band effects, from the interaction of the electron with the lattice. From (4.33) we see that in such a case the force will accelerate a particle in the opposite direction to that in which it would accelerate a free particle. This is because the reaction from the lattice overwhelms the applied force. (A simple example of this in everyday life is the observation that a cotton reel pulled by a length of cotton will roll away from the puller because the reaction at the floor exceeds the applied force.)

Finally, a property of a hole which is of great interest is its apparent charge. The electric current in a band is given by

$$\mathbf{j} = e\sum \boldsymbol{v} \tag{4.39}$$

where the sum is over all occupied electron states. For a full band the total current is zero. Thus

$$\mathbf{j}_h = -\mathbf{j}_e \tag{4.40}$$

and the hole current is equal and opposite to the current in the unoccupied states. From (4.37) and (4.39) we see *that the hole has an effective charge opposite to that of the electron* (i.e. positive). This is borne out by the effect of an electric field. A field $\mathbf{F}$ produces a force $e\mathbf{F}$ which is therefore opposite in sign for holes and electrons, in agreement with (4.31) since $\boldsymbol{k}_h = -\boldsymbol{k}_e$ and with (4.33) since $m_h^* = -m_e^*$ but $\boldsymbol{v}_h = \boldsymbol{v}_e$.

4.7 Electrons in magnetic fields

4.7.1 Free electrons (AD)

When a magnetic field $\mathbf{B}$ is applied to an electron its motion takes on certain characteristic features. By studying the changes which occur in a magnetic field it is possible to deduce a number of important properties of the electrons. In particular the effective mass in semiconductors and the detailed structure of the Fermi surface in metals can sometimes be determined. In order to understand what happens in a real solid in a magnetic field we begin by considering a free electron.

The electron experiences the Lorentz force perpendicular to its velocity and perpendicular to the field so that the classical equation of motion is

$$m(\mathrm{d}^2\boldsymbol{r}/\mathrm{d}t^2) = e(\mathrm{d}\boldsymbol{r}/\mathrm{d}t) \times \mathbf{B} \tag{4.41}$$

The component of this along the direction of **B** (taken as the z-direction) gives free steady motion described by

$$m(\mathrm{d}^2 z/\mathrm{d}t^2) = 0 \tag{4.42}$$

but in the plane perpendicular to **B** the Lorentz force will just balance the centrifugal force to give motion in a circle of radius a_c and at frequency ω

$$m\omega^2 a_c = e\omega a_c \mathrm{B} \tag{4.43}$$

The characteristic frequency of this circular motion is then

$$\omega_c = e\mathrm{B}/m \tag{4.44}$$

which is called the cyclotron frequency.

The energy levels of such electrons should be deduced from wave mechanics but the essential results can be easily seen from (4.42) and (4.43). Along the z-axis the particle behaves as if it were free. It will therefore be described by a wave of some wave vector, k_z say. For a specimen of length $\mathscr{L}$ in the z-direction the wave function has the form

$$\mathscr{L}^{-1/2} \exp(\mathrm{i}k_z z) \tag{4.45}$$

and the energy of this motion is

$$E_z = \hbar^2 k_z^2/2m \tag{4.46}$$

The classical motion in the x–y plane is circular. This will be quantised in quantum mechanics to give energies

$$E_\perp = (\mathrm{n} + \tfrac{1}{2})\,\hbar\omega_c \tag{4.47}$$

where n is an integer.

This circulation of electrons at a fixed frequency can be observed experimentally. They will absorb radiation of frequency ω_c (as they do in a real cyclotron) and, in terms of (4.47), make transitions between the nth and (n + 1)th state. Such experiments in solids are called cyclotron resonance—the details of the method differ in semiconductors and in metals (see § 7.2).

In order to generalise our discussion of the cyclotron effect to real metals and semiconductors it must be extended from free electrons to electrons in a general band. However, we shall continue to assume that the electrons move without scattering. In real crystals there is a finite lifetime for the electron state which broadens the energy levels (§ 2.7). If this broadening

$$\hbar/\tau > \hbar\omega_c \tag{4.48}$$

the resonant features will not be observed. Alternatively one can say that in this situation $\omega_c\tau < 1$ and hence the electrons only describe a small part of their circular orbits before being scattered. This normally means that experiments to measure ω_c must be performed at low temperatures, in large magnetic fields and in very pure materials, since impurities and thermal fluctuations cause scattering. We return to this point in § 7.2.

4.7.2 Band electrons (AD)

The extension from free to band electrons is best achieved by examining the variation of electron wave vector in a magnetic field. An electron in a cyclotron orbit is continuously accelerated towards the centre so that its wave vector changes continuously. We can rewrite (4.41) as the rate of change of momentum $\hbar\boldsymbol{k}$

$$\hbar\frac{\mathrm{d}\boldsymbol{k}}{\mathrm{d}t} = e\frac{\mathrm{d}\boldsymbol{r}}{\mathrm{d}t} \times \mathbf{B} \tag{4.49}$$

For the special case of free particles $\mathrm{d}\boldsymbol{r}/\mathrm{d}t = \hbar\boldsymbol{k}/m$ and (4.49) gives component equations equivalent to (4.42) and (4.43). k_z is constant while $k_\perp$ describes a circle with frequency ω_c. In the motion therefore the end point of $\boldsymbol{k}$ traces out a circle whose area is $\pi k_\perp^2$. In quantum mechanics the energy associated with this motion in the plane perpendicular to $\mathbf{B}$, $\hbar^2 k_\perp^2/2m$, is quantised as in (4.47). Thus the area $\mathscr{D}$ of the orbit in $\boldsymbol{k}$ space is quantised

$$\mathscr{D} = (\mathrm{n} + \tfrac{1}{2})\, e\mathbf{B}/h \tag{4.50}$$

It can be said that the field constrains the motion so that the allowed $\boldsymbol{k}$ values end on a set of cylinders of area given by (4.50) (see figure 4.23).

Further consideration of the motion in $\boldsymbol{k}$ space gives a generalisation to band electrons. The simplest extension is to an isotropic band with effective mass m^*. The velocity in (4.49) should now be written as $\hbar\boldsymbol{k}/m^*$ and everything follows through as before with

$$\omega_c = e\mathbf{B}/m^* \tag{4.51}$$

This could also be obtained by using m^* in the $\boldsymbol{r}$ equation (4.41).

For general bands, however, it is not clear what to do with (4.41) but (4.49) still holds. The allowed motions remain closed orbits (now not necessarily circular) on the surfaces of constant energy in $\boldsymbol{k}$ space. The period of this motion is the time taken to complete an orbit

$$\mathrm{t_c} = \oint \delta t \tag{4.52}$$

where the integral is round the orbit in the plane perpendicular to $\mathbf{B}$. In this plane we get from (4.49)

$$\delta t = \hbar\, \delta k / e v_\perp \mathbf{B}$$

and using (4.28)

$$\delta t = \frac{\hbar^2 \, \delta k \, \Delta k_\perp}{e\mathrm{B} \, \Delta E}$$

Now $\oint k_\perp dk = \mathscr{D}$, the area of the orbit and hence

$$t_c = \frac{\hbar^2}{e\mathrm{B}} \frac{\partial \mathscr{D}}{\partial E} \quad \text{or} \quad \omega_c = \frac{2\pi e\mathrm{B}}{\hbar^2} \frac{\partial E}{\partial \mathscr{D}} \tag{4.53}$$

Since the total energy is quantised in units of $\hbar\omega_c$ it follows that $\mathscr{D}$ is quantised and

$$\mathscr{D} = \frac{e\mathrm{B}}{h}(\mathrm{n} + \gamma) \tag{4.54}$$

This is identical to (4.50) except that the constant γ is not necessarily $\frac{1}{2}$ as it is in the free-electron case.

As long as the function $E(\boldsymbol{k})$ remains parabolic, although dependent on the direction of $\boldsymbol{k}$, $\mathscr{D}$ depends on k^2 and hence E is a linear function of $\mathscr{D}$. $\partial E/\partial \mathscr{D}$, which defines ω_c in (4.53), is then constant throughout this region of the band. For example in the case of the spheroids (4.23) if the field is placed in the x–z plane at an angle θ to the z-direction, the ellipse perpendicular to this direction has the form

$$E = \frac{\hbar^2}{2}\left[\frac{k_1^2}{m_t^*} + k_2^2\left(\frac{\cos^2\theta}{m_t^*} + \frac{\sin^2\theta}{m_l^*}\right)\right] \tag{4.55a}$$

and

$$\frac{E}{\mathscr{D}} = \frac{\hbar^2}{2\pi}\left[\frac{\cos^2\theta}{m_t^{*2}} + \frac{\sin^2\theta}{m_t^* m_l^*}\right]^{1/2} \tag{4.55b}$$

In this case, which applies to the conduction bands of Si and Ge, ω_c depends on the direction of **B**. Measurements of ω_c against angle allow m_t^* and m_l^* to be determined directly.

At a general point in the band $\partial E/\partial \mathscr{D}$ will depend on the value of $\boldsymbol{k}.\mathbf{B}$ as well as direction. The cylinders in figure 4.23 will not now have uniform cross section. However, as discussed below the value of ω_c appropriate to $k_z = 0$ is normally observed. As an example we consider the situation in InSb where the conduction band is not parabolic though it is spherical. E is given by

$$E = \frac{\hbar^2 k^2}{2m_0} - \eta\left(\frac{\hbar^2 k^2}{2m_0}\right)^2 + \ldots \tag{4.56a}$$

If the second term is small, expansion in series gives

$$\omega_c = \frac{e\mathrm{B}}{m_0}\left(1 - \frac{\eta E}{2}\right) \tag{4.56b}$$

and the apparent effective mass increases as we go up the band.

4.7.3 Metals (AD)

As has been stressed before in this chapter the Pauli Exclusion Principle has important consequences for the properties of electrons in metals. In the case of free electrons in a magnetic field the energy levels are given by (4.46) and (4.47). There are a large number of states associated with each quantised energy for motions in the x–y plane, this degeneracy corresponding to the fact that in the classical motion the centre of the orbit is not defined. In quantum mechanics the number of these states can be determined from the uncertainty principle (see problem 10.9) to be $\mathrm{B}e\mathscr{A}/h$ if the area of the cross section in the x–y plane is $\mathscr{A}$.

We have shown (2.44) that the density of states for free particles confined to the x–y plane in the absence of a field is constant and equal to $\mathscr{A}m/2\pi\hbar^2$. The number of states in the energy interval $\hbar\omega_c$ is then

$$\mathscr{A}m\omega_c/2\pi\hbar = \mathrm{B}e\mathscr{A}/h \tag{4.57}$$

which is the same as the degeneracy given above. Thus the effect of the magnetic field is to cause the free electron levels to coalesce into groups with energy $(\mathrm{n}+\frac{1}{2})\hbar\omega_c$ (see figure 4.20). These groups are often called the *Landau levels.* The existence of free one-dimensional motion in the direction of **B** means that a band of energy levels (usually termed a sub-band) arises from each of these Landau levels. The density of states in one dimension varies as $E^{-1/2}$ from the bottom of each sub-band (cf. (2.45)) so that the total density of states is a saw-toothed function

$$\mathrm{S}(E) = \frac{\mathscr{L}\mathscr{A}e\mathrm{B}}{\pi h}\left(\frac{2m}{\hbar^2}\right)^{\frac{1}{2}} \sum_{\mathrm{n}} (E - (\mathrm{n}+\tfrac{1}{2})\hbar\omega_c)^{-1/2} \tag{4.58}$$

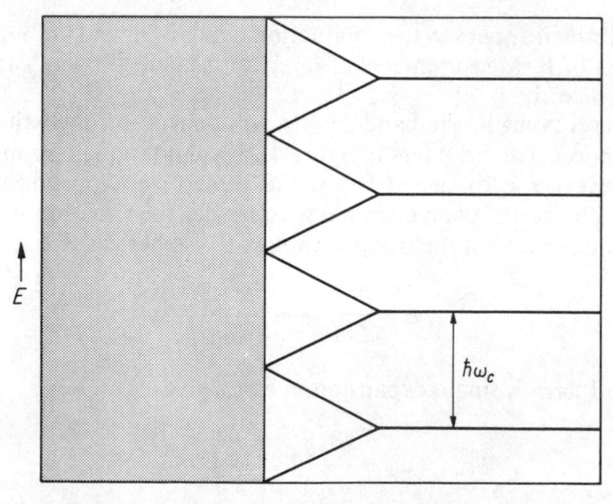

Figure 4.20 Change in density of states in two dimensions induced by a magnetic field (schematic)

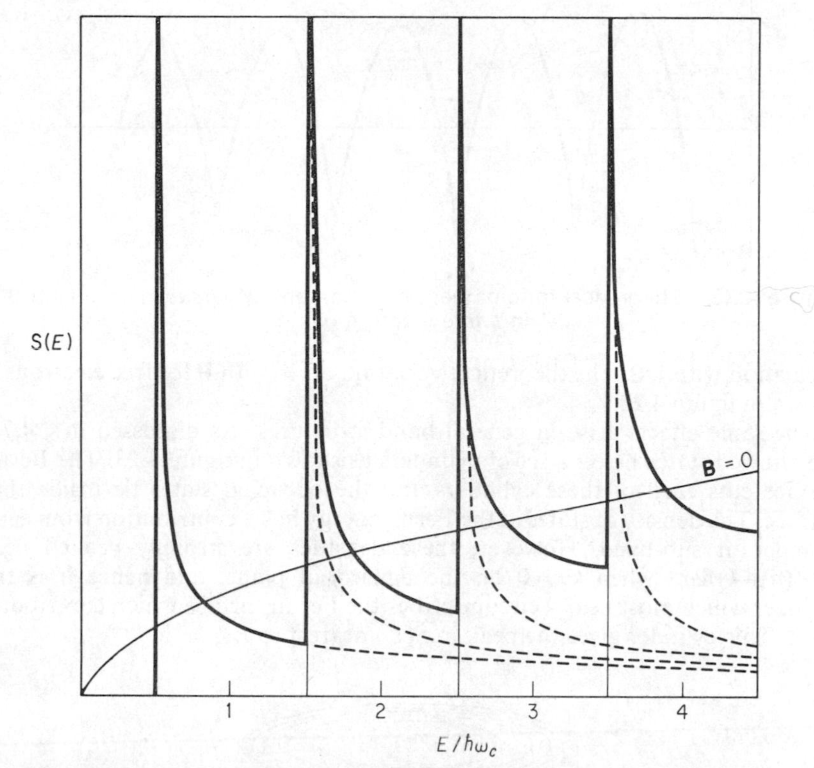

Figure 4.21 Density of states for free electrons in a magnetic field. It comprises the sum of identical contributions from each sub-band. The density of states with **B** = 0 (shown light) is an average of the peaked curves

as plotted in figure 4.21. This change from a smooth to a peaked density of states leads to important changes in many thermodynamic properties of metals. The Fermi level must move with B to ensure that the same total number of filled states lie below it. There is clearly a discontinuity in the properties each time one of the peaks in the density of states curve passes through the Fermi level and the states in this sub-band become depopulated. This occurs whenever

$$(\mathrm{n}+\tfrac{1}{2})\,\hbar\omega_c = E_F \tag{4.59}$$

approximately and for large n leads to a periodic variation in the properties of the metal with $\mathbf{B}^{-1}$. The Fermi energy and the total energy of the system become periodic functions in $\mathbf{B}^{-1}$.

Almost all properties of metals show this oscilliatory behaviour to some degree. We shall not discuss the details here though some further reference to them is made in Chapter 8. The most important type of experiment is named after its discoverers as the de Haas–van Alphen effect. They demonstrated that the magnetic moment of the system $\mathscr{M} = -\partial\langle E\rangle/\partial\mathbf{B}$ shows a strong

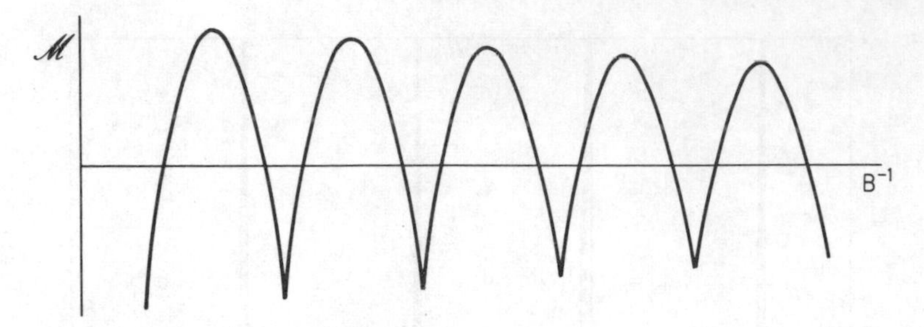

Figure 4.22 Theoretical form of magnetic moment $\mathscr{M}$ *versus* inverse field B^{-1} in a free electron gas

oscillation with 1/B. The theoretical variation of $\mathscr{M}$ with B for free electrons is shown in figure 4.22.

The same effects arise in general band structures. As discussed in § 4.7.2 the allowed states lie on a set of cylinders as shown in figure 4.23. The Fermi surface cuts each of these cylinders and the occupied states lie inside that surface. The density of states at the Fermi energy has a contribution from each cylinder or sub-band. However, these densities are strongly peaked near $E = (\mathrm{n} + \frac{1}{2})\hbar\omega_c$, when $k_z = 0$ on the equatorial plane, and hence it is the cylinder which most nearly circumscribes the Fermi surface which contributes most. This cylinder gives the peak in S(E) nearest to E_F.

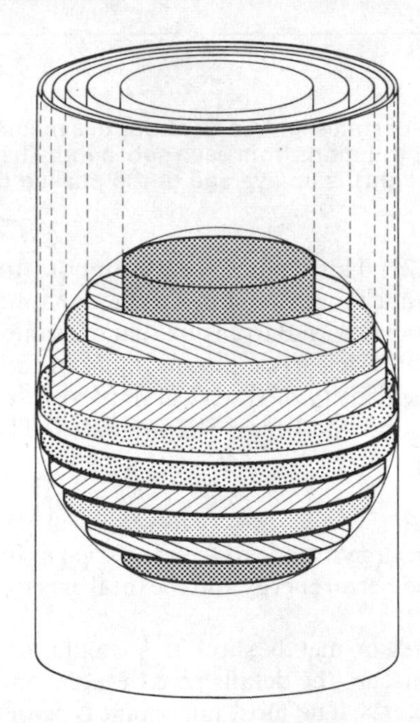

Figure 4.23 Occupied regions of ***k***-space for a free electron gas in a magnetic field. The allowed states lie on cylinders, cut off by the Fermi sphere

The de Haas–van Alphen effect measures directly the area of the circumscribing cylinders defined by (4.58) since the oscillations arise as the peaks in the density of states at each Landau level pass through the Fermi energy and become depopulated. $\mathscr{M}$ therefore gives direct information about the size of the Fermi surface. Figure 4.24 shows oscillations observed in Cu for **B** in the (1, 1, 1) direction. From (4.54)

$$\Delta(1/\mathbf{B}) = e/h\mathscr{D} \tag{4.60}$$

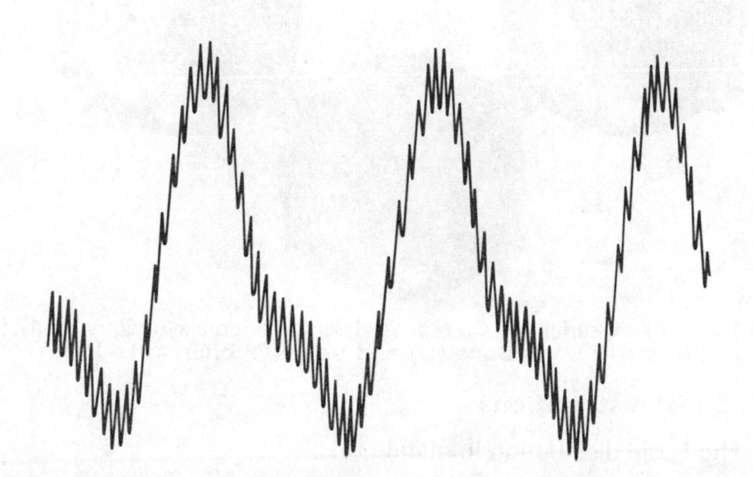

Figure 4.24 Measurement of de Haas–van Alphen effect in Cu (after A. S. Joseph *et al., Phys. Rev.,* **148**, 569 (1966)) for **B** near the (1,1,1) direction. The long period corresponds to the belly orbit (B) and the short period to the neck orbit (N). These are marked in figure 4.25

The rapid oscillations correspond to an orbit of large $\mathscr{D}$ around the 'belly' of the surface. The slow oscillations correspond to orbits around the small neck (cf. figure 4.12). In other directions more complicated orbits can be observed, for example the 'dog's bone' orbit shown in figure 4.25 for **B** in the (1, 1, 0) direction.

In some directions it is impossible to close the orbit on a repeating zone picture, as shown in figure 4.25. These 'open orbits' with infinite area do not affect the properties discussed here, but they can have strong effects on magnetoresistance (see § 8.8.4).

Cyclotron resonance and de Haas–van Alphen experiments have given much information and allowed details of several Fermi surfaces to be determined. They have been supplemented by other techniques and some of these other experiments will be considered in later chapters. In this chapter we are only concerned to show how basic information about bands can be definitively determined.

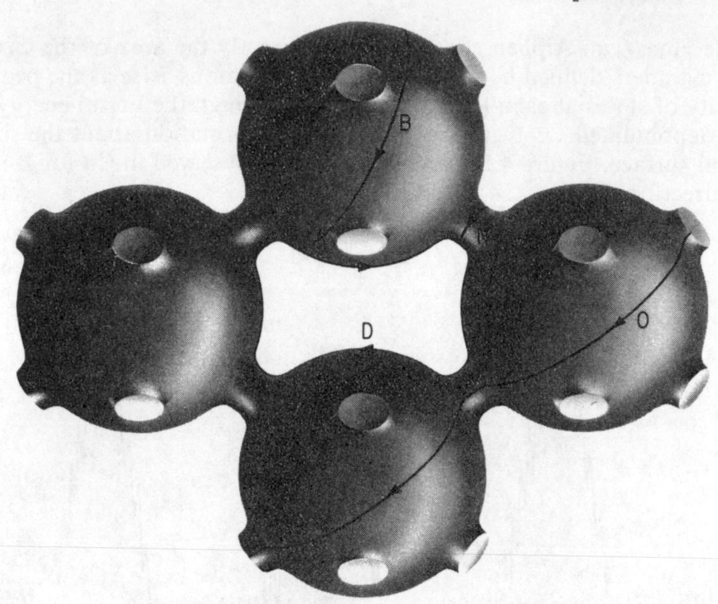

Figure 4.25 Fermi surface of Cu repeated using periodic symmetry. Belly (B), Neck (N), Dog's Bone (D) and Open (O) orbit marked

4.8 Thermal distributions

4.8.1 The Fermi distribution in metals (a)

We have so far discussed the electronic properties of systems in a single energy state—usually the ground state. In this case there is a well-defined Fermi surface in a metal dividing occupied and unoccupied electron states and a clear division between full and empty bands in a semiconductor. At finite temperatures, however, higher energy states may be occupied and these divisions are blurred. The probability of occupation of states, given by the Fermi–Dirac distribution (2.37), is shown in figure 4.26. At temperature T states of total energy about kT above that of the ground state become occupied. In these states individual electrons are promoted in energy by about kT. In a metal this can only occur in states near the Fermi level since only there are there empty states available for the promoted electrons and it leads to a smearing of the Fermi level by about kT. Now the position of the Fermi level itself is determined by the fixed total number of particles, N (2.38). If, as in the free-electron approximation illustrated in figure 4.27, the density of states is a rising function of E the number of states emptied below E_F in the smaller area marked v is smaller than that in area f of occupied states. To keep N constant E_F must therefore be slightly reduced. If $\mathrm{S}(E)$ were a decreasing function of E near the top of a band E_F would increase slightly with T. In fact

$$E_F(T) = E_F - \frac{\pi^2}{6}\frac{(kT)^2}{\mathrm{S}(E_F)}\left(\frac{\partial \mathrm{S}}{\partial E}\right)_{E=E_F} \tag{4.61}$$

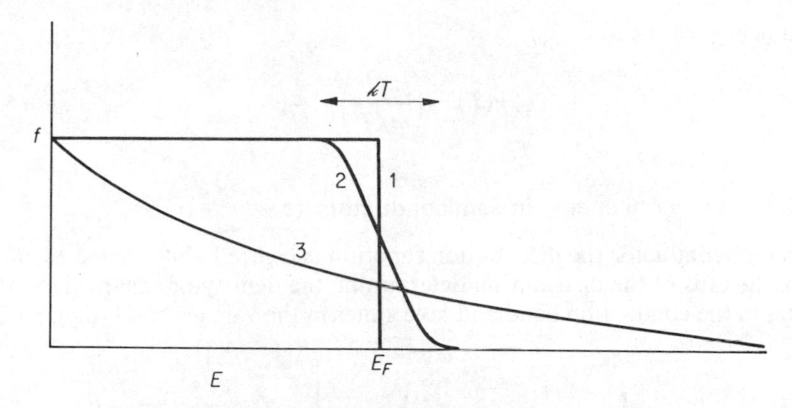

Figure 4.26 The Fermi–Dirac distribution function (1) $\mathscr{k}T = 0$, (2) $\mathscr{k}T \ll E_F$, (3) $\mathscr{k}T \gg E_F$. (The last is the Maxwell–Boltzmann distribution $e^{-E/\mathscr{k}T}$ arbitrarily normalised.)

In metals $E_F \gg \mathscr{k}T$ so the second term in (4.61) is relatively small. If the temperature becomes comparable to $E_F/\mathscr{k}$ it becomes more difficult to find the temperature dependence of $E_F(T)$ analytically. However, if $\mathscr{k}T \gg E_F$, $E_F(T)$ becomes large and negative and the distribution function becomes a Boltzmann one

$$\mathscr{f}(E) = g(T)\exp(-E/\mathscr{k}T) \tag{4.62}$$

If $S(E)$ is given by (2.43), and $N = v\mathscr{N}$ is fixed, it follows from (2.38) that

$$\frac{\mathscr{N}\Omega}{(2\pi)^2}\left(\frac{2m}{\hbar^2}\right)^{3/2}\int g(T)\,E^{1/2}\exp(-E/\mathscr{k}T)\,\mathrm{d}E = v\mathscr{N} \tag{4.63}$$

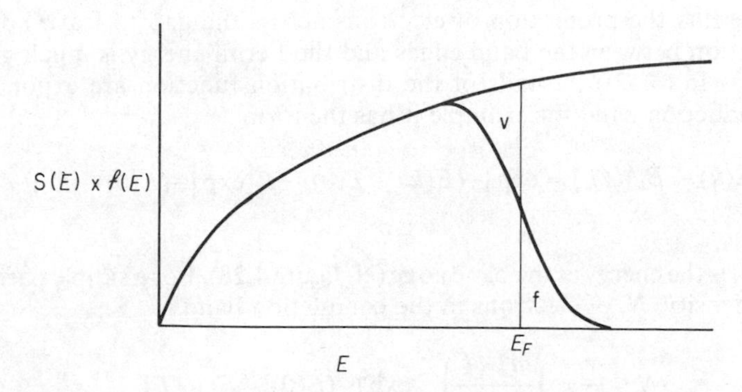

Figure 4.27 Occupied states $S(E) \times \mathscr{f}(E)$ for a free electron gas at low T

and hence

$$g(T) = \left(\frac{\pi\hbar^2}{m\ell T}\right)^{3/2} \frac{\nu}{\Omega} \tag{4.64}$$

4.8.2 The Fermi energy in semiconductors (a)

In a semiconductor the distribution function is centred about some E_F in the gap, the tails of the distribution determining the density of occupied electron states in the conduction band and hole states in the valence band (figure 4.28).

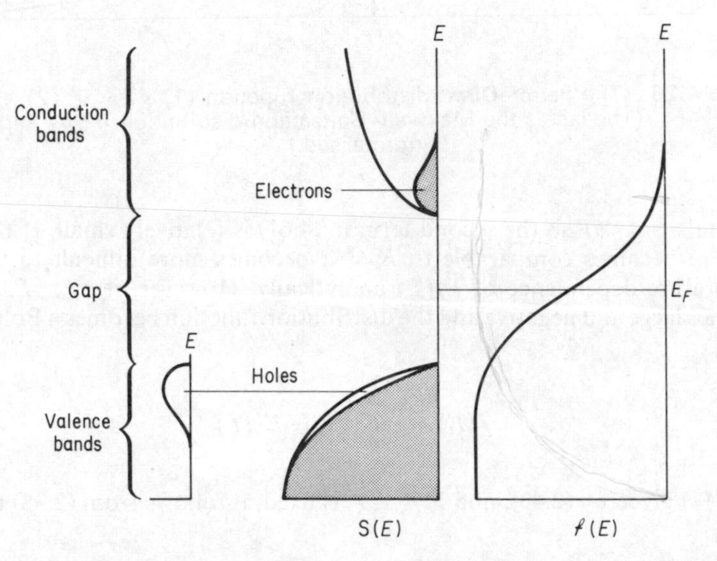

Figure 4.28 Occupied states $S(E) \times \ell(E)$ in a semiconductor. The unshaded portion of the valence band corresponds to a hole distribution

This represents the promotion of electrons across the gap to leave holes. If the separation between the band edges and the Fermi energy is much greater than ℓT (as in (4.62)) the tails of the distribution function are exponential. In the conduction band for example it has the form

$$\exp\left[-(E(\boldsymbol{k}) - E_F)/\ell T\right] = \exp\left[-(E(\boldsymbol{k}) - E(0))/\ell T\right] \exp\left[-(E(0) - E_F)/\ell T\right] \tag{4.65}$$

where $E(0)$ is the energy at the band edge (cf. figure 4.28). For a simple parabolic band the density, N, of electrons in the conduction band is

$$N = \frac{\nu}{\Omega} = \left(\frac{m_e^* \ell T}{\pi\hbar^2}\right)^{3/2} \exp\left[-(E(0) - E_F)/\ell T\right] \tag{4.66}$$

provided the material contains no impurities (see Chapter 5). Since $\mathcal{N}\Omega$ is the volume of the crystal, $N\mathcal{N}\Omega$ is the total number of electrons in the band. The number of holes is given by a distribution of the form

$$1 - \{\exp[(E(\boldsymbol{k}) - E_F)/kT] + 1\}^{-1} = \frac{\exp[-(E_h(\boldsymbol{k}) + E_F)/kT]}{\exp[-(E_h(\boldsymbol{k}) + E_F)/kT] + 1}$$

$$\sim \exp[-(E_h(\boldsymbol{k}) + E_F)/kT] \qquad (4.67)$$

and again the density of holes in the valence band is

$$P = \left(\frac{m_h^* kT}{\pi\hbar^2}\right)^{3/2} \exp[-(E_h(0) - E_F)/kT] \qquad (4.68)$$

In a perfect and pure semiconductor each of these holes has appeared by the promotion of an electron so that

$$N = P$$

or

$$(E(0) - E_F)/kT - \tfrac{3}{2}\log(m_e^*/m) = (E_h(0) + E_F)/kT - \tfrac{3}{2}\log(m_h^*/m)$$

whence:

$$E_F = \tfrac{1}{2}E_g + \tfrac{3}{4}kT\log(m_h^*/m_e^*) \qquad (4.69)$$

where E_g is the energy gap. Thus the Fermi level in a perfect semiconductor at $T = 0$ lies half-way between the conduction and valence bands. It shifts linearly with increasing temperature towards the band with the lower effective mass and smaller density of states. If the effective mass depends on direction as in (4.23) and (4.24) an average is taken as required to give the correct density of states.

Semiconducting crystals will normally contain impurities and by techniques to be described in Chapter 5 these can be specially introduced and controlled. We shall see in Chapter 5 that these impurities modify the band states and may introduce new levels in the forbidden gap. They may also introduce extra electrons or holes if the valence of the impurity is respectively greater or less than that of the host. Leaving aside the details of the new levels we see that if extra electrons are added they must go into the lowest conduction band states, so that at $T = 0$, E_F must lie near the bottom of that band. If holes are added, empty electron states will appear at the top of the valence band and E_F will lie nearer there. There will be a Fermi–Dirac distribution of electrons or holes with low E_F relative to the appropriate band edge. As T increases from low temperatures the distribution will become a Boltzmann distribution when $kT > (E_F - E(0))$. Provided the temperature is not too high the electron (hole) density is determined by the density of impurities and is substantially independent of temperature. As T increases further the promotion of electrons from the valence band to the conduction band becomes important and the number of thermally promoted electrons and holes rises to exceed the number donated by impurities. E_F then moves towards the centre of the energy gap, the electron and hole densities tend to equalise and both increase rapidly with T as in (4.66) and (4.68). Thus at high enough temperatures the impurity content is of relatively little importance.

4.8.3 The electronic contribution to the specific heat of metals (a)

The way in which the Fermi distribution in a metal is smeared allows us to understand the thermodynamic properties of an electron gas of this type. In the smeared distribution some of the electrons have increased their energy by about kT, but because of the exclusion principle only that fraction of the electrons in states within kT of the Fermi level can be promoted in this way, the fraction being about kT/E_F of the whole number. Thus the increase in energy is given roughly by

$$\langle E \rangle \sim \mathcal{N} v(kT)^2/E_F$$

and hence

$$\partial\langle E \rangle/\partial T = C_V \sim \mathcal{N} v(k^2 T/E_F) \tag{4.70}$$

where C_V is the electron contribution to the specific heat of the metal. In a proper calculation allowance must be made for the change (4.61) in E_F with temperature and the actual result for the electronic contribution to the specific heat is

$$C_V = \frac{\pi^2}{3} S(E_F) k^2 T = \gamma T \tag{4.71}$$

Experimentally the results for alkali metals are very close to the values calculated on this free-electron model. In fact $S(E_F)$ (2.43) can be written

$$S(E_F) = \left(\frac{m^*}{\pi^2 \hbar^2}\right) k_F \mathcal{N} \Omega \tag{4.72}$$

and using (4.19) the only unknown is m^*. Experimentally, specific heat data gives

	Na	K	Rb	Cs
(m^*/m)	1.25	1.27	1.31	1.56

4.9 Electron–electron interactions (AD)

So far the interactions between the electrons have been completely neglected, except in so far as they contribute to the average potential used in calculating the bands. Remarkably, the results of the independent particle model are in excellent agreement with experiment, both for details like the geometry of the Fermi surface and generally for thermodynamic properties like the specific heat. We must now try to give some indication of the theoretical justification for this fact.

The Coulomb interaction is very long range because, although it falls off as r^{-1}, the number of particles in a shell of constant thickness at distance r increases as r^2. Thus one particle is strongly affected by all the other particles, even those a large distance away. This relatively strong coupling between the particles allows them to take part in collective motions. A gas of charged particles (e.g. electrons) interacting in this way is called a plasma and the collective density fluctuations are called plasma oscillations. The frequency of plasma oscillations can be obtained from a simple argument. If the particles move a distance X, a polarisation $\mathbf{P}$ is set up relative to the static positive charges.

$$\mathbf{P} = NeX \tag{4.73}$$

where N is the number of particles of charge e per unit volume. The electric field arising from this polarisation is $-\mathbf{P}/\varepsilon$ (where ε is the dielectric constant) so the equation of motion of the particles is

$$m^*(\mathrm{d}^2 X/\mathrm{d}t^2) = -Ne^2 X/\varepsilon \tag{4.74}$$

This leads to simple harmonic oscillations of frequency

$$\omega_p = (Ne^2/\varepsilon m^*)^{1/2} \tag{4.75}$$

At the electron densities appropriate to metals this frequency is very high. For nearly-free electrons, assuming $m^* \sim m$ and $\varepsilon = \varepsilon_0$, $\hbar\omega_p$ is >1 eV. For example:

	Li	Na	K	Mg	Al
$\hbar\omega_p$(obs.) (eV)	7.1	5.8	3.8	10.6	15.3
$\hbar\omega_p$(calc.) (eV)	8.0	6.0	4.3	10.9	15.8

Experimentally, the generation of such large quanta can be observed by passing fast electrons through metal foils and recording how much energy they lose in creating such oscillations. The optical properties of metals and semiconductors also show big changes at $\omega \sim \omega_p$ (cf. Chapter 7). The zero-point energy of the plasma oscillations contains the main contribution of the Coulomb repulsion to the total energy of electrons.

Although the major part of the interaction between the electrons manifests itself in these collective oscillations some residual interaction remains. This arises because the plasma oscillations involve the motion of individual particles which do not all have the same velocity. In a Fermi–Dirac distribution the most energetic of the particles move with a velocity $v_F \sim \hbar k_F/m^*$. In the period of one oscillation they move a distance w given by

$$w^2 \sim v_F^2/\omega_p^2 \sim \varepsilon E_F/Ne^2 \tag{4.76}$$

so that we expect oscillations with shorter wavelengths than this to be smeared out. We also expect single-particle properties to exist over a distance of about

w. This is borne out by a more rigorous calculation based on Poisson's equation. The extra electrostatic potential ϕ due to a density fluctuation ΔN may be written

$$\varepsilon \nabla^2 \phi = e \Delta N \tag{4.77}$$

N is the equilibrium density of electrons. Such a density fluctuation due to electrons at the Fermi level caused by an energy $e\phi$ is

$$\Delta N = e\phi \mathrm{S}(E_F)/\mathcal{N}\Omega \tag{4.78}$$

Thus

$$\nabla^2 \phi = \kappa^2 \phi$$

$$\phi = \frac{e}{r} \exp(-\kappa r) \tag{4.79}$$

where

$$\kappa^2 = e^2 \mathrm{S}(E_F)/\mathcal{N}\Omega\varepsilon \tag{4.80}$$

Using (4.20) we see that $w^2\kappa^2 = 3/v^2 \sim 1$ so that the screening length κ^{-1} is similar to the length w, which in a typical metal is $\sim 10^{-8}$ cm.

For electrons in a Boltzmann distribution (as in a semiconductor) similar considerations apply. The mean velocity now is

$$v^2 \sim kT/m^* \tag{4.81}$$

so that (4.76) is replaced by

$$w^2 \sim \varepsilon kT/Ne^2 \tag{4.82}$$

Similarly (4.77) gives

$$\Delta N = Ne\phi/kT \tag{4.83}$$

and again $w^2\kappa^2 \sim 1$. This is sometimes known as Debye–Hückel screening.

We can say that the long range part of the Coulomb interaction gives rise to the collective motions. Only a residual short range interaction, screened by the plasma, is left to give electron–electron interactions. In a metal the effect of these on the important electrons in states at the Fermi level is even further reduced by the influence of the Fermi–Dirac statistics. If a pair of electrons with wave vectors $\boldsymbol{k}_1$, $\boldsymbol{k}_2$ interact and are scattered to $\boldsymbol{k}_1'$, $\boldsymbol{k}_2'$ the conservation rules (§ 2.6) require

$$\boldsymbol{k}_1 + \boldsymbol{k}_2 = \boldsymbol{k}_1' + \boldsymbol{k}_2' \tag{4.84}$$

and

$$E(\boldsymbol{k}_1) + E(\boldsymbol{k}_2) = E(\boldsymbol{k}_1') + E(\boldsymbol{k}_2') \tag{4.85}$$

Further at $T = 0$, $E(\boldsymbol{k}_1)$ and $E(\boldsymbol{k}_2)$ must be less than E_F while $E(\boldsymbol{k}_1')$ and $E(\boldsymbol{k}_2')$ must be greater than E_F if the former states are to be occupied and the latter unoccupied, but (4.85) cannot then be satisfied. It is only when the Fermi distribution is smeared over the range kT that the processes are allowed and then only the electrons in states in this region can be scattered. This is a fraction $\sim kT/E_F$ of the electrons. Since the cross section for scattering *via* the interaction (4.81) would normally be about π/κ^2 the statistics reduce this to

$$\sigma_{\text{el.}} \sim (kT/E_F)^2(\pi/\kappa^2) \tag{4.86}$$

At room temperature in a metal where $(kT/E_F) \sim 10^{-2}$ and $\sigma_{\text{el.}} \sim 10^{-19}\ \text{cm}^2$ the mean free path between electron–electron collisions, l_f, is given by making $\sigma_{\text{el.}} l_f$ equal to the volume for an electron (cf. § 2.7)

$$l_f = 1/N\sigma_{\text{el.}} \tag{4.87}$$

and is $\sim 10^{-4}$ cm or 10^4 atomic spacings. Thus interactions between individual electrons in most metals are negligible and the whole effect is absorbed by the collective properties of the plasma. In transition metals which have large electron densities and small E_F in the d bands the effect is more important and affects the resistivity at low T.

4.10 Insulators (A)

We have so far only considered the theory of crystalline metals and semiconductors for which band theory forms a good approximation. There are other crystals—one might call them insulators, although the definition is not a precise one—where this is an unsatisfactory description.

In simple ionic crystals like the alkali halides, the concept of band structure remains of some value in discussing small numbers of electrons and holes, but as we shall see in § 6.5, it must be modified to discuss optical properties associated with excitations across the energy gap where the electron–hole interaction is of paramount importance and *excitons* are formed.

There are other ionic crystals where band theory is an even more unsatisfactory starting point. Consider a crystal like MnO which has an NaCl-type structure. Ionic Mn^{2+} has a $3d^5$ configuration so there are an odd number of electrons in the unit cell. Thus they cannot fill an integral number of bands, and if band theory held, the crystal would be a metal! (Actually MnO is antiferromagnetic at low T (see Chapter 10) and so appears to have two different Mn ions per cell with spins pointing in opposite directions. Strictly speaking this might give an integral number of filled bands, but the gap should disappear as the magnetic order is lost, while in fact no significant change in electrical conductivity does take place at that temperature.) In fact MnO is very like NiO in properties though the latter has an even number of electrons per cell.

The reason for the insulating properties of a crystal like MnO lies elsewhere—in the electron–electron interaction. The d bands for such a material must be very flat since the overlap and hence the transfer energies (the E_δ of (4.13)) will be small. The kinetic energy of such electrons would be small and the energy gained by sharing electrons between atoms small. Against this, any transfer of an electron would change a Mn^{2+}–Mn^{2+} pair to Mn^{1+}–Mn^{3+}. The energy of the latter is considerably greater because of the extra repulsion energy of the six d electrons on the Mn^{+}. In fact the question now is whether this repulsive energy is larger or smaller than the band width

$$U \lesseqgtr 6E_\delta$$

For very narrow bands the correlation energy will win and the crystal retain its atomic Mn^{2+} form throughout.

Mott has argued that one may in some crystals have a transition from insulating to conducting band states as the thermal energy helps to overcome the repulsion U. In fact some transition metal oxides do show transitions where the conductivity leaps by several orders of magnitude and although they are accompanied by lattice distortions they may be thought of as 'Mott transitions'.

It should be noted that such systems only become insulators if they are free of impurities. If some electrons are removed from MnO to leave a few Mn^{3+} ions the hole may transfer from site to site without introducing a repulsive energy. This is particularly true in crystals containing atoms like Fe which have two common valence states (Fe^{2+} and Fe^{3+}).

The electron–electron interactions may also destroy the conducting state in a slightly different situation. If for instance we have a relatively small number of electrons in a band, the kinetic energy will be low but the Coulomb repulsion will fall off less rapidly because of its long range nature. If we define a characteristic length r_s which gives the volume per electron as

$$\frac{4\pi}{3} r_s^3 = \Omega/v = 1/N \tag{4.88}$$

then the zero point kinetic energy is about $\hbar^2/mr_s^2$ while the mean Coulomb repulsion is about $e^2/4\pi\varepsilon_0 r_s$. The latter dominates if

$$(e^2/4\pi\varepsilon_0 r) > (\hbar^2/mr_s^2) \tag{4.89}$$

i.e. if

$$r_s > 4\pi\varepsilon_0 \hbar^2/me^2 \; (= a_0, \text{ the Bohr radius}) \tag{4.90}$$

If the density is small enough Wigner showed that the electrons would set into a lattice. The characteristic oscillations of this lattice are at the plasma frequency. It might be expected that the electron lattice would become unstable and change into a gas if the mean electron displacements due to the vibrations (the zeropoint motion) exceeded r_s, the electron spacing. The mean displacement of such an oscillator is $(\hbar/m\omega_p)^{1/2}$ so that the condition for an insulator is

$$r_s > (\hbar/m\omega_p)^{1/2} \tag{4.91}$$

and since from (4.75) and (4.88) $\omega_p^2 = 3e^2/4\pi\varepsilon_0 mr_s^3$ the conditions (4.90) and (4.91) are the same within a small numerical factor.

4.11 Effects of electron–lattice coupling (A)

In all the discussion of electron states so far, we have assumed a static, perfectly regular lattice. The motion of the heavy ions was discussed in Chapter 3 on the basis of the separation allowed by the adiabatic approximation. This separation is not strictly accurate and the residual coupling between the moving ions and the valence electrons (electron–phonon coupling) leads to a number of effects. The most important is the finite lifetime and mean free path of the electrons which gives a width to the energy states. We remarked in § 4.7 that this width might smear out the Landau levels (cf. (4.48)). It will be more effective than the temperature in smearing out all the features associated with the cut-off in the Fermi–Dirac distribution if $\hbar/\tau > kT$. The effect of a finite mean free path on transport properties will be discussed in detail in Chapter 8.

In addition to these scattering effects we saw in § 2.7 that a coupling of this type leads to a modification of electron states and a change in their energy (by the addition of the 'self-energy'). Physically this can be regarded as a local distortion of the lattice around the electron which moves with it. This makes the electron motion slower and increases the effective mass. This is a simple example of a concept commonly used in field theory—the electron can be said to be dressed in a cloud of phonons which constitute the lattice distortion—and the properties of the system discussed in terms of the dressed quasi-particles.

In metals the change in the effective mass due to this interaction has been estimated for the alkalis. It is of about the same magnitude as the residual electron–electron interaction discussed in § 4.9.

$(\delta m/m)$ Interaction and m^*/m	Na	K	Rb
$(\delta m/m)$ elec.-phonon	0.18	0.15	0.17
$(\delta m/m)$ elec.-elec.	0.06	0.11	0.13
(m^*/m) Band	1.00	1.07	1.18
(m^*/m) Total	1.24	1.33	1.48

These effects lead to results which are not negligible and go a long way towards explaining the effective masses obtained from the interpretation of specific heat data (§ 4.8).

In ionic crystals the electron–phonon coupling is much stronger—particularly to the longitudinal optic phonon modes which set up strong electric fields (cf. § 3.4). The quasi-particle in this case is called a polaron. The coupling is conveniently specified by a dimensionless parameter α_p which measures the number of phonons necessary to produce the local distortion and polarisation. The estimated value of α_p in a few ionic crystals are

	KCl	KI	AgCl	InSb	GaAs
α_p	3.6	2.5	1.9	0.014	0.06

Semiconductors have small values but the alkali halides have much larger ones. For relatively small α_p, the polaron mass is

$$m_p = m^*(1 + \alpha_p/6) \tag{4.92}$$

In fact m^*/m for the conduction band in KCl is estimated by band theory to be about 0.5 while the value of m^*/m found in cyclotron resonance experiments is ~0.9; this is about 10% larger than predicted by (4.92). The mean free path of such polarons becomes very small once kT is large enough to allow appreciable population of longitudinal optic phonons. The strong coupling of these phonons to the electrons has also been demonstrated by observing the effect on cyclotron resonance when $\omega_c \sim \omega_L$. Holes in ionic crystals have a large effective mass and the polaron effects can cause large local distortions.

PROBLEMS

(Some answers, where appropriate, are given on page 481.)

4.1 An N type semiconductor of dielectric constant $16\varepsilon_0$ contains 10^{19} electrons per cm^3 which may be assumed to occupy states at the bottom of the band where $m^* = m/10$. Estimate the plasma frequency of the electrons.

4.2 Obtain expressions for the electron densities at which the free electron spherical Fermi surface first touches the zone boundary in

(i) a body-centred cubic metal,

(ii) a face-centred cubic metal.

4.3 In a two-dimensional square lattice draw the free electron Fermi circles for two and three electrons per unit cell. Obtain the shape of the Fermi curves for a nearly free electron model in the reduced zone scheme.

4.4 Using the tight-binding method with a single band in a body-centred cubic lattice and only nearest neighbour interaction energy E_1, show that the energy bands are spherical near $\boldsymbol{k} = 0$ and deduce the effective mass.

4.5 Assuming the free electron theory calculate the Fermi energy of body-centred cubic Na and face-centred cubic Al. The dimensions of the unit cube are 0.43 nm and 0.40 nm, respectively.

4.6 Show that, at low temperatures, the Fermi energy varies with temperature as

$$E_F(T) = E_F(0) - \frac{(\pi k T)^2}{6S(E_F)}\left(\frac{\partial S}{\partial E}\right)_{E=E_F}$$

(equation (4.61)).

4.7 Derive equation (4.71) for the specific heat of an electron gas using the identity

$$\int_0^\infty \frac{x\,\mathrm{d}x}{1+\exp(x)} = \frac{\pi^2}{12}$$

4.8 Show that for a free electron gas at zero temperature the internal energy $U = (3/5)E_F(0)$. By considering the variation of E_F with volume show that there is a relation between pressure and volume of the gas

$$PV = (2/3)\,U$$

and show that the bulk modulus, $(1/V)(\mathrm{d}V/\mathrm{d}P)$, is given by $9V/10U$.

5

Imperfections in crystals

5.1 Introduction (a)

The translational symmetry of a perfect crystal lattice allows us to describe the excitations (electrons, phonons, etc.) of a crystal in terms of a wave vector and requires that the energies of the excitations lie in bands. It is on this basis that we are able to discuss the detailed behaviour of many types of crystals, but imperfections in crystals destroy this symmetry. Provided the density of imperfections is small we might assume that they would modify the crystal properties to only a small extent. For many crystal excitations this is true and imperfections can, for example, be considered as only slightly perturbing the motion of electrons or phonons through the crystal. (We shall discuss the scattering of electrons and phonons by imperfections in Chapter 8.) On the other hand there are some properties of crystals, typically those that are null in a perfect crystal, which are grossly affected by the presence of imperfections. The most important example is probably the electrical conductivity of semiconductors: the number of mobile charge carriers (electrons and holes) is very small in the pure, intrinsic, state, particularly at low temperatures but we shall see that the carrier density can be considerably increased by the addition of suitable impurities.

Crystal imperfections can be crudely classified on the basis of size. A single atom or ion out of place, or a single foreign impurity atom substituted for one of the host lattice atoms are examples of *point defects.* Larger imperfections can occur when two or more point defects cluster to form a complex. A simple example arises from missing lattice atoms which alone constitute point defects; these are referred to as vacancies or Schottky defects. Such defects may coalesce to form macroscopic voids or bubbles in the crystal.

Macroscopic structural defects, referred to as dislocations, also occur in crystals. The simplest example of these defects, called an edge dislocation, can be visualised with the aid of figure 5.1. Imagine an otherwise perfect cubic crystal bent as shown in the figure. If both upper and lower surfaces are

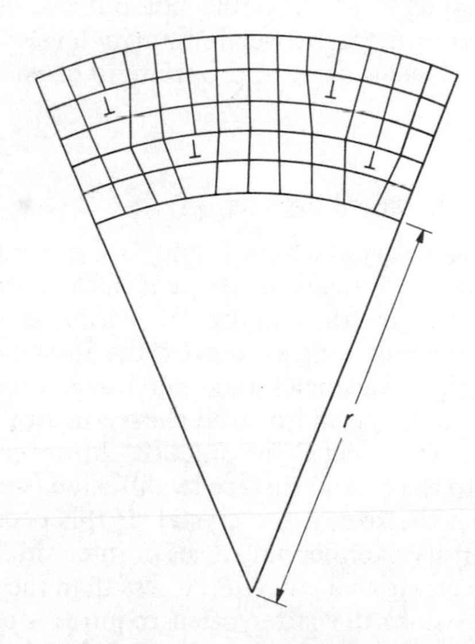

Figure 5.1 Bending of a simple cubic lattice produces edge dislocations, indicated symbolically by $\perp$. If r is the radius of curvature of the bend and **b** is the Burger's vector (defined in § 5.4.1) the density of edge dislocations produced by the bend is $1/r\mathrm{b}$. (After F. L. Vogel, *J. Met.*, **8**, 946 (1956).)

to exhibit the correct lattice spacing there must be more atomic planes on the upper surface than on the lower surface and this means that one or more planes of atoms must terminate between the surfaces to accommodate the bend. The termination of such a plane constitutes an edge dislocation; it extends, as a line imperfection, through the crystal in a direction normal to the direction of bend.

The properties of imperfections are important in all aspects of solid state physics and particularly its applications. We shall, therefore, need to refer to the properties of imperfections in most succeeding chapters of this book. The structure of this chapter differs somewhat from the preceding ones (and indeed those that follow) for it is desirable to describe the grosser consequences of the presence of imperfections before attempting a more detailed theoretical treatment of the imperfections themselves. We shall start, therefore, by considering such matters as ion diffusion, impurity doping in semiconductors and the influence of dislocations on the mechanical properties of crystals. We continue with a description of how crystals are grown and manipulated to control the density and nature of the imperfections they contain. These techniques are important in the fabrication of solid state, and particularly semiconductor, devices. With the main experimental situation thus established we shall discuss the electron energy states associated with particular types of imperfection (chiefly impurities) and for experimental confirmation refer mainly to the optical absorption associated with such imperfections. Here some overlap with Chapter 6 is inevitable but optical absorption due, for example, to electronic transitions between impurity levels is so characteristic of impurity effects that it seems more appropriate to consider them here.

5.2 Defects

5.2.1 Vacancies and interstitial defects (a)

We assume initially that the crystal contains no foreign (impurity) atoms and, if a compound, that the constituents are present in the correct stoichiometric proportions. Point defects are then limited to *vacancies* (missing atoms or ions) or *interstitials* (atoms or ions at sites in the interstices of the lattice). If the crystal had no surface, vacancies and interstitials would have to occur in pairs to maintain stoichiometry and overall charge neutrality. Such pairs are sometimes called Frenkel defects. In practice, however, interstitials and vacancies can migrate to the crystal surface by diffusion (see below) and either add to or subtract from the size of the crystal. If this occurs the crystal can contain a net excess density of either vacancies or interstitials. In most crystals the energy necessary to create a vacancy, E_v, is less than the energy E_I required to create an interstitial since the latter often requires a considerable lattice distortion to be accommodated. If one type of defect (say, vacancies) predominates, charge neutrality can be maintained in an ionic crystal by equal anion and cation vacancy densities. Alternatively a positively charged vacancy may capture a free electron to achieve local charge neutrality, or a negative ion vacancy capture a hole. A vacancy with a trapped electron constitutes a new type of defect and it is important to notice that this electron may be excited, and even freed (i.e. excited to the conduction band), by the absorption of a photon of appropriate energy. Thus electrons trapped at defects can lead to optical absorption and, if the photon energies involved correspond to visible light (2 or 3 eV), give a colour to the crystal. The coloration provides a convenient way of detecting this type of defect.

Even in the absence of deliberate damage to a crystal some vacancies and interstitials will exist in thermal equilibrium. Since a formation energy is involved the number of, for example, vacancies in equilibrium at a temperature T will be of the form

$$\mathcal{N}_v \sim \mathcal{N}_0 \exp(-E_v/kT) \tag{5.1}$$

where $\mathcal{N}_0$ is the number of atoms in the crystal and E_v is typically 1 or 2 eV. For $E_v = 1.0$ eV the fraction $\mathcal{N}_v/\mathcal{N}_0$ is about 10^{-16} at $T = 300$ K, rising to about 10^{-6} at 1000 K. The number can be estimated experimentally by measuring the specific heat of the crystal (energy being absorbed by the creation of defects) and sometimes by measurement of changes in crystal size.

A defect density in excess of the thermal equilibrium number can be obtained easily—for some purposes, all too easily. A relatively large density can be obtained by rapidly cooling ('quenching') a crystal heated to a temperature near its melting point. Such a crystal is not, of course, in an equilibrium state but the defects can only move to the surface very slowly by diffusion at the lower temperature and can be considered as 'frozen in'.

Diffusion of defects, like any other diffusion process, arises from a gradient in the concentration. As explained before, defects will diffuse to the surface to build up new crystal layers. The current of atoms crossing unit area in unit time is proportional to the gradient in the concentration, N_d, and is

$$J = -D \operatorname{grad} N_d \tag{5.2}$$

where the diffusion constant, D, varies with temperature as

$$D = D_0 \exp(-E_0/kT) \tag{5.3}$$

since the jump of an atom into an adjacent site requires an activation energy E_0. The value of E_0 is of the order of 1 eV for vacancies in many crystals, which move as an adjacent atom jumps into the vacant site (see figure 5.2). The vacancy can then be considered to have moved one lattice spacing in the opposite

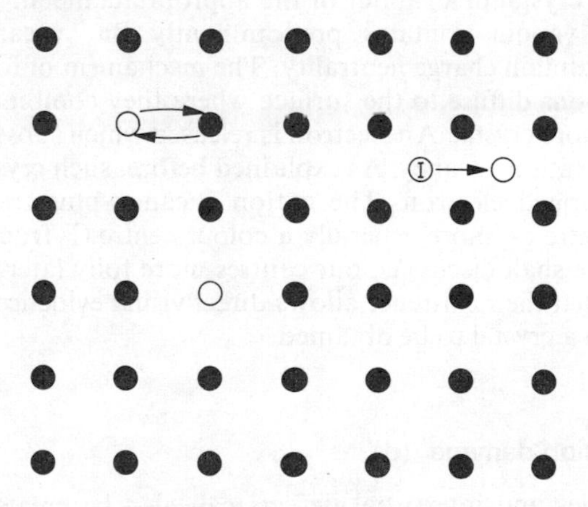

Figure 5.2 Planar array of atoms showing vacancies and interstitial. The vacancy moves into a neighbouring site as an atom moves in the reverse direction

direction. The activation energy for the motion of interstitials is often much smaller and may be as low as 0.1 eV. Hence interstitials diffuse out of (or in to, if the concentration gradient is reversed) a crystal much more readily than vacancies. Appreciable diffusion of interstitials occurs in some crystals even at room temperature. As a consequence heating and quenching followed by a relatively low temperature anneal favours the generation of vacancies with a consequent expansion of the crystal.

Frenkel defects (vacancy–interstitial pairs) in an ionic crystal can maintain charge neutrality but the existence of an isolated vacancy implies that the crystal is locally charged. Pairs of vacancies of opposite charge may therefore form to avoid this. The number of vacancy pairs in thermal equilibrium will be of the form

$$\mathcal{N}_p \sim \mathcal{N}_0 \exp(-E_p/2kT) \tag{5.4}$$

where E_p is the energy required to form a vacancy pair. The number of Frenkel pairs on the other hand is given by

$$\mathcal{N}_F = (\mathcal{N}_0 \mathcal{N}')^{\frac{1}{2}} \exp[-(E_v + E_I)/2kT] \tag{5.5}$$

where $\mathcal{N}'$ is the number of interstitial sites that can accommodate an ion. In the alkali halides (NaCl, etc.) it is found experimentally that $E_p < (E_v + E_I)$ and is about 2 eV. Hence vacancy pairs are the predominant defect in equilibrium. But in the silver halides $E_p > E_v + E_I$. The existence of interstitial silver ions is important in the photographic process; they diffuse easily to form macroscopic and consequently visible aggregates.

Single cation vacancies can be made predominant in alkali halide crystals by heating the crystal in a vapour of the appropriate metal. Thus KBr heated in potassium vapour contains predominantly Br^- vacancies which trap electrons to maintain charge neutrality. The mechanism of formation requires that bromine ions diffuse to the surface where they combine with potassium and build up more crystal. An electron is released which subsequently becomes trapped at the cation vacancy. As explained before, such crystals are coloured due to the trapped electron. The cation vacancy plus trapped electron is called an F centre or, more generally a colour centre (F from Farbe, German for colour). We shall discuss colour centres more fully later; for the moment we note only that their existence allows direct visual evidence for the diffusion of vacancies in a crystal to be obtained.

5.2.2 Radiation damage (d)

Excess vacancies and interstitial defects can also be created as a result of irradiation by energetic particles. A fast incident particle from an accelerator or radioactive source may strike an atom in the crystal and give up part of its energy. The incident particle may go on to make several such collisions. The atoms which are struck may also acquire a large kinetic energy and plough through the lattice causing further atomic displacements. The latter effect and the range of the incident particle will, in general, depend on crystal orientation, a phenomenon known as 'channelling'. In any case disorder will extend over a considerable region along the track of the incident particle. The energy required to form a vacancy-interstitial pair by irradiation is typically about 5–10 eV (i.e. appreciably more than required thermally) so that an incident particle in the MeV range will produce many such pairs, either directly or indirectly. Lighter particles, such as electrons, do proportionately less damage, much of their energy being dissipated by exciting crystal electrons to higher energy states.

Energetic photons—γ-rays, X-rays and short-wavelength ultraviolet light—can also produce radiation damage. The process here is usually indirect, the incident photon imparting its energy to a lattice electron which in turn creates defects. The subject of radiation damage is complex, the pattern of damage to a crystal being a complicated function of the energy, mass and charge of the incident particle; the crystal nature, orientation and temperature. Annealing effects occur during irradiation even at room temperature due to interstitial diffusion. Complicated though radiation damage effects are, they have been much studied recently in connection with nuclear reactor materials. The generation of vacancies in neutron-irradiated materials, for example, leads to a substantial increase in the size of reactor components and aggregation of vacancies into large bubbles.

Alkali halide crystals (and indeed many transparent insulators) become coloured on irradiation due to the formation of F and similar centres. This coloration has found application in dosimetry and in information storage and display. As dose meters such crystals are relatively insensitive, requiring for example around 3×10^3 roentgens of 1 MeV gamma rays to produce reasonably measurable coloration. Smaller doses can be detected using thermoluminescence, as discussed in § 6.7.4.

Crystal coloration by electron irradiation was first used for information display during World War II. The device was the Skiatron or 'dark-trace' tube used in ground-based radar installations. The tube employed a potassium iodide screen and conventional cathode ray tube electron gun, the display being projected episcopically by reflection off the tube face. Since the electrons trapped at the vacancies can be freed by absorbing light, the projection lamps also served to clear the coloration produced by the incident electrons. The chief deficiency of these tubes was the build-up of persistent coloration during use which could only be cleared by prolonged heating (annealing). Recently interest in dark-trace tubes has revived. A number of new materials have been found, for example the mineral sodalite (an aluminium silicate) which can be produced synthetically and which appears to suffer much less from persistent coloration. Coloration is still by high voltage electrons (~10 KeV), erasure and display by illumination. There is also some interest in the use of colour centres created by electron irradiation for data storage in computers but at present this seems unlikely to compete with other available methods.

5.2.3 Ionic conduction (a)

The existence of lattice vacancies permits the motion of lattice ions by providing sites into which they can jump. As interstitial ions are similarly free to move, defects in ionic crystals permit the movement of ions and hence the transport of charge, i.e. ionic conduction. We have seen in Chapter 4 that many ionic crystals are insulators, containing relatively few mobile electrons or holes even at elevated temperatures. Such crystals conduct at high temperatures due to ion motion. We would expect the conductivity to be proportional to the number of vacancies, vacancy pairs or interstitials, whichever predominates, and to the diffusion rate. If therefore the crystal has not been deliberately damaged and is in thermal equilibrium we expect the conductivity, σ, to be of the form

$$\sigma = A_0 \exp[-(E + E_0)/kT] \tag{5.6}$$

where $E = E_p$ or $(E_v + E_I)$ whichever is the smaller and E_0 is the activation energy for ion motion (5.3). Using the Einstein relationship between the velocity per unit field and the diffusion constant (see § 9.1.2) one can show that

$$A_0 = D_0 e^2/kT \tag{5.7}$$

a result which is in adequate agreement with experiment.

The coloration of crystals by the introduction of excess vacancies at which electrons may be trapped permits the migration of vacancies under the influence of an electric field to be observed directly. Perhaps one of the most striking experiments of solid state physics, performed over 40 years ago by Gudden and Pohl, is the migration of colour through KBr crystals. Their own illustration is reproduced as figure 5.3.

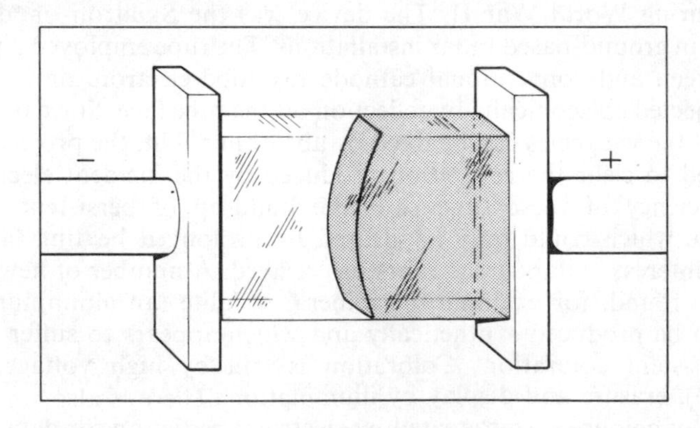

Figure 5.3 Migration of F centres in an applied electric field KBr at 580°C. (After R. W. Pohl, *Proc. Phys. Soc.*, **49**, 3 (1937).)

5.3 Impurities (a)

Impurities, i.e. foreign atoms or ions, can take up both substitutional or interstitial positions in a host lattice, depending on the temperature and the activation energy required. They may diffuse through a lattice in much the same way as the defects discussed in the previous sections and can contribute to ionic conduction. The diffusion of impurities at elevated temperatures and their subsequent freezing-in is one of the most widely used techniques of semiconductor device fabrication (see § 5.5.4).

A substitutional impurity will in general differ from the replaced atom in mass, electronic configuration and valency. A mass difference alone (i.e. an isotopic impurity) will affect the lattice vibrations locally, as will be evident from our discussion in Chapter 3. A different electronic configuration and valency will lead to new electronic energy levels and increase or decrease the total electron concentration in the crystal. We discuss the electronic energy levels associated with impurities in various types of host lattice in later sections. In the next section we shall consider only the grossest effect of impurity addition, namely its effect on the density of mobile electrons or holes in semiconductors and the consequent changes in conductivity. Impurities can also affect the conductivity of solids in other ways, notably by interrupting the periodicity of the lattice and reducing the ease with which electrons can move through the crystal. This effect is relatively small and, as mentioned, will be discussed in Chapter 8.

5.3.1 Impurities in semiconductors (a)

Impurities may increase the electron concentration or decrease it, the latter being equivalent to the addition of holes. If, as in most insulators, the additional electrons (or holes) are tightly bound to the impurity centres or trapped at other defects no substantial change in conductivity results. In metals and covalent crystals on the other hand, the additional electrons may be only weakly bound and free to move through the crystal at ordinary temperatures. In metals containing $\mathscr{N}_I$ impurities, however, the fractional increase in free electron density is of the order of $\mathscr{N}_I/\mathscr{N}$ which is small as the initial density is so large. It is in the covalent or partially covalent semiconductors, where the initial electron and hole densities are small, that conductivity changes of many orders of magnitude can be obtained. A good example of the mechanism involved is the addition of Group V or Group III element impurities into the Group IV elemental semiconductors, i.e. germanium and silicon, though similar remarks apply to Group VI and Group II impurities in the III-V compound semiconductors. At low concentrations these impurities take up substitutional positions in the host lattice. If atoms of a Group V element (e.g. As) are added to germanium, four of the five valence electrons of each impurity atom are taken up in covalent bonding to the impurity's nearest neighbours. The fifth electron is superfluous for bonding and only very weakly bound to the impurity atom (see figure 5.4). If the binding energy is much less than kT, which it is at room temperature for the example chosen,

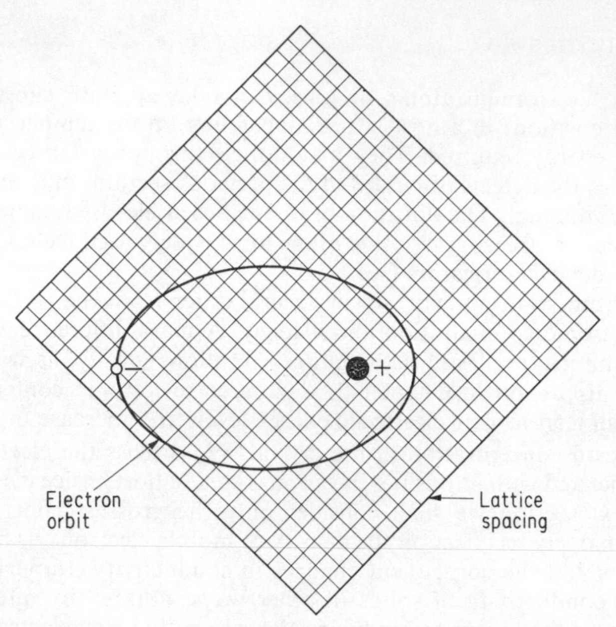

Figure 5.4 Schematic representation of a donor impurity electron orbit in Ge. The impurity atom is shaded. The classical elliptic orbit covers many atoms

almost all the impurity centres become ionised. Each carries a net positive charge and an equal number of electrons are free to contribute to electronic conduction. Similarly the substitution of Group III atoms for germanium or silicon atoms leads to enhanced conduction, this time by the addition of holes. Each impurity contributes three valence electrons, which are taken up in covalent bonds leaving a vacant electron site (a hole) which may move through the crystal by successive electron motions, as described in Chapter 4. When acting in this way a Group V impurity is called a *donor* centre, as it has donated a free electron, and a Group III impurity an *acceptor*.

We saw in Chapter 4 that in a pure, intrinsic semiconductor the free electron and hole densities are equal; the Fermi energy lies midway between the valence and conduction band edges at $T=0$ and moves slowly with increasing temperatures towards the band with the smaller density of states. On the addition of donors or acceptors this is no longer true. We can distinguish three temperature regions. At very low temperatures (e.g. below about 20 K for arsenic donors in germanium) kT is less than the binding energy so that electrons (holes) are bound to donor (acceptor) impurities and make relatively little contribution to conduction. The mobile charge carriers (electrons and holes) can be said to be 'frozen out'. When kT is greater than the carrier binding energy but still much less than the energy gap between valence and conduction bands the impurities are ionised and the carrier concentration almost independent of temperature. In this, the exhaustion region, the density of electrons (holes) is almost equal to the donor (acceptor) density unless both are present,

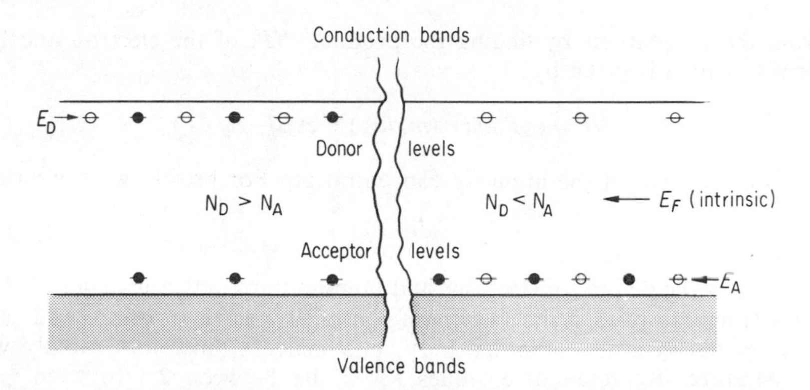

Figure 5.5 Energy levels for donors and acceptors relative to the valence and conduction band in a semiconductor. If $N_D > N_A$ all the acceptor states and some of the donor states are filled with electrons so E_F lies near E_D at low T. If $N_A > N_D$ all the donor states will be empty, and there will be holes on the acceptors, so E_F lies near E_A at low T. Full circles indicate electrons in the impurity states, open circles indicate holes or missing electrons in these states

when the density and nature of the charge carriers is determined by the excess of donors over acceptors or *vice versa* (see figure 5.5). A semiconductor sample in which the free electron density exceeds the hole density is called N type; conduction is primarily by negatively charged electrons. Conversely in a P type semiconductor there is an excess of acceptors over donors and current is carried primarily by positive holes. Many semiconductor devices rely on the properties of junctions between P and N type material. We discuss the fabrication of such junctions in § 5.5.4 and their properties in Chapter 9.

As the temperature is increased an increasing number of electrons are promoted across the energy gap between valence and conduction bands. Eventually the equilibrium density of thermally generated electrons and holes will greatly exceed the excess of electrons or holes due to the impurities and the material will be an intrinsic semiconductor. The temperature at which this occurs obviously depends on the initial carrier surplus. In this third temperature region the conductivity of the material increases very rapidly with temperature. We indicate how electron and hole densities in semiconductors can be measured experimentally in § 8.1.

The Fermi level in an N type semiconductor lies near and, if the impurity concentration is very high, within the conduction band. Similarly the Fermi level in a P type sample lies near or in the valence band. As the temperature increases and the intrinsic region is reached the Fermi level moves towards the centre of the energy gap (figure 5.6(a)). When the density of intrinsic carriers greatly exceeds the impurity density the variation of the Fermi level with temperature follows that of pure material (Chapter 4).

If the impurity density is not too high and the Fermi energy E_F lies well within the energy gap the density of electrons and holes in a semiconductor sample are given by (4.66) and (4.68). The Fermi energy can be eliminated from these equations by finding the product, NP, of the electron and hole density which is given by

$$NP = (kT/\pi\hbar^2)^3 (m_e^* m_h^*)^{3/2} \exp(-E_g/kT) \tag{5.8}$$

and independent of the impurity concentration. For brevity we can write

$$NP = n_i^2 \tag{5.9}$$

where n_i is the density of electrons and holes in pure, intrinsic material at the same temperature. E_g is the energy difference between valence and conduction band edges, typically of order 1 eV, and hence n_i^2 increases rapidly with temperature. Representative values for n_i lie between 2×10^{16} cm^{-3} for InSb and 10^{10} cm^{-3} for Si at room temperature. The value of n_i for germanium is 2.4×10^{13} cm^{-3} at 300 K. A donor or acceptor concentration appreciably greater than n_i increases the conductivity and implies conduction predominantly by electrons or holes respectively.

A donor centre, having lost an electron, is positively charged. Similarly an ionised acceptor is negatively charged. If a fraction f_D of the donors are occupied by electrons and a fraction f_A of the acceptors occupied by electrons, charge neutrality requires

$$N + N_A f_A = P + N_D(1 - f_D) \tag{5.10}$$

from which P or N can be eliminated using (5.9). N_A is the acceptor concentration, N_D the donor concentration and f the Fermi function. Thus f_D is given by

$$f_D = \{1 + \exp[(E_D - E_F)/kT]\}^{-1}$$

where E_D is the energy of an electron bound to a donor. For most donors and acceptors in silicon and germanium E_D and E_A differ only by ~0.01 eV

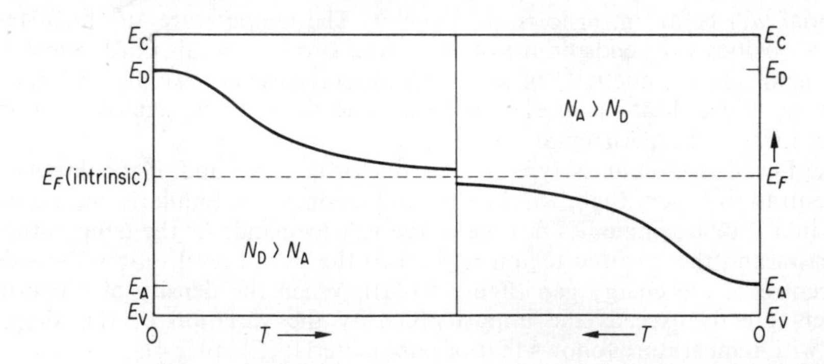

Figure 5.6 (a) Variation of E_F with T in N type and P type semiconductors. At low T the position is determined by the position of the impurity levels, at high T it tends to the intrinsic value in the middle of the gap

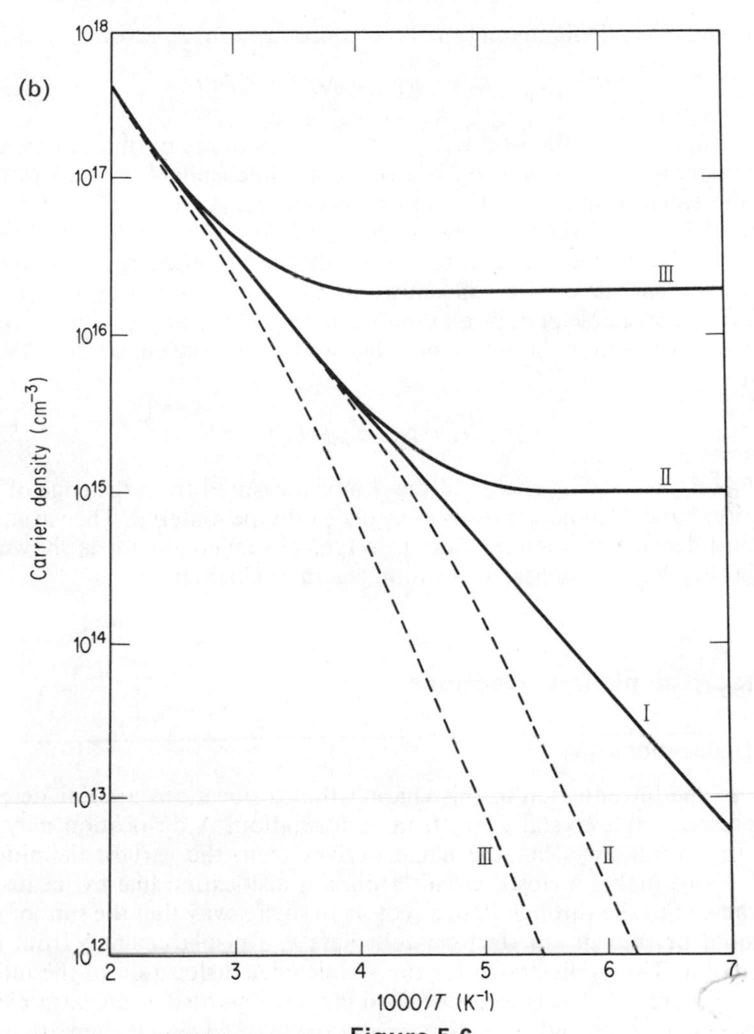

Figure 5.6

(b) Temperature dependence of majority electron density (solid curve) and minority hole density (broken curve) in a semiconductor where the intrinsic density $n_i = 1.45 \times 10^{15}\ T^{3/2} \exp(-1850/T)$ cm^{-3} is plotted as curve I. Different dopings $N_D - N_A = 10^{15}$ cm^{-3} and 2×10^{16} cm^{-3} are given by curve II and III respectively. At low temperature the number of carriers becomes equal to $N_D - N_A$: this is called the exhaustion range. At high T all samples give the intrinsic result (After J. S. Blakemore, *Semiconductor Statistics*.)

from the conduction and valence band energies respectively so that, except for very high impurity concentrations or very low temperatures, it is clear from figure 5.6(a) that $(E_D - E_F)$ is large and positive, and $(E_A - E_F)$ large and negative. Hence, $f_D \sim 0$ and $f_A \sim 1$. To be definite we assume the material is N type at temperatures below the intrinsic range. Inserting these approximations into (5.10) and eliminating the hole concentration, P, gives

$$2N = (N_D - N_A) + [(N_D - N_A)_i^2 + 4n_i^2]^{1/2} \tag{5.11}$$

so that at high temperatures ($n_i \gg N_D - N_A$), $N = n_i$ in the intrinsic range and at low temperatures $N = N_D - N_A$ is a constant, independent of temperature. This is the exhaustion range. The approximation $(E_D - E_F) > kT$ and hence equation (5.11) is not valid at very low temperatures when kT is comparable to the donor binding energy and consequently the Fermi energy approaches E_D. The free charge carrier concentration then falls as electrons become trapped for appreciable periods on donor centres. At the lowest temperatures another approximation becomes possible and the electron density, N, is given by

$$N \sim N_D \exp(-E_{DO}/2kT) \tag{5.12}$$

where E_{DO} is the binding energy of the donor measured from the edge of the conduction band. Similar expressions apply to P type material. The variation in electron density with temperature in a typical semiconductor is shown in figure 5.6(b). We shall consider this topic again in Chapter 8.

5.4 Macroscopic imperfections

5.4.1 Dislocations (a)

We saw in the introduction to this chapter that dislocations are line defects whose presence in a crystal arises from deformation. A dislocation may be defined in several ways but the name derives from the earliest definition, namely: if one makes a closed circuit around a dislocation line by means of steps from site to site through lattice vectors in such a way that the sum of the steps would be zero in a perfect crystal, there is a displacement **b** from the starting point. This is illustrated for the simple edge dislocation of the introduction in figure 5.7. The edge dislocation may be regarded as an extra plane of atoms which terminated at the line, x, and the marked circuit shows that in this case **b**, called the Burgers vector, is perpendicular to the line.

A second simple type of dislocation is the screw dislocation shown in figure 5.8. Here **b** is parallel to the line of the dislocation. It may be thought of as a spiral arrangement of atomic planes, with a step up for each revolution.

In general a dislocation line will be curved and will either form a closed loop or end at the crystal surface. The Burgers vector **b** may be at a general angle to the loop. A dislocation loop may be thought of as if the crystal had been cut over the area of the loop. The crystal on one side is then displaced by **b** and sufficient atom planes inserted to fill the space. A simple example of this would be a raft of atoms surrounded by edge dislocations. A part of this can be visualised from figure 5.9. This can be obtained when a perfect crystal is cut in a vertical plane and the crystal then moved apart by **b** so that

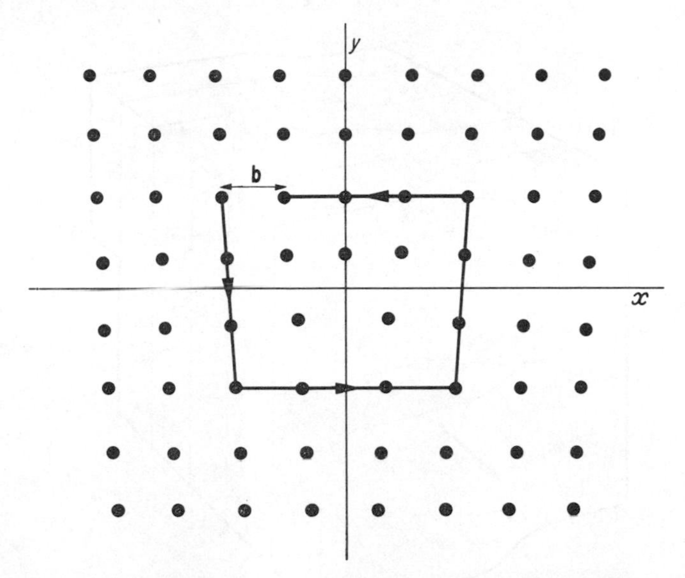

Figure 5.7 A section through an edge dislocation which runs normal to the plane through the origin of co-ordinates. The Burger's vector **b** is shown. The dislocation may be thought of as a cutting of the crystal along the positive OY axis, a displacement by **b** of the positive octant and the insertion of an extra plane. The upper half of the crystal is compressed and the low half is extended. The strain is largest near the origin

a new plane is inserted. A similar description of the screw dislocation in figure 5.10 would cut the crystal over the plane and move the top half parallel to **b**.

If a dislocation terminates at a crystal surface its presence can be detected by chemical etching. Suitable mixtures of chemicals—usually based on strong acids—have been devised for many crystalline substances, the choice frequently being based on guess work and folklore. A moderate rate of dissolution (say 10–100 μm per minute) is desirable and if the correct conditions are achieved the dislocations etch more rapidly than the undisturbed crystal so that 'pits' appear. Occasionally the reverse process occurs and dislocations meeting the surface appear as hillocks. Etch pits are frequently regular in shape (e.g. pyramidical) and bounded by low index crystal planes. Because etch pit counting is widely used as a method of estimating dislocation densities in crystals, densities are usually quoted on a unit area basis. In practice a good single crystal contains about 10^2 dislocations per cm^2 but densities of 10^6 cm^{-2} are common in metal crystals and occasionally densities as high as 10^{12} cm^{-2} may occur. For a few materials (e.g. silicon) techniques have been developed which enable large dislocation free crystals to be prepared.

The misorientation of adjacent crystal planes implied by the presence of dislocations can be detected by X-ray scattering if the dislocation density exceeds about 10^4 cm^{-2}, which allows 'absolute' calibration of a given etching technique. In transparent crystals dislocation lines can be made visible (at

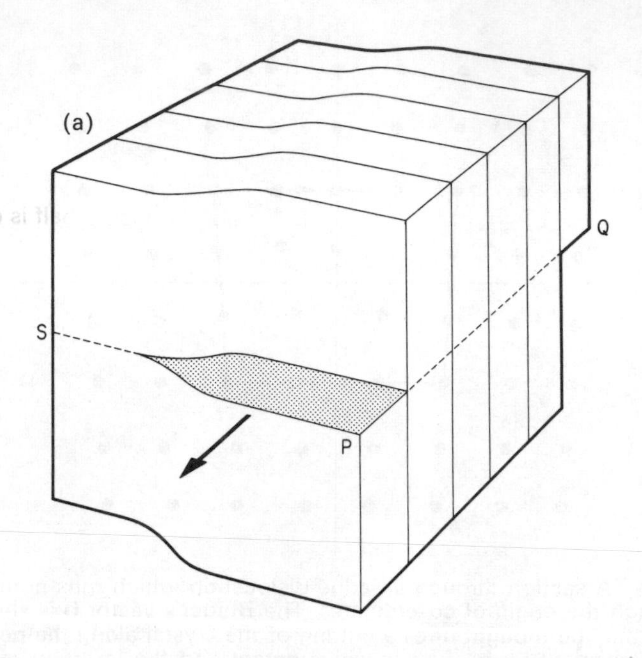

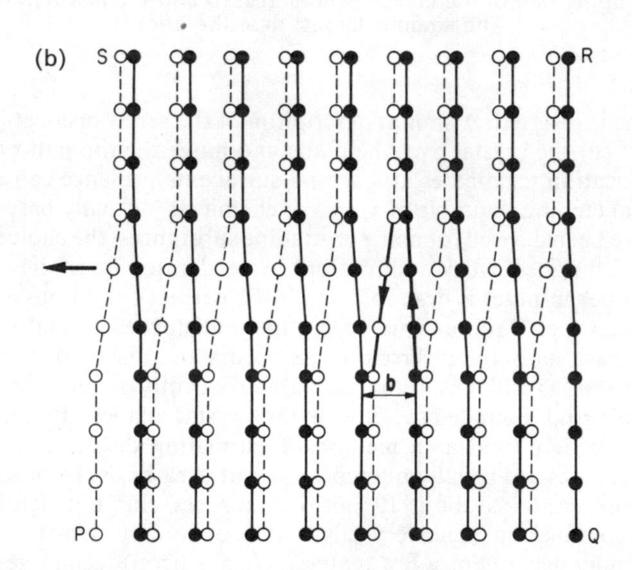

Figure 5.8 A screw dislocation. (a) showing planes of atoms. On going once round the axis of the dislocation marked with an arrow there is a displacement parallel to PQ. (b) shows the atomic planes immediately above (block circles) and below (white circles) the plane PQRS. The Burger's vector is obtained by a path along a line in the upper plane and returning in the lower plane

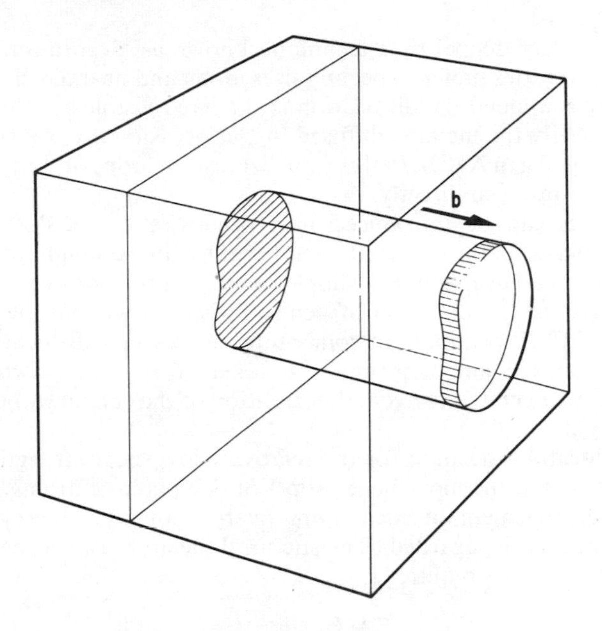

Figure 5.9 Dislocation loop caused by displacement of a region of crystal by **b** perpendicular to the plane of the loop

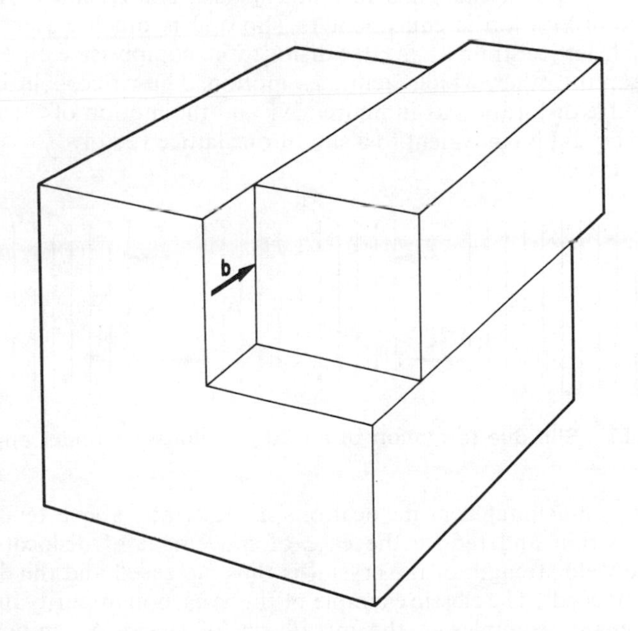

Figure 5.10 Screw dislocation as displacement **b** of part of crystal parallel to axis (compare figure 5.8(a))

least under a microscope) by a technique known as 'decoration'. It is found that many impurities prefer to occupy sites along and near a dislocation and if the impurity is a metal the dislocation is rendered visible as a line of opaque material. Usually the metal is diffused in but occasionally is a constituent of the crystal (e.g. Ag in AgCl). In this way dislocation loops of the type described above can be observed directly.

Dislocations can be introduced into otherwise low dislocation density material by plastic deformation at a temperature high enough for the material to be ductile. We have seen that simple bending introduces edge dislocations at right angles to the direction of bend. Twisting a crystal introduces both screw and edge dislocations. In solid state devices low dislocation densities are usually desired and care must be taken, for example when diffusing impurities into a crystal, to avoid deformation of the crystal by bending under its own weight.

Edge dislocations account for the relatively low shear strength of crystals. In a perfect crystal the movement ('slip') of one plane of atoms over another requires the movement of each atom over a potential energy barrier (cf. § 5.2). If the strain is regarded as elastic until the atom is, say, halfway across (at $x = a/2$), the stress required is

$$\Xi = G_s x/a \sim G_s/2$$

where G_s is the shear modulus. In practice the elastic limit at which slip begins is very much smaller than this in real crystals. The reason is that motion of an edge dislocation is equivalent to slip and is much easier. There is a considerable lattice strain at an edge dislocation, compressive on one side and extensive on the other, which facilitates motion. Thus it costs little energy to move an edge dislocation as in figure 5.11 and the motion of the extra plane across the crystal is equivalent to a slip of one lattice vector.

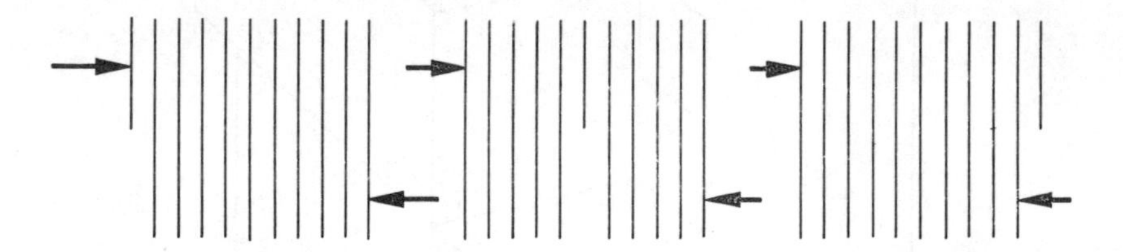

Figure 5.11 Slip due to motion of an edge dislocation under applied stress

Impurities precipitated at dislocations as described above tend to relieve the lattice strain and reduce the ease of movement of dislocations under stress. The yield strength of the crystal is thus increased and the dislocations said to be 'locked'. The classic example of this is carbon impurity in α-iron but there are many examples in the metallurgy of steels. An impurity-locked dislocation will, of course, still move under sufficient stress and a sharp yield point is observed when the dislocations are dragged off the impurities. The material then deforms easily. Annealing of a crystal after yielding allows the impurities to diffuse back to the dislocations and relock them, when a further yield point can be observed.

5.4.2 Planar defects (a)

Low-angle boundaries between otherwise relatively dislocation-free regions often occur in crystals. Such a boundary may be thought of as a planar array of edge dislocations as illustrated in figure 5.12. If a new vertical plane of atoms is added every few lattice spacings a well-defined tilt boundary develops.

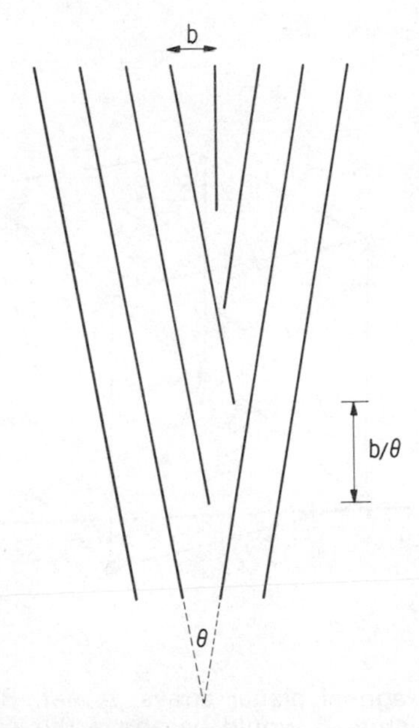

Figure 5.12 Section of low angle grain boundary as a planar array of dislocations. The average distance between successive dislocations is b/θ

Annealing of a crystal initially containing a random array of edge dislocations often leads to the development of low-angle boundaries due to the coalescence of dislocations. The total energy of the crystal is reduced by this movement, the crystal preferring to be free of dislocations over as large a volume as possible.

Another type of planar defect, common in face-centred cubic and hexagonal close-packed crystals, is called a stacking fault. Here there is a direct mismatch between adjacent crystal planes. A f.c.c. crystal, viewed along the (1,1,1) directions, is built up of three types of plane lattices A, B and C, as shown in figure 5.13. The arrangement of planes is ABCABC.... An h.c.p. crystal is rather similar, with packing ABAB.... In fact the face-centred cubic crystals often contain a stacking fault so successive planes might be ABCBCABC, i.e. an A plane missing. The adjoining planes have local h.c.p. symmetry.

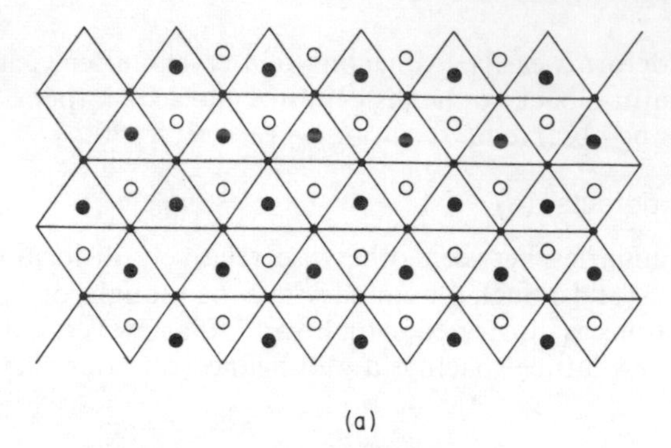

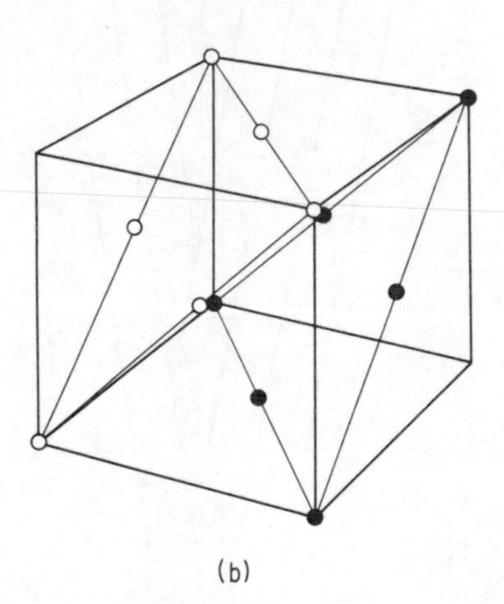

Figure 5.13 (a) Hexagonal planar arrays, A net, B black circles, C white circles. For a f.c.c. structure B would be above the A plane and C below it; for a h.c.p. structure B planes occur above and below A
(b) Cubic f.c.c. cell showing adjacent B and C planes. A planes run through the upper left and lower right-hand corners. (The energy of interaction is clearly the same between any pair of neighbouring unlike planes. A stacking fault requires extra energy to compensate the longer range forces.)

5.5 The growth of crystals of controlled perfection (d)

Imperfections in crystals can most profitably be studied experimentally when their nature is known and their density controlled. This requires high purity and high quality starting materials to which imperfections might be added, for example by the addition of foreign atoms (called doping) or by mechanical deformation to introduce dislocations. The production of reproducible solid state devices, too, requires a substantial degree of control over the position, density and state of ionisation of doping impurities. In this section we will briefly describe some of the more important techniques used in the fabrication of crystals for use in research and device production. It is convenient to distinguish three main steps: purification, crystal growth and incorporation of impurities, though the boundaries are not particularly distinct.

5.5.1 Purification (d)

Initial purification is of course by chemical methods, particularly distillation. Most metals and many semiconductors can then be further purified by a technique called zone melting. Zone melting relies on the fact that the solubility of an impurity is different (usually greater) in the liquid than the solid phase of a solvent. A familiar example of this effect is provided by icebergs which contain a lower concentration of salt (impurity) than sea water. The ratio of the solubilities in the solid and liquid phases is called the segregation constant, S_0 (or sometimes 'distribution coefficient') of the impurity. If the presence of the impurity lowers the melting point of the mixture, S_0 is less than unity and the impurity more soluble in the liquid phase. This is the most common situation. Zone refining then consists essentially of drawing a solid rod through one or more hot zones where melting occurs (see figure 5.14). From the point of view of the rod molten zones travel along its length. Each zone sweeps

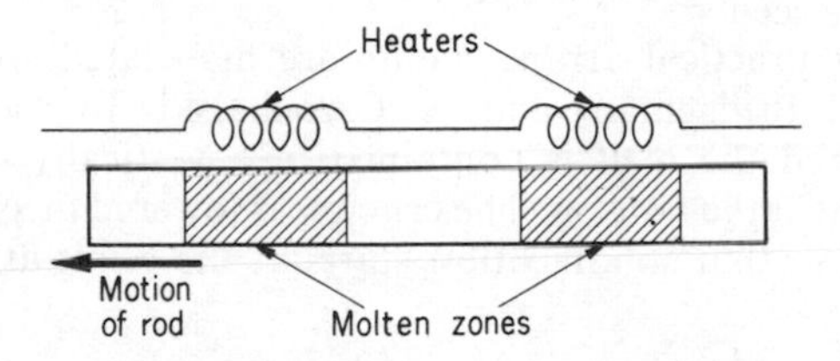

Figure 5.14 Schematic representation of the zone refining technique

impurities with $S_0 < 1$ with it to the end of the bar. Impurities with $S_0 > 1$ tend to migrate to the other end. Purification of the central region of the bar proceeds until equilibrium is reached when the rate at which impurities are being swept out equals the rate of incorporation from the boat in which the material is supported, or from the surrounding gas or simply by back diffusion from the very impure ends.

Naturally there are a number of technical refinements to which we should give brief consideration. The segregation constant defined above assumes an equilibrium situation in melt and solid, valid only if the molten zone moves sufficiently slowly for the impurity in it to be uniformly distributed. At finite zone speeds the effective segregation constant, S, will be more nearly unity than S_0. There is a tendency for material to be physically transported by the zone: this can be offset with the help of gravity by operating with the boat inclined to the horizontal. Contamination by impurities from the boat itself can be extremely troublesome with high melting point materials. This can be eliminated, in exchange for other problems, by dispensing with a boat and mounting the rod vertically. Provided that only a small zone is melted in the rod the liquid zone stays in position due to surface tension. Vertical zone refining has been used to purify metals with high melting points such as tungsten.

5.5.2 Crystal growth from the melt (d)

Crystals may be grown by a variety of methods: from solution (e.g. salt crystals), from a melt or from the vapour phase. Unless the segregation constant is orders of magnitude different from unity, solution growth inevitably leads to contamination by the solvent and we shall not consider it further. Vapour phase growth has some important advantages which we discuss below but growth rates are often very slow. To grow crystals of large size, melt growth is usually the most suitable method when it is applicable.

To grow a crystal from a melt one needs a small crystal to act as a seed. The basic features of the crystal growing system are then as shown in the sketch, figure 5.15, and the only essential condition is that the solid–liquid interface be at the melting temperature of the crystal. If we now adjust the temperature or the temperature gradient so as to move the interface into the molten region sufficiently slowly, the crystal will grow with the same orientation as the seed.

The two most practical arrangements are associated with the names of Stockbarger and Bridgeman, and of Czrochalski. In the Stockbarger or Bridgeman method the melt is contained in a vertically mounted crucible with a sharp point at the bottom. The crucible is lowered mechanically through a freezing plane so that solidification starts at the point at the bottom. The

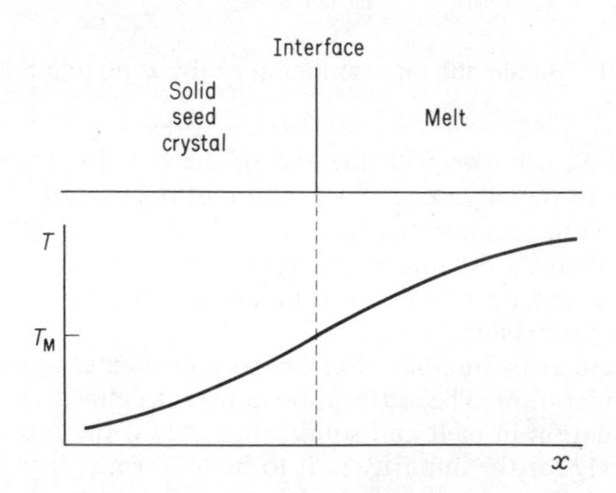

Figure 5.15 The temperature at the solid–liquid interface of a crystal growing from the melt must be at the melting temperature T_M. The crystal will grow if the average temperature is lowered, thus moving the interface to the right, or as in the Stockbarger and Bridgeman method, physically moving the crystal and melt to the left while the temperature distribution remains unaltered

shape of the pointed end is such that a small single crystal often forms and provides a seed for the remainder of the melt. This technique is remarkably successful for some materials, e.g. alkali halides, and crystals up to 30 cm diameter and 30 cm long have been successfully grown this way. The crucible is normally sealed off at the top to prevent atmospheric contamination but clearly this cannot be done when the constituents of the melt have high vapour pressures. A more fundamental difficulty with the Stockbarger method arises from the constraint imposed on the growing crystal by the crucible. Some materials expand on solidification (ice is again an example) and this can lead to strain in the crystal or even fracture of the crucible. The differential contraction of the crucible and crystal on cooling must also be considered. These difficulties are to a large extent overcome by the Czrochalski method.

In the Czrochalski method an existing seed crystal attached to the bottom of a vertical rod is dipped into a melt of the material to be grown. The seed is now slowly and steadily withdrawn, liquid adhers by surface tension and if sufficient heat loss occurs through the seed crystal and pulling rod the seed crystal grows. No mechanical constraint is imposed on the growing crystal and the diameter of the rod grown is dependent primarily on the pull rate and melt temperature, which are under the operator's control. Consequently it is possible to reduce the crystal diameter to a thin neck at some point when many structural defects (e.g. dislocations) originating in the seed but faithfully reproduced in the bulk crystal, may grow out to a surface. Continued growth and expansion of the crystal diameter yields a crystal with fewer defects than the original seed: clearly a desirable feature. A few materials can be grown totally free from dislocations by this method.

If a doped crystal is required, impurities must be added to the melt from which the crystal is to be grown. Typical concentrations may be indicated by a numerical example: N type germanium with a resistivity of 5 Ωcm contains about 3×10^{14} donor impurities per cm^3 which corresponds to a doping concentration of one part in 1.6×10^8. For convenience impurity is normally added to a pure melt in the form of pellets already diluted by germanium. Impurity segregation occurs during growth so that, unless steps are taken to avoid it, the impurity concentration varies through the crystal in the growth direction. The variation of concentration c, can be shown quite simply to be given by

$$\mathrm{c} = S\mathrm{c}_0(M_0/M_\mathrm{M})^{1-S} \tag{5.13}$$

where c_0 is the initial concentration, M_0 the initial mass and M_M the mass of the melt when the impurity concentration is c. The effective segregation constant, S can, however, be varied by varying the growth rate: as the growth rate increases, S changes from S_0 towards unity. If the variation of S with growth rate is known the growth rate can be programmed to keep the impurity concentration constant over a substantial fraction of the grown crystal. Other factors are important in determining the radial variation of impurity across the (assumed planar) solid–liquid interface. For example it is found that the segregation constant is frequently dependent on the type of crystal plane, (1,0,0), (1,1,1), etc., which lies parallel to the solid–liquid interface.

The latter may be flat at the centre of the crystal but it is difficult to avoid some radial temperature gradient so a flat interface cannot be maintained to the edges. A curved interface, exposing other crystal planes for growth, consequently leads to a non-uniform radial impurity distribution.

It is frequently important to ensure not only that the required impurity element is incorporated but that it is in the required state of ionisation. The state of ionisation of an incorporated impurity is usually determined by the position it takes up in the lattice. Thus an impurity substituted for a divalent cation is likely to be doubly charged. The state of ionisation can be modified by the simultaneous addition of a charge-compensating impurity. As an example, consider the incorporation of Nd^{3+} ions into $CaWO_4$ (a laser material). Charge compensation is achieved by the addition of a Group I metal ion (Na^{1+}, K^{1+}, Li^{1+}, etc.) which substitutes for a Ca^{2+} on a site near a Nd^{3+} which has also substituted for Ca^{2+}. The experimentally determined segregation constant of Nd^{3+} ions in $CaWO_4$ is shown in figure 5.16.

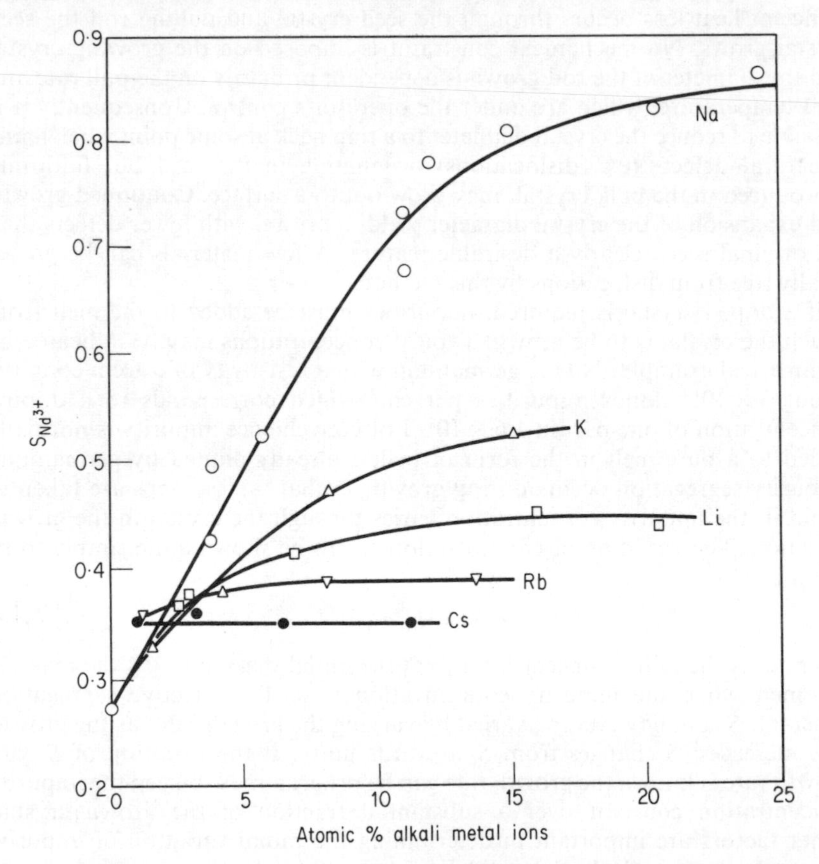

Figure 5.16 Segregation constant or distribution coefficient S for Nd^{3+} ions in $CaWO_4$ in the presence of alkali metal atoms (after K. Nassau, *Proc. Symp. on Optical Masers*, Brooklyn Poly Inst. (1963))

The growth rate that can be achieved in growth from the melt is substantially dictated by the need to incorporate as uniform a density of the desired impurity as possible. Slow growth rates and, when technically feasible, stirring of the melt give the maximum opportunity for a uniform impurity concentration to be maintained in the melt and ensure that the segregation constant is near its equilibrium value. If equilibrium is not achieved, impurity segregation tends to build up a high concentration of impurity in the melt just below the growing interface. In heavily doped and poorly stirred melts the build-up of impurity in this way can significantly modify the melting point of the material, making the melting point a function of position. Unstable growth may then result. Typical growth rates lie in the range of 0.1 to 10 cm per hour, the slower rates being employed for heavily doped melts.

During the past few years the Czrochalski technique has been extended to materials with appreciable vapour pressures (up to a few atmospheres) at their melting points. To achieve this (a) the melt is covered with a moderately thick layer of Borax (~1 cm) which generally forms a complete layer round the melt and (b) the vessel containing the crucible and crystal pull-rod is pressurised by an inert gas. The seed crystal is lowered through the molten Borax layer to make contact with the melt. This technique, known as liquid encapsulation, has been successfully applied to GaAs, GaP and a number of other compound semiconductors.

5.5.3 Crystal growth from the vapour (d)

Crystal growth from the vapour, also known as epitaxial ('on the same axis') growth, is used when a high melting point or high vapour pressure precludes melt growth (e.g. CdS) or when a thin epitaxial film, with different doping from the bulk, is required. The vapour may be either vapour of the material itself (e.g. CdS) or a gas which dissociates on the heated substrate (e.g. $SiCl_4$ on Si). In either case the substrate or seed crystal must be maintained at an elevated temperature. Normal practice is to use a flowing gas stream and to heat the substrate until evaporation occurs to clean the surface. The substrate temperature is then lowered to permit growth. The growth rate, which is typically around 100 μm per hour, is determined either by the diffusion rate in the gas or the surface reaction rate, whichever is the slower. Reaction rates vary rapidly with temperature and it is usually desirable to operate under gas flow limited conditions, i.e. at high temperatures, not much below the melting point of the crystal.

5.5.4 Spatial control of impurities (d)

In semiconductor devices it is frequently necessary to introduce abrupt discontinuities and graded distributions of different types of doping impurity. The obvious examples are the fabrication of PN junctions (i.e. single crystal blocks, part P type, part N type) and PNP or NPN transistors. It is clearly possible to introduce additional impurities of any type during crystal growth by the Czrochalski method. The production of junctions by this method is not, however, commercially successful. The most widely used techniques are (a) alloying and (b) diffusion, which we now describe. Direct implantation of ions from high voltage accelerators is also possible and valuable in some specialised contexts.

Alloying is, in effect, growth from solution of one part of a PN junction. A fortunate feature of PN junctions (see Chapter 9) is that their electrical properties are determined by the least heavily doped side if the doping concentrations on each side are markedly different. Hence the concentration of impurity in the alloyed region does not matter, provided only that it is high enough. The alloying process is used in the fabrication of some germanium devices. As an example consider a pellet or disc of indium (a Group III acceptor impurity in germanium) placed on a slice of N type germanium. On heating to about 500°C the indium melts and dissolves some germanium: the maximum solubility of germanium in indium increases with temperature. When cooled with the germanium cooler than the indium, germanium heavily doped with indium regrows on the original germanium crystal as a seed. Excess indium finally solidifies on the surface and a P^+N junction (P^+ = heavily P type) results. The main shortcoming of the alloying technique is the inability to control the area of 'wetting' by liquid alloy and consequently the depth of dissolution of germanium because the temperature only determines the mass of germanium dissolved. For junction transistors, which require two junctions to be formed extremely close together, this is a serious limitation.

Introduction of impurities by diffusion usually proceeds in two stages. We refer to silicon as an example. In the initial deposition stage the slice of silicon is heated to about 1000°C in a stream of vapour, usually an oxide of the desired impurity. P_2O_5 (phosphorous impurity, N type) and B_2O_3 (boron impurity, P type) are frequently used. A vapour pressure high enough to saturate the solubility is normally used. Some diffusion occurs which follows the diffusion equations (5.2) and (5.3). The variation of the concentration N_0 in one dimension, x, may be found from these equations together with the boundary conditions

$$N_0(x=0) = \mathrm{c}_0; \; N_0(x=\infty) = 0; \; N_0(t=0) = 0$$

the solution being

$$N_0 = \mathrm{c}_0 \,\mathrm{erfc}\left(\frac{x}{2\sqrt{Dt_p}}\right) \tag{5.14}$$

where t_p is the time of deposition and erfc, is the complementary error function. Following deposition the sample is usually further heated in an oxidising (but undoped) atmosphere to permit the impurity to diffuse further to the required depth. The boundary conditions to be applied to the diffusion equation now include the above solution (5.14) as the distribution at $t = 0$. We also have $N_0(x = \infty) = 0$ and $(\mathrm{d}N/\mathrm{d}x)_{x=0} = 0$. With these conditions and assuming $t \gg t_p$, which is usually the case, we find

$$N_0 = (2\mathrm{c}_0/\pi)\exp(-x^2/4Dt) \tag{5.15}$$

namely a Gaussian distribution of impurities. If an N type impurity has been diffused into a P type slice and the impurity centres are fully ionised a junction will occur when N_0 is equal to the initial acceptor concentration provided, of course, that $N_0 > P$ for $x = 0$.

The values of the diffusion constants of many impurities in solids of technical interest have been determined over wide temperature ranges and are available in the literature. The above model is somewhat over simplified, and ignores, for example, the effect on the diffusion constant of one species of the presence of another. In general, however, the solutions of the diffusion equation given are well obeyed.

The great virtue of solid state diffusion of impurities is the high accuracy and reproducibility obtainable when high quality single crystal substrates are used, though diffusion constants generally vary with crystal orientation and diffusion down dislocation lines can be very rapid. The area of, say, a silicon slice into which diffusion is required can be delineated by oxidising the silicon surface elsewhere; a sufficiently thick oxide film is effectively impervious to the diffusant. Selective doping of a silicon surface may be achieved by the following procedure, which is illustrated in figure 5.17:

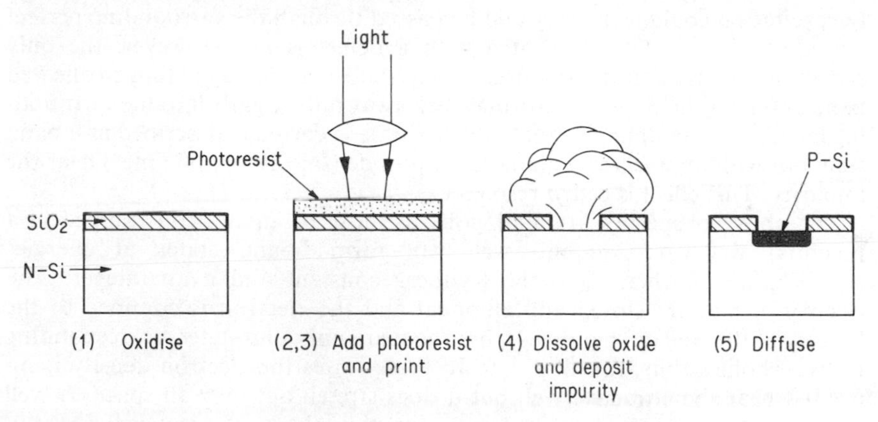

Figure 5.17 Doping of Si surface (see text)

(1) oxidise all surface by heating in O_2 or H_2O vapour;
(2) apply to the silicon surface one of the many proprietary brands of 'photoresist', a chemical similar to photographic emulsion;
(3) expose the photoresist to a pattern of ultraviolet light, usually using a microscope;
(4) dissolve with acid the printed area, and dissolve the oxide film underneath with hydrofluoric acid;
(5) diffuse impurity. As the diffusion progresses sideways as well as into the bulk, the junction between P and N regions is covered near the surface by the oxide film, which protects the junction from ambient gases during the working life of the device.

This procedure forms the basis of the field of microelectronics: the fabrication of complete electronic circuits in and on one slice of silicon. We discuss other aspects of this type of structure in Chapter 9.

5.6 Energy states associated with imperfections (a)

We now probe more deeply into the theory of impurity states. We have seen that an excitation of wave vector $\boldsymbol{k}$ in a perfect crystal has the same amplitude in each unit cell and spends only $1/\mathcal{N}$th of its time in any particular cell. In the presence of an impurity or other imperfections the amplitude may be expected to change over the affected cell or cells but the change will be localised and only affect about $1/\mathcal{N}$th of the whole. Hence we would expect only slight modification of the bulk excitations of a crystal. However, it is also possible that some excitations will be confined to the region of the defect and only have appreciable amplitude there. If so they will be totally unlike the perfect lattice excitations and the wave vector will be indeterminate, indeed irrelevant. These energy states will in general be 'bound' or localised states and must have energies outside the range of allowed band energies, or otherwise the excitation could leak away and be passed through the surrounding perfect lattice. A bound state associated with a defect is not, however, the only possibility. If an impurity produces a local state at an energy within an allowed band of the crystal the excitation may leak away only slightly into the surrounding lattice if the coupling is weak. Such a state is correctly described as a band state but with appreciably enhanced amplitude (say, 10 or 100 times) near the impurity. This effect is called resonance.

The above properties are analogous to those of an electron trapped in a potential well. A Coulomb well will form bound states at energies $E = -E_{\mathrm{H}}/\mathrm{n}^2 < 0$ where E_{H} is the Rydberg constant and n an integer. The electron orbits are closed and elliptical and the electron is confined to the locality of the well (cf. figure 5.4), but there are unbound states, corresponding to hyperbolic orbits, for all $E > 0$. In these states the electron density is increased near the attractive well but it does stretch out over all space. A well need not always bind a localised state: for example, a spherical well does not unless it exceeds a critical depth. Potentials of more complex shape (see figure 5.18) may give a resonance. In the central well alone there can be a localised state which can tunnel to the outside through the potential barriers to give a resonant state.

In general, therefore, we conclude that in the presence of impurities the excitations may be affected in various ways. There can be highly localised and tightly bound impurity states with energies very different from the band states of the host lattice, which in the main reflect the properties of the impurity. Weakly bound states on the other hand are best envisaged as crystal states somewhat perturbed by the impurity. This applies to the band states which persist in the presence of impurities and are only slightly modified. This modification can be regarded as giving a finite mean free time and path to the band states. Even in a resonance this picture retains its usefulness. The most important effects arise from the localised states and these will be discussed in some detail below.

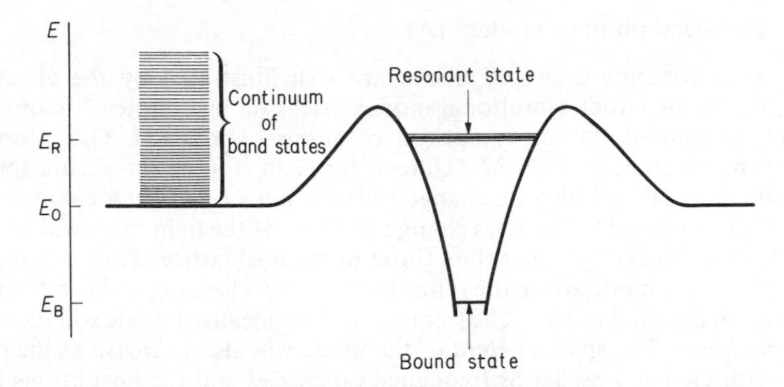

Figure 5.18 Potential well showing bound, resonant and band states. The central well alone would have a bound state at E_R but it may leak through the barriers to form a resonance

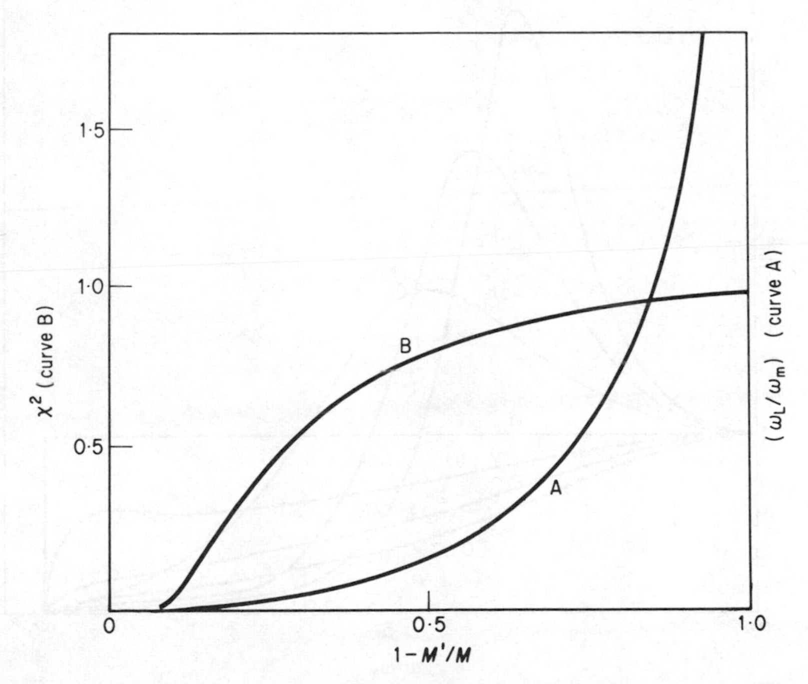

Figure 5.19 (A) Characteristic frequency ω_L of a localised mode due to a light mass relative to the maximum frequency ω_m of the host lattice. For light masses ω_L varies like $M'^{-\frac{1}{2}}$

(B) Amplitude squared $|\chi^2|$ of localised mode on defect atom. This rises to unity for the highly localised modes when $M'/M \ll 1$ and falls for the extensive modes when $M'/M \sim 1$. Note that $1 - M'/M$ must exceed a critical value before a localised mode appears (after R. J. Elliott and P. G. Dawber, *Proc. Roy. Soc.*, **A273**, 222 (1963))

5.6.1 Localised phonon modes (A)

The general concepts discussed above are well illustrated by the effect of impurities on the lattice vibrational modes discussed in Chapter 3. Consider a simple monatomic lattice comprised of atoms of mass M. One atom is replaced by an atom of mass M'. Unless this is an isotope the elastic spring constants Λ (3.24) will also be changed locally but in fact such changes are often small compared to the mass change. If $M' < M$ the light mass will have a characteristic frequency higher than those of the host lattice. Thus it can give rise to a localised mode above the range $0 < \omega < \omega_m$ where ω_m is the maximum frequency of the band modes. The frequency of the localised mode will increase as M' decreases. The spatial extent of the mode will also decrease as the mismatch with the band oscillator frequencies increases and the host atoms find it harder to respond. On the other hand as M' approaches M the mode will become more extensive. These effects are shown in figure 5.19. The band modes

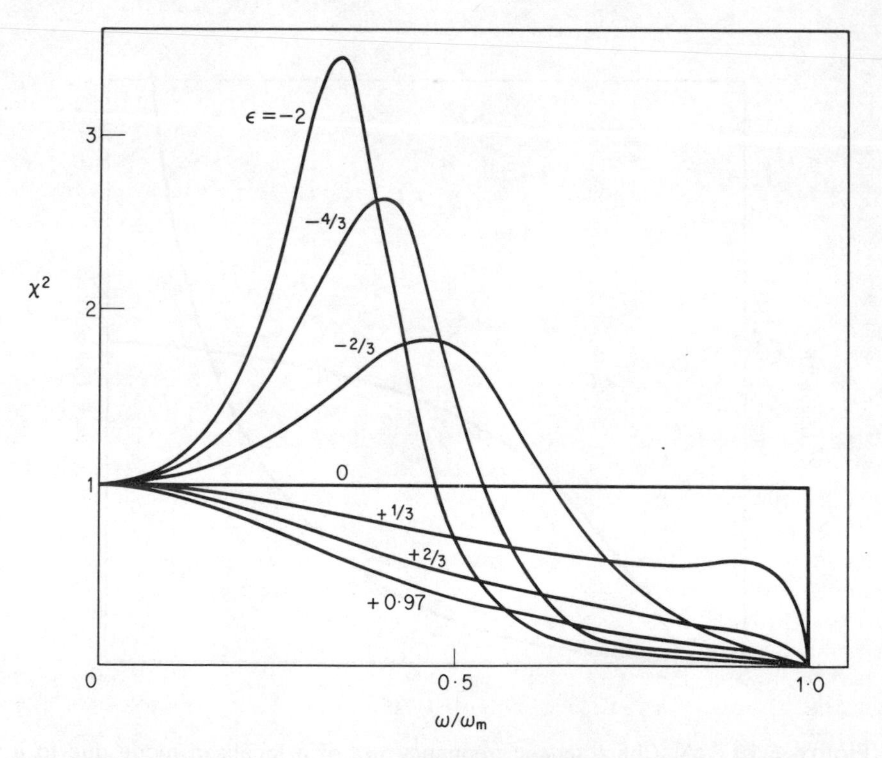

Figure 5.20 Relative squared amplitude $|\chi|^2$ of band modes at a defect site as a function of frequency in a simple Debye model. Note that for all light defects $M'/M < 1$ the amplitude is attenuated at all frequencies. For heavy atoms $M'/M > 1$ there is enhanced amplitude in a resonance which moves to lower frequencies and sharpens at M' increases (after R. J. Elliott and P. G. Dawber, *loc. cit.*)

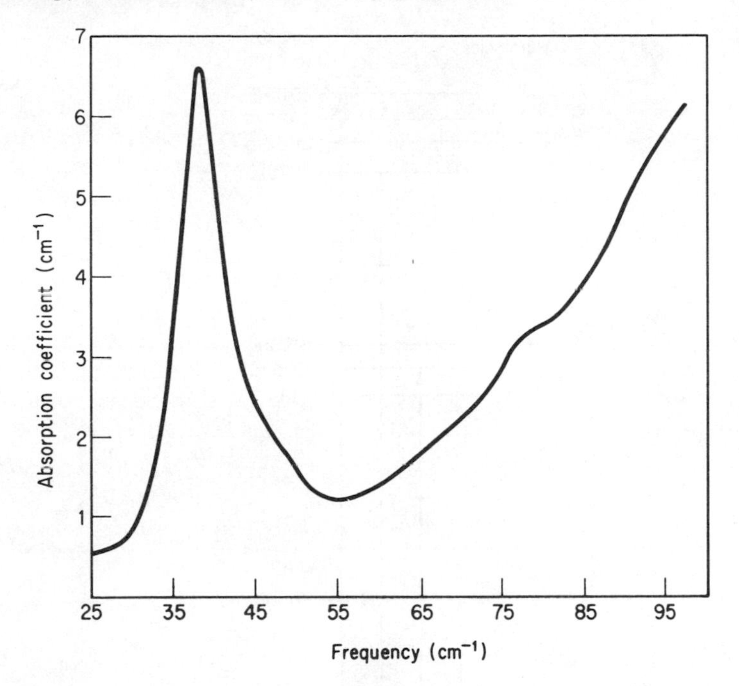

Figure 5.21 Optical absorption in the far infrared showing a resonance due to Ag impurities in KCl (after A. F. Sievers, *Localised Excitation in Solids*, p. 37, Plenum Press)

which all have frequencies below the characteristic frequency of the defect will be attenuated at the defect site.

If $M' > M$ on the other hand there is no tendency to form a localised mode. The characteristic frequency of thc impurity atom now lies in the band. Near this frequency the band modes will be enhanced near the defect by a resonance, at other frequencies they will be attenuated there. This effect is shown in figure 5.20. Since phonon frequencies are comparable to those of infrared photons these localised vibrational modes can give rise to absorption bands in the infrared (see § 6.6). As an example, figure 5.21 shows absorption due to a sharp low frequency resonance mode due to Ag impurity in KCl. Here the force constants are weaker and this helps to decrease the frequency. H^- ions in substitutional positions in ionic crystals give rise to very high frequency, highly localised modes in which the light H^- ions vibrate almost alone in a nearly stationary lattice. The potential here is markedly anharmonic. For example, for the case of H^- in CaF_2, the potential energy when expanded in powers of the displacement $R = (X, Y, Z)$ takes the form

$$V = \tfrac{1}{2}M\omega^2 R^2 + \Lambda' XYZ + \Lambda_1'' \Sigma X^4 + \Lambda_2'' \Sigma Y^2 Z^2 \tag{5.16}$$

which reflects the local tetrahedral symmetry. By measuring the $\Delta n = 1, 2, 3$

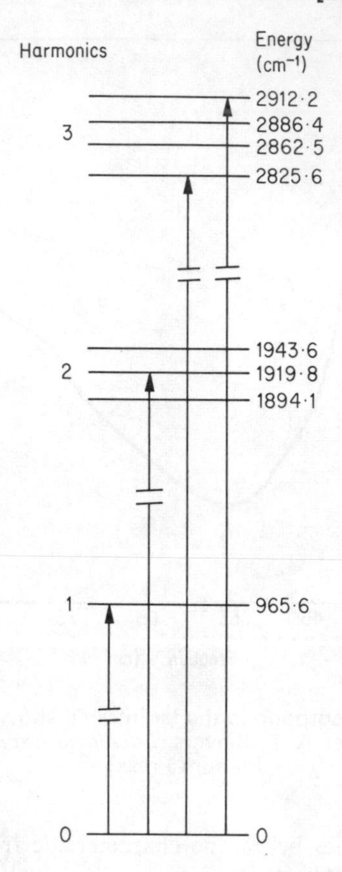

Figure 5.22 The vibrational energy levels of H^- ions in CaF_2 showing anharmonic effects. The arrows show allowed observed transitions (after R. J. Elliott *et al., Proc. Roy. Soc.* **A289**, 1 (1965))

transitions of this anharmonic oscillator the actual form of the potential well can be determined (see figure 5.22) as in molecular spectroscopy. Another example of local mode spectra is given in § 6.6.5.

Thus apart from demonstrating the general form of impurity excitations (as outlined in § 5.6), the measurement of localised and resonant phonon modes can give detailed information about force constants in crystals. As a technique it can be useful in identifying defects. Clusters of defects can give rise to complex spectra.

5.6.2 Impurity atomic states (AD)

(a) *Rare Earth Ion Impurities.* In an ionic or Van der Waals bonded crystal where the electrons retain their atomic character, the behaviour of electrons in impurity states is often also dominated by the character of the impurity atom. This is particularly true of the electrons in the partially filled f or d shells of rare earth or transition metal impurities since electrons in these levels are screened from the lattice by the outer electrons. The surrounding atoms will exert a small influence which can, to a good approximation, be described by an electric field—referred to as the crystal field.

The states of the free ions are usually described by a configuration f^n or d^n whose degenerate levels are split by the inter-electronic Coulomb interaction. If this is neglected the independent electron states have spherical symmetry. The uncertainty principle therefore allows precise determination of their angular momentum ($l = 2$ for d electrons, etc.). Various values of the l_z-component correspond to different orbitals which concentrate in different parts of space (figure 5.23). When the inter-electronic interaction is included

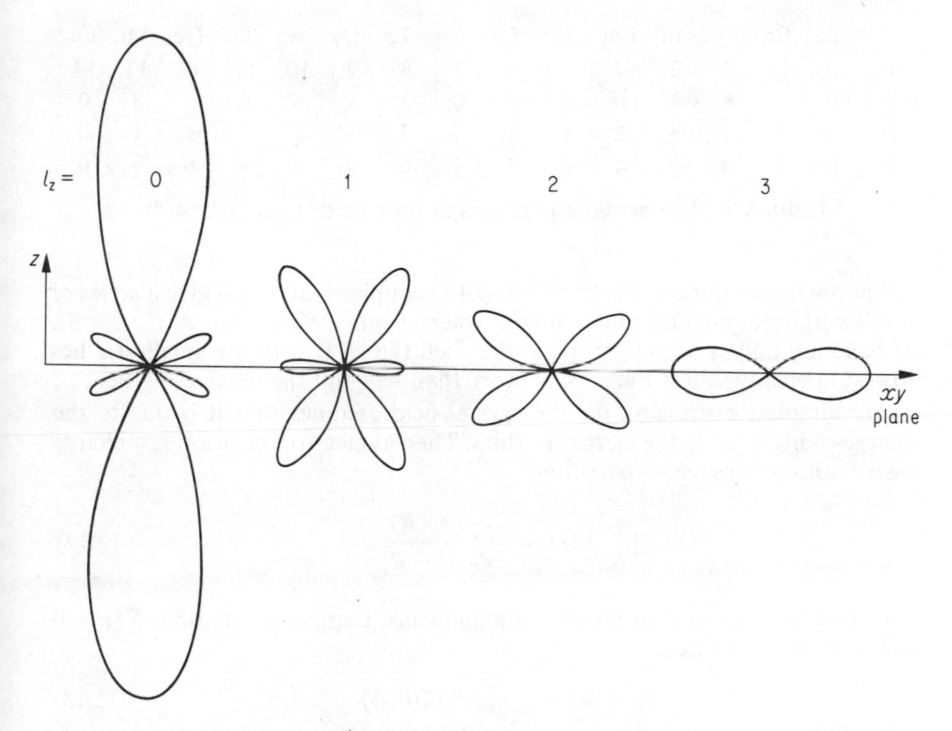

Figure 5.23 Angular dependence of the charge density $|\psi|^2$ for f orbital wave functions of different azimuthal quantum number l_z as a function of polar angle θ. For the various l_z values

$$\psi_0 \sim 5\cos^3\theta - 3\cos\theta$$
$$\psi_{\pm 1} \sim (5\cos^2\theta - 1)\sin\theta\exp(\pm i\phi)$$
$$\psi_{\pm 2} \sim \cos\theta\sin^2\theta\exp(\pm 2i\phi)$$
$$\psi_{\pm 3} \sim \sin^3\theta\exp(\pm 3i\phi)$$

The charge density $|\psi|^2$ is independent of ϕ and has circular symmetry about the $\theta = 0$ axis. For $l_z = 0$ it is concentrated in a pillar near that axis but for larger $|l_z|$ it is concentrated in annuli of large radius. As $|l_z|$ increases so does the circular current around these annuli the individual electron orbit no longer has spherical symmetry but depends on the position of the other electrons. The whole atom, however, may be rotated without energy change. In this situation the uncertainty principle ensures that the whole atom has a fixed value of total angular momentum L and total spin momentum S. The lowest energy state is one in which the electrons are kept apart in separate orbitals so as to minimise the repulsive energy. Electrons in different orbitals may all have the same spin component without violating the Pauli principle. The lowest state then obeys Hund's rule and has the maximum value of S and the maximum L for that S. The values of L and S for transition metal and rare earth ions are given in tables 5.1 and 5.2.

	La	Ce	Pr	Nd	Pm	Sm	Eu	Gd	Tb	Dy	Ho	Er	Tm	Yb	Lu
n	0	1	2	3	4	5	6	7	8	9	10	11	12	13	14
L	0	3	5	6	6	5	3	0	3	5	6	6	5	3	0
S	0	$\frac{1}{2}$	1	$\frac{3}{2}$	2	$\frac{5}{2}$	3	$\frac{7}{2}$	3	$\frac{5}{2}$	2	$\frac{3}{2}$	1	$\frac{1}{2}$	0
J	0	$\frac{5}{2}$	4	$\frac{9}{2}$	4	$\frac{5}{2}$	0	$\frac{7}{2}$	6	$\frac{15}{2}$	8	$\frac{15}{2}$	6	$\frac{7}{2}$	0

Table 5.1 Lowest Energy Levels of Rare Earth Ions (R^{3+} $4f^n$)

Spin-orbit coupling in the atom (cf. § 4.8) couples L and S to give a series of levels with total angular momentum J where $J = L + S, L + S - 1, \ldots |L - S|$. If there are only a few electrons in the shell the level with the smallest J lies lowest in energy, but if the shell is more than half-full the reverse is true.

The simplest expression for the crystal field assumes that it is due to the charges lying outside the electron orbits. Then at electron position $\boldsymbol{r}$, a charge distribution $\rho(\boldsymbol{R})$ gives a potential

$$V(\boldsymbol{r}) = -\sum \frac{e\rho(\boldsymbol{R})}{|\boldsymbol{r} - \boldsymbol{R}|} \tag{5.17}$$

This may be expanded in powers of r and since Laplace's equation, $\nabla^2 V = 0$ holds, it can be written

$$V(\boldsymbol{r}) = \sum_{n,m} A_n^m r^n y_n^m(\theta, \phi) \tag{5.18}$$

in terms of spherical harmonics $r^n y_n^m(\theta, \phi)$, for an electron position given by spherical polar coordinates r, θ, ϕ. Each harmonic gives a multipole of the electron distribution. For example, $n = 2$, $m = 0$ gives

$$r^2 y_2^0(\theta, \phi) = \tfrac{1}{2}(3z^2 - r^2) \tag{5.19}$$

the quadrupole moment of the electron cloud. This is acted on by the gradient of the electric field along the z-direction represented by A_2^0. Such a potential energy changes the energy of electrons in different orbitals. For example those of figure 5.23 have different quadrupole moments. If positive charges are placed on the z-axis they will have different energies, decreasing as l_z increases.

The crystal field thus breaks the degeneracy of the spherically symmetric states. The detailed behaviour depends on the size of the effect. If the crystal field energies are less than the spin-orbit energy as happens in rare earths there are groups of levels around each free ion J-level position. If the crystal field has low symmetry then states will be singly degenerate for even numbers of electrons and doubly degenerate for odd numbers. In higher symmetries, two- and three-fold states may occur with even numbers, four-fold with odd numbers.

Thus the energy level scheme for rare earth ions is largely similar in any host matrix. The positions of the main groups of levels are shown later (see figure 6.23). When studied spectroscopically, these groups are found to

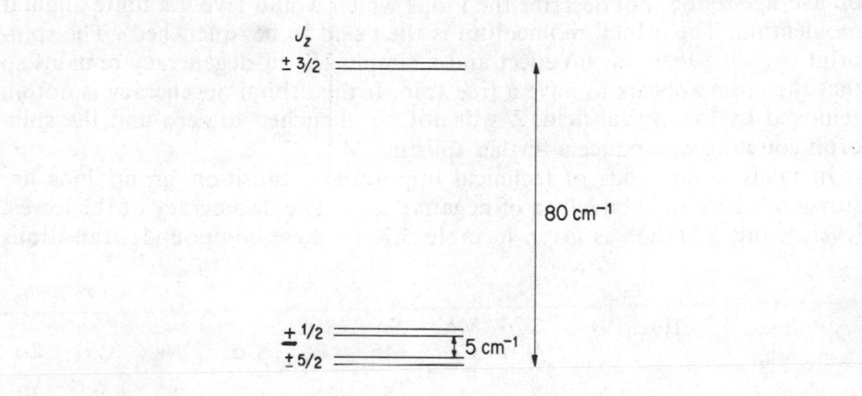

Figure 5.24 Low-lying energy levels of Ce^{3+} ions in cerium ethyl sulphate, showing crystal field splitting of the J manifold of states. The field has approximately axial symmetry and the levels are described by values of the axial component J_z

have crystal field splittings of around 10^{-2} eV (figure 5.24). The characteristic features of these ions is that they have many low-lying energy levels, the energy separation between many of them corresponding to photon energies in the visible. As a consequence doping with these impurities colours otherwise transparent crystals. The absorption and fluorescence emission is generally weak since the transitions are between levels of the same electronic configuration and are therefore forbidden. If, however, the impurity lies away from a centre of inversion symmetry the crystal field will weaken the selection rules (see § 6.2.3).

The sharp fluorescence lines associated with rare earth ion impurities make them of interest for solid state lasers (see Chapter 6). The impurity Nd^{3+} is particularly valuable. The Nd^{3+} lines near 1.16 eV (1.06 μm wavelength) show strong laser action and are at the same energy to within 2 or 3 per cent in host matrices as different as $CaWO_4$, glass and yttrium aluminium garnet (see figure 6.24).

(b) *Transition Metal Ions.* In transition metal ions the d electron shells lie farther toward the outside of the atom than do the f electrons in the rare earths. In consequence the crystal field has a larger effect. For most of these ions it is a good approximation to assume that the energy splitting associated with the crystal field is much smaller than the inter-electronic interaction but much larger than the spin-orbit coupling. The electric field acts on the orbital motion of the degenerate levels with a particular value of L. The spin motion remains unaffected and each level keeps a $(2S+1)$ degeneracy. If the symmetry is low enough the orbital levels will be singlets in which the average value of L is zero. The orbit of the electron is then so restricted by the electric fields that on average it does not describe the loops which would give it a finite angular momentum. The orbital momentum is then said to be 'quenched'. The spin-orbit coupling then has no effect and a simple $2S+1$ degeneracy remains so that the atom appears to have a free spin. If the orbital degeneracy is not all removed by the crystal field, L will not be quenched to zero and the spin-orbit coupling can induce a further splitting.

In many compounds of technical importance transition group ions are surrounded by an octahedron of negative ions. The degeneracy of the lowest level in this situation is given in table 5.2. In these compounds transitions

3+	Ti	V	Cr	Mn	Fe	Co				
2+			V	Cr	Mn	Fe	Co	Ni	Cu	Zn
n	1	2	3	4	5	6	7	8	9	10
L	2	3	3	2	0	2	3	3	2	0
S	$\frac{1}{2}$	1	$\frac{3}{2}$	2	$\frac{5}{2}$	2	$\frac{3}{2}$	1	$\frac{1}{2}$	0
Octahedral degeneracy	3	3	1	2	1	3	3	1	2	1

Table 5.2 Lowest Energy Levels of Transition Metal Ions ($3d^n$)

between different levels again often correspond to photon energies in the visible region. Transition group ions therefore colour crystals to give absorption and fluorescence (§ 6.7) in lines of well-defined frequency. Al_2O_3 with Cr^{3+} impurity (ruby) is a famous example whose absorption spectrum is shown in figure 5.25. The ground state is a singlet orbital state with four-fold spin degeneracy due to $S=\frac{3}{2}$. This is slightly split by the crystal field. The two strong lines then arise from transitions between the split ground state and an excited state. Thus Cr^{3+} inclusion gives Al_2O_3 (sapphire, which is transparent) the characteristic colour of ruby. Ruby was the first material in which laser action was observed and ruby rods can be made to 'lase' on both the strong lines of figure 5.25.

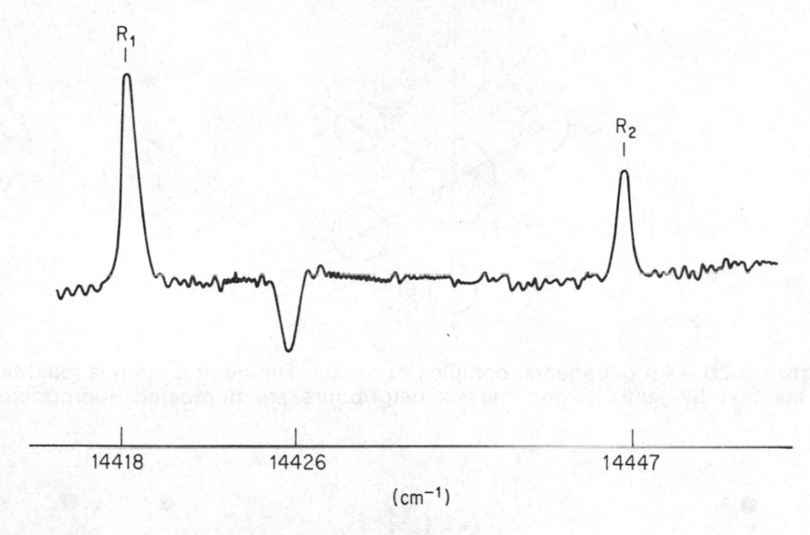

Figure 5.25 The R absorption lines in the spectrum of Cr^{3+} impurities in Al_2O_3 (ruby). (After S. Sugano and I. Tsujikawa, *J. Phys. Soc. Japan*, **13**, 899 (1958).)

5.6.3 Ligand field theory for impurities (A)

For electrons moving on the outside of impurity atoms or ions it is not satisfactory to represent the effect of the host lattice by an electric field. We may consider that this is because the effective crystal field becomes so large that the shapes of the electron orbits are modified. An alternative picture is that of an electron in a moderately large orbit spending an appreciable amount of time on the immediately neighbouring ions. Either way the electron orbit is modified significantly by the environment and indeed the change in orbit can often be detected experimentally by observation of the magnetic fields set up at neighbouring nuclei in a nuclear magnetic resonance spectrum. A defect centre of this type is best considered as a complex consisting of the impurity and its immediate neighbours. The electrons are shared between the atoms of the complex forming covalent bonds. Such bonds are called 'ligands':—hence the title. This situation is intermediate between the tightly bound electron states discussed in the last section and the very extended states discussed below.

We shall consider the example of the F-centre in alkali halides, an impurity which has been extensively studied. The results are typical of many more complex defects in such crystals. The F-centre consists of a negative ion vacancy in which an electron has been trapped to maintain charge neutrality (see § 5.2.1). There is no atom at the centre of the defect and no deep potential well to bind the electron. There are, however, six surrounding positive ions at the corners of an octahedron as in figure 5.26. These do have deep potential wells. We may approximate by considering the electron to move in this complex well and

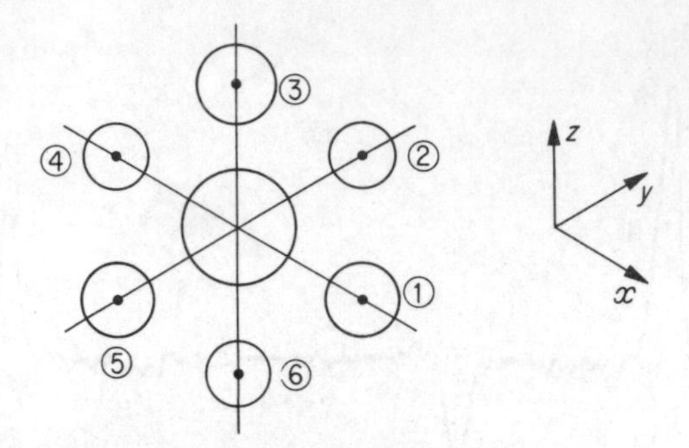

Figure 5.26 An octrahedral complex of atoms. The central atom is identified in the text by suffix 0 and the six neighbours are numbered appropriately

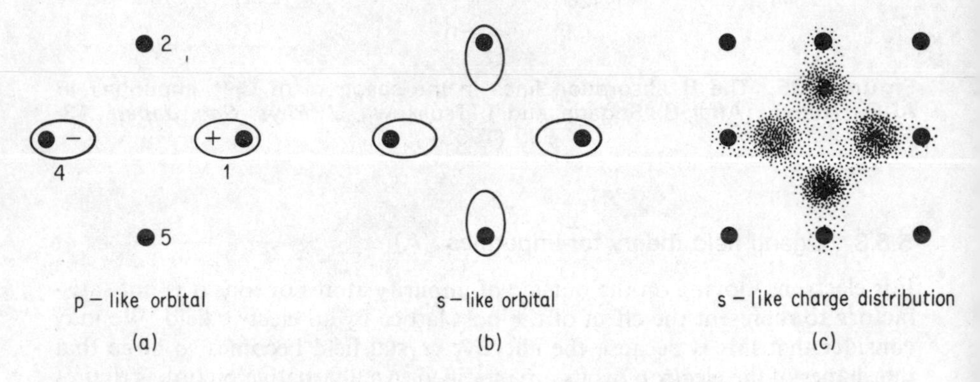

Figure 5.27 A section in the x–y plane (through atoms 1, 2, 4 and 5 of figure 5.26) of orbitals of the F centre. The central atom is now missing. Orbitals of p- and s-like symmetry are shown schematically in (a) and (b). The charge density of the latter is sketched in (c)

neglect the rest of the lattice. When near these atoms the electron wave function will be like those of the atom in question. These alkali ions have closed shells of atomic electrons and so this extra electron must occupy one of the next vacant states because of the Pauli principle. These are 3s and 3p states in Na, 4s and 4p in K and so on. Since the electron moves from atom to atom the total functions will be made up by combining the atomic functions on the different atoms. In addition the atomic functions will be distorted by the vacancy. This can be allowed for by taking on each atom a mixture of s and p states so that the charge density is greater in the vacancy than outside as shown in figure 5.27. For atom 1 of figure 5.26 this will have an axis in the x-direction, and a wave function

$$\psi_x(1) = \alpha\phi_s(1) + \beta\phi_x(1) \tag{5.20}$$

For atom 2 the axis will be in the y-direction and so on.

The combination of these states with lowest energy is one which gives the same weight to all six atoms.

$$\psi = \frac{1}{\sqrt{6}} [\psi_x(1) - \psi_x(4) + \psi_y(2) - \psi_y(5) + \psi_z(3) - \psi_z(6)] \qquad (5.21)$$

The charge density $|\psi|^2$ of such a state is shown schematically in figure 5.27. Excited states of the defect will be described by different combinations, for example

$$\psi = \frac{1}{\sqrt{2}} [\psi_x(1) + \psi_x(4)] \qquad (5.22)$$

which, because the x-, y- and z-directions are all equivalent, is degenerate with two similar states. The electric field of light waves will excite transitions between states like (5.21) and (5.22) which give a characteristic absorption which imparts a blue colour to alkali halides. The F absorption bands of several crystals are shown in figure 5.28. The most striking feature of the

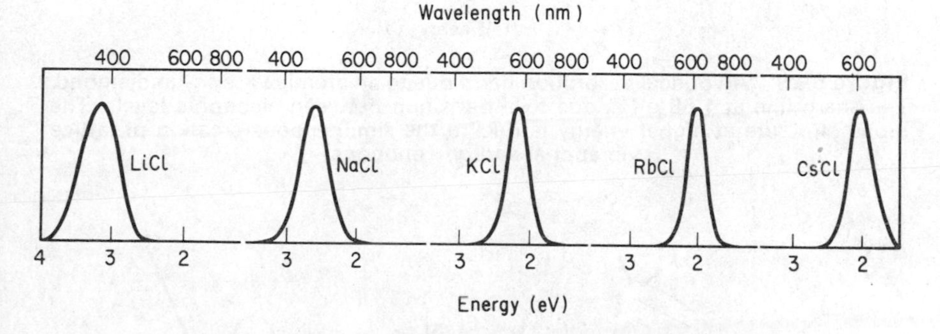

Figure 5.28 Absorption bands due to F centres in several alkali halides. Note that all are very broad. (After R. W. Pohl, *Proc. Phys. Soc.*, **49**, 3 (1937).)

absorption is the very considerable width in energy which decreases only slowly with decreasing temperature. This is due to a strong coupling to the lattice vibrations which will be discussed in § 5.8.

A vacancy in a covalently bonded crystal must also be thought of in terms of the electrons on the surrounding atoms. In the diamond structure each of the four neighbours is left with a spare electron on a bond 'dangling' into the vacancy. A considerable lattice distortion may be expected as the neighbours modify their positions to rearrange their bonds. The four electrons will have appreciable interaction and a fairly complex level scheme results. The optical absorption thought to arise from a vacancy in diamond is shown in figure 5.29.

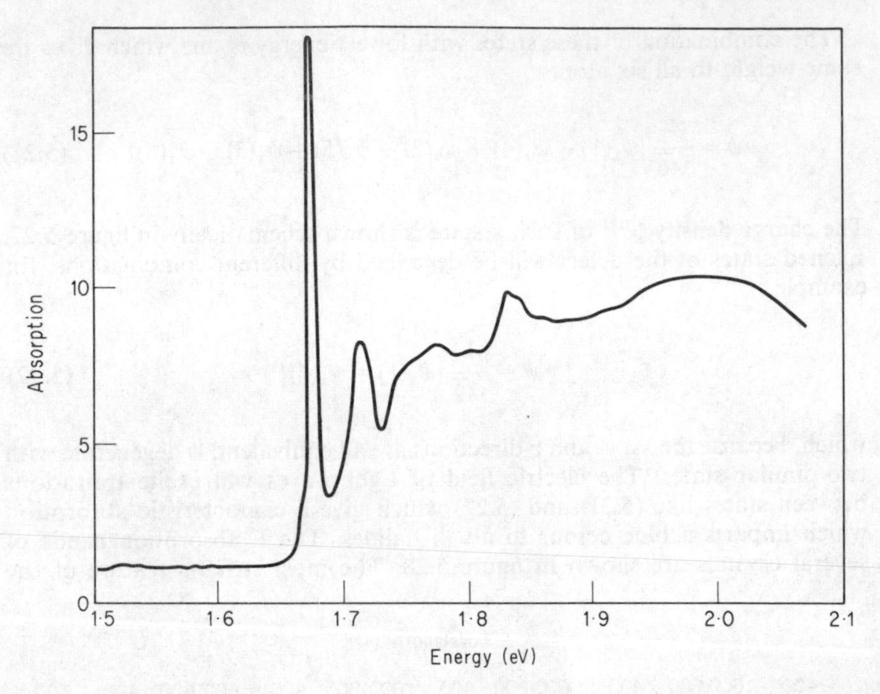

Figure 5.29 An optical absorption band due to an atomic vacancy in diamond. The sharp line at 1.68 eV is due to a transition between electronic levels. The broad structure at higher energy is due to the simultaneous creation of lattice vibrational energy (phonons)

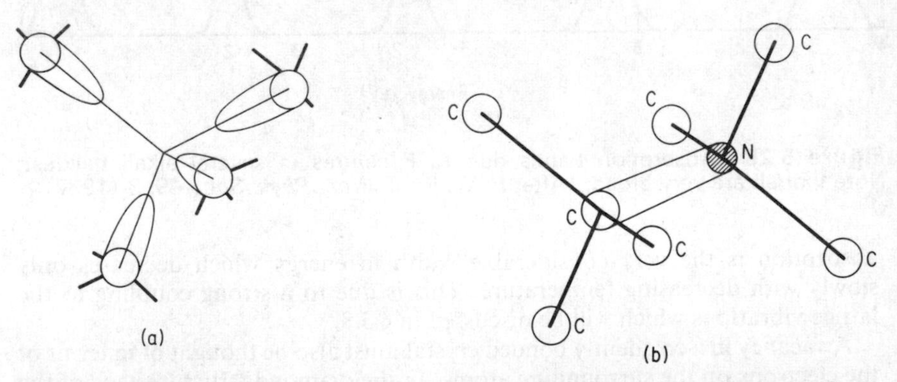

Figure 5.30 (a) Schematic representation of a vacancy in diamond. Each surrounding atom has three electrons shared in covalent bands with their neighbours and one electron which occupies an orbital pointing towards the vacant site. These four electrons in 'dangling' bands interact to give impurity levels
(b) Representation of N centre in diamond. The extra electron contributed by the N is found to be largely confined in an anti-bonding orbital with one neighbouring C atom

Another simple defect arises from nitrogen, a common impurity in diamond, which in some circumstances occurs as an isolated substitutional atom. The extra electron appears to share one bond with a single C neighbour, and to elongate this bond considerably (figure 5.30).

5.6.4 Aggregate centres in alkali halides (A)

The F band is by no means the only impurity-induced optical absorption band that can be observed in alkali halide crystals although the F centre is the simplest and serves as a model for more complex structures. The numerous other bands are each designated by a capital letter (sometimes with subscripts) which reflect either the discoverer's name or his whim. Thus the M band commemorates Mollwo. Other bands are designated K, L_1, L_2, L_3, F′, F_A, N, R_1, R_2, U, V_K, H, etc. Considerable effort has been devoted to elucidating the origin of these bands but some controversial points still remain.

The K, L_1, L_2 and L_3 bands, which are observed at low temperature, are associated with transitions to progressively higher excited states of the F centre. For example, in KBr the F centre transition energies are

F	K	L_1	L_2	L_3
2.07	2.36	3.31	3.93	4.5 eV

In the higher excited states the orbit of the electron extends over many lattice ions and the hydrogenic model of the impurity centre, to be discussed in the next section in respect of impurities in semiconductors, becomes applicable. A highly excited F centre is easily ionised thermally, when the electron escapes from the field of the vacancy, and this is why the K and L bands can be observed only at low temperatures.

The F′ band—a very broad absorption band on the low-energy side of the F band—is due to the capture, by an F centre, of a second electron. Naturally this second electron is less strongly bound than the first. The F′ band is formed at room temperature during prolonged illumination in the F band, which decreases in strength ('bleaches', see § 5.2.2) at the same time. The mechanism of formation is then clearly the excitation of F centres, so that electrons are ionised into the conduction band and eventually captured by other unionised F centres. Illumination in the F′ band reverses the process.

Surprisingly, the positive hole equivalent of an F centre does not appear to exist. A hole trapped on a halogen ion is apparently shared preferentially with a neighbouring halogen ion to give a negatively charged halogen molecule. This entity gives rise to the V_K band.

The remaining bands are probably due to aggregates or clusters of defects of the type referred to in § 5.1. Two adjacent negative-ion vacancies with two trapped electrons form an M centre, and three such vacancies an R centre. An F_A centre is a vacancy adjacent to an impurity. More complex centres certainly form but their characteristic absorption remains to be identified with certainty. Well-established models for some centres are illustrated in figure 5.31.

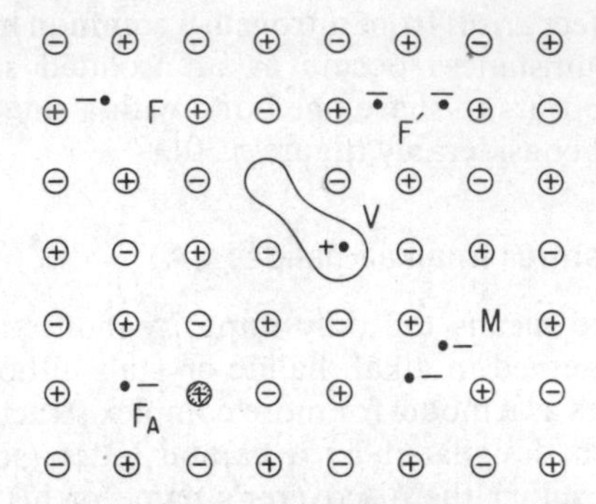

Figure 5.31 Models of some centres in alkali halides. Electrons are marked by •– and holes by+•. F denotes an electron trapped at a negative ion vacancy. F_A a similar centre when one neighbour is an impurity (shaded). F′ has two electrons at a negative ion vacancy. M has two electrons at two adjacent negative ion vacancies. V has a hole shared between two adjacent halides—like a Cl_2^+ molecule

Finally, we recall (§ 5.6.2) that impurities in alkali halides show essentially their atomic spectra unless they have electrons in large orbits. A famous example is the U centre formed by an interstitial hydrogen atom. In KBr, for example, the main absorption line corresponds to a 1s to 2p transition and occurs at 5.5 eV. In free atomic hydrogen it is at 10.2 eV. The large difference is due to the spread of the electrons on to the surrounding anions.

5.6.5 Impurity centres in covalent semiconductors (AD)

(a) *Shallow Donors and Acceptors*. In § 5.3 we described how simple donor and acceptor centres formed from Group V and Group III impurities could raise the electron or hole concentration in covalent semiconductors. We noted that the electron (hole) binding energies were small and the centres thermally ionised above about 20 K. In the conduction band of these materials electrons move very readily from atom to atom. The small binding energy indicates that the impurity states spread widely in the crystal. A trapped electron then spends very little time near the impurity and the atomic character of the impurity is of small importance. It is then possible to treat the impurity as producing a small perturbation of the lattice states and use the so-called *effective mass approximation* which leads to a hydrogen-like centre.

In the presence of a small perturbation the electron wave functions can be written

$$\psi = \sum_n F(\boldsymbol{R}_n)\,\phi(\boldsymbol{r} - \boldsymbol{R}_n) \tag{5.23}$$

i.e. as a linear combination of the atomic-like functions appropriate to this band. For the free electron case $F(\boldsymbol{R}_n) = \exp(i\boldsymbol{k}.\boldsymbol{R}_n)$ as in § 4. 8 but here it will peak about the impurity site. $F(\mathbf{R}_n)$ in equation (5.23) represents the large-scale motion of the electron through the medium. For a simple band this may be regarded, following (4.33), as similar to the motion of an electron in a hydrogen atom except that here it has an effective mass m^* while the medium has a high dielectric constant ε due to the polarisability of the atoms. The latter reduces the Coulomb force and $F(\mathbf{R}_n)$ is approximately the solution of a Schroedinger-type equation

$$\left(-\frac{\hbar^2 \nabla^2}{2m^*} - \frac{e^2}{4\pi\varepsilon r}\right) F(\mathbf{R}_n) = EF(\mathbf{R}_n) \tag{5.24}$$

The centre therefore has the characteristics of a hydrogen atom with a fundamental 'Bohr radius' given by

$$\frac{1}{a_I} = \frac{e^2 m^*}{4\pi\varepsilon\hbar^2} = \frac{1}{a_0}\left(\frac{m^*}{m}\right)\left(\frac{\varepsilon_0}{\varepsilon}\right) \tag{5.25}$$

where a_0 is the Bohr radius of hydrogen. The binding energy, equivalent to the ionisation energy of hydrogen is given by

$$(e^2/2a_I)(\varepsilon_0/\varepsilon) = E_{\mathrm{H}}\left(\frac{m^*}{m}\right)\left(\frac{\varepsilon_0}{\varepsilon}\right)^2 \tag{5.26}$$

where E_{H} is 13.54 eV.

For $(m^*/m) \sim 0.1$ and $(\varepsilon/\varepsilon_0) \sim 10$, values which are approximately appropriate for germanium and silicon, the orbit radius is about 25 lattice spacings so the impurity centre does spread widely in the crystal. The atomic character of the impurity is not apparent and the effect of crystal structure is only represented through m^* and ε. (cf. figure 5.32). We expect the optical absorption due to an electron (hole) trapped at such an impurity to consist, like the spectrum of atomic hydrogen, of a series of lines corresponding to transitions into excited states with a series limit at the binding energy given by (5.26). An example of this is given by the absorption of arsenic donors in germanium at 4 K which is shown in figure 5.33.

On the basis of the above model the ionisation energies of all shallow donors should be the same in a given host lattice. Using the known values of m^* and ε for germanium the donor ionisation energy should be about 0.01 eV. Experimentally observed values are

As	0.014 eV
Sb	0.0098 eV
P	0.0128 eV

the small differences being largely due to the atomic character of the impurity which the model neglects. A similar model is used for Wannier excitons in § 6.5.1.

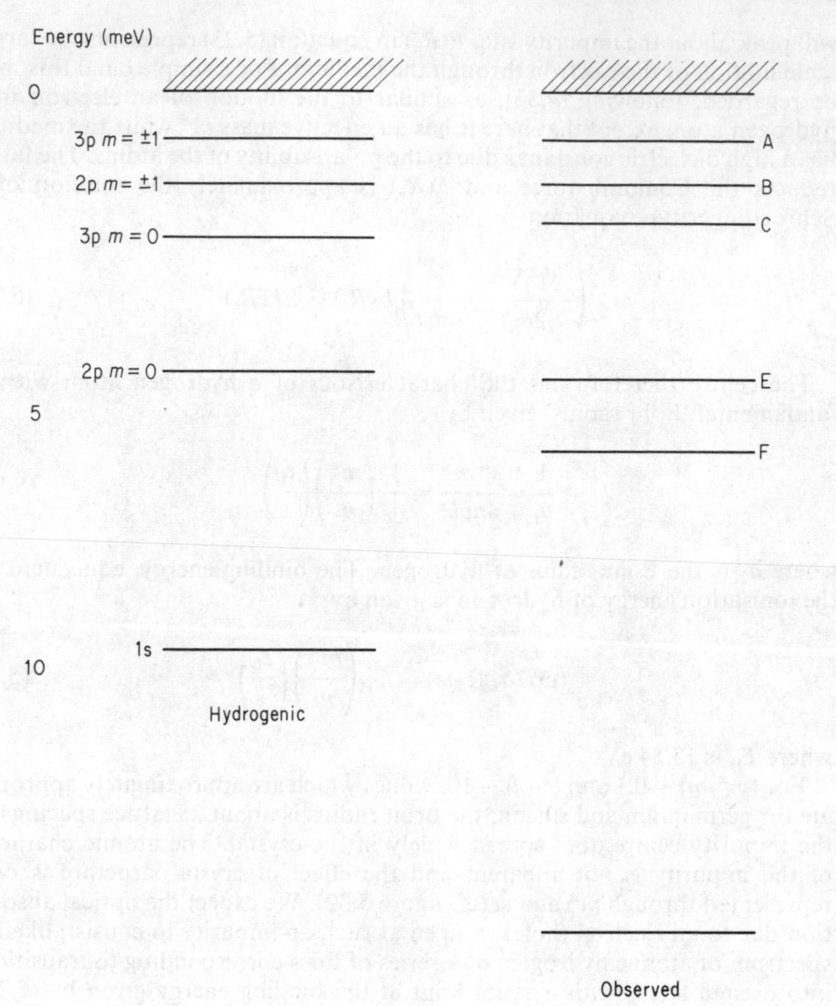

Figure 5.32 Hydrogenic energy levels of a donor compared with the levels observed for As in Ge (see figure 5.33). Transitions from ground to various excited states are identified by letters

In germanium and silicon a further complication arises from the multi-valley nature of the conduction bands (§ 4.5). In Si for example there are six impurity states observed from the six minima. These interact so that the final pattern of levels observed (figure 5.34) is not precisely hydrogenic. Similarly for the hole levels the degeneracy of the valence band (cf. § 4.5.4) modifies the result.

(b) *Deep Levels in Semiconductors*. Impurities from groups other than III and V produce energy levels in the energy gap between valence and conduction bands. These levels are often well separated from the band edges and the

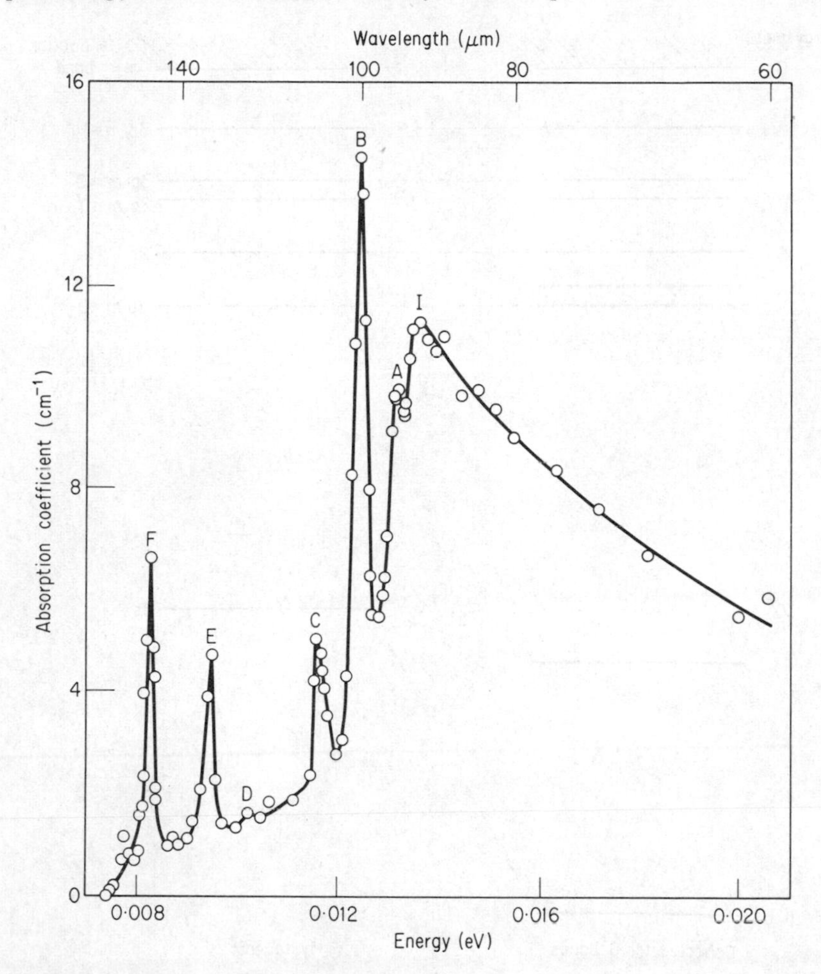

Figure 5.33 Observed absorption spectrum from As impurities in Ge. Transitions are between levels shown in Figure 5.32 (after H. Y. Fan and P. Fisher, *J. Phys. Chem. Solids*, **8**, 270 (1959))

electron orbits are smaller so that the preceding theory is less satisfactory. In some cases an approach similar to that of § 5.6.3 provides a better description. The most carefully studied are the divalent transition elements Mn, Fe, Co, Zn and Cd. They all behave as elements from Group II and act as double acceptors. Thus they can trap two holes. The first of these is always tightly bound. Then the centre attracts a further hole rather weakly. For Zn and Cd the level sequence of this second hole is not unlike a simple acceptor centre; but the other elements with d electrons always form deep levels. Cu in Ge is a triple acceptor and it behaves like a Group I element. The final hole is only weakly bound.

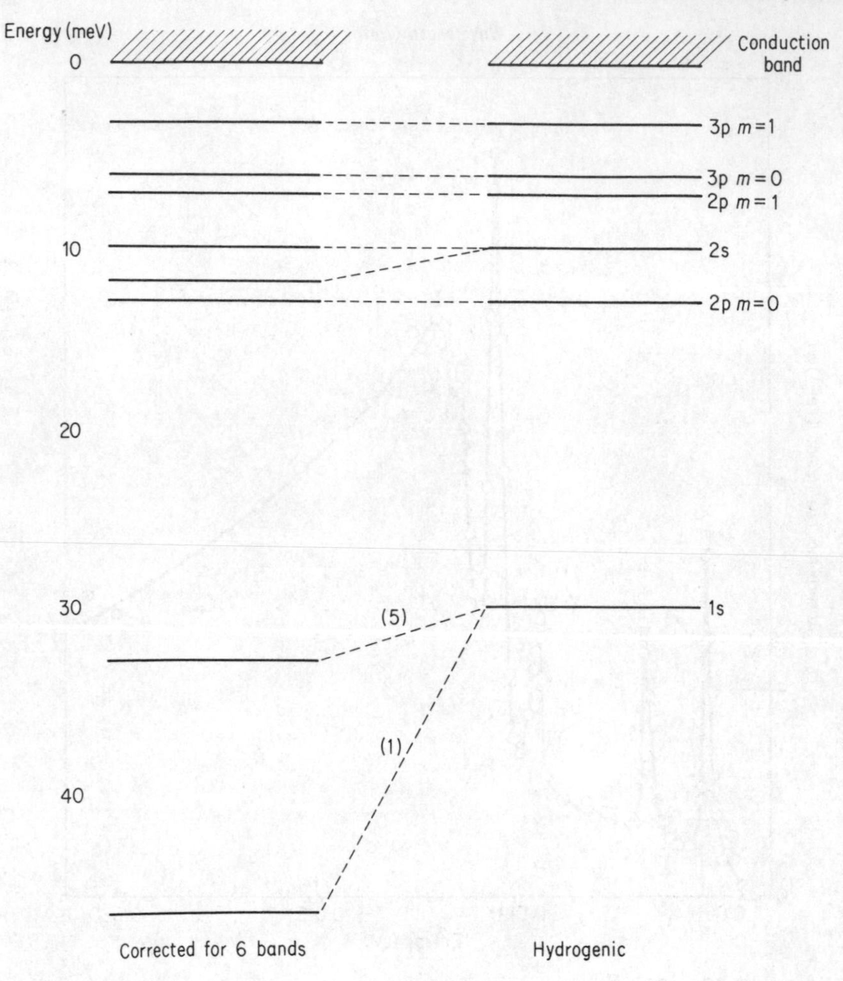

Figure 5.34 Energy levels for P impurities in Si calculated for a simple hydrogenic model but including the effects which arise because the conduction band has six valleys

It is possible for some elements to act as donors and as acceptors—Au in Ge is an example. It can trap up to three holes with varying energy and can also contribute an electron to the conduction band from a very deep level. Examples of these energies are given in figure 5.35.

Illumination by photons of energy equal to or greater than the impurity ionisation energy frees an electron (or hole) which contributes to conduction. This phenomenon is called photoconductivity and impurity-induced photoconductivity forms the basis of many infrared detectors. Clearly the detector must be operated at a temperature low enough for the electrons (holes) to be trapped in the dark, so the operating temperature is reduced as the ionisation

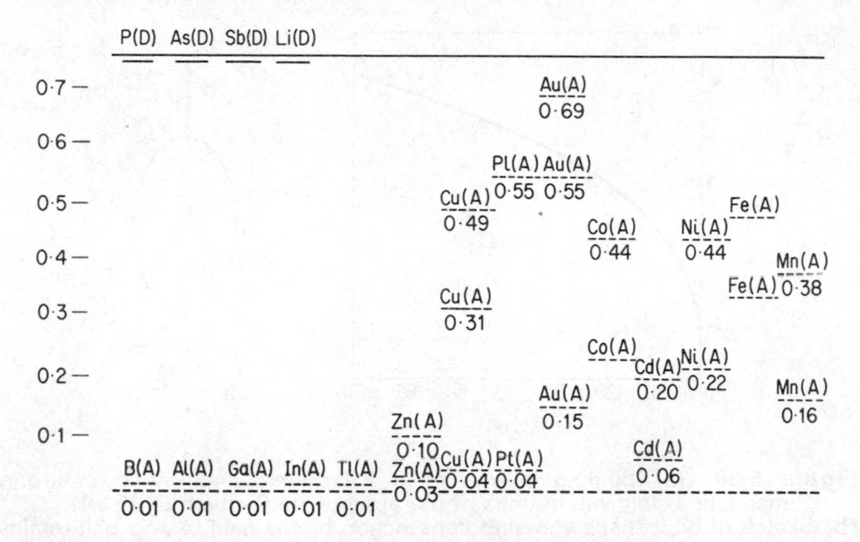

Figure 5.35 Energy levels of various impurities giving deep levels in Ge. Accepter levels (A) are measured from valence band edge, donor levels (D) from conduction band edge

energy decreases and the long wavelength limit of detection increases. If a photoresponse at very long wavelengths is not required the inconvenience of operation at, say, liquid helium temperature can be avoided by the use of deep level impurities. Thus germanium doped with Cu or Zn is used in detectors for use in the near infrared region (>0.1 eV), the state of ionisation of the Cu or Zn being adjusted by simultaneous doping with a shallow donor (e.g. As). At the other extreme donors due to Te impurity in InSb are very shallow ($\sim 10^{-4}$ eV) since $(m^*/m) \sim 0.01$ and $(\varepsilon/\varepsilon_0) \sim 25$. No significant binding of electrons to the donor centres occurs even at 1 K. The binding energy can, however, be increased by applying a magnetic field. This reduces the orbit by constraining it in the plane perpendicular to the field to have a radius given in § 4.7. For large enough magnetic fields this is smaller than the Bohr radius (5.25) (see figure 5.36). This procedure enables an impurity-type photoconductive detector to be made which operates at wavelengths as long as 1 mm (see also § 8.7.2).

Finally, we note that deep levels in semiconductors increase the recombination rate of electrons and holes if both are present simultaneously. Equally they assist generation since the equilibrium density is independent of the rates and the two must be equal under equilibrium conditions. The actual rates are, however, important when the equilibrium electron and hole densities are disturbed by external forces, as they are in many semiconductor devices (see Chapter 9). The two-stage process of electron (hole) capture at a deep level followed by hole (electron) capture is considerably more likely than direct interband recombination, particularly if the level involved is near the Fermi level and approximately 50 per cent occupied.

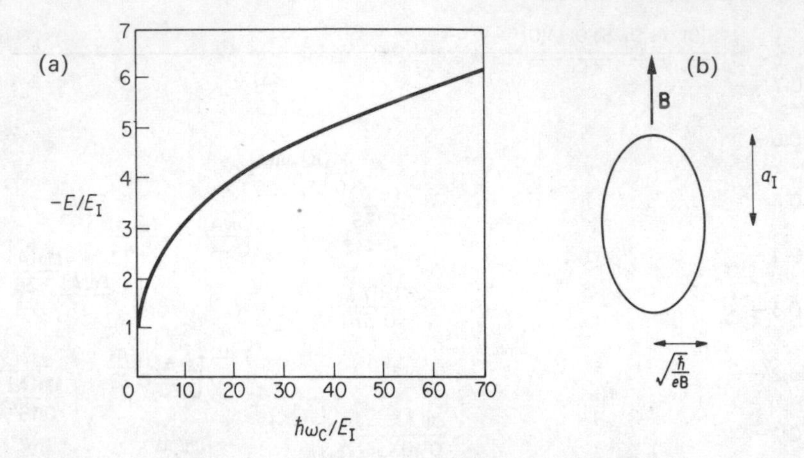

Figure 5.36 (a) Binding energy of a hydrogenic impurity in a strong magnetic field given in units of the appropriate Rydberg E_I (5.26)
(b) Sketch of orbit shape showing constriction by the field. Along **B** it retains roughly the same radius as the unperturbed hydrogen atom (a_I given by 5.25)) but perpendicular to **B** it is more constricted

5.6.6 Impurity centres in metals (A)

An impurity, vacancy or interstitial in a metal will change the wave function of all the electrons locally. We saw in § 4.9, however, that because of the electron–electron interaction the electric field of such an impurity will be modified, particularly at long distances. For example, if a divalent atom like Zn is added to a monovalent nearly-free electron metal like Cu it contributes an extra electron and an extra positive charge. However, because of the screening effect of the other electrons the effective potential may be expected to be of shorter range than a Coulomb potential. The approximate theory of § 4.9 gave

$$V(r) = \frac{e^2}{4\pi\varepsilon r} \exp(-\kappa r) \tag{5.27}$$

The effect of such a short-range potential is to modify each wave function in the band as discussed in § 5.6. It may or may not give a bound state. The occupation of the modified states is of course determined by the Fermi–Dirac distribution. The charge density in the region near each impurity will also be modified by the existence of these states. Freidel has stressed the important point, which he used to obtain a self-consistent expression for the effective potential (e.g. to determine κ in (5.27)), that since this is a local potential the local excess charge density must satisfy charge neutrality and completely screen out the impurity charge in a finite distance. Moreover the electron states will be unaffected at large distances from the impurity—in particular k_F, the value of k at the Fermi level, will be unchanged.

In fact the extra local charge density will be an oscillating function of distance since it is compiled from occupied states with $k < k_F$. To prove this we assume a weak potential $V(r)$ with a Fourier transform

$$V(\boldsymbol{q}) = (\mathcal{N}\Omega)^{-1/2}\int V(\boldsymbol{r})\exp(\mathrm{i}\boldsymbol{q}.\boldsymbol{r})\mathrm{d}\boldsymbol{r} \tag{5.28}$$

Using perturbation theory this mixes states $\boldsymbol{k}$ and $\boldsymbol{k}+\boldsymbol{q}$ into the form

$$\psi(r) = (\mathcal{N}\Omega)^{-1/2}\left\{\exp(\mathrm{i}\boldsymbol{k}.\boldsymbol{r}) + \sum_{\boldsymbol{q}} \alpha_k(\boldsymbol{q})\exp[\mathrm{i}(\boldsymbol{k}+\boldsymbol{q}).\boldsymbol{r}]\right\} \tag{5.29}$$

where

$$\alpha_k(\boldsymbol{q}) = \left(\frac{1}{\mathcal{N}\Omega}\right)\frac{V(\boldsymbol{q})}{E(\boldsymbol{k}) - E(\boldsymbol{k}+\boldsymbol{q})} \tag{5.30}$$

The charge density, $e|\Psi(r)|^2$, is perturbed by

$$\sum_{\boldsymbol{q}}\left(\frac{1}{\mathcal{N}\Omega}\right)\alpha_k(\boldsymbol{q})\exp(\mathrm{i}\boldsymbol{q}.\boldsymbol{r}) \tag{5.31}$$

Summing over all states in the Fermi distribution ($k < k_F$) gives a total change in the charge density

$$\rho(r) = \frac{1}{(2\pi)^6}\int_{k<k_F}\mathrm{d}\boldsymbol{k}\int\frac{V(\boldsymbol{q})\exp(\mathrm{i}\boldsymbol{q}.\boldsymbol{r})}{E(\boldsymbol{k}) - E(\boldsymbol{k}+\boldsymbol{q})}\mathrm{d}\boldsymbol{q} \tag{5.32}$$

The sum over $\boldsymbol{k}$ gives

$$\frac{2}{3E_F}\left\{1 + \frac{4k_F^2 - q^2}{4k_F q}\log\left|\frac{2k_F - q}{2k_F + q}\right|\right\} \tag{5.33}$$

This has infinite slope at $2k_F$ as shown in figure 5.37. If we choose a simple form for the potential with a very small range $\mathrm{V}(\boldsymbol{q}) = \mathrm{V}$ independent of $\boldsymbol{q}$ the integral (5.32) gives

$$\rho(R) = \frac{\mathrm{V}}{E_F}\left(\frac{\sin 2k_F R}{2k_F R} - \cos 2k_F R\right)\frac{1}{(2k_F R)^3} \tag{5.34}$$

which falls off as R^{-3} and oscillates, reflecting the singularity in (5.33) at $2k_F$. The form of (5.34) is shown in figure 5.38. The oscillations in the charge density about an impurity can be inferred from nuclear resonance experiments.

Impurities with unfilled d shells are of particular interest in deriving the magnetic properties of metals. As we saw in band structures like those illustrated in figure 4.15 the d levels lie in the middle of the s band. Thus the d

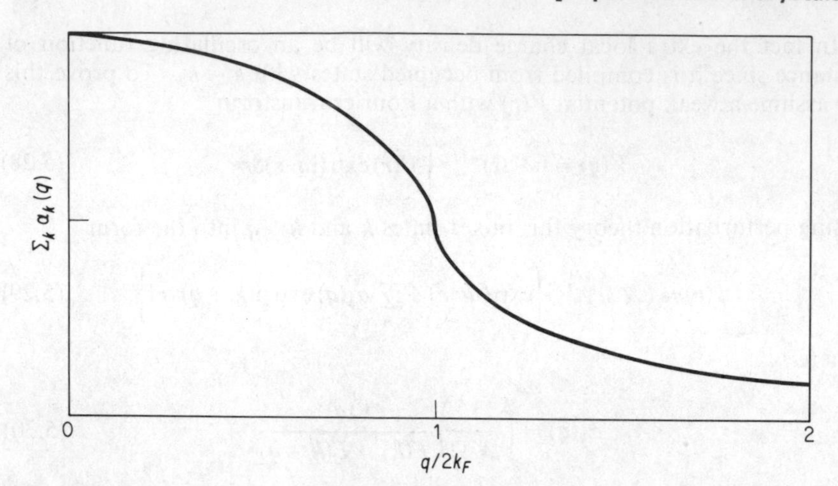

Figure 5.37 Variation of the function $\Sigma_k\ \alpha_k(\boldsymbol{q})$ defined by (5.33) with q. The strong variation at $q = 2k_F$ reflects the cut-off in the Fermi–Dirac distribution of electrons

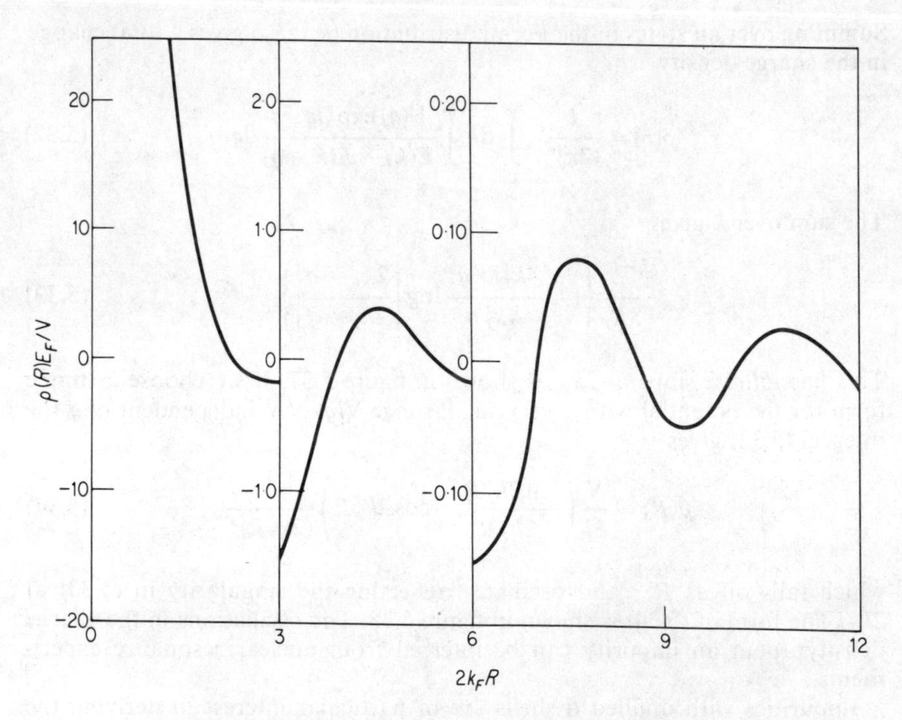

Figure 5.38 Radial electron density in a metal due to an impurity. The oscillations arise because of the cut-off in the Fermi–Dirac distribution

electrons will not be completely localised but will mix with the s levels in a resonance. Here, however, the density of states due to the d level is high over a very narrow band, and for many properties can be regarded as effectively localised. Such levels are sometimes referred to as quasi-localised. Thus Mn in Cu exhibits magnetic properties not unlike those of a Mn ion with an effectively localised magnetic moment. If a resonance lies close to the Fermi level there is greatly increased scattering of the conduction electrons.

5.7 Alloys (A)

Alloys can be considered as metals containing substantial concentrations of metallic impurity. The crystal excitations can be expected to be significantly modified and even the crystal structure may vary.

The simplest situation occurs with metals that are adequately described by the free-electron model. With these metals the lattice has little influence on the electronic band structure and hence the electron states should be little influenced by a random array of impurities. In fact it is found that, provided the atomic sizes are compatible, such metals form alloys over the whole composition range. The chief effect of alloying is to increase or decrease the Fermi energy as the electron density is increased or decreased. In addition the ease of movement of the free electrons is reduced due to increased scattering, as discussed in Chapter 8.

In many alloys of this type, however, the crystal structure is observed to change with composition. Moreover, ordered structures form at particular compositions which are given by a set of empirical rules first enunciated by Hume-Rothery. These ordered structures can now be explained as follows. It was shown in Chapter 4 that the energy states near a zone boundary are split, the lower ones decreasing and the upper increasing in energy to form an energy gap. If the Fermi energy lies in the gap all the occupied states are lowered and hence so is the total energy. Consequently the crystal tends to take up a crystal structure which makes the Fermi surface approach the zone boundary thus minimising the total energy. When the Fermi surface touches the zone boundary the number of electrons per unit cell, ν, is 1.36 for face-centred cubic structures, 1.48 for body-centred and 1.69 for hexagonal close-packed structures. For example Cu is face-centred cubic with $\nu \sim 1$. As Zn is added ν increases and the body-centred cubic phase of CuZn alloys is observed to occur near the 50:50 concentration ($\nu = 1.5$). Around 20:80 ($\nu = 1.8$) the hexagonal close packed phase is observed. The ordered alloys Cu_3Al, AgZn, Cu_5Sn all form in body-centred cubic structures and all have $\nu = 1.5$.

More complicated effects occur if there is a substantial change in the potential energy of electrons near an impurity. At low concentrations localised bound states may appear. As the concentration of impurity increases these states broaden to form what may be called an impurity band. Consequently in a binary alloy there may be two separate conduction bands, one associated with the host and the other with the impurity. Whether or not they are separated in energy depends on the difference in electron potential energy, $|E_A - E_B|$, when the electron is on the two species of atom. If E_A and E_B are sufficiently different two bands are formed at all concentrations, as shown in figure 5.39. If the difference is smaller, separate impurity bands form at low concentrations of both constituents but a single broad band is found around the 50:50 concentration. For very small values of $|E_A - E_B|$ no separate impurity band appears.

Even if impurity bands do appear, all the energy bands retain a finite width and energy gaps remain. If $|E_A - E_B|$ is small the energy band widths are not much larger in alloys than in pure crystals. Thus semiconductor alloys such as Si–Ge and GaAs–GaP remain semiconductors with well-defined energy gaps.

Finally we note that the phonon bands of alloys show similar properties, as shown in figure 5.40 for a simple alloy with two constituents of significantly

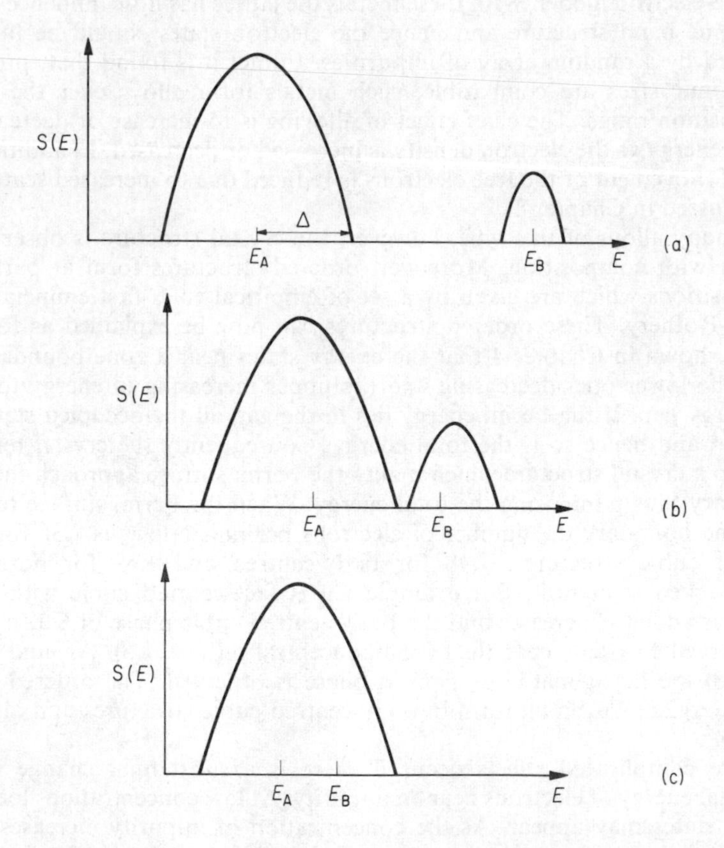

Figure 5.39 Schematic density of electron states in an A–B random alloy with 85% of A atoms. (a) $E_A - E_B \gg \Delta$, (b) $E_A - E_B \sim \Delta$, (c) $E_A - E_B \ll \Delta$ where E_A, E_B are the atomic energy levels and Δ the band width

different mass. At low concentrations of the light mass constituent new, localised, high energy states are formed as described in § 5.6.1. As the concentration increases a phonon impurity band forms. At the 50:50 concentration there is a single broad band and finally at low concentrations of the heavy mass constituent a resonance is seen at low frequencies.

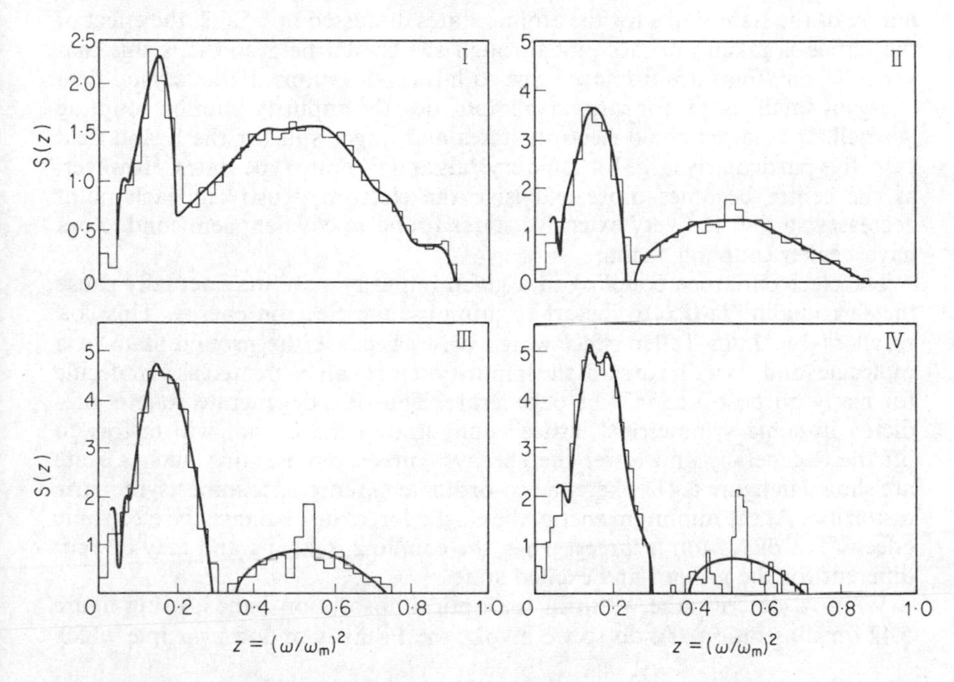

Figure 5.40 Density of phonon states as calculated for a simple model of an alloy of heavy A atoms and light B atoms (where $M_A = 3M_B$) by two different methods (curve and histogram respectively) I 25% A shows a resonance due to the A atoms. II 50% A, III 75% A. IV 86% A shows an impurity band due to the B atoms. (After D. W. Tayler, *Phys. Rev.*, **156**, 1017 (1971).)

5.8 The coupling of impurity states to the lattice (AD)

We have seen that the optical absorption and emission spectra due to electronic transitions between bound impurity states consist of narrow lines if the levels are well shielded from the lattice (e.g. rare earth impurities) but the lines become broader as the bound electron orbits increase in size. The absorption of F centres is characterised by a very broad band. For the very large electron orbits associated with shallow donors and acceptors in semiconductors, however, the absorption bands become narrower again. We shall now argue that this result reflects the degree of coupling between the impurity states and the vibrations of the lattice. We first discuss the magnitude of the impurity-phonon coupling and secondly show how it leads to spectral broadening.

Within the adiabatic approximation (§ 2.1) the energy of electronic states depends on the positions and relative displacements of the atoms. For a state localised on an impurity the energy depends on the local atomic displacements in much the same way as it does in a small molecule and line splitting due to vibrational motion is familiar in molecular spectroscopy. The extent to which the energy of a state varies with local atomic displacement depends on the nature of the state. Thus for the atomic states discussed in § 5.6.2, the effect of the lattice is taken into account through the crystal field and it is this that varies when atoms are displaced due to lattice vibrations. If the crystal field energy is small, as it is for rare earth impurities, the impurity-phonon coupling is small. It is larger for d electron states and larger still for the ligand field case. It is particularly large for ionic crystals and F centre type states. However, as the centre becomes more extensive the electron density at each point decreases so that the very extensive states found in covalent semiconductors have weaker coupling again.

The electron-lattice coupling in a given impurity state may actually cause the surrounding lattice to distort to minimise the electron energy. This is a result of the Jahn–Teller effect which arises because the ground state in a molecule (and, as we have said, the impurity centre can be treated as a molecule for many purposes) cannot be degenerate. Thus if a degenerate state is predicted in some symmetrical lattice configuration the crystal will distort to lift the degeneracy and lower the energy. Curves representing such a state are shown in figure 5.41, where the co-ordinate Q represents some asymmetric distortion. At the minimum energy the elastic forces just balance the electronic effects. The distortion is largest when the coupling is strong and may be very different for the ground and excited states.

We now describe the width of the optical absorption bands, using figure 5.42 for illustration. To do so we invoke the Frank–Condon principle which

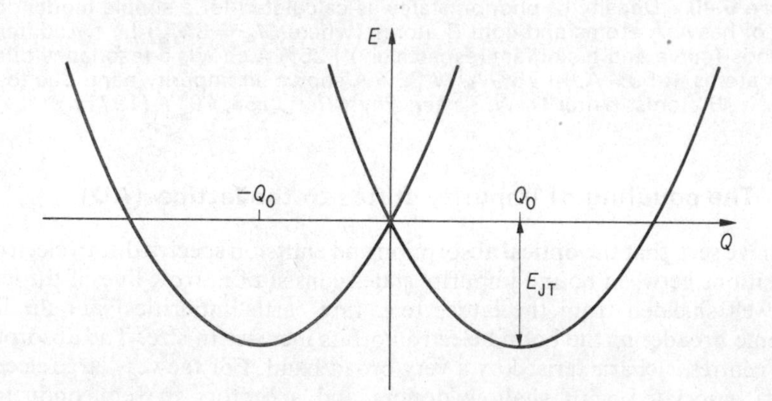

Figure 5.41 Diagram showing splitting of a degenerate electronic state by lattice displacement: the Jahn–Teller effect. The vibrations of the lattice take place around a different equilibrium distortion $\pm Q_0$ depending on the electronic state. The energy gained by distortion is E_{JT}

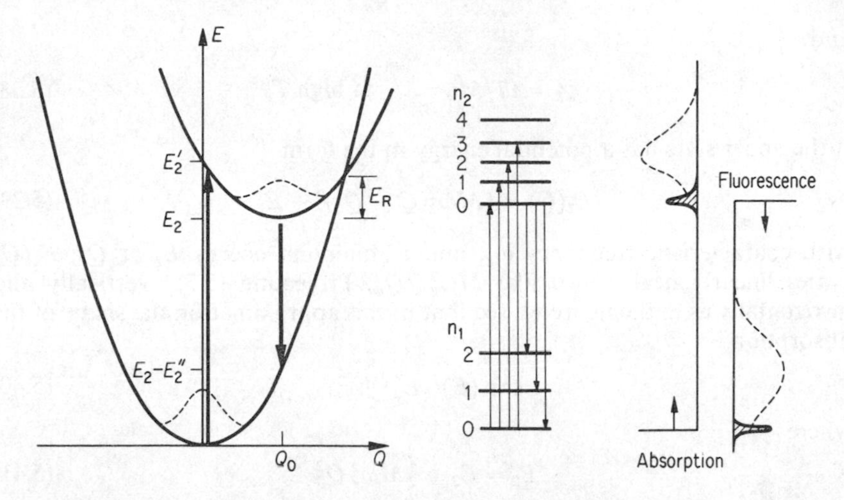

Figure 5.42 Illustration of absorption and emission of light from an electronic centre in a solid assuming the Frank–Condon principle (see text). In the ground state the system has a distribution of Q values (shown dashed) because of zero-point motion of the lattice. Absorptive transitions take place vertically and hence have a distribution of energy centred about E_2' as shown on right. Fluorescent transitions are also distributed about E_2''. If the vibrations are quantised the transitions take place between n_1 and n_2. The zero phonon line $n_1 = 0 \rightleftarrows n_2 = 0$ (shown shaded) occurs at the same energy in absorption and fluorescence

states that because the motion of atoms is so sluggish an optical transition takes place instantaneously by comparison, i.e. the atoms are stationary and Q does not change during the transition. Hence optical transitions are vertical in figure 5.42 and it is only after a transition is completed that the lattice distorts so as to minimise the electron energy. Inspection of figure 5.42 then shows that the broad absorption spectrum arises as follows. When the electron is in the ground state there is a distribution of Q values around the minimum of the curve. Over this range of Q values the excited state energy changes substantially and there is a correspondingly large range of energies over which a transition can take place, so the absorption band will be broad. To put this argument on to a more quantitative basis we note that in the ground state the system will have a Gaussian distribution over the motion of Q, like

$$\exp(-Q^2/\xi_T^2) \tag{5.35}$$

If the potential energy curve is written

$$E_1(Q) = \tfrac{1}{2}M\omega_1^2 Q^2 \tag{5.36}$$

where M is the atomic mass and ω_1 the characterististic frequency then the mean square displacement of the oscillator is

$$\xi_0^2 = \tfrac{1}{2}\hbar/M\omega_1 \quad \text{at } T=0 \tag{5.37}$$

and

$$\xi_T^2 = kT/M\omega^2 \qquad \text{at high } T \tag{5.38}$$

If the upper state has a potential energy in the form

$$E_2(Q) = \tfrac{1}{2}M\omega_2^2(Q - Q_0)^2 + E_2 \tag{5.39}$$

with characteristic frequency ω_2, and a minimum energy E_2 at Q_0, $E_2(Q)$ varies linearly near $Q = 0$ like $M\omega_2^2 QQ_0$. Projecting (5.35) vertically and horizontally as in the figure we see that in this approximation the shape of the absorption is

$$\exp(E - E_2')^2/\Delta^2 \tag{5.40}$$

where

$$E_2' = E_2 + \tfrac{1}{2}M\omega_2^2 Q_0^2 \tag{5.41}$$

and

$$\Delta^2 = M^2\omega_2^4 Q_0^2 \xi_T^2 \tag{5.42}$$

The width increases with the displacement Q_0.

Actually the vibrational energy will be quantised. Considering only a single mode described by the variable Q, the quantised levels are as shown in figure 5.42. Transitions will take place from the lowest level to each of the upper levels with an intensity which depends on the range of Q values covered by the motion in each level. Assuming that in the upper levels the most probable value of Q is at the turning points of the oscillator but that the lower level has the distribution (5.35) the envelope of the absorption will still be given by (5.40). The lowest energy transition to the lowest of the upper levels is called the zero phonon line and has an intensity $\exp(-Q_0/\xi_0)^2$. The most intense line will be that corresponding to the maximum of (5.40) when

$$n_2\hbar\omega_2 = \tfrac{1}{2}M\omega_2^2 Q_0^2 \tag{5.43}$$

In fact a real lattice has a spectrum of vibrational frequencies and each mode will couple differently to the defect. The zero-phonon line will remain sharp, but the others will be broad and reflect the density of states. For a simple atomic centre n_2 will be small, but as the coupling increases and Q_0 increases n_2 will grow. For the absorption curve of the centre in diamond shown in figure 5.29, $n_2 \sim 2$. For the F centre in alkali halides $n_2 \sim 30$, and the absorption shown in figure 5.28 (§ 5.6.3) is very broad. No zero-phonon line is visible.

Electronic transitions accompanied by optical fluorescent emission can take place from the minimum of the upper curve. Classically the fluorescence band, too, is a Gaussian centred at

$$E_2'' = E_2 - \tfrac{1}{2}M\omega_1^2 Q_0^2 \tag{5.44}$$

If Q_0 and hence n_2 are small the strong zero-phonon line will be at almost the same energy in both absorption and emission (cf. figure 5.42). The rest of the fluorescence emission will be at lower frequencies. If Q_0 is large as in an F centre the shift in energy between the maximum of the fluorescence and the maximum in the absorption is enormous.

It may happen that curves representing the ground and excited states of a centre actually cross, as indicated in figure 5.42. If so an excited electron can return to the ground state without light emission, the energy simply being given off in a series of vibrational quanta. This radiationless emission is thought to occur in some centres. For an electron in an excited state thermal energy is required to allow it to reach the crossover point so the efficiency of fluorescence might be expected to decrease at high temperatures and be of the form

$$\eta = \frac{1}{1 + \text{const.} \exp(-E_R/kT)} \tag{5.45}$$

where E_R is the energy difference between the excited state minimum and the crossover point. Figure 5.42 provides a model for the F centre and many commercially important luminescent materials based on impurity doped ionic crystals such as ZnS. We discuss such materials in § 6.7. The efficiency of luminescence in these materials certainly falls off at high temperatures in the manner indicated above but there is positive evidence that for some of these at least the mechanism is not that just described.

Finally we note a further consequence of the Frank–Condon principle which is evident from figure 5.42. If the energy separation between ground and excited states is measured by thermal activation the observed energy difference is the separation between the two minima. This is always less than the photon energy at peak absorption but greater than the fluorescence emission peak energy.

PROBLEMS

(Answers to some problems are given on page 481.)

5.1 The elastic displacement round an isolated dislocation of Burger's vector $\boldsymbol{b}$ is approximately $\boldsymbol{b}\theta/2\pi$, where θ is an angular co-ordinate around the dislocation line. Show that two parallel dislocations with Burger's vectors $\boldsymbol{b}_1$ and $\boldsymbol{b}_2$, a distance r apart in a crystal of shear modulus μ repel each other with a force per unit length of approximately $\mu b_1 b_2/2\pi r$.

5.2 A fine X-ray beam 2×10^{-3} cm in diameter is used to obtain a diffraction pattern from a single crystal of copper. From the broadening of the spots it can be deduced that over the area of the beam there is a change in lattice orientation of 3 minutes of arc. Assuming this change is due to a homogeneous array of edge dislocations of Burger's vector 2.5×10^{-8} cm, estimate the minimum dislocation density.

5.3 Describe the pair of dislocations which together are equivalent to

(i) a line of vacancies,
(ii) a line of interstitial atoms.

5.4 Estimate the binding energy of donors in InSb ($m^*/m = 0.01$, $\varepsilon/\varepsilon_0 = 25$ in zero applied magnetic field). Using Figure 5.36, estimate the binding energy with B = 1 tesla.

5.5 A semiconductor of energy gap $E_g = 0.76$ eV and an intrinsic density n_i of 3×10^{13} cm^{-3} at 300 K contains 2×10^{16} donor impurities per cc of binding energy 0.01 eV. The acceptor density is negligible. Estimate

(i) the temperature at which there are 1×10^{16} cm^{-3} electrons in the conduction band;
(ii) the position of the Fermi level at that temperature;
(iii) the electron density at 300 K;
(iv) the temperature at which the electron density reaches 3×10^{16} cm^{-3}.

5.6 From the positions of the low-lying energy levels in figure 6.23 estimate the spin-orbit coupling constant for the rare-earth ions Nd^{3+}, Eu^{3+}, Dy^{3+} and Yb^{3+} and show that it varies smoothly along the series.

5.7 In KCl the F absorption band is centred at 2.3 eV (photon energy) and has a width of 0.16 eV at 4 K. At the same temperature the fluorescence emission is centred at 1.2 eV and has a width of 0.26 eV. From these data estimate the parameters describing the model in figure 5.42. What is the average number of phonons created with this F centre absorption?

At 300 K the width of the absorption band is 0.36 eV. Is this compatible with the simple model?

5.8 The octahedral complex shown in figure 5.26 has a transition metal ion at the centre. The possible d electron states have wave functions in the form $\psi_{xy} \sim xy\mathcal{f}(r)$, $\psi_{yz} \sim yz\mathcal{f}(r)$, $\psi_{zx} \sim zx\mathcal{f}(r)$ and $\psi_{x^2-y^2} \sim (x^2 - y^2)\mathcal{f}(r)$, $\psi_{3z^2-r^2} = (3z^2 - r^2)\mathcal{f}(r)$. By considering a polar diagram show that the first three point their charge clouds between the surrounding ions while the last two point towards them.

Show that the mixture of p states on the surrounding atoms with the same symmetry as ψ_{xy} is

$$\Phi_y(1) - \Phi_y(4) + \Phi_x(2) - \Phi_x(3)$$

and find the combinations with the symmetry of the other functions.

In such complexes the six states like ψ_{xy} (each has a double spin degeneracy) have lower energy and fill with electrons first. Show that the effective spin value expected for $[MCl]^{2-}$ complexes where M represents the 5d transitions elements Re, Ir, Pt is $S = \frac{3}{2}$, $\frac{1}{2}$ and 0, respectively.

6

The optical properties of solids

In the preceding chapter we described how the occurrence of lattice defects or the inclusion of impurities in a solid can give rise to optical absorption. The origin of such absorption is either the excitation of vibrational modes or the promotion of electrons from a given set of energy states to others of higher energy. Analogous effects occur in perfect crystals. Simply because the density of impurities and defects in a crystal is usually small compared with the density of lattice atoms the most prominent features of the absorption spectrum of a solid are characteristic of the host lattice and impurity absorption bands are usually observable only if they occur in those spectral regions in which the host material is transparent. The emission of light from solids, on the other hand, is usually dominated by impurities even when their density is as low as 1 in 10^6. Thus a solid can absorb photons of one energy due to electronic transitions characteristic of the host lattice and in consequence emit photons of a different (usually lower) energy corresponding to a transition between impurity energy levels. A description of light emission processes in solids follows the sections on absorption.

The term 'optical properties' usually implies that the dimensions of the sample under examination are large compared with the wavelength. The mean free path of photons can then be large and their wave vector well defined. Hence in an interaction with a non-localised excitation of the crystal the wave vector conservation rule applies, as explained in § 2.6. We shall use this rule extensively. Arbitrarily we have chosen a frequency of 3×10^{11} Hz, corresponding to a wavelength of 1 mm (photon energy 1.23×10^{-3} eV) as the lower limit of the optical frequency range. Phenomena normally studied at lower frequencies are discussed in later chapters. At the high frequency end of the spectrum technical difficulties become severe in the ultra violet above about 3×10^{15} Hz (100 nm wavelength, photon energy 12.3 eV) and this provides an effective upper limit. This wavelength is, however, still large compared with atomic dimensions which allows the optical properties of a solid to be described in terms of macroscopic 'optical constants' (the refractive index and the extinction coefficient) which are, of course, not really constants at all as they vary with frequency.

The optical properties of solids are of considerable technological importance. One of the many striking developments of the last two decades has been the extension of optical techniques and relatively high resolution spectroscopy throughout the spectral range from visible light to microwaves. Solid state physics has both contributed to, and profited from, this development. The main contribution has been in the field of photoconductive detectors: semiconducting elements that decrease in resistance on illumination. We have seen in § 5.6.5 that photoconductivity can arise from the promotion, on absorption of a photon, of an electron from a bound state of an impurity to the conduction band where it is free to move through the lattice and carry a current. Photoconductivity can also result from the promotion of electrons from a filled energy band to the conduction band, two mobile charges (an electron and a hole) being produced per photon in this case. The spectral response of a photoconductive detector is an optical property of the solid, being determined by the photon energy necessary to promote an electron. However, the magnitude of the resultant photocurrent and the rate at which it can respond to changes in illumination level depend respectively on the ease of electron (hole) movement through band states and the probability of electron recapture by an impurity or hole. These aspects are discussed in subsequent chapters (particularly Chapter 9). At this stage we simply note that by suitable choice of doping impurity, fast and sensitive detectors can now be constructed for use at any desired wavelength throughout the infrared spectral region. Light detectors and light emitters are the most important optical solid state devices but solids also provide filters, polarisers, modulators, non-reciprocal devices, frequency changers, etc., not to mention the more familiar lenses and prisms. Solids exhibit a wealth of optical phenomena, the study of which gives information on the fundamental properties of the material under examination. The purpose of this chapter is to describe these phenomena and their origin, with particular reference to those of technical importance.

Before discussing the observed optical properties of solids we shall describe briefly how such properties are measured and interpreted. Fuller accounts of these aspects can be found in standard textbooks of optics, electromagnetic theory and quantum mechanics. In § 6.1 the macroscopic optical constants of a solid are described and § 6.2 shows how these can be related to microscopic quantities. § 6.3, to which the reader familiar with the theory of optics may turn immediately, summarises the various optical absorption processes that occur in solids. These are described in more detail in §§ 6.4, 6.5 and 6.6. In § 6.7 we consider the emission of light from solids. Finally, in § 6.8 we discuss the interaction of light with light, coupled through a solid as intermediary.

6.1 The measurement of the optical properties of a solid

6.1.1 The definition of optical constants (a)

Macroscopically, the optical properties of any medium can be fully specified by two quantities: the refractive index n_0 and the extinction coefficient k_0. It is useful to use both these quantities although they are not independent (see § 6.1.3). They are defined by a set of equations of the form

$$F_y = F_{y_0} \exp\left\{-i\omega\left[t - \frac{z}{c}(n_0 + ik_0)\right]\right\} \tag{6.1}$$

which describes the electric field component in the y-direction of a transverse electromagnetic wave propagating in the z-direction through a medium described by n_0 and k_0. In (6.1) ω is the angular frequency of the radiation and c is the velocity of light *in vacuo*. Similar expressions apply to the other components of the electromagnetic wave, but only in an isotropic crystal are the values of n_0 and k_0 independent of the direction of polarisation with respect to the axes of the crystal.

6.1.2 Reflection and absorption coefficients (a)

Experimentally, n_0 and k_0 are usually determined indirectly although the refractive index of transparent media ($n_0 \gg k_0$) can be measured directly by fabricating a prism or by interferometry. A more widely applicable technique is to measure simultaneously the transmission and reflection of a parallel sided slab of material. The radiation intensity, I, which is proportional to the square of the amplitude, falls off with distance through the sample as

$$I = I(0)\exp(-Kz) \tag{6.2}$$

where K is called the absorption coefficient. Comparison of (6.2) with (6.1) shows that

$$K = \frac{2k_0\omega}{c} = \frac{4\pi k_0}{\lambda} \tag{6.3}$$

Equation (6.2) does not fully describe the intensity of radiation transmitted through a sample of finite thickness $\mathscr{L}$ in the z-direction. The effects of reflection at the incident and exit surfaces cannot be neglected. At a boundary between vacuum and a medium specified by n_0 and k_0, and at normal incidence, the intensity reflection coefficient, $\mathscr{R}$, is given by

$$\mathscr{R} = \frac{(n_0 - 1)^2 + k_0^2}{(n_0 + 1)^2 + k_0^2} \tag{6.4}$$

and reflection is accompanied by a change in phase given by

$$\tan\theta = \frac{2k_0}{n_0^2 + k_0^2 - 1} \tag{6.5}$$

Equations (6.4) and (6.5) are derived in most textbooks of optics from classical wave theory. An alternative approach which uses our discussion of wave vector conservation in Chapter 2, is to note that the velocity of the wave in the medium is $c/(n_0 + ik_0)$ and use the fact that the total photon momentum must be conserved at the interface. If unit photon flux is incident on the sample, a fraction $\mathscr{R}'$ reflected and $(1 - \mathscr{R}')$ transmitted, the momentum balance gives

$$1 = -\mathscr{R}' + (1 - \mathscr{R}')(n_0 + ik_0)$$

from which

$$\mathscr{R}' = \frac{n_0^2 + k_0^2 + 2ik_0 - 1}{(n_0 + 1)^2 + k_0^2}$$

where $\mathscr{R}'$ is the amplitude reflection coefficient. From a statistical point of view the power reflection coefficient $\mathscr{R}$ is the probability of finding a reflected photon and so is equal to the product of $\mathscr{R}'$ and its complex conjugate (just as $\psi\psi^*$ gives electron density). Equation (6.5) is then the ratio of the real and imaginary terms while (6.4) is the sum of the squares of the real and imaginary parts. It should be noted that $\tan\theta$ is zero if k_0 is either zero or infinity. These solutions correspond to $\theta = 0$ and π respectively, on reflection from a non-absorbing medium and on reflection from a highly absorbing material such as a metal.

Equations (6.4) and (6.5) describe the reflection from a real sample only if reflection from the exit surface can be neglected. In general this must be included, together with multiple reflections inside the sample. The results are quite complicated. (See J. A. Stratton, *Electromagnetic Theory*, McGraw-Hill (1941).) In practice, however, it is often permissible to neglect multiple reflections, particularly if n_0 and k_0 are varying rapidly with frequency and all that is required is an indication of the frequency at which absorption maxima occur. Multiple reflections can be ignored if

$$\mathscr{R}\exp(-K\mathscr{L}) \ll 1$$

and

$$\sin^2 \theta \ll 1$$

where $\mathscr{L}$ is the sample thickness. If these approximations are valid the incident, transmitted and reflected intensities, $I(0)$, $I(T)$ and $I(R)$, are given by

$$I(T)/I(0) = (1 - \mathscr{R})^2 \exp(-K\mathscr{L}) \tag{6.6}$$

and

$$I(R)/I(0) = \mathscr{R} \tag{6.7}$$

Experimentally, simultaneous measurement of transmission and reflection is the most common (and most reliable) method of measuring the optical constants of a material. Care must obviously be taken in reflection measurements to ensure that the surfaces are clean and their optical properties characteristic of the bulk material.

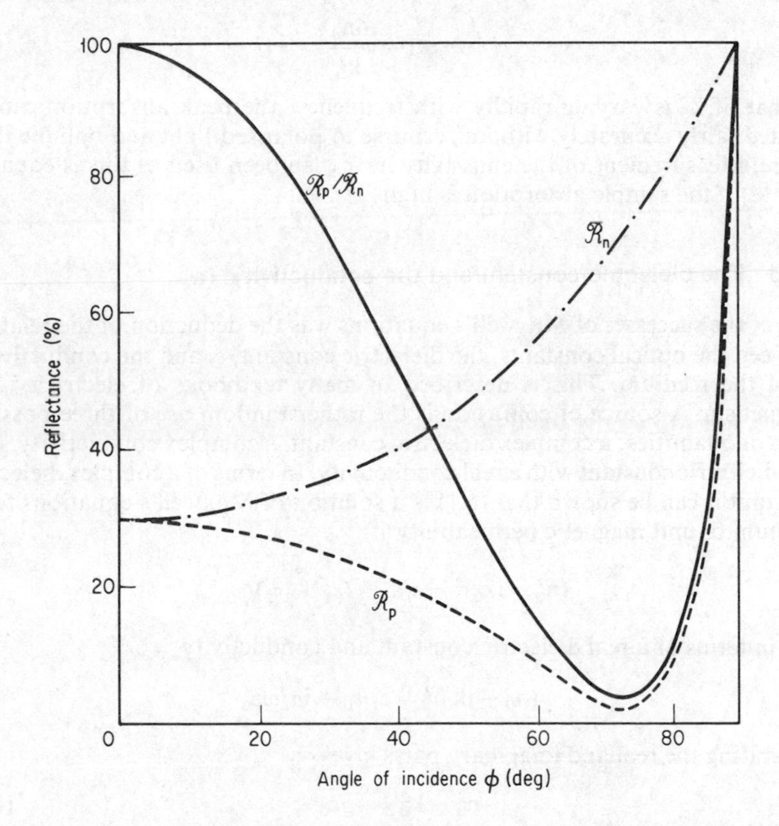

Figure 6.1 Theoretical reflectance curves for $n_0 = 3$, $k_0 = 1$. (After T. S. Moss, *The Optical Properties of Semiconductors*, Butterworth.)

If the absorption is very high it may be impossible to detect any transmitted radiation. The absorption coefficient of solids can equal or even exceed 10^6 cm^{-1} in some frequency ranges and it is more or less impossible to polish bulk samples down to much less than 10^{-4} cm thick. Some solids, notably the alkali halides, can be prepared as thin crystals by vacuum evaporation on to a heated substrate of another alkali halide crystal. In general, however, only reflectivity measurements can be made if K is very large. Reflection at normal incidence (6.4) gives only one piece of information, and two are needed. A solution is to measure the reflectivity of polarised light at non-normal incidence. The relevant equations are illustrated in figure 6.1. $\mathscr{R}_n$ is the reflectivity for a wave with its electric vector in the plane of the reflecting surface, and $\mathscr{R}_p$ refers to the orthogonally polarised component. ϕ is the angle of incidence relative to the normal. If $k_0 = 0$, $\mathscr{R}_p$ passes through zero at $\tan\phi = n_0$ (Brewster's angle). It will be seen that n_0 and k_0 can be deduced either from measurements of $\mathscr{R}_n$ and $\mathscr{R}_p$ at one angle of incidence or from their ratio at two angles.

It is possible to use simple normal incidence reflection if the absorption is sufficiently high for $k_0^2 \gg (n_0 + 1)^2$. Equation (6.4) can then be expanded to give

$$\mathscr{R} \approx 1 - \frac{4n_0}{k_0^2} \tag{6.8}$$

so that, if k_0 is varying rapidly with frequency, the peak absorption can be located fairly accurately without recourse to polarised light and oblique incidence. Measurement of the emissivity have also been used as this is equal to $(1 - \mathscr{R})$ if the sample absorption is high.

6.1.3 The dielectric constant and the conductivity (a)

One of the successes of Maxwell's equations was the deduction of the relation between the optical constants, the dielectric constant, ε, and the conductivity, σ, of the medium. This is described in many textbooks of electricity and magnetism. A source of confusion is the rather random use of three possible pairs of quantities; a complex dielectric constant, a complex conductivity, or a real dielectric constant with a real conductivity. In terms of a complex dielectric constant it can be shown that (6.1) is a solution of Maxwell's equations for a medium of unit magnetic permeability if

$$(n_0 + ik_0)^2 = \varepsilon/\varepsilon_0 = (\varepsilon_1 + i\varepsilon_2)/\varepsilon_0$$

and in terms of a real dielectric constant and conductivity

$$(n_0 + ik_0)^2 = \varepsilon_1/\varepsilon_0 + i\sigma/\omega\varepsilon_0$$

Separating the real and imaginary parts gives

$$n_0^2 - k_0^2 = \varepsilon_1/\varepsilon_0 \tag{6.9}$$

$$2n_0 k_0 = \varepsilon_2/\varepsilon_0 = \sigma/\omega\varepsilon_0 \tag{6.10}$$

The constants in these equations are defined by

$$\varepsilon_1 = D/F \qquad \sigma = J/F$$

where D is the electric displacement, F is the electric field, J is the current density and ε_0 is the permittivity of free space. It is convenient to use ε_2 rather than σ when the absorption process is not due to mobile electrons since the use of σ tends to imply the possibility of d.c. conduction. In metals, on the other hand, the use of ε_1 and σ is appropriate when absorption is due to the promotion of mobile electrons to states above the Fermi level.

ε_1 and ε_2 (or σ) determine the magnitudes of the currents in and out of phase with the electric field. They are therefore related quantities and through them n_0 and k_0 can be related. The exact relationship between ε_1 and ε_2 depends on the nature of the absorbing centres in the medium. There are, however, two perfectly general relationships, called the Kramers–Kronig relations, which are valid under all conditions. These are derived in many textbooks of optics and are

$$\varepsilon_1(\omega) = 1 + \frac{2}{\pi} \mathcal{P} \int \frac{\omega' \, \varepsilon_2(\omega) \, d\omega'}{\omega'^2 - \omega^2} \tag{6.11}$$

and

$$\varepsilon_2(\omega) = -\frac{2\omega}{\pi} \mathcal{P} \int \frac{\varepsilon_1(\omega') \, d\omega'}{\omega'^2 - \omega^2} \tag{6.12}$$

where $\mathcal{P}$ indicates the Cauchy principal value of the integral (there is a singularity at $\omega = \omega'$ which is omitted from the integration). Thus, in principle at least, ε_1 can always be obtained if ε_2 is known over a sufficiently wide (ideally infinite) frequency range. The reverse is also true. A 'sufficiently wide frequency range' is difficult to define but it must cover the feature of the spectrum of interest and often one decade of frequency is sufficient, particularly if the asymptotic values on each side of the feature can be deduced from a knowledge of the physical processes involved.

6.2 Absorption, emission and dispersion

6.2.1 Classical dispersion theory (a)

In the classical theory of the dielectric constant electrons bound to nuclei are assumed to behave like harmonic oscillators. The force acting on an electron due to a radiation field is

$$e(\mathbf{F}(t) + \boldsymbol{v} \times \mathbf{B}(t))$$

where $\boldsymbol{v}$ is the electron velocity, $\mathbf{F}$ and $\mathbf{B}$ the electric and magnetic fields of the radiation. The term in $\mathbf{B}$ is small and is neglected in the 'dipole approximation'.

The equation of motion of an electron is then

$$\frac{d^2 \boldsymbol{r}}{dt^2} + \frac{1}{\tau}\left(\frac{d\boldsymbol{r}}{dt}\right) + \omega_0^2 \boldsymbol{r} = \frac{e\mathbf{F}(t)}{m} \tag{6.13}$$

where $\boldsymbol{r}$ is the displacement, ω_0 the resonant frequency and $1/\tau$ a measure of the damping of the motion. The solution of (6.13) when $\mathbf{F}(t) = \mathbf{F}_0 \exp(i\omega t)$ is

$$\boldsymbol{r} = \frac{e\mathbf{F}(t)}{m}\left[\frac{1}{\omega_0^2 - \omega^2 + i\omega/\tau}\right] \tag{6.14}$$

The displacement of charge produces an electric dipole moment $e\boldsymbol{r}$ and a total polarisation $\mathbf{P}$ for N oscillators per unit volume equal to $Ne\boldsymbol{r}$. The polarisation is in turn related to the dielectric constant by

$$\varepsilon = \varepsilon_1 + i\varepsilon_2 = 1 + \mathbf{P}/\mathbf{F}(t)$$

and hence

$$(n_0 + ik_0)^2 = 1 + \frac{Ne^2/\varepsilon_0 m}{\omega_0^2 - \omega^2 + i\omega/\tau} \tag{6.15}$$

The variation in absorption and refractive index with frequency predicted by (6.15) is of the form given in (2.32) and illustrated in figure 2.5.

If all the oscillators are identical and non-interacting the classical model gives a good description of the shape of an absorption line. The same form is obtained from a quantum mechanical treatment. Thus (6.15) describes the absorption lines observed in gases at low pressures. Doppler broadening in a gas and the broadening of impurity lines due to crystal field effects in a solid can be considered to arise from a distribution in ω_0 values. The line shape is then often Gaussian. Coupling of identical classical oscillators (e.g. coupled tuned circuits) leads to shifts in resonant frequency so we would expect the atoms of a crystal lattice to show broad absorption bands, as indeed they do.

The classical model was unable to give an adequate account of the factors determining ω_0 and τ and, of course, unable to account for the multiplicity of absorption lines observed in gases which we now ascribe to a multiplicity of electronic energy levels. Also (6.15) usually predicts too large an absorption coefficient. To allow for this, N was multiplied by an adjustable parameter called the oscillator strength, denoted by f where $f < 1$. This term is still in use.

6.2.2 The Einstein *A* and *B* coefficients (a)

In 1917 Einstein appreciated that a quantum description of the interaction of radiation and matter required the inclusion of a stimulated emission process as well as absorption and spontaneous emission. Stimulated emission is, in fact, implicit in the classical Rayleigh–Jeans description of black body radiation since it is based on thermodynamics. Einstein's argument was essentially that the (then) new quantum picture should be consistent with statistical mechanics.

We consider two isolated energy levels separated by an energy $\hbar\omega$. The number of oscillators having the lower energy is N_1 and the number with the higher energy is N_2. Transitions take place between the levels as shown in figure 6.2 and in the presence of radiation of frequency ω and energy density

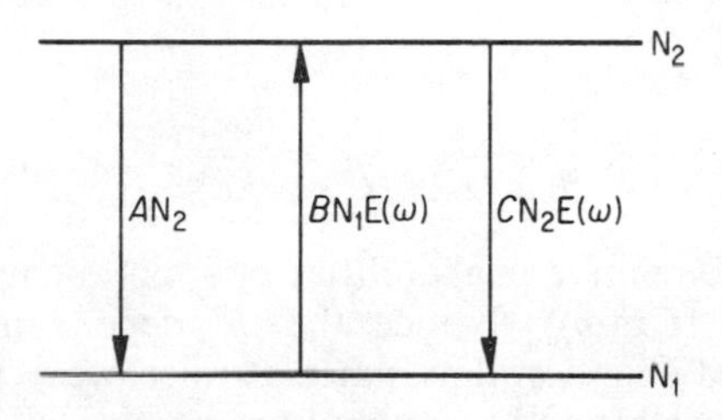

Figure 6.2 Stimulated and spontaneous transitions between a pair of isolated energy levels with populations N_1 and N_2

$E(\omega)$ per unit bandwidth the rate of change of the population (N_2) of the upper state is given by

$$\frac{dN_2}{dt} = \underbrace{-AN_2}_{\text{spontaneous emission}} + \underbrace{BN_1 E(\omega)}_{\text{stimulated absorption}} - \underbrace{CN_2 E(\omega)}_{\text{stimulated emission}} \qquad (6.16)$$

where A, B and C are constants. Under steady state conditions $dN_2/dt = 0$.

When a steady state is achieved two conditions must be met by (6.16). First, if $E(\omega)$ is infinite, which is equivalent thermodynamically to an infinitely high temperature, we must have

$$N_2/N_1 = g_2/g_1$$

where g_2 and g_1 are the statistical weights (degrees of degeneracy) of the levels involved. Inserting this condition into (6.16) gives

$$Bg_1 = Cg_2 \qquad (6.17)$$

so that in a non-degenerate system the probabilities of stimulated emission and absorption are equal.

The second condition applies when the external radiation intensity is zero so that $E(\omega)$ equals $E_0(\omega)$, the black body radiation arising from the finite temperature of the medium. The ratio of N_1 to N_2 must then be given by the Boltzmann distribution

$$\frac{N_1}{N_2} = \left(\frac{g_1}{g_2}\right) \exp(\hbar\omega/kT) \qquad (6.18)$$

Inserting this into the steady state form of (6.16) and using (6.17) gives

$$\mathrm{E}_0(\omega) = \frac{A}{B(g_1/g_2)}\left[\frac{1}{\exp(\hbar\omega/kT)-1}\right] \tag{6.19}$$

for the black body radiation at temperature T. The factor in the square brackets is the Bose distribution function. Comparison with Planck's radiation law then gives

$$\frac{A}{B(g_1/g_2)} = \frac{A}{C} = \frac{\hbar\omega^3}{\pi^2 c^3} \tag{6.20}$$

which determines the relative probabilities of spontaneous emission (A) and stimulated emission ($C\mathrm{E}(\omega)$). Notice the ω^3 dependence. For 'ordinary' intensities, stimulated emission dominates at microwave frequencies but the reverse is true in the visible and throughout most of the infrared.

Among the consequences of Einstein's formulation are the following:

(a) The net rate of absorption of photons is proportional to the stimulated absorption minus the stimulated emission, or

$$B\mathrm{N}_1 - C\mathrm{N}_2 = B(\mathrm{N}_1 - (g_1/g_2)\,\mathrm{N}_2)$$

In a bandwidth $\mathrm{d}\omega$ the energy density $\mathrm{E}(\omega)\,\mathrm{d}\omega$ of the incident radiation equals I/c where I is the incident intensity. Assuming that the intensity is constant over the absorption band it then follows that the absorption coefficient, K, is given by

$$\int \frac{\mathrm{K}\,\mathrm{d}\omega}{\omega} = \frac{B\hbar}{c}(\mathrm{N}_1 - (g_1/g_2)\,\mathrm{N}_2) \tag{6.21}$$

Since in thermodynamic equilibrium $\mathrm{N}_1 \gg (g_1/g_2)\,\mathrm{N}_2$, from (6.18), the absorption coefficient predicted by (6.21) is positive. If, however, it were possible to invert the populations by some external agency so that $g_1\,\mathrm{N}_2$ became greater than $g_2\,\mathrm{N}_1$, (6.21) shows that the absorption would be negative. This would give rise to Light Amplification by Stimulated Emission of Radiation (Laser action) which we discuss later in this chapter.

(b) The concepts of stimulated and spontaneous emission, and stimulated absorption, apply equally to transitions induced by any Bose particle and we will, for example, apply them to electron scattering by phonons (Chapter 8). Suppose an electron is in state j and may be promoted to state j + 1 or demoted to state j − 1, both the latter states having equal statistical weights and being equally spaced in energy from state j. In thermal equilibrium $\mathrm{E}_0(\omega)$ represents the energy density of Bose particles (e.g. phonons) of appropriate energy. then the probability of promotion is $B\mathrm{E}_0(\omega)$ which, using (6.19) equals $An(\omega)$ where $n(\omega)$ has been written for the Bose distribution. Similarly the probability of demotion is

$$C\mathrm{E}_0(\omega) + A = B\mathrm{E}_0(\omega) + A = A(n(\omega)+1) \tag{6.22}$$

We conclude, as shown previously (§ 3.5) in connection with the scattering of neutrons by phonons, that the probabilities of absorption and emission of Bose particles are proportional to $n(\omega)$ and $(n(\omega)+1)$ respectively (cf. (3.48)).

6.2.3 The quantum mechanical treatment of absorption and emission (A)

In quantum mechanics the interaction of radiation and matter is treated by time dependent perturbation theory. A derivation of the results which are quoted below is outside the scope of this book and can be found in standard textbooks of quantum mechanics. The essential steps in the argument are as follows. Schrödinger's equation is written in time dependent form with a Hamiltonian that consists of the sum of the Hamiltonian of the unperturbed system and a term representing the perturbation due to the radiation field. The latter is assumed to be small and the total Hamiltonian is expanded to first, second or at most third order. Since the perturbation is time dependent the solutions for the wave functions and consequently the probability of finding an electron in a particular state are also time dependent. If $a_j a_j^*$ represents the probability of occupation of the jth state and at $t=0$ $a_i=1$ and $a_j=0$ the probability of a transition from state i to state j in time t is taken to be $(a_j a_j^*)/t$. For transitions induced by radiation, the Hamiltonian of the perturbation is a periodic function of time. Its average over the wave functions of the states involved, ψ_i and ψ_j, has the form

$$\mathscr{H}_{ij} = \mathscr{H}_{ij}^0 \exp(i\omega t) = \int \psi_i^* \mathscr{H} \psi_j \, d\mathscr{V}$$

where the integration is over all space and $\mathscr{H}_{ij}$ is called the matrix element of the perturbation since terms like $\mathscr{H}_{ij}$ can be written out in matrix form. When the expansion of the Hamiltonian is to first order only the value of a_j at time t with the above boundary conditions is found to be

$$a_j = \frac{\mathscr{H}_{ij}^0}{2i\hbar}\left[\frac{\exp[i(\omega_{ij}+\omega)t]-1}{\omega_{ij}+\omega} - \frac{\exp[i(\omega_{ij}-\omega)t]-1}{\omega_{ij}-\omega}\right] \tag{6.23}$$

where the two terms correspond to stimulated emission and stimulated absorption respectively. Each is large if the appropriate energy conservation condition

$$\hbar\omega_{ij} = E_i - E_j = \pm\hbar\omega \tag{6.24}$$

is met, as the denominators go to zero.

To obtain an expression for the transition probability in a practical situation we note that the frequency of the incident radiation cannot have a discrete value if t is finite (a consequence of the uncertainty principle) so that (6.23) must be summed over a finite range of ω values. Concentrating on the second (absorption) term we therefore assume that the final states are distributed in energy and write $\rho(j)$ for the density of final j states per unit energy interval. We further assume that $\mathscr{H}_{ij}$ does not vary significantly over this finite range of frequencies. The transition rate ν_{ij} per unit time then becomes

$$\nu_{ij} = \hbar\rho(j)\int a_j a_j^* \, d\omega = \frac{2\pi}{\hbar}\rho(j)|\mathscr{H}_{ij}^0|^2 \tag{6.25}$$

In a similar way the absorption line shape for an oscillator with two energy levels can be shown to be Lorentzian (cf. (6.15)) as there is an energy spread of $\hbar/\tau$ due to the Uncertainty Principle.

To find expressions for the Einstein B coefficient and the oscillator strength f we must find an explicit value for $\mathscr{H}_{ij}$ in terms of the energy density of the radiation. The simplest approximation, valid for isolated atoms, is to assume that the interaction is between the electric dipole of the system and the electric field of the radiation, as in § 6.2.1. Then

$$\mathscr{H}_{ij} = e\mathbf{r}.\mathbf{F}(t) \tag{6.26}$$

and if the radiation is plane polarised the energy density is

$$\mathrm{E}(\omega)\,\mathrm{d}\omega = I/c = \varepsilon_0\,\mathbf{F}_0^2/2 \tag{6.27}$$

An expression for the B coefficient can then be found and is

$$B = \frac{e^2\pi}{\hbar^2\varepsilon_0}|R_{ij}|^2 \tag{6.28}$$

where

$$R_{ij} = \int \psi_i^* r \psi_j \,\mathrm{d}\mathscr{V} \tag{6.29}$$

Comparison with (6.15) including a scaling factor f as explained above gives the relationship between the oscillator strength and the Einstein B coefficient:

$$\mathrm{f} = \left(\frac{2m\varepsilon_0\,\hbar\omega}{\pi e^2}\right)B \tag{6.30}$$

Though the approximation used above is usually a good one for isolated atoms and some impurities in solids it is not generally valid for energy band states in solids where r can have the dimensions of the crystal. A more appropriate approximation for $\mathscr{H}_{ij}$ in this case is

$$\mathscr{H}_{ij} = \mathbf{j}.\mathbf{A} \tag{6.31}$$

where $\mathbf{j}$ is the current operator ($\mathbf{j} = ep/m$) and $\mathbf{A}$ is the vector potential of the electromagnetic field.

In principle, equations (6.26), (6.31) and (6.32), as appropriate, allow the Einstein B coefficient, and hence the absorption from (6.21) to be calculated from microscopic properties of the system. In practice it is very difficult to make quantitative predictions though, for example, (6.31) usually predicts transitions with $\mathrm{f} \sim 1$ and for solid state densities $\sim 10^{22}$ atoms per cc this implies absorption coefficients in excess of 10^5 cm^{-1} over wide spectral bandwidths, as observed. Perhaps the most important conclusions to be drawn from the quantum mechanical picture are the general confirmation of the concepts of § 6.2.2, the interpretation of resonant frequencies such as ω_0, the 'golden rule' (6.25) and the prediction of selection rules.

Equation (6.29) illustrates how selection rules arise. Since the displacement $\boldsymbol{r}$ is an odd function (of opposite sign on each side of the origin) R_{ij} will vanish and the absorption be zero if either both ψs are odd or both even functions. Such transitions are said to be forbidden. Since, however, the theory is only approximate a finite but small transition probability may be obtained when second order effects are taken into account. For example the magnetic field of the radiation will interact with the magnetic dipole of the system to give a matrix element of the interaction Hamiltonian

$$\mathscr{H}_{ij} = \mathscr{m}\,.\,\mathbf{B} \tag{6.32}$$

where $\mathscr{m}$ is the magnetic dipole operator. Thus 'forbidden' transitions are usually characterised by small ($\sim 10^{-4}$) but non-zero f values. Transitions within the 4f shell of rare earth ion impurities are frequently of this type. The symmetry of the wave functions associated with an impurity in a solid will generally reflect the symmetry of the crystal field, and also depend on the degree of coupling to the lattice (§ 5.6). As a consequence the absorption coefficient may depend on the polarisation direction of the electric vector of the radiation with respect to the crystal axes. Equation (6.31) has similar properties and in anisotropic crystals the interband absorption may depend on the direction of polarisation.

6.3 A survey of photon absorption processes in solids (a)

Before discussing how photons interact with the various possible excitations of a solid we now give a brief survey to put each process in perspective. Figure 6.3 shows the absorption spectrum of a hypothetical solid: it is a semiconductor to which a magnetic field may be applied and is antiferromagnetic at some low temperature. Semiconductors show all the optical properties of insulators and metals though not, of course, to the same degree. The main features are as follows:

(a) In the ultraviolet, and sometimes extending into the visible and infrared, is a region of intense absorption which arises from electronic transitions between the valence and conduction bands. Such transitions generate mobile electrons and holes, resulting in photoconductivity. The absorption coefficient is typically in the range 10^5 to 10^6 cm^{-1} but some structure is normally observable in reflection measurements. The origin of this structure is discussed in § 6.4. On the high energy side of this band (typically around 20 eV) there is often a smooth fall in absorption over a range of several electron volts. On the low energy side, on the other hand, the absorption coefficient falls more rapidly and may fall by as much as six orders of magnitude within a few tenths of an eV. In semiconductors this low energy boundary of the fundamental absorption is often the most striking feature of the spectrum and is referred to as *the* absorption edge.

(b) The limit of the absorption edge corresponds to the photon energy required to promote electrons across the minimum energy gap, E_g. As it usually

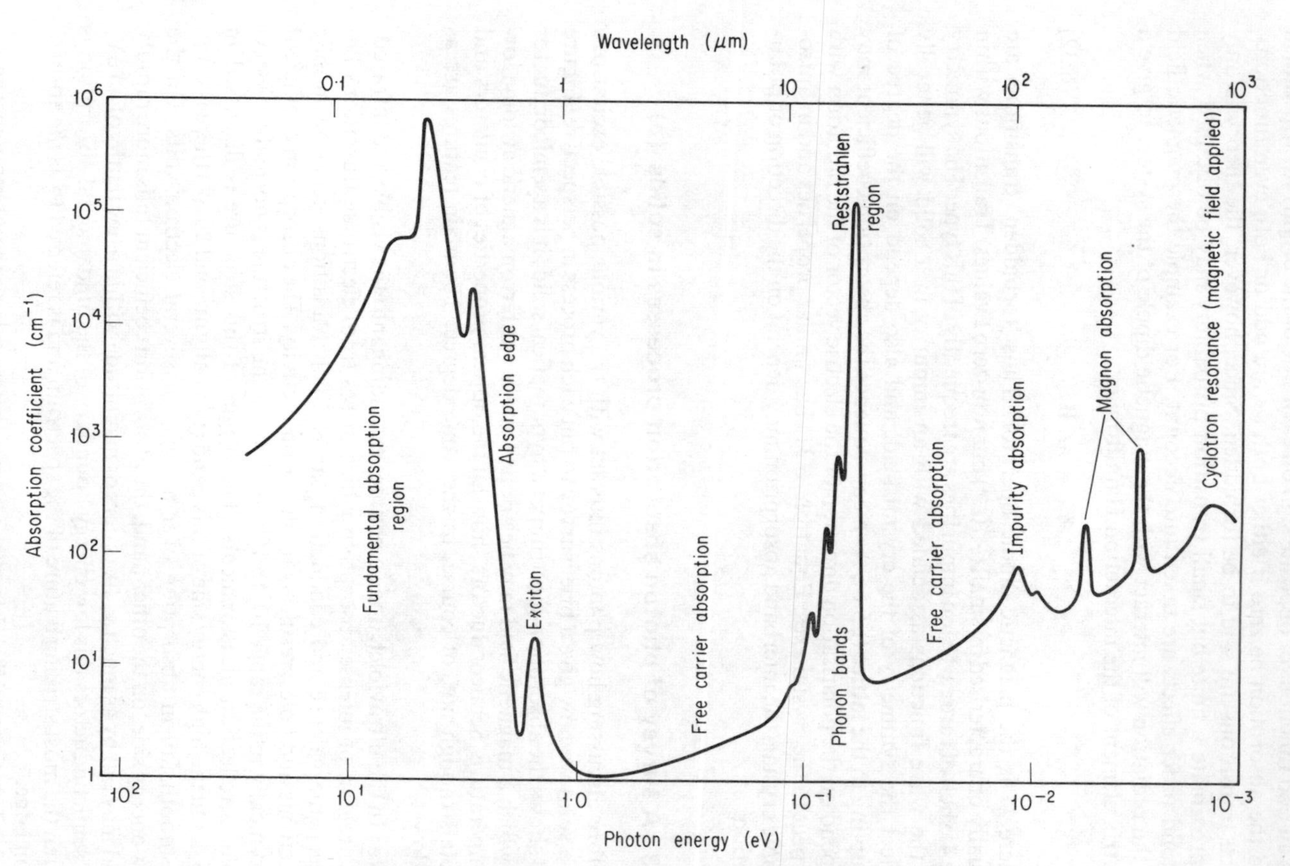

Figure 6.3 Absorption spectrum of a hypothetical semiconductor

occurs in a technically convenient frequency range and the absorption coefficient over much of the edge is sufficiently low for transmission measurements to be made, the absorption edge region is much studied in non-metallic solids. The edge region often shows some structure, in particular that due to *excitons*. Exciton absorption is more pronounced in insulators, particularly ionic crystals, than in semiconductors, and can lead to strong narrow-line absorption as in atomic spectra. We return to more detailed discussion of the edge region in § 6.5.

(c) As the wavelength is increased beyond the absorption edge the absorption starts to rise slowly again. This increase is due to electronic transitions within the conduction or valence bands (the material being a semiconductor) and is referred to as free-carrier absorption. It extends throughout the infrared and microwave regions of the spectrum. Its magnitude is, of course, a function of the mobile electron (or hole) density. This is very high in metals, when the free-carrier absorption may mask all other features of the spectrum, at least in the infrared. We shall postpone discussion of free-carrier absorption until the next chapter.

(d) At photon energies between 0.02 and 0.05 eV (50 to 20 μm wavelength) a new set of absorption peaks appears. These are due to interaction between the incident photons and the vibrational modes of the lattice (Chapter 3). If the crystal is ionic the absorption coefficient may reach 10^5 cm^{-1} and strong reflection occurs (6.4). Homopolar crystals show similar features but with much lower absorption coefficients, around 10^1 or 10^2 cm^{-1}. Phonon spectra are considered in § 6.6.

(e) As discussed in Chapter 5, impurities give rise to additional absorption. Impurity absorption is included in figure 6.3 only for completeness. To construct the figure we have assumed a shallow impurity with an ionisation energy ~0.01 eV. Absorption due to such an impurity would be observable only at low temperatures, such that kT was less than the ionisation energy (see § 5.3.1).

(f) Absorption may occur in solids due to electron spin reversal. Thus a solid containing paramagnetic impurities will show absorption lines in the presence of an external magnetic field which splits otherwise degenerate spin levels. There may, additionally, be crystal field splitting. Very large magnetic fields are usually required before the energy difference between spin states corresponds to the energies of infrared photons. However, in an antiferromagnetic material where there is a substantial internal field, the spins are coupled together and there is the possibility of collective excitations of the spins. Such collective excitations are called spin waves or magnons and are discussed in Chapter 10. Again, absorption due to magnon generation is included in figure 6.3 for completeness.

(g) Finally the absorption peak at the long wavelength end of figure 6.3 is intended to indicate the possibility of enhanced absorption by mobile carriers in the presence of a magnetic field. This phenomenon, referred to as cyclotron resonance, is discussed in § 4.7 and the next chapter.

6.4 Interband absorption in the fundamental region (a)

As an example of interband absorption, we show in figure 6.4 the reflectivity of germanium at normal incidence and in figure 6.5 the value of $\varepsilon_2/\varepsilon_0$ deduced from the reflectivity measurements. It should be noticed that the latter is on a linear scale which, compared with the logarithmic scale of figure 6.3, has the effect of compressing the absorption edge into a tiny step around 0.7 eV.

Figure 6.6 shows the energy band structure of germanium. Electronic transitions can take place between filled and empty bands subject to the conservation of energy and wave vector. If single photons and electrons are involved, the conservation rules require

$$E_{\mathrm{f}} - E_{\mathrm{i}} = \hbar\omega(\boldsymbol{\beta}) \tag{6.33}$$

and

$$\boldsymbol{k}_{\mathrm{f}} - \boldsymbol{k}_{\mathrm{i}} = \boldsymbol{\beta} \tag{6.34}$$

where the subscripts f and i refer to the final and initial electron states and $\boldsymbol{\beta}$ is the wave vector of a photon of energy $\hbar\omega(\boldsymbol{\beta})$. Since the wavelength of the radiation is very much greater than the lattice spacing $\boldsymbol{\beta}$ is very much less than a reciprocal lattice vector. Hence on the scale of figure 6.6 $\boldsymbol{\beta}$ can be neglected, $\boldsymbol{k}_{\mathrm{f}} \approx \boldsymbol{k}_{\mathrm{i}}$ and only vertical transitions between points separated in energy by $\hbar\omega(\boldsymbol{\beta})$ are allowed. Given this restriction and noting that the transition probability is proportional to the density of final states (§ 6.2.3) we examine the band structure in figure 6.6 to see where a large number of transitions with energies

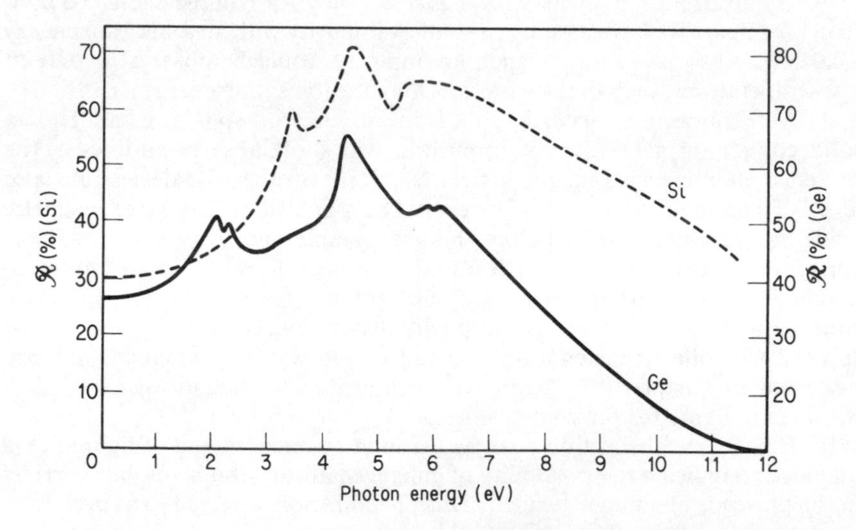

Figure 6.4 Reflectivity, $\mathscr{R}$, at room temperature and normal incidence of germanium and silicon in the fundamental absorption region. Note displacement of germanium scale. (After D. L. Greenaway and G. Harbeke, *Optical Properties and Band Structure of Semiconductors*, Pergamon (1968).)

between $\hbar\omega(\boldsymbol{\beta})$ and $\hbar[\omega(\boldsymbol{\beta})+d\omega(\boldsymbol{\beta})]$ might be possible. This occurs whenever an empty band runs parallel to a filled band and there is an extended region of $\boldsymbol{k}$ space over which the energy separation of the bands is nearly constant. There is then a large density of initial and final electron states available for a small range of photon energies. This condition is expressed by

$$\nabla_{\boldsymbol{k}} E_v(\boldsymbol{k}) = \nabla_{\boldsymbol{k}} E_c(\boldsymbol{k}) \tag{6.35}$$

where $\nabla_{\boldsymbol{k}} E_v(\boldsymbol{k})$ and $\nabla_{\boldsymbol{k}} E_c(\boldsymbol{k})$ are respectively the gradients of the full (valence) and empty (conduction) band energies in $\boldsymbol{k}$ space. In addition we have seen in Chapter 2 that the density of states increases monotonically with k. This biases the absorption in favour of transitions between states of large k but (6.35) remains the main source of structure shown in figure 6.5.

The numbered arrows in figure 6.6 indicate the transitions in germanium for which (6.35) is met and the numbers on figure 6.5 the features probably associated with these transitions. A feature is not necessarily an absorption peak. The exact shape of the feature depends on the sign of both gradients away from the condition (6.35) and the consequent rate of variation in state density with E (see § 2.9).

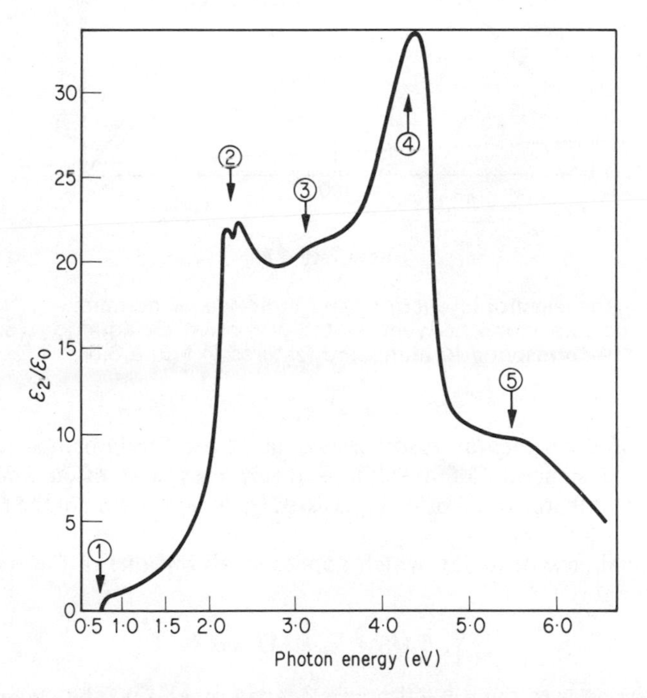

Figure 6.5 The value of ε_2 for germanium at room temperature deduced from reflectivity data. Numbered features referred to in text. Curve constructed from D. Brust, J. C. Phillips and G. F. Bassani, *Phys. Rev. Lett.*, **9**, 94 (1962) and G. Harbeke, *Z.f. Naturforsch.*, **19a**, 548 (1964)

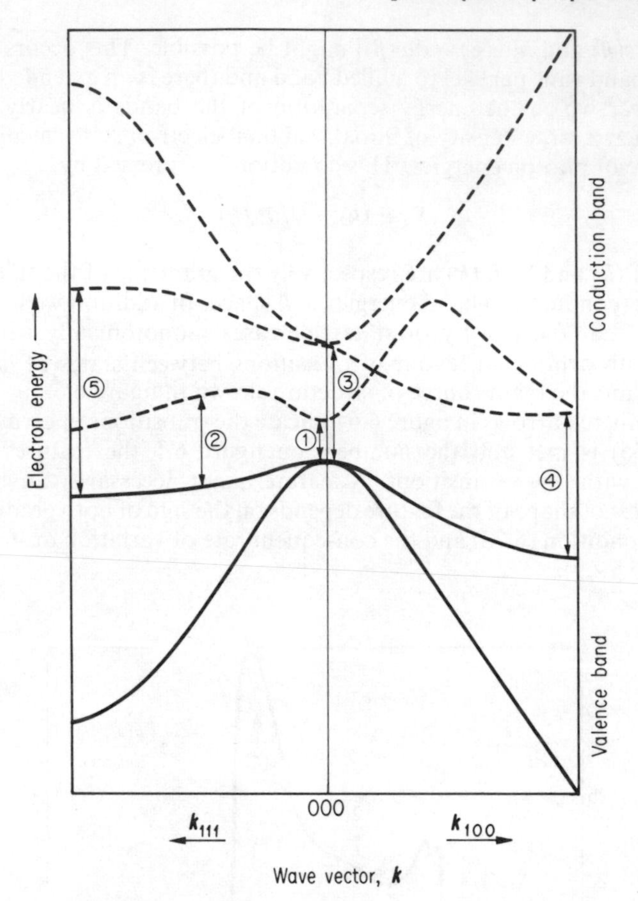

Figure 6.6 The electronic energy band structure of germanium. Numbered arrows indicate transitions between states which satisfy equation (6.35) and correspond to numbered features of figure 6.5

Analysis of experimental absorption data in the fundamental region, in terms of band structure of a solid, is rarely easy and often ambiguous. Identification of features can often be assisted by consideration of the following points.

(a) A special case of (6.35), which applies to all but one of the transitions indicated in figure 6.6, is

$$\nabla_{\boldsymbol{k}} E_v(\boldsymbol{k}) = \nabla_{\boldsymbol{k}} E_c(\boldsymbol{k}) = 0 \tag{6.36}$$

which corresponds to the critical points defined by (2.47). The occurrence of such points can be predicted on symmetry grounds.

(b) As in atomic spectroscopy, solids can be arranged in isoelectronic series through which spectra change quantitatively but remain qualitatively similar.

In the fundamental absorption region series may be found which run both vertically and horizontally through the Periodic Table: a horizontal series could start from a Group IV element, followed by a III-V compound, II-VI compound, I-VII compound. An example is Ge, GaAs, ZnSe, CuBr. The preparation of alloys can provide similar information.

(c) Due to thermal expansion and other causes, absorption peaks and other features shift with temperature, typically by 10^{-4} to 10^{-3} eV K^{-1}. Features arising from the same type of transition in an isoelectronic series usually have the same temperature dependence.

The main features of the fundamental absorption spectrum of many non-metallic solids have been accounted for along the lines indicated above and illustrated by germanium. Spectroscopy in the fundamental region is one of the best methods of obtaining data on the electronic band structure of solids over a wide range of E and $\boldsymbol{k}$ values. It is sometimes desirable to insert spectroscopic data into a band structure calculation *ab initio* and see if the programme can correctly predict any remaining unused data. The technique is not, generally, applicable to metals as the free-carrier absorption may still be appreciable even in the fundamental absorption region, and in any case a large fraction of the conduction band states are occupied and hence inaccessible to electrons excited from lower bands.

6.5 The absorption edge region (a)

For most solids the minimum energy at which interband transitions can take place lies between 1 and 10 eV. It is found that the absorption edge is rarely a simple, monotonic fall in absorption coefficient. Structure in this region of the spectrum is due mainly to excitons, which we consider first. In addition, absorption processes that might be dismissed as second order can be important in regions of low absorption when first-order processes do not occur. In particular, transitions in which the initial and final wave vectors are unequal can be observed.

6.5.1 Excitons (a)

The excitation of electrons to the conduction band leads to photoconductivity and is equivalent to the photo-ionisation of isolated atoms. It is natural to expect that excited but bound electron states will exist at energies below the photo-ionisation limit. Particularly for those solids for which an 'atomic' picture is a good description (e.g. the solidified rare gases) there will be bound states, derived from atomic levels, into which electrons can be excited by the absorption of photons. Such transitions will give rise to absorption lines, similar to those of isolated atoms, at photon energies less than the ionisation energy, E_g.

As noted in Chapter 2, there is an important difference between the excitation of solids and that of isolated atoms. Excited states in solids are excitations of the whole crystal, are not localised and have a well defined wave vector. Such excitations are called excitons.

When discussing electrons in solids in Chapter 4 we showed it was convenient to use two extreme approximations: the nearly free electron model appropropriate to metallic and covalent crystals (§ 4.2.1) and an atomic picture for ionic and molecular crystals (§ 4.2.2). In a similar way there are two models of the exciton, named after Wannier and Frenkel, respectively. Frenkel envisaged an excited electron describing an orbit of atomic dimensions around an atom with a vacant valence state. Since the excitation can move from atom to atom the empty valence state, though instantaneously on one atom, constitutes a mobile hole. Thus a Frenkel exciton is like an electron bound in a relatively deep impurity state (Chapter 5) except that it is mobile. As with impurities, however, this model is inappropriate for covalent semiconductors. We saw in § 5.6.5 that shallow impurity levels in covalent solids can be described by analogy with hydrogen atoms. The Wannier picture of an exciton is similar, the electron and hole being bound by their mutual Coulomb attraction but separated by many lattice distances. Exactly as with impurities, the Coulomb force is reduced by the dielectric constant of the intervening medium. Following the arguments of § 5.6.5 we can deduce the exciton binding energy, E_B, by replacing the effective mass by the reduced mass of the electron-hole pair constituting the exciton. By comparison with (5.25) and (5.26) we obtain

$$a_{\text{ex}} = \text{n}^2 \left(\frac{m}{m_r}\right)\left(\frac{\varepsilon_1}{\varepsilon_0}\right) a_0 \tag{6.37}$$

and

$$E_B = -\frac{1}{\text{n}^2}\left(\frac{m_r}{m}\right)\left(\frac{\varepsilon_0}{\varepsilon_1}\right)^2 E_{\text{H}} \tag{6.38}$$

where E_{H} is the ionisation energy of hydrogen (13.54 eV), a_{ex} is the exciton 'orbit radius' and a_0 is the Bohr radius of the ground state of hydrogen. $\text{n} = 1, 2, 3, \ldots$, etc., corresponding to a series of excited states. The reduced mass, m_r is given by

$$m_r = m_e^* m_h^* / (m_e^* + m_h^*)$$

where m_h^* and m_e^* are the effective masses in the valence and conduction bands, respectively (Chapter 4).

Since on this model we can identify the energy gap of the solid, E_g, with the photo-ionisation energy we can write the total energy of an exciton, including a term for its motion through the lattice, as

$$E_{\text{ex}} = E_g - \frac{m_r e^4 \varepsilon_0^2}{2\hbar^2 \text{n}^2 \varepsilon_1^2} + \frac{\hbar^2 \boldsymbol{K}_{\text{ex}}^2}{2(m_e^* + m_h^*)} \tag{6.39}$$

where the second term is the binding energy, E_B, $\boldsymbol{K}_{\text{ex}}$ is the exciton wave vector and we have assumed simple parabolic bands. When excitons are generated by photon absorption and no other particles are involved, energy conservation requires $\hbar\omega(\boldsymbol{\beta}) = E_{\text{ex}}$ and wave-vector conservation requires $\boldsymbol{K}_{\text{ex}} = \boldsymbol{\beta}$. Since $\boldsymbol{\beta}$ is small (§ 6.5) the term in $\boldsymbol{K}_{\text{ex}}^2$ can normally be neglected in such circumstances.

It is convenient to distinguish between the Frenkel and Wannier models on the criterion of orbit size (cf. impurities, § 5.6). If a Frenkel exciton is to have approximately the same energy as its atomic counterpart the electron orbit must be of similar size. On the other hand, the use of a macroscopic dielectric constant in the Wannier model involves the average over many lattice atoms so the model can hardly be valid unless a_{ex} is an order of magnitude greater than a_0. The Frenkel picture is applicable to the solidified rare gases and some ionic crystals, the Wannier model to the Group IV semiconductors and many partially ionic, partially covalent crystals (e.g. III-V compounds). There is a range of orbit size for which neither is a good description. Unfortunately this includes many alkali halides, which show particularly strong and well-resolved absorption peaks. Excitons in alkali halides are thought to be similar to U centres (§ 5.6.4) with the hole located on a halide ion and the electron orbit encompassing only the nearest neighbour metal ions.

Exciton energies predicted from (6.39) correspond quite well to the values observed experimentally, even in alkali halides. Only rarely, however, is an extended hydrogenic-type series of lines, corresponding to different n values, detected. A good example is Cu_2O. At very low temperatures (necessary to avoid thermal ionisation of the exciton, which occurs if $E_B \ll kT$) lines up to

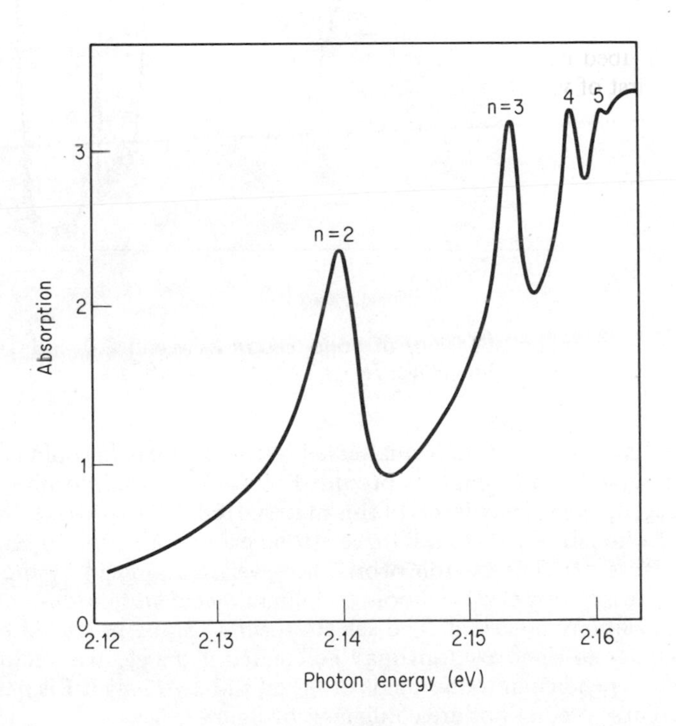

Figure 6.7 Exciton absorption lines in Cu_2O at 77 K. (After Baumeister, *Phys. Rev.*, **121**, 359 (1961).)

n = 11 have been observed. The absorption spectrum of Cu_2O in the region of its absorption edge is shown in figure 6.7. The n = 1 line is missing: this transition is forbidden.

Figure 6.8 shows the absorption spectrum of solid xenon. The two strong peaks at 8.4 eV and 9.55 eV correspond to spectral lines in gaseous xenon and consequently are clearly Frenkel excitons. The smaller peaks at 9.1 and 9.2 eV can be described by a Wannier type model as the n = 2 and n = 3 lines of a series, the first of which is the 8.4 eV line. It would thus appear that we have both types of exciton in one crystal. This is not particularly surprising since the condition on orbit size for the Wannier model can be met for $n \geqslant 2$ but not n = 1.

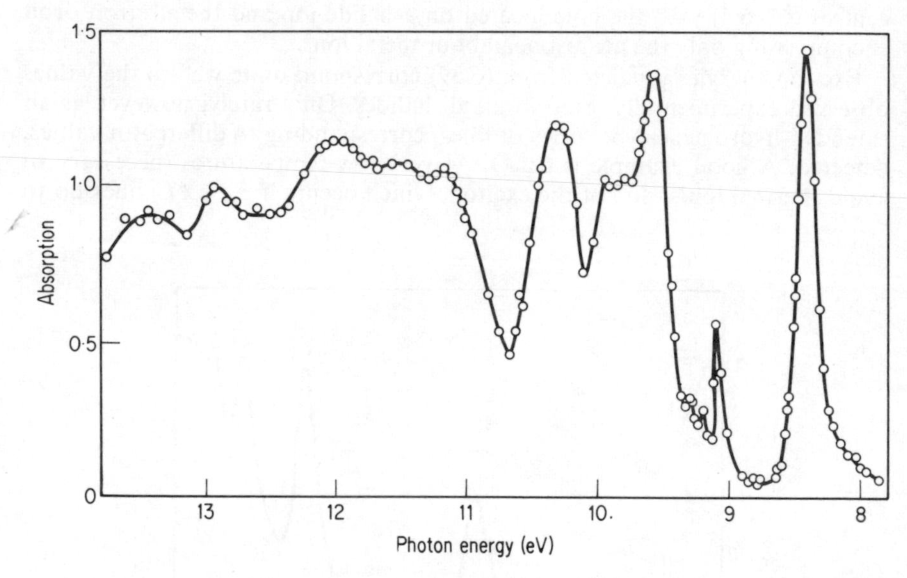

Figure 6.8 Absorption spectrum of solid xenon between 8.0 and 14.0 eV. (After Baldini, *Phys. Rev.*, **128**, 1562 (1962).)

A Frenkel exciton orbit totally unaffected by the crystal field could not occur in an ionic crystal. An example is provided by CdS. The spin orbit coupling splits the degenerate valence levels of this material (§4.2.2) to provide three hole bands. Correspondingly there are three strong exciton absorption series,* as shown in figure 6.9. The exciton orbits, however, are aligned by the crystal field and there is strong electric dipole coupling to incident photons only if the radiation is suitably polarised. The selection rules can be predicted on symmetry grounds: all three excitons may be created if the electric vector has a component perpendicular to the c-axis of the crystal, two only if **F** is parallel to the c-axis. These predictions are confirmed by figure 6.9.

* We note in passing that there are two series in Cu_2O for the same reason.

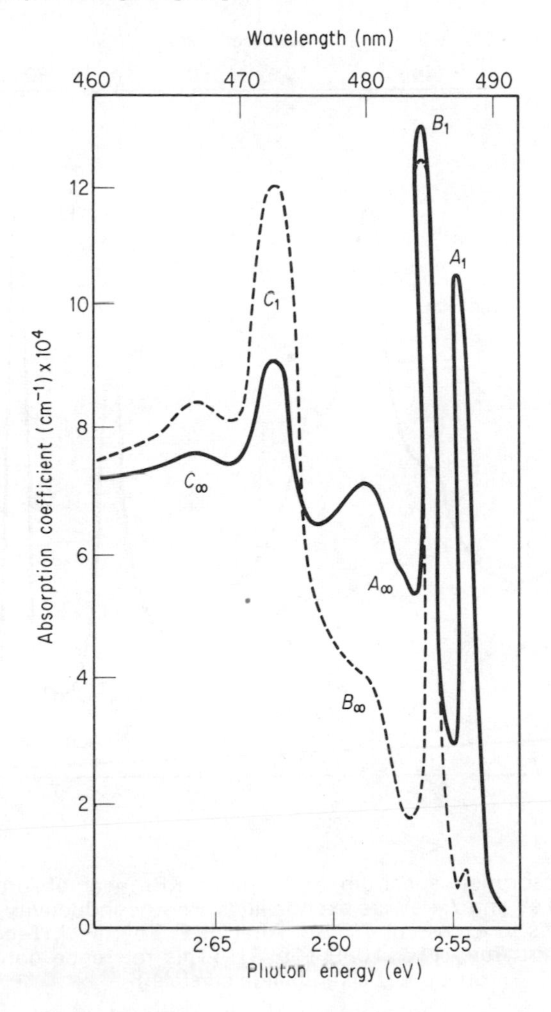

Figure 6.9 Absorption of CdS at 77 K showing exciton lines. – – – Electric field parallel to c-axis,—— F perpendicular to c-axis. A_1, B_1 and C_1, first members of series. A_∞, B_∞ and C_∞ limits of series. Note that the A series is missing for electric vector parallel to c-axis

As already remarked, the alkali halides provide excellent examples of strong exciton absorption lines with peak absorption coefficients as large as 10^6 cm^{-1}. The spectrum of KBr is shown in figure 6.10. One problem in analysing such a spectrum is to determine where interband absorption begins, and hence the value of E_g. Information on this can be obtained from photoconductivity data. Since an exciton is an electrically neutral particle (electron + hole) it does not contribute to photoconduction, which begins at $\hbar\omega(\boldsymbol{\beta}) = E_g$. By this criterion it would appear that interband absorption in KBr begins at 7.8 eV, where there is a small and almost insignificant step in the absorption spectrum.

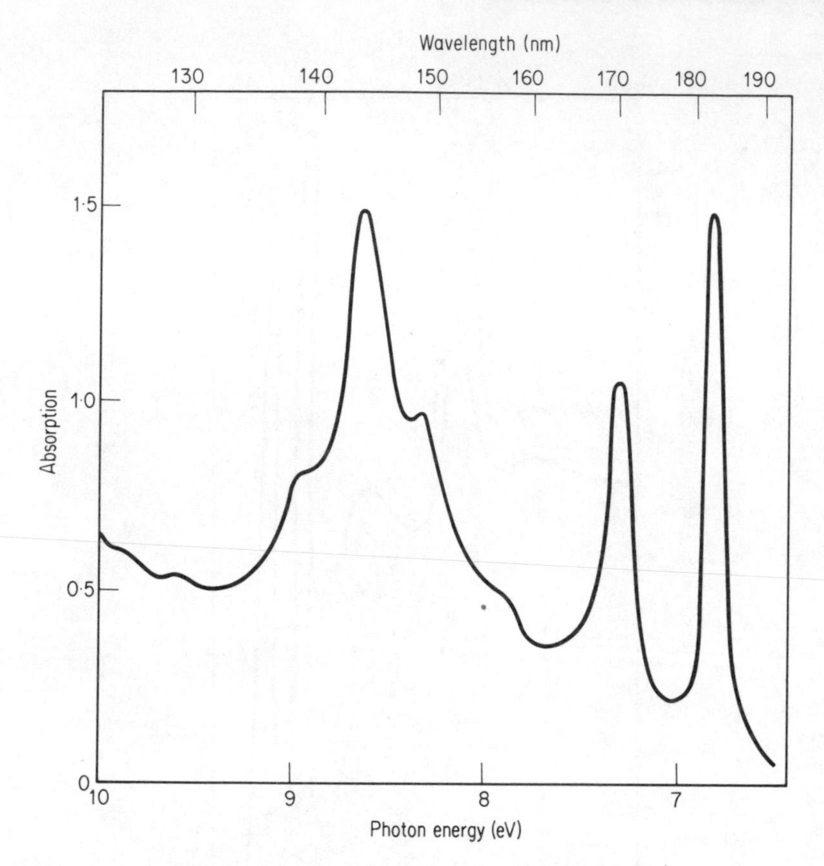

Figure 6.10 Absorption spectrum at 77 K of KBr near absorption edge. Peaks at about 6.8 eV and 7.4 eV are exciton lines. Photoconductivity is observed for photon energies in excess of 7.8 eV. After J. E. Eby, K. J. Teegarden and D. B. Dutton, *Phys. Rev.,* **116**, 1099 (1959). (This reference contains many spectra of alkali halide crystals.)

A puzzling feature of exciton absorption lines in many ionic crystals (including alkali halides) is the shape of the line. It is frequently not Lorentzian ((6.15) and (2.32)), as it would be if excitons had a well-defined lifetime, or Gaussian, as it would be if excitons had a distribution of energies. Empirically it is found that the absorption coefficient, K, is given by the expression

$$K = K_0 \exp[\eta(\omega - \omega_0)] \tag{6.40}$$

where K_0 and η are constants and ω is less than the peak frequency ω_0. Equation (6.40) is known as Urbach's rule and in many materials it is obeyed over an enormous range of K values (up to 10^6:1). The origin of Urbach's rule and the considerable width of some exciton lines is probably 'phonon-assisted transitions'. These are interactions in which three particles are involved: photon, phonon and exciton. Although three-particle interactions are less likely than two they may be important when the direct transitions are weak, far from $\omega = \omega_0$. The energy and wave vector conservation rules now become

$$E_{ex}(\boldsymbol{K}) = \hbar\omega(\boldsymbol{\beta}) \pm \hbar\omega(\boldsymbol{q}) \tag{6.41}$$

and

$$\boldsymbol{K}_{ex} = \boldsymbol{\beta} \pm \boldsymbol{q} \tag{6.42}$$

where $\hbar\omega(\boldsymbol{q})$ is the energy of a phonon of wave vector $\boldsymbol{q}$. The $\pm$ signs indicate that a phonon may either be absorbed by the exciton (+ sign) or emitted (− sign). We are now no longer constrained to direct transitions and the creation of at least some excitons of finite wave vector is possible. Since $\boldsymbol{q}$ can have a range of values there will be a tendency for the absorption line to be broadened but the exact form of (6.40) is difficult to justify.

6.5.2 The limit of interband absorption (a)

Except at very low temperatures exciton lines are not observed in materials of high dielectric constant and low effective mass since $E_B \ll kT$ and the exciton levels are indistinguishable from the conduction band. Excitons are also absent in materials containing a high density of electrons (or holes), such as metals and heavily doped semiconductors, since the free charge-carriers adjust their position to screen out the field between electron and hole (§ 4.9). Hence it is possible to study the absorption edge of many substances without the complexity of exciton structure. This we now do, though noting that the most common situation is a blend of the two extremes.

Figure 6.11 shows the simplest situation, that in which the bands are parabolic and the minimum conduction band energy occurs at the same $\boldsymbol{k}$ value as the maximum valence band energy. Wave vector conservation permits only near vertical transitions to take place (§ 6.4) between states separated in energy by

$$\hbar\omega(\boldsymbol{\beta}) = E_g + \frac{\hbar^2 \boldsymbol{k}^2}{2m_h^*} + \frac{\hbar^2 \boldsymbol{k}^2}{2m_e^*} \tag{6.43}$$

If the valence band states are fully occupied and the conduction band states empty and if the matrix element of the transition is independent of $\boldsymbol{k}$ the absorption will simply be proportional to the density of states at $\boldsymbol{k}$. For parabolic bands this varies as $E^{1/2}\,\mathrm{d}E$ if E is measured from a band edge (see (2.43)). Consequently the absorption will be proportional to

$$[\hbar\omega(\boldsymbol{\beta}) - E_g]^{1/2} \qquad \text{for} \qquad \hbar\omega(\boldsymbol{\beta}) > E_g \tag{6.44}$$

and zero otherwise. This expression is obeyed over a substantial frequency range by, for example, InSb, which is a direct gap semiconductor (table 4.1) and which shows no exciton lines even at moderately low temperatures since

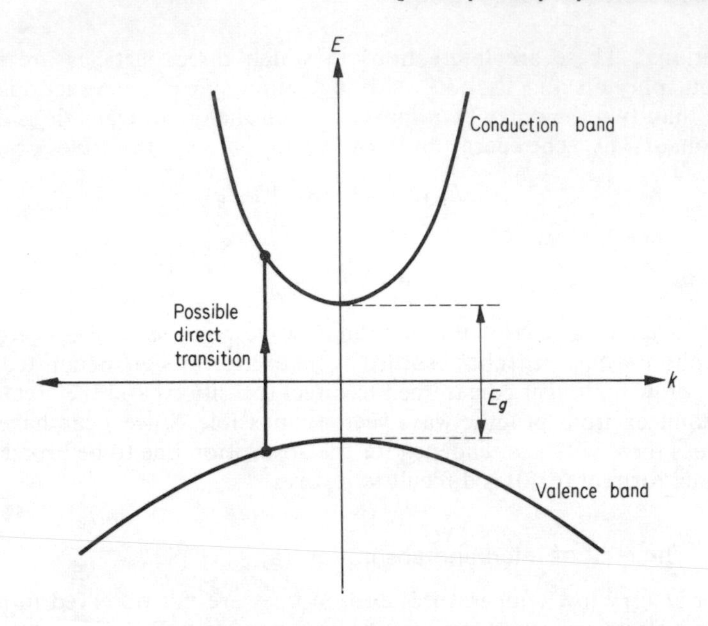

Figure 6.11 The valence and conduction band edges of an insulating solid, assuming the minimum energy of electrons and holes occurs at zero wave vector

$E_B \sim 10^{-3}$ eV. At low absorption levels a long tail, extending to photon energies less than E_g, is observed in InSb. This is due to phonon assisted transitions and photoconductivity has been observed with photon energies as low as $E_g - 2\hbar\omega_0$, where $\hbar\omega_0$ is the optical phonon energy.

Equation (6.44) is modified if the matrix element varies with $\boldsymbol{k}$. In particular the transition may be forbidden on symmetry grounds near high symmetry points like $\boldsymbol{k} = 0$ but the matrix element may increase linearly with k due to second order effects (§ 6.2.3). If so, the absorption varies as $[\hbar\omega(\boldsymbol{\beta}) - E_g]^{3/2}$. In general, then, we expect a smooth curve following some power law with exciton peaks at the lower end. A good example, GaSe, is shown in figure 6.12. GaSe is a direct gap semiconductor, as illustrated in figure 6.11, in which two Wannier type excitons can be seen at the lower temperature of measurement.

In indirect gap semiconductors (e.g. Ge, Si and GaP; see table 4.1) the maximum valence band energy and minimum conduction band energy do not occur at the same $\boldsymbol{k}$ value: see figure 6.13(b). The lowest energy interband transitions must then be phonon assisted to conserve wave vector and, since the separation of the band extrema is about half the width of the Brillouin zone, only phonons at the zone edge have a sufficiently large wave vector to contribute (cf. (6.41) and (6.42)). Ge and Si have two atoms per unit cell and hence four characteristic phonon frequencies at the zone edge (figure 3.9). The absorption edge now shows structure, with a separate feature for each of the four phonon energies.

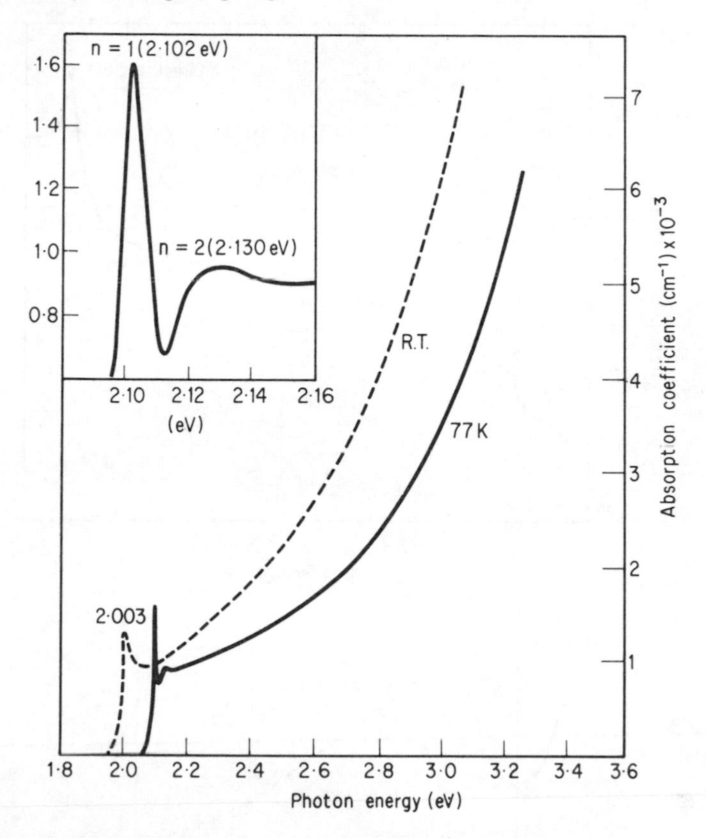

Figure 6.12 The absorption edge of GaSe. After F. Bassani, D. L. Greenaway and G. Fischer, *Proceedings of the 7th International Conference on the Physics of Semiconductors,* p. 51, Dunod, Paris (1964)

The phonon energies deduced from a detailed experimental study of the absorption edge of germanium agree with those found from neutron scattering (see § 3.5). It is worth noting that, since several different zone edge phonons are involved, phonon assistance leads to structure in this case and not to broadening. The absorption edge spectrum of germanium at 20 K is shown in figure 6.13(a). At this temperature exciton effects also appear since $E_B \sim 10^{-2}$ eV. Figure 6.13(b) shows the relevant part of the band structure (cf. figure 6.6). At high photon energies direct transitions as in figure 6.11 are possible and there is a direct exciton line at 0.888 eV. At lower energies only the indirect, phonon assisted, transitions can occur, as described above. The exciton levels now provide a secondary absorption edge, E_B below the indirect interband absorption edge. These features can be seen in figure 6.13(a) but the scale of this diagram is too small to show up the phonon induced structure.

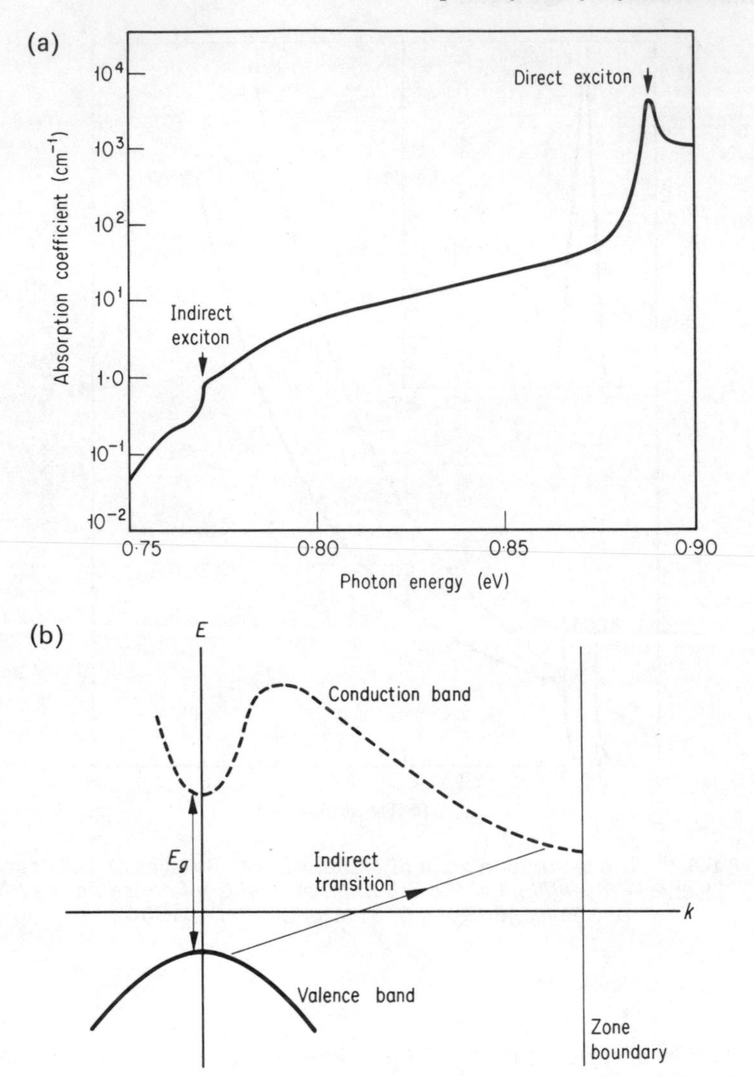

Figure 6.13 (a) The absorption edge of germanium at 20 K (after T. P. McLean, *Progress in Semiconductors*, Volume 5)
(b) Valence and conduction band edges of a material for which the minimum energy gap corresponds to an indirect transition (e.g. Ge and Si)

6.6 Photon-phonon interactions (a)

All solids show a region of absorption in the infrared, usually between 0.1 eV and 0.01 eV, due to the interaction of photons and lattice vibrations. The energies and wave vectors of phonons are well defined as they have appreciable lifetimes and their wave functions are not localised. Hence the energy and wave vector conservation rules apply to photon–phonon interactions. The wave vectors of phonons range up to π/a, where a is a lattice constant, and are large compared with photon wave vectors, which may be neglected (cf. § 6.4).

Electric dipole coupling between photons and phonons arises from the motion of charged ions. Strong electric dipole fields are generated by lattice vibrations when oppositely charged ions move in opposite directions, i.e. in the optic mode vibrations of ionic crystals. There is no comparable polarisation in covalent crystals. Hence we expect strong absorption in ionic crystals and weak absorption in elemental semiconductors like germanium and silicon. This is indeed observed.

6.6.1 The reststrahlen band (a)

Crystals with two atoms per unit cell and hence both acoustic and optic phonon branches show all the phenomena of interest and additional atoms only increase the complexity. As an example we quote results on GaAs. The theoretical phonon spectrum of GaAs, which is partially ionic, is shown in figure 6.14. This may be compared with the spectrum obtained experimentally from neutron scattering and shown in figure 3.15.

If only one photon and one phonon are involved in the interaction the conservation rules

$$\hbar\omega(\boldsymbol{\beta}) = \hbar\omega(\boldsymbol{q}) \qquad \text{and} \qquad \boldsymbol{q} = \boldsymbol{\beta} \approx 0$$

indicate immediately that only optic phonons near $\boldsymbol{q} = 0$ can be involved.

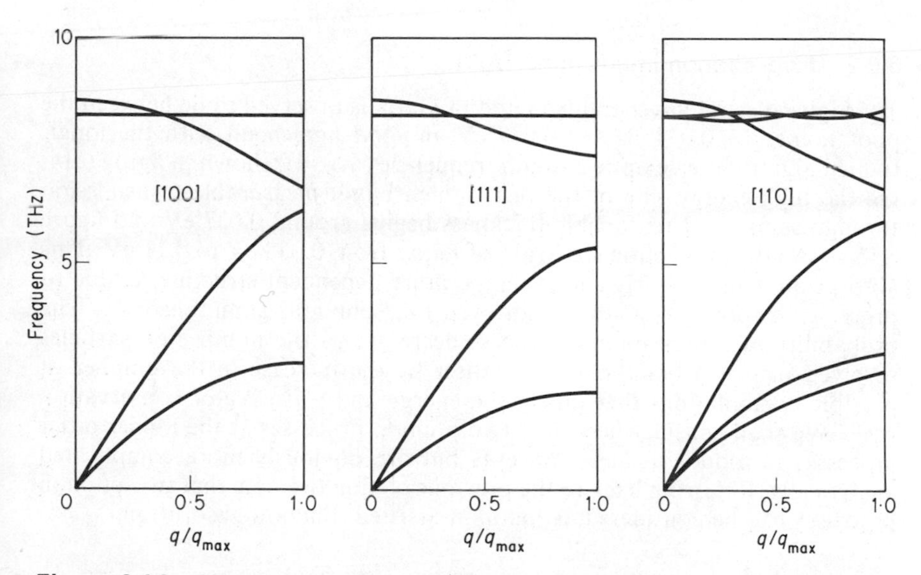

Figure 6.14 The phonon spectrum of GaAs in three principle crystal directions, based on the shell model. (After F. A. Johnson and W. Cochran, *Proceedings of the 6th International Conference on the Physics of Semiconductors*, p. 498, Inst. of Phys. and Phys. Soc., Exeter (1962).)

As the photon represents a transverse electromagnetic wave it is most strongly coupled to the transverse optic phonon particularly when both waves travel at the same phase velocity ($\omega(\beta)/\beta = \omega(q)/q$). In ionic and partially ionic crystals (e.g. GaAs) the absorption coefficients at the transverse optic mode frequency are very high and comparable with those in the fundamental absorption region. The high absorption results in a high reflectivity (6.4) but experimentally the reflectivity is greater and extends over a larger frequency range than this argument would indicate. The reflection coefficient is found to be almost unity for all frequencies between the longitudinal and transverse optic mode frequencies at $\boldsymbol{q} = 0$, a range called the reststrahlen band.* The origin of this band has in fact already been described in § 3.4 in connection with the Lyddane, Sachs and Teller relation. From (3.37) and (3.38) we find

$$\frac{\varepsilon_1(\omega)}{\varepsilon_1(\infty)} = \frac{\omega_L^2 - \omega^2}{\omega_T^2 - \omega^2} \tag{6.45}$$

so that the dielectric constant is negative for $\omega_T < \omega < \omega_L$. Reference to (6.9) shows that if ε_1 is negative the extinction coefficient and reflectivity must be large. The dielectric constant is zero at $\omega = \omega_L$. In an ionic crystal a longitudinal optic mode can be considered as an electric as well as a mechanical wave and a longitudinal electromagnetic wave is a solution of Maxwell's equations only if the dielectric constant is zero. We shall see that an analogous result is obtained for plasma oscillations of mobile electrons in solids (Chapter 7).

6.6.2 Two-phonon interactions (AD)

The high-reflectivity reststrahlen band of GaAs is observed to lie between the photon energies 0.034 eV and 0.037 eV, in good agreement with the longitudinal and transverse optic phonon frequencies at $\boldsymbol{q} = 0$ shown in figure 6.14. On the high energy side of the reststrahlen band, measurable transmission through samples of reasonable thickness begins around 0.037 eV and figure 6.15 shows the absorption spectrum of GaAs from 0.038 eV to 0.11 eV. This absorption, which clearly shows temperature-dependent structure, is due to processes involving two, three and even four phonons simultaneously. The probability of a multiphonon process decreases as the number of particles involved increases but this can be offset by the increase in the number of possible combinations that satisfy the energy and wave vector conservation rules. We shall consider here only two-phonon processes as the higher order processes introduce no new concepts but are obviously more complicated analytically. Referring back to the previous section, we note that two-phonon processes can be considered as 'phonon assisted' phonon absorption.

* From the German for residual rays. Before the development of modern infrared monochromators, reflection from ionic crystals was used to select particular infrared frequencies. After a few reflections, only the 'residual rays' were left from the broad band of frequencies emitted by a hot body.

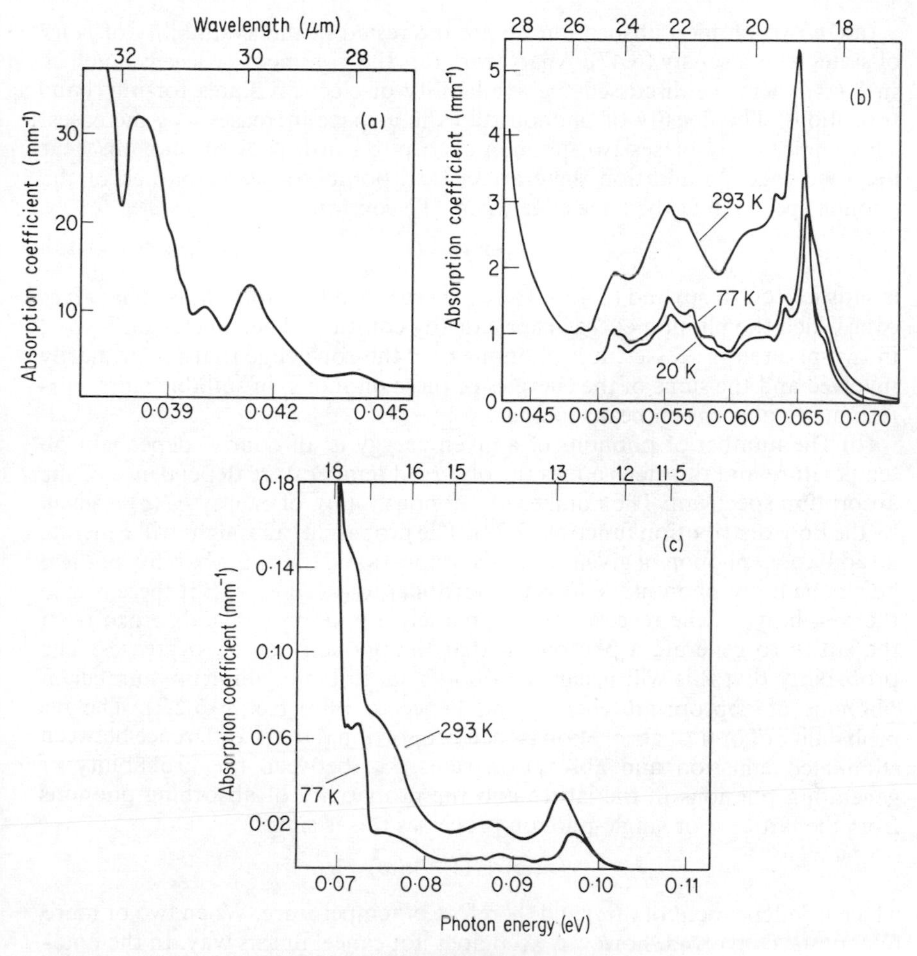

Figure 6.15 (a), (b) and (c). The absorption spectrum of GaAs from 0.038 eV to 0.11 eV. (After W. Cochran *et al., J. Appl. Phys.,* **32**, 2102 (1961).) Note changes in scale

If a single photon generates two phonons, the conservation rules require

$$\hbar\omega(\boldsymbol{\beta}) = \hbar\omega(\boldsymbol{q}_1) + \hbar\omega(\boldsymbol{q}_2) \tag{6.46}$$

and

$$\boldsymbol{\beta} = \boldsymbol{q}_1 + \boldsymbol{q}_2 \tag{6.47}$$

As before, $\boldsymbol{\beta}$ is small so that $\boldsymbol{q}_1 \approx -\boldsymbol{q}_2$, i.e. the two phonons emitted have approximately equal but oppositely directed wave vectors. The strength of the absorption at a particular frequency will depend primarily on three factors: the density of available phonon states, the phonon distribution and the matrix element of the transition. We now discuss these factors separately.

(a) In two-phonon absorption we are interested in the availability of *pairs* of states which satisfy (6.47). Apart from this the argument proceeds much as in § 6.4 where we discussed the availability of electron states for interband transitions. The density of phonon states in $\boldsymbol{q}$-space increases as $\boldsymbol{q}$ increases. This immediately biases two-phonon absorption in favour of phonons near the zone edge. In addition there are critical points for every branch of the phonon spectrum at the zone edge where the condition

$$\nabla_{\boldsymbol{q}}\, \omega(\boldsymbol{q}) = 0$$

is satisfied (cf. (6.36) and (2.47)). Hence there is a substantial range of $\boldsymbol{q}$ values over which the phonon energy is practically constant. The conclusion is that, in two-phonon processes, it is phonons near the zone edge that are primarily involved and the sums of the energies of these phonons, in suitable pairs, correspond to absorption peaks in figure 6.15.

(b) The number of phonons of a given energy is, of course, dependent on temperature and this determines the observed temperature dependence of the absorption spectrum. The number of phonons, $n(\omega)$, of energy $\hbar\omega(\boldsymbol{q})$ is given by the Bose distribution function (2.39). The probability that light will generate an additional phonon of given energy is proportional to the probability of there being one more phonon, i.e. it is proportional to $(n(\omega) + 1)$. But there is also the possibility of the reverse process, namely a phonon being absorbed from the lattice to generate a photon, so that the phonon density decreases. The probability that this will occur is proportional to the equilibrium number of phonons of appropriate energy, and hence to $n(\omega)$ (see § 6.2.2). The net probability of optical absorption is then proportional to the difference between stimulated emission and absorption rates, i.e. between the probability of generating phonons in the lattice and the probability of absorbing phonons from the lattice. For single-phonon processes this is simply

$$(n(\omega) + 1) - n(\omega)$$

which is independent of $n(\omega)$ and therefore of temperature. When two or more phonons are involved, however, $n(\omega)$ does not cancel in this way. In the notation of (6.46) we write the probabilities of emitting phonons of energy $\hbar\omega(\boldsymbol{q}_1)$ and $\hbar\omega(\boldsymbol{q}_2)$ as $(n_1(\omega) + 1)$ and $(n_2(\omega) + 1)$, respectively. The probability that both phonons are emitted *simultaneously* is proportional to

$$(n_1(\omega) + 1)(n_2(\omega) + 1)$$

The net optical absorption is then proportional to

$$(n_1(\omega) + 1)(n_2(\omega) + 1) - n_1(\omega)\, n_2(\omega) = 1 + n_1(\omega) + n_2(\omega)$$

$$= \frac{\exp\,[\hbar(\omega(\boldsymbol{q}_1) + \omega(\boldsymbol{q}_2))/kT] - 1}{[\exp\,(\hbar\omega(\boldsymbol{q}_1)/kT) - 1]\,[\exp\,(\hbar\omega(\boldsymbol{q}_2)/kT) - 1]}$$

(c) For certain combinations of phonon pairs the matrix element of the transition vanishes on symmetry grounds. Near such points the matrix element for the transition will vary with $\boldsymbol{q}$ in much the same way as that discussed in connection with forbidden electronic transitions (§ 6.5.2). The matrix element for allowed transitions may also vary with $\boldsymbol{q}$ but the effect of this on the spectrum is usually small compared with the variation in the density of states and it can often be neglected.

6.6.3 The analysis of phonon absorption spectra in ionic crystals (A)

Taking all these considerations into account the analysis of a set of data such as that in figure 6.15, usually requires some *a priori* knowledge of the phonon spectrum. If this is not available from neutron scattering experiments (Chapter 3) some clues may be obtained from the elastic constants of the material. The energies of the optic phonon branches at $\boldsymbol{q} = 0$ can be obtained from the limits of the reststrahlen band. For materials like GaAs there are only four other phonon energies of importance (all at the zone edge) and all other absorption peaks must occur at photon energies equal to the sum of two or more of these six. There can be no two-phonon lines at energies greater than $2\hbar\omega_L$; this 'two-phonon cut off' can usually be identified by a marked drop in the general level of the absorption. Finally a tentative assignment of given phonon combinations to particular peaks can be checked by measuring the temperature dependence. Even then unambiguous assignment is usually only possible if neutron scattering data is available. After assignment, however, phonon energies at the critical points can usually be determined with three-figure accuracy. As an example, the following table lists the major peaks in figure 6.15, and their origin, as deduced by Cochran *et al.*

Peak energy (eV)	Assignment
0.0955	3TO
0.0648	2TO
0.0548	TO + LA
0.0510	LO + LA
0.0413	TO + TA
0.038	LO + TA

6.6.4 Phonon absorption in non-polar crystals (AD)

Pure elemental crystals like silicon do not show absorption due to single-phonon generation as there is no electric dipole coupling. Two-phonon and higher order processes are observed in these materials. The coupling between the phonons and photons can be accounted for by the 'shell model' discussed in § 3.3.3 and illustrated in figure 3.10. One phonon wave can be considered as polarising the lattice by displacing and distorting the charge clouds around the atoms, thus inducing a charge on them. The second phonon and photon now interact *via* the induced electric dipole moment. Analysis of the resulting two, three and multi-phonon spectra proceeds in exactly the way described above for ionic and partially ionic crystals.

6.6.5 Localised modes (A)

The effect of defects on the normal modes of lattice vibrations was discussed in § 5.6.1. Because the defects destroy the lattice periodicity all the normal modes may cause absorption of radiation provided there is a coupling mechanism. The local charges and local polarisation induced by charge cloud distortion near a defect is the most important cause of this coupling.

The most striking optical effects are the sharp lines arising from localised modes or narrow resonances. An example of the former is seen in figure 6.16 and the latter in figure 5.21.

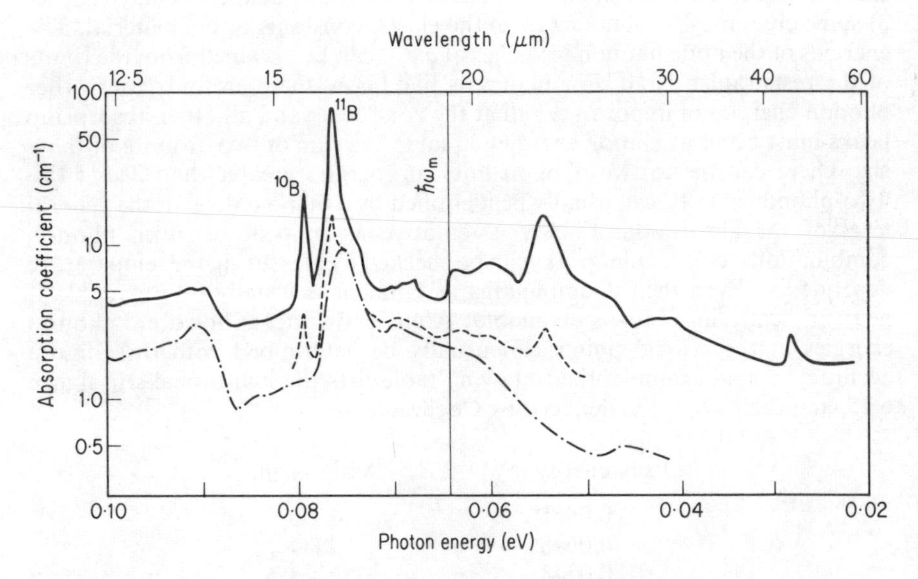

Figure 6.16 Local phonon modes due to boron (isotopes 10 and 11) in silicon. Full line —— 5×10^{19} cm^{-3} boron impurities; Dashed – – – 5×10^{18} cm^{-3}; Dashed and dotted –·–·– nominally pure silicon. (After J. F. Angress, A. R. Goodwin and S. D. Smith, *Proc. Roy. Soc.*, **287**, 64 (1965).)

6.7 Optical emission from solids (a)

The return of, say, an excited electron to lower energy states may be accompanied by the emission of photons, phonons or both. The rate of spontaneous emission from a two level system, as described in § 6.2.2, is given by AN_2. Using equations (6.17), (6.18), (6.20) and (6.21) we find

$$AN_2 = \frac{\omega^3 \displaystyle\int \frac{K}{\omega} d\omega}{\pi^2 c^2[\exp(\hbar\omega/kT) - 1]} \tag{6.48}$$

so that it might be concluded that nothing could be learnt from the study of photon emission which was not already available from absorption data. This would be false. In fact (6.48) describes the emission only if the absorption and emission processes considered are thermodynamically reversible. Thus absorption accompanied by phonon emission must be compared only with photon emission accompanied by the absorption of a similar phonon. In general absorption and emission form part of an irreversible cycle and emission measurements can give information not available from absorption data. In any case there are more compelling practical reasons for interest in emission. Firstly emission is often easier to measure than absorption and secondly the applications of emission from solids (e.g. fluorescent lighting and television) are far more numerous and important than the applications of absorption.

In the following sections we discuss four main aspects of optical emission from solids, namely (a) the many types of transition observed; (b) excitation mechanisms; (c) the mechanisms that delay emission and give rise to luminescence 'afterglow'; and (d) Laser action. Consequent on their high absorption at almost all frequencies, metals do not emit radiation other than black body radiation so we will be concerned exclusively with semiconductors and insulators.

6.7.1 A survey of photon emission processes in solids (a)

An electron excited to a high energy state in the conduction band will eventually recombine with a hole in the valence band. Some of the possible electronic transitions which may take place with the emission of radiation are shown schematically in figure 6.17 and detailed below.

Processes A and B. An electron high in the conduction band of an insulator comes into thermal equilibrium with the lattice ('thermalises') very rapidly. The energy relaxation rate of electrons is discussed in Chapter 8. When thermalised, electrons lie in a region $\sim kT$ wide at the bottom of the conduction band. Thermalisation may be achieved by photon emission, normally phonon assisted, or entirely by phonon emission, the latter being most likely. Photons generated in such transitions will, therefore, usually cover a broad spectral range. If their energy exceeds E_g they may be reabsorbed to generate a further electron (process A, figure 6.17).

The observation of such *intraband* emission requires rather special excitation conditions. It can be observed in reverse-biased PN junctions in which electron avalanches, analogous to gas discharges, occur (see Chapter 9). Under these conditions, *interband* transitions also occur. An example is shown in figure 6.18, the spots corresponding to avalanche regions.

Processes C and D. Interband recombination of electrons and holes with photon emission is obviously possible as the reverse process leads to strong absorption. Compared with other, competing, processes it is, however, relatively unlikely and only observed in high purity, high quality, single crystals. Interband emission from germanium is shown in figure 6.19, together with the spectrum to be expected from absorption data (6.48). The two peaks are associated with the indirect (phonon assisted) and direct transitions respectively (see § 6.5.2).

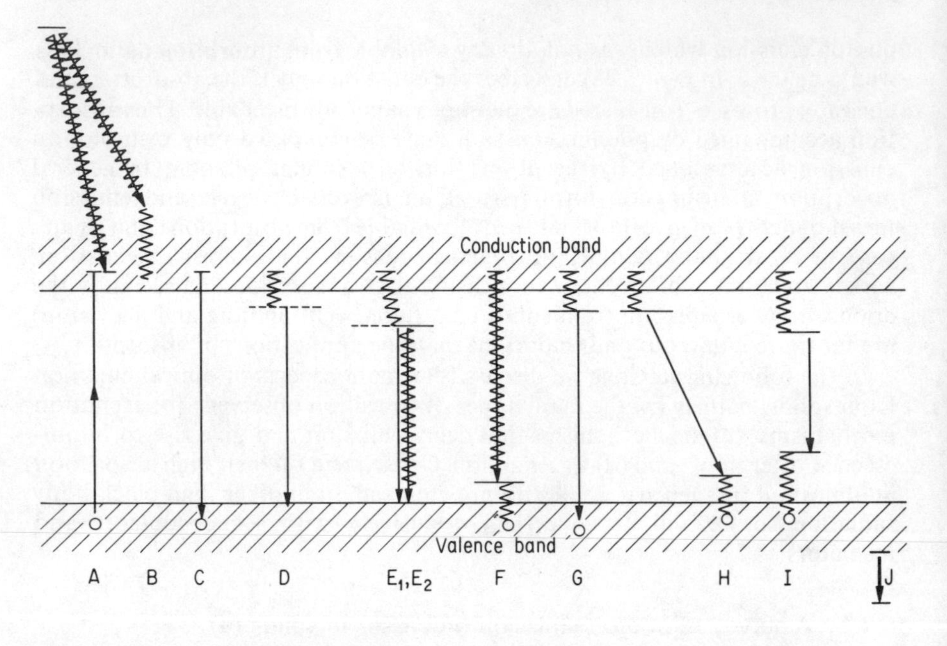

Figure 6.17 Examples of transitions leading to photo emission from solids.

photon emission   phonon emission   phonon emission, photon assisted

A and B Intraband transitions, phonon assisted to conserve wave vector
C Interband transition
D Recombination *via* exciton states
E Recombination *via* bound exciton states
F, G Conduction band to neutral acceptor, neutral donor to valence band transitions
H Donor level to acceptor level transitions
I Recombination *via* impurity excited state(s)
J Transitions between well screened energy levels of impurities

The width of the emission bands should be $\sim kT$ since the thermalised electrons and holes lie within this range of the band edges.

Superficially the agreement between the emission spectrum and that computed from absorption data is excellent but more detailed examination, particularly at low temperatures, reveals that the emission bands are narrower than, and slightly shifted in energy from, their predicted values. This is found to be due to the capture of electrons to form excitons and subsequent annihilation of excitons with the emission of photons (process D of figure 6.17). Exciton binding energies are about 10^{-2} eV in germanium so the energy shift is small. On the other hand exciton levels are well defined and sharp emission lines are to be expected. At low temperatures emission due to excitons is completely dominant in high purity germanium and hence, if excitation is direct to the conduction band, absorption and emission are irreversible.

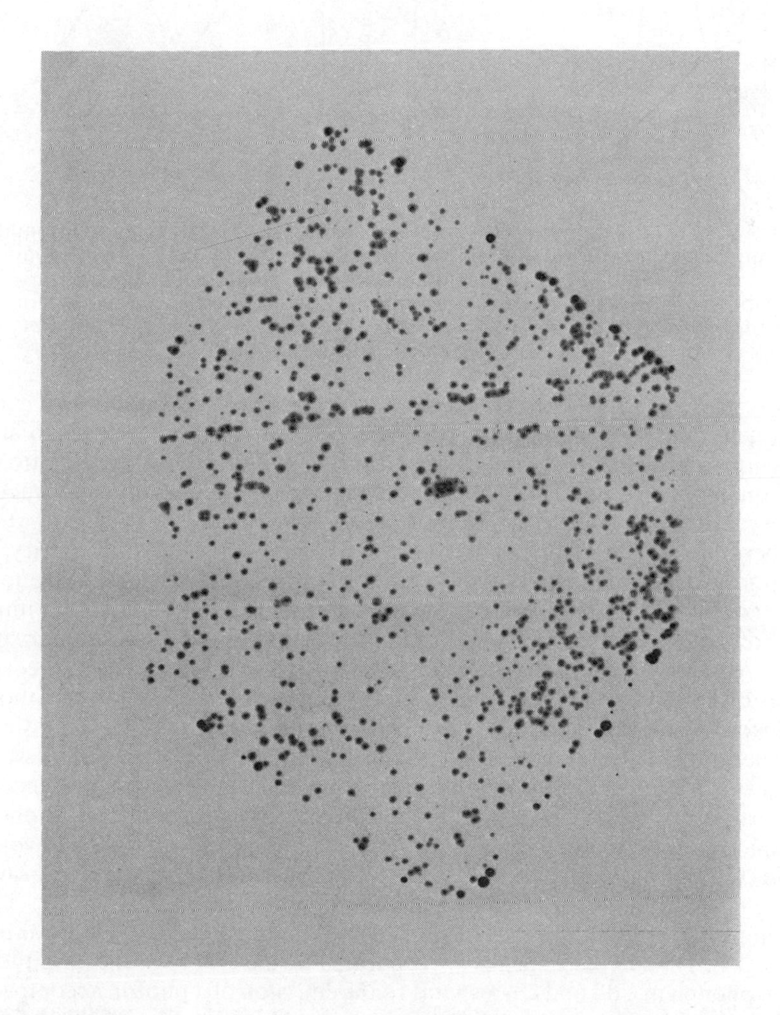

Figure 6.18 Light emission for a reverse biased silicon PN junction 2.7 μm below N type surface. (Photograph kindly provided by K. G. McKay, Bell Telephone Laboratories, Murray Hill, N.J., U.S.A.)

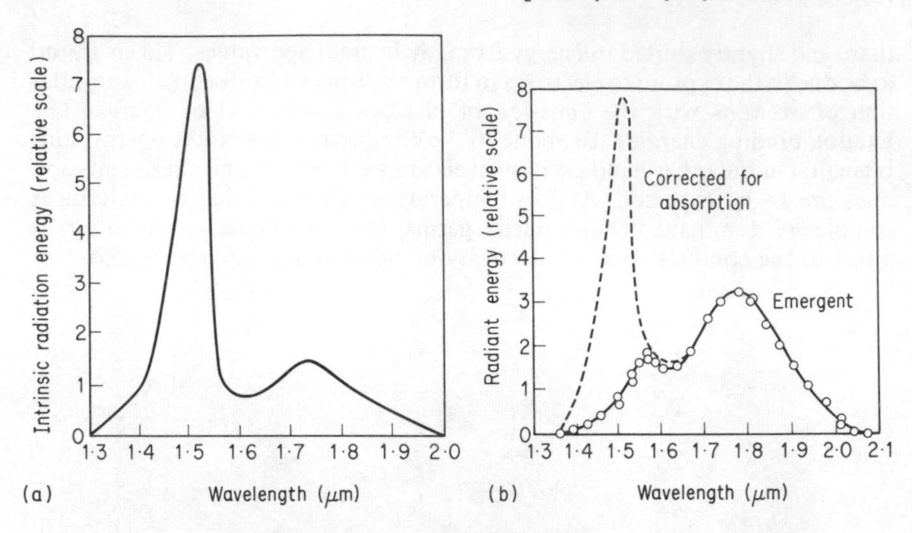

Figure 6.19 (a) Theoretical curve of emission due to interband recombination of 'thermalised' electrons and holes (process C, figure 6.17) in germanium, deduced from absorption data and equation (6.48)
(b) Observed emission from germanium. Dashed curve corrected for self absorption (i.e. process A of figure 6.17). (After J. R. Haynes, *Phys. Rev.*, **98**, 1866 (1955).)

Processes E_1 and E_2. Since germanium and silicon are not polar and can be obtained with very high purity they are excellent media in which to study emission associated with impurities. Emission due to the annihilation of excitons bound to impurities (the analogue of hydrogen molecules) was first observed in these materials. Figure 6.20 shows emission from silicon at low temperatures. The main peak at 1.10 eV, which is independent of purity, is a phonon-assisted free exciton line. As silicon is an indirect gap material, phonon assistance is essential to conserve wave vector. An exciton bound to an impurity, however, is localised in space and consequently wave vector conservation is not required. The very sharp peak which appears at 1.149 eV on the incorporation of arsenic is then interpreted as a bound exciton line *without* phonon emission. The line at 1.091 eV is displaced with respect to the 1.149 eV line by the energy of a TO phonon in silicon: it is a bound exciton line, phonon assisted.
Processes F and G. Transitions between band edges and donor and acceptor impurities are quite common in solids. Frequently observed transitions are conduction band to neutral acceptor (process F) and neutral donor to valence band (G). They may be phonon assisted. For illustration process F is shown in figure 6.17 as phonon assisted and process G as not assisted.

Figure 6.21 shows emission from CdS which is almost certainly an example of process F. The successive peaks are separated by the energy of the longitudinal optic phonon in CdS and correspond to the emission of a photon accompanied by the simultaneous emission of 0, 1, 2, 3, etc. phonons. Since CdS is a polar material electrons trapped at impurities can be strongly coupled to optic phonons, as discussed in § 5.8.

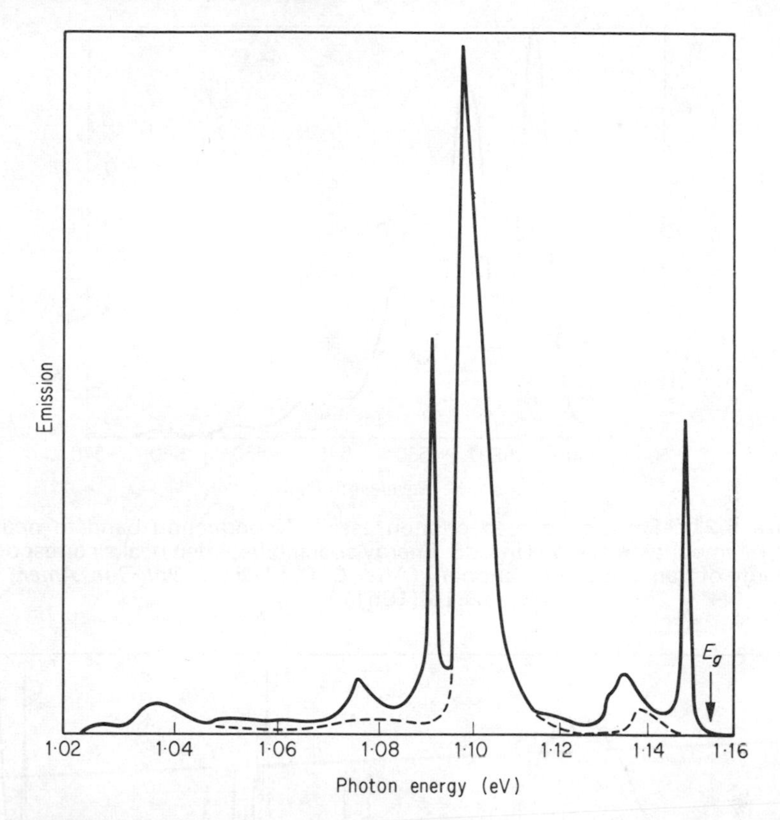

Figure 6.20 Emission from optically excited silicon at 25 K. Dashed curve, pure silicon; full curve, doped with 8×10^{16} arsenic impurities per cc. For origin of peaks, see text. (After J. R. Haynes, *Phys. Rev. Lett.*, **4**, 361 (1960).)

Process H. Electrons captured at donor impurity levels can recombine with holes captured at acceptor levels. Due to the Coulomb potential well of an impurity centre the energy of the emitted photons depends on the separation, r, of the two centres as

$$\hbar\omega(r) = E_g - (E_A + E_D) + e^2/4\pi\varepsilon r \tag{6.49a}$$

where E_A and E_D are the acceptor and donor ionisation energies. In a crystal lattice r can only have discrete values so that (6.49a) predicts a series of lines.

Figure 6.22 shows emission from gallium phosphide containing sulphur (donors) and silicon (acceptors), both occupying substitutional phosphorous sites. The possible values of r are

$$r = (\mathrm{m}/2)^{1/2}\, a \tag{6.49b}$$

where a is the lattice constant and m an integer. It will be seen that (6.49) is in excellent agreement with the data. It is particularly interesting to note that some

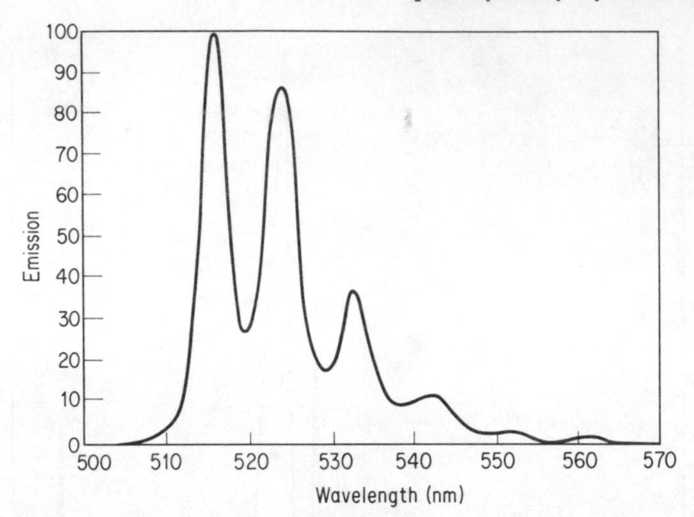

Figure 6.21 Emission due to phonon assisted conduction band to neutral acceptor impurity transitions in CdS. Energy spacing between peaks corresponds to energy of transverse optic phonon. (After C. C. Klick., *J. Opt. Soc. Amer.*, **41**, 816 (1951).)

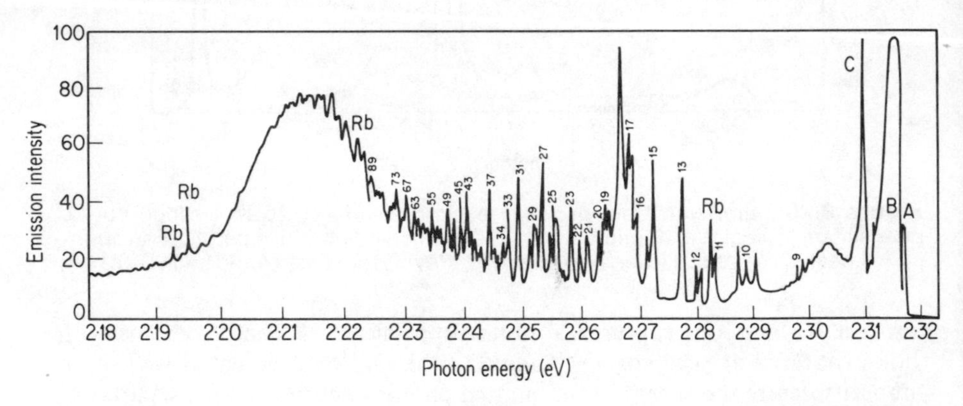

Figure 6.22 Emission due to neutral donor to neutral acceptor transitions in gallium phosphide at 1.6 K. Crystal excited by ultraviolet illumination. Peaks marked Rb, rubidium calibration lines; A, B and C are bound exciton lines (cf. figure 6.20). Digit on emission peaks indicates value of m in equation (6.49b). $E_A + E_D = 0.140$ eV. (After D. C. Thomas, M. Gershenzon and F. A. Trumbore, *Phys. Rev.*, **133**, 269 (1964).)

values of m are not possible in the GaP lattice, e.g. m = 14, 30 and 46: the gap at 2.275 eV (m = 14) is easily distinguished.

Emission spectra of this type have been observed in doubly doped Si, Ge, GaAs and other III–V compounds, with emission from donor–acceptor pairs separated by as much as 25 unit cells. The large permissible separation arises of course, from the large orbits of impurity bound electrons and holes in these high dielectric constant materials (§ 5.6.5).

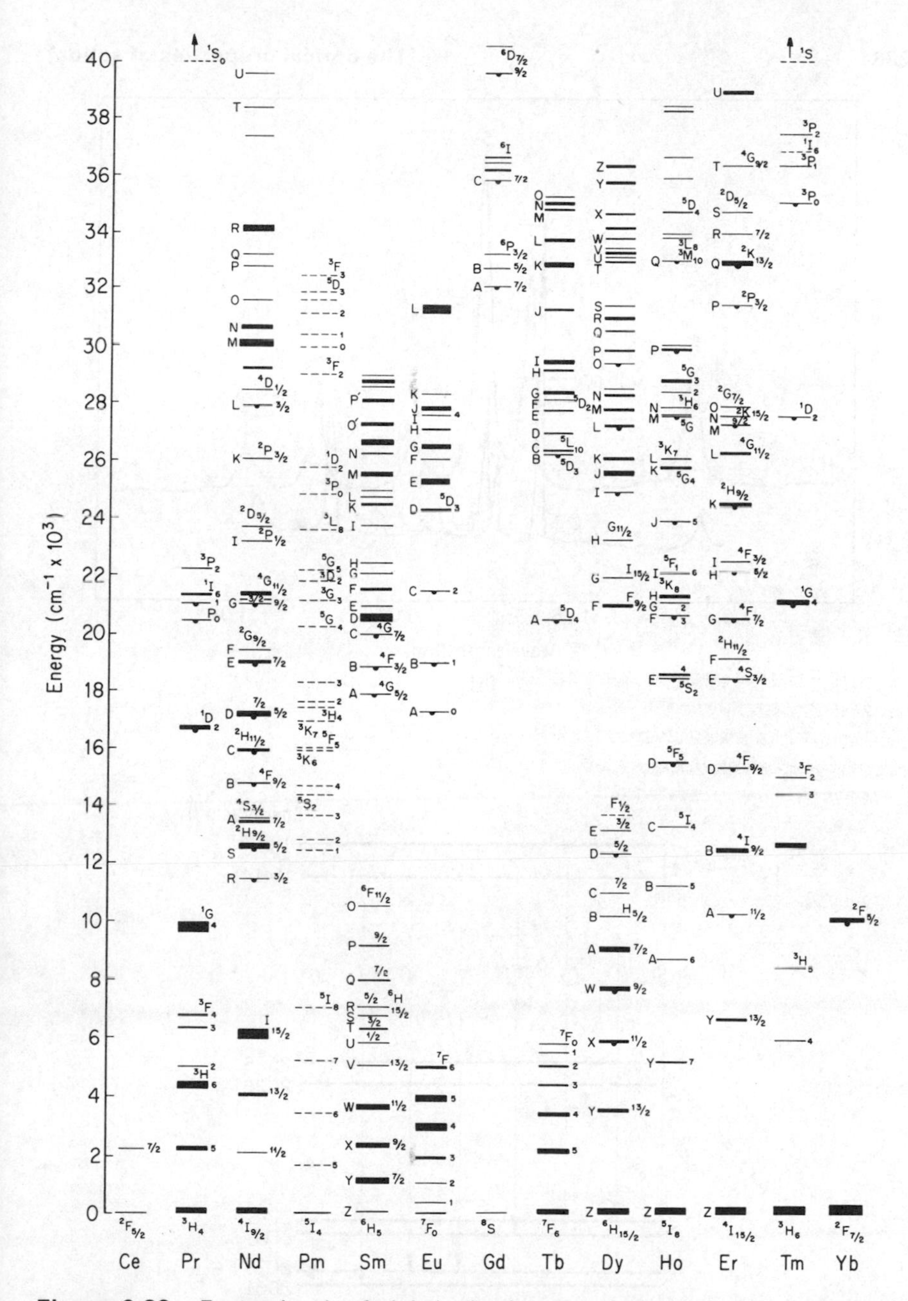

Figure 6.23 Energy levels of triply ionised rare earth ions in $LaCl_3$, determined primarily from fluorescence studies. Pendent under energy level indicates strong emission from this level. All these ions are of interest in respect of laser action (§ 6.7.6) and particularly Nd^{3+}. Strong laser action can be obtained due to transitions between the $^4F_{3/2}$(R) and the $^4I_{11/2}$ levels of Nd^{3+} in many host lattices. (Diagram from G. H. Dieke and B. Pandey, *Proceedings of the Symposium on Optical Masers*, p. 327, Brooklyn Polytechnic Institute Press (1963).)

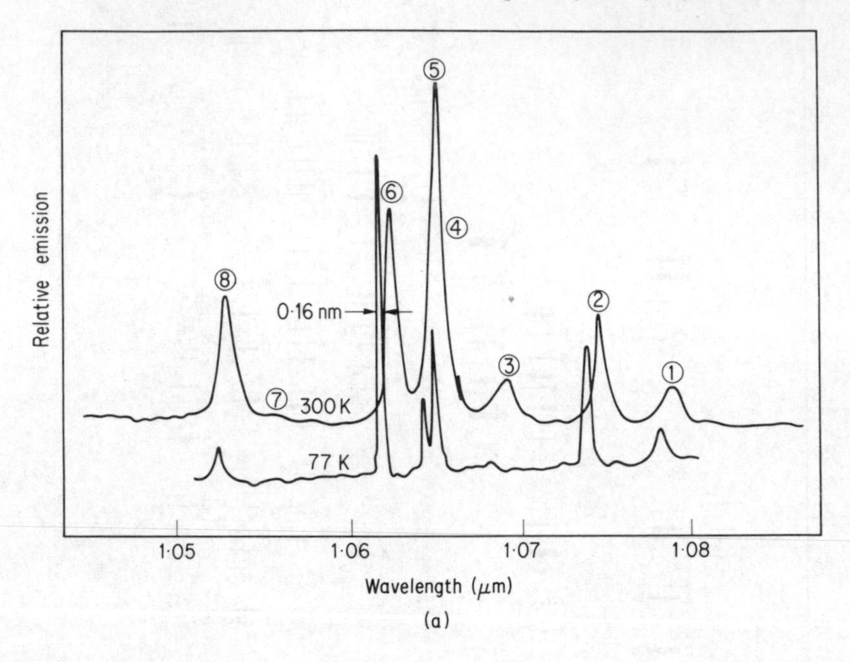

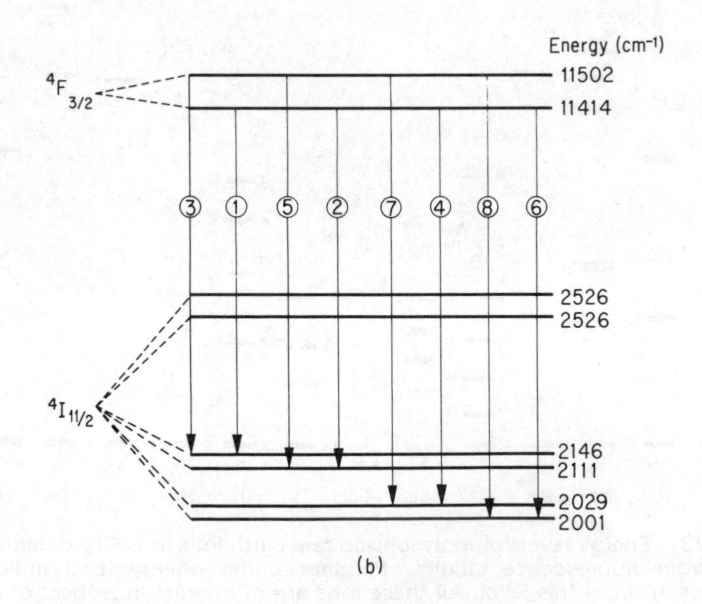

Figure 6.24 Emission spectrum of $Y_3Al_5O_{12}:Nd^{3+}$ in the 1.06 μm wavelength region and the transitions involved. Strong laser action can be obtained on lines 5 and 6. (After L. G. Van Uitert. *Luminescence of Inorganic Solids*. P. Goldberg (ed.), Academic Press (1966).)

Processes I and J. Optical transitions from excited states to the ground state of single impurities are clearly possible. I and J of figure 6.17 are intended to symbolise two extreme cases: I, impurity levels strongly coupled to the lattice and to which figure 5.42 applies and J, deep lying levels well shielded from the lattice which only slightly perturbs the atomic levels.

Transitions between levels well shielded from the crystal field produce emission spectra consisting of sharp lines related to the equivalent undisturbed atomic lines (§ 5.6.2). The energy levels of triply ionised rare earth ion impurities in $LaCl_3$, as deduced from fluorescence data, are shown in figure 6.23. The rare earth ions, particularly Nd^{3+}, are of importance in connection with lasers. One of the most efficient solid state laser materials known is $Y_3Al_5O_{12}$ (yttrium aluminium garnet, colloquially known as YAG) containing Nd^{3+} ions. Part of the emission spectrum of YAG-Nd^{3+} is shown in figure 6.24, strong laser action occurring on the lines near 1.06 μm wavelength.

In contrast, figure 6.25 shows the broad, structureless bands obtained from impurities in ZnS, a polar crystal which is the archetype of the commercially important inorganic luminescent phosphors. The broad bands probably arise as discussed in connection with figure 5.42 but further broadening may arise from the relatively high impurity densities employed, the presence of additional 'co-activator' or sensitising impurities, and crystal defects. Most commercially important luminescent materials are used in the form of microcrystalline powders.

Non-radiative Processes. Finally it should be pointed out that recombination of holes and electrons can also take place non-radiatively. Indeed in the great majority of solids non-radiative recombination with the emission of phonons is more likely than photon emission. The probability of a large energy transition, involving the simultaneous emission of many phonons, obviously falls off as the number of phonons involved increases. Hence non-radiative recombination processes are often associated with impurities and/or crystal defects which provide intermediate energy levels between conduction and valence bands. In another mechanism, which is known to occur in some solids and is referred to as the Auger process, an electron may make an interband transition on giving up energy to another conduction band electron, which is promoted to a higher state in the same band. The second electron then percolates down to the bottom of the band, the near continuum of states permitting it to do so by the successive emission of single phonons. In most solids, however, the mechanism of non-radiative recombination is not known with any certainty. If unknown impurities are involved, they are sometimes collectively described as 'deathnium'.

6.7.2 Photoluminescence (d)

Optical emission (luminescence) in solids can be excited in a variety of ways and luminescent materials are sometimes classified by the preferred mode of excitation. Excitation on irradiation by light, known as photoluminescence, can obviously take place. Thus absorption of light in the fundamental band

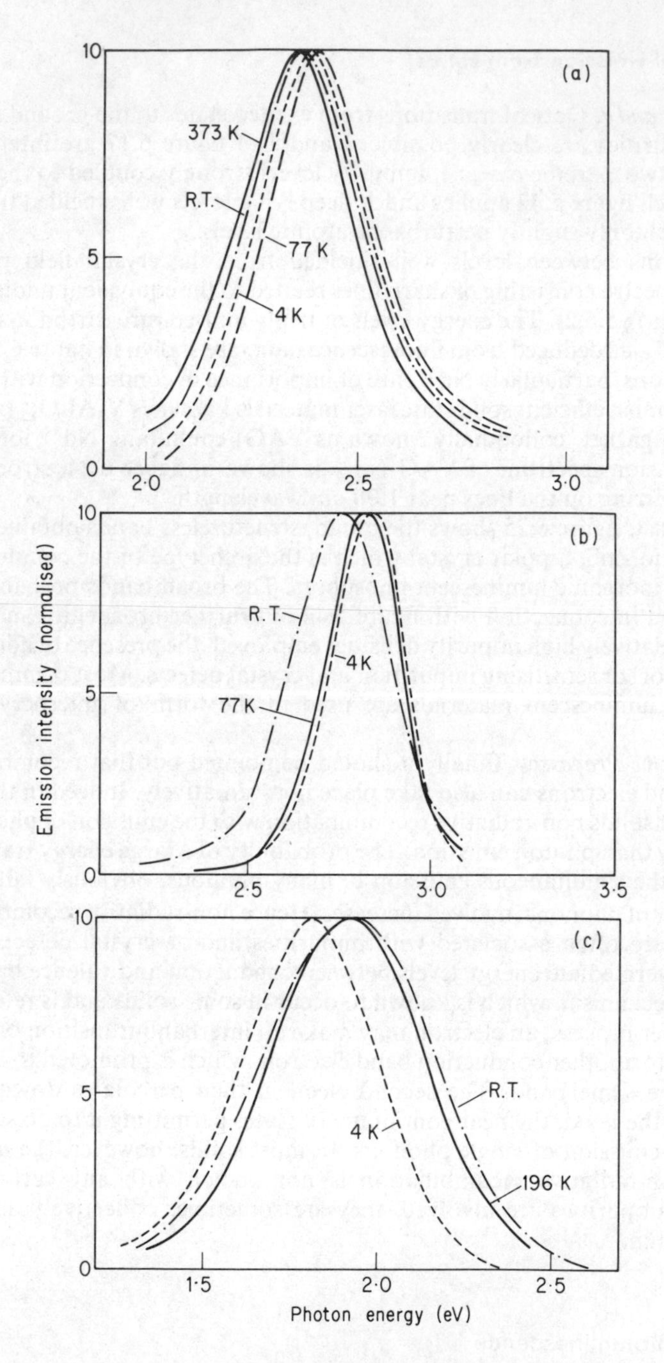

Figure 6.25 Emission spectra of ZnS phosphors under ultraviolet excitation.
(a) ZnS. Cu activator, Al coactivator;
(b) ZnS. Cu activator, I coactivator;
(c) ZnS. Cu activator, In coactivator.
(After S. Shionoya *et al., J. phys. Soc. Japan*, **19**, 1157 (1964).)

leads to the generation of holes and electrons, photoconductivity and optical emission by any of the processes A to J (figure 6.17). Photoconductivity is not, however, a necessary adjunct to luminescence and the absorption may be entirely within the 'luminescence centre' (i.e. an impurity and its immediate surroundings). Processes I and J need not be accompanied by photoconductivity and hence photoconductivity measurements can give information on the processes involved. In a centre of type I (figure 6.17) the occurrence of photoconductivity may depend on the energy of the exciting photons. Thus ZnS containing Mn^{2+} impurity shows the same broad structureless emission band centred on the orange part of the visible spectrum when excited by blue or ultraviolet light. Blue or near ultraviolet excitation does not, however, produce photoconductivity: only when the photon energy exceeds 3.9 eV (the energy gap of ZnS) does photoconductivity accompany luminescence. As we shall see, the occurrence of photoconductivity may also influence the rate of decay of luminescence after excitation.

If excitation is by absorption in the fundamental region, the absorption coefficient is so high that the exciting light penetrates only a short distance into the crystal ($\sim 10^{-5}$ cm). If, on the other hand, absorption only occurs at the luminescence centres, the absorption coefficient may be too low for the material to absorb most of the exciting light in the available crystal thickness. Both situations may result in the inefficient conversion of excitation energy to luminescence. When excitation is confined to a thin surface layer of the crystal there are, however, mechanisms capable of transferring the absorbed energy further into the crystal. These mechanisms include the migration of the electrons and holes generated by the primary excitation and the migration of excitons. The range of free migration of electrons, holes and excitons in luminescent materials is generally not known with any certainty, but is probably $\sim$0.1 mm or less. At the other extreme, when excitation is confined to the luminescence centres, too low an absorption coefficient can obviously be offset by an increased concentration of active impurities. It is found, however, that luminescence efficiency falls at high concentrations due to interaction between adjacent centres in too close proximity. Typical optimum concentrations range from about 0.01 % for type I centres in, say, ZnS up to about 5 % for well screened (type J) centres (e.g. YAG–Nd^{3+}).

It has been found empirically that the inclusion of more than one species of impurity can enhance the efficiency of photoluminescence. If the emission spectrum associated with, say, species A is only slightly changed on inclusion of species B we can identify A as the luminescence centre or 'activator' impurity and B is then called a co-activator or sensitiser. Among the numerous examples of this behaviour is that of certain rare earth ions in CaF_2. Ce^{3+} acts as a sensitiser for Tb^{3+}, Eu^{2+} for Eu^{3+} and Tb^{3+} for Eu^{3+}. Note that more than one species may sensitise a given activator and an ion like Tb^{3+} may play the role of activator or sensitiser. This behaviour is further illustrated in figure 6.26. The effect of some sensitising impurities in ZnS was illustrated in figure 6.25. Here the sensitiser does materially change the emission spectrum and the classification into activator and sensitiser arises because the latter alone produces negligible luminescence.

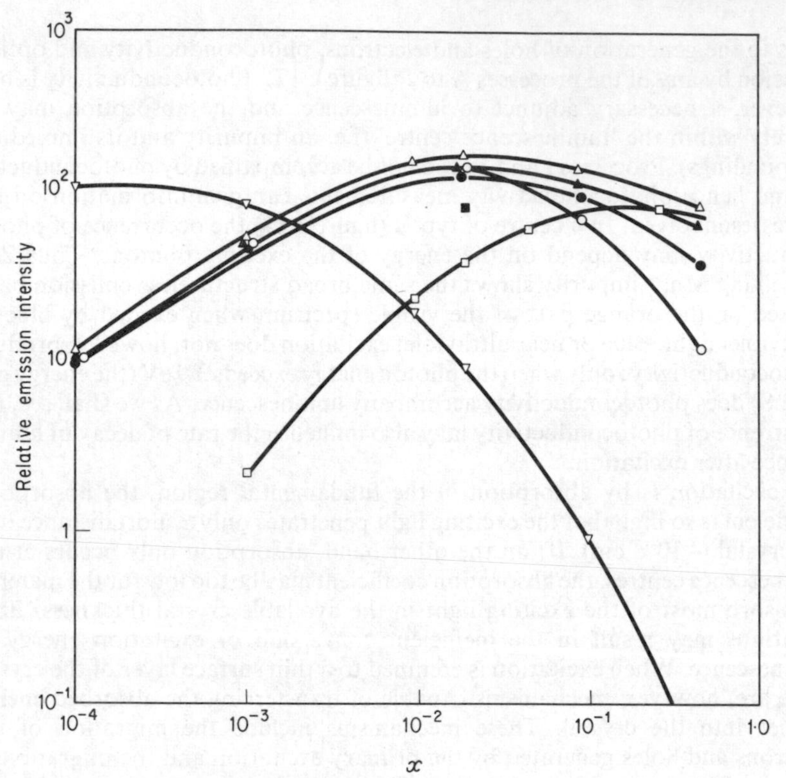

Figure 6.26 Relative emission intensities of Tb^{3+} or Eu^{3+} *versus* Eu^{3+} concentration, *x*. Excitation by (A) 253.7 nm radiation, (B) 366 nm radiation and (C) broadband long wavelength ultraviolet.

Symbol	Series	Excitation	Emission
△	$Li_{0.5}Tb_{0.5-x}Eu_xWO_4$	B	Eu^{3+}
○	$Na_{0.5}Tb_{0.5-x}Eu_xWO_4$	A	Eu^{3+}
▲	$Na_{0.5}Tb_{0.5-x}Eu_xWO_4$	B	Eu^{3+}
●	$Na_{0.5}Tb_{0.5-x}Eu_xWO_4$	C	Eu^{3+}
▽	$Na_{0.5}Tb_{0.5-x}Eu_xWO_4$	A, B or C	Tb^{3+}
□	$Na_{0.5}Y_{0.5-x}Eu_xWO_4$	B	Eu^{3+}
■	$Na_{0.5}Y_{0.24}Tb_{0.24}Eu_{0.02}WO_4$	B	Eu^{3+}

(After L. G. Van Uitert and R. R. Soden, *J. chem. Phys.*, **36**, 1289 (1962).)

The action of sensitisers is far from fully understood. Some undoubtedly increase and widen the absorption band(s) of the activator and hence can increase the efficiency of excitation. Examples are shown in figure 6.27. Some provided charge compensation (e.g. Cl in ZnS–Cu) as discussed in § 5.5.2 and may increase the fraction of activator impurities in favourable lattice sites. Others appear to do neither of these things and their action is obscure. Not surprisingly, commercial development of luminescent materials has proceeded along semi-empirical lines. In spite of the problems photoluminescent materials can be made which convert the ultraviolet emission of the mercury discharge

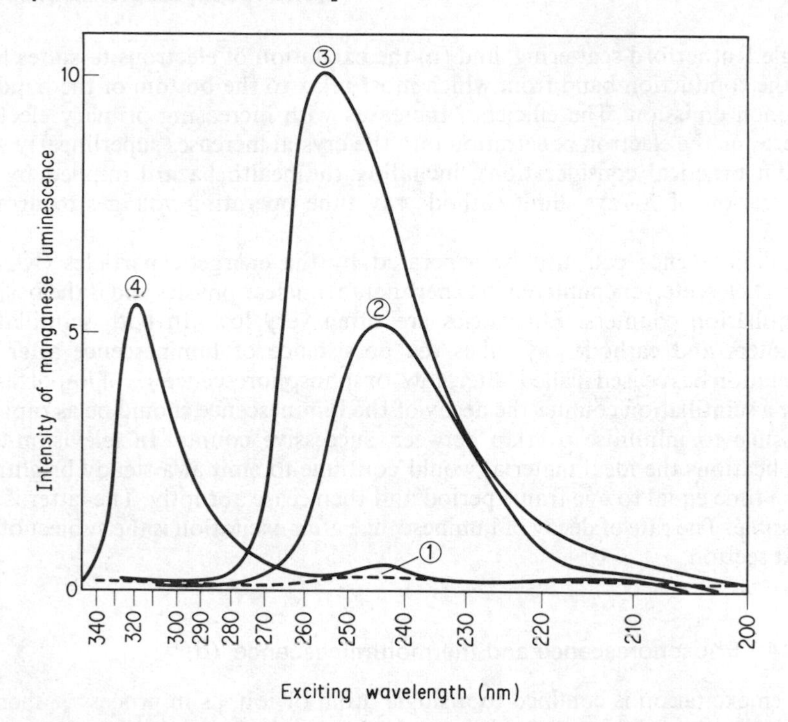

Figure 6.27 Intensity of luminescence emission from manganese impurity in $CaCO_3$ with various sensitisers as a function of wavelength of exciting radiation.
(1) No sensitising impurity;
(2) Tl sensitised;
(3) Pb sensitised;
(4) Ce sensitised.
(After J. H. Schulman *et al., J. appl. Phys.*, **18**, 732 (1947).)

to visible light with efficiencies as high as 80%. The high efficiency of fluorescent lamps is well known; the luminescent materials used in such lamps are based on Mn and Sb activated halophosphates ($Ca_5(PO_4)_3X - Sb^{3+}, Mn^{2+}$, where X is a halogen ion). The use of two activators with broad, overlapping emission bands allows nearly white light to be produced.

6.7.3 Cathodo-luminescence (d)

Electron-beam excited luminescence is called cathodo-luminescence. The obvious example is the television tube where the primary electrons reach the luminescent screen with energies of about 15 keV and there produce a shower of electron-hole pairs. Emission takes place by the processes already described.

Cathodo-luminescence is considerably less efficient than photo-luminescence and efficiencies in the range 5–15% are common. The reasons for the low efficiency include (a) backscatter of electrons at the entry surface, due to large angle Rutherford scattering, and (b) the excitation of electrons to states high in the conduction band from which most relax to the bottom of the band by phonon emission. The efficiency increases with increasing primary electron energy as the electron penetration into the crystal increases superlinearly with E but practical considerations, including the health hazard implied by the generation of X-rays, limit cathode ray tube operating voltages to around 15 kV.

Luminescence can also be generated by the energetic particles (X-rays, α-particles, etc.) encountered in experimental nuclear physics and is the basis of scintillation counters. Efficiencies are often very low. In both scintillation counters and cathode ray tubes the persistence of luminescence after the excitation has ceased (called 'afterglow' or 'phosphorescence') is of importance. For a scintillation counter the decay of the luminescence should be as rapid as possible to minimise overlap between successive counts. In television-type applications the ideal material would continue to emit at a steady brightness for a time equal to one frame period and then cease abruptly. The latter is not possible. The rate of decay of luminescence after excitation is the subject of the next section.

6.7.4 Phosphorescence and thermoluminescence (d)

When excitation is confined to a single atom or ion, as in process J, there is always an empty state available into which an excited electron may decay with the emission of radiation. The number of such transitions per second, and hence the intensity of emission, is then simply proportional to the density, N, of excited electrons. Hence

$$\frac{\mathrm{d}N}{\mathrm{d}t} = -AN \tag{6.50}$$

The decay of luminescence after cessation of excitation is exponential. Interband transitions rates, on the other hand, are proportional to the probability of both a free electron and a free hole having the appropriate E and $\boldsymbol{k}$ values simultaneously. Assuming the electron and hole densities to be equal and equal to N

$$\frac{\mathrm{d}N}{\mathrm{d}t} = -GN^2$$

or

$$N = 1/G(t + t_0) \tag{6.51}$$

where G and t_0 are constants. The decay is hyperbolic and the light emission, in photons per second, is proportional to N^2.

All the emission processes discussed in connection with figure 6.17 should have exponential or hyperbolic decays. Some do not. Moreover, the time constants predicted are of the order of 1 μs or less—possibly up to 1 ms for a forbidden transition—while it is well known that some phosphorescent materials will continue to glow for many hours. It appears that the reaction rates just described determine the initial decay but that the long term luminescence is due to the release of electrons (or holes) from 'traps'.

We postulate the existence of isolated states (due to impurities or other crystal defects) lying below the conduction band and which may trap electrons from the conduction band. We suppose that the probability of electronic transitions from these states to holes in the valence band is negligibly small but thermal excitation to the conduction band is possible. The probability of the latter transition per unit time will be of the form $\Gamma \exp(E_t/kT)$ where E_t is the energy depth of the trap below the conduction band and Γ is some constant. Experimentally Γ is found to be about 10^{-8} s^{-1} in many luminescent solids. The traps store electrons for a time determined by their depth and the temperature. On escape from a trap an electron may be either retrapped or captured at a luminescent centre with the emission of radiation. These processes are illustrated schematically in figure 6.28.

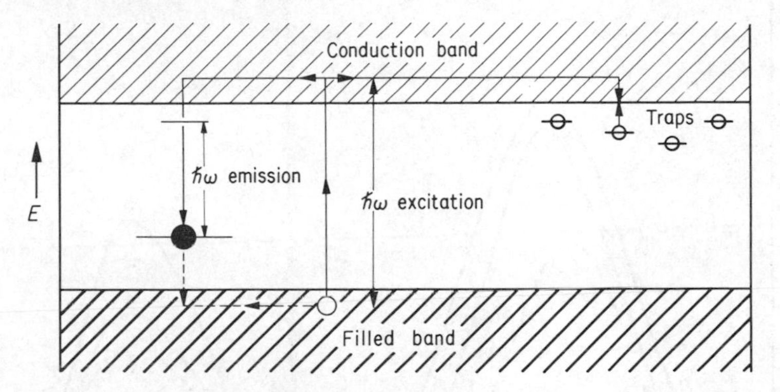

Figure 6.28 A possible electron trap mechanism for long-term phosphorescence and thermoluminescence

On this hypothesis long afterglow luminescent materials should, after excitation and some decay, produce a burst of luminescence on heating due to thermal excitation from traps. This they do, the phenomenon being known as 'thermoluminescence'. If the initial excitation is at low temperatures and the temperature is subsequently raised rapidly and uniformly in the dark, the thermoluminescence should be bright whenever a group of traps of given depth are thermally emptied. In this way thermoluminescence measurements are capable of indicating the distribution of traps as a function of depth which can be correlated with the observed luminescence decay rate at various fixed temperatures. Some representative thermoluminescence curves are shown in figure 6.29.

Thermoluminescence has some interesting scientific applications. The storage of electrons in traps provides a mechanism for integrating the effects of prolonged excitation by, for example, high energy particles. Provided the

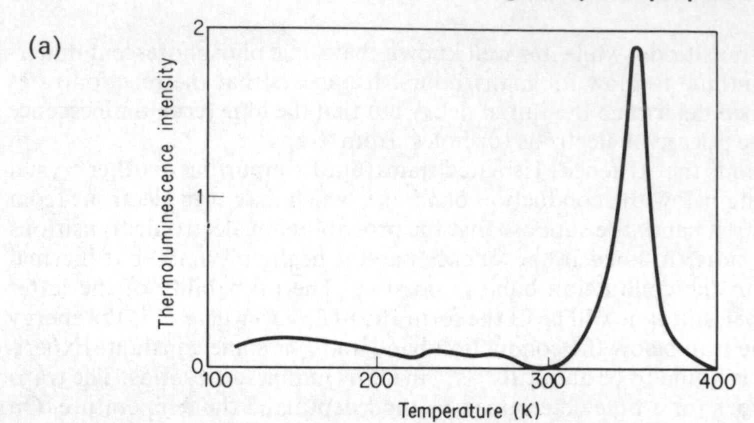

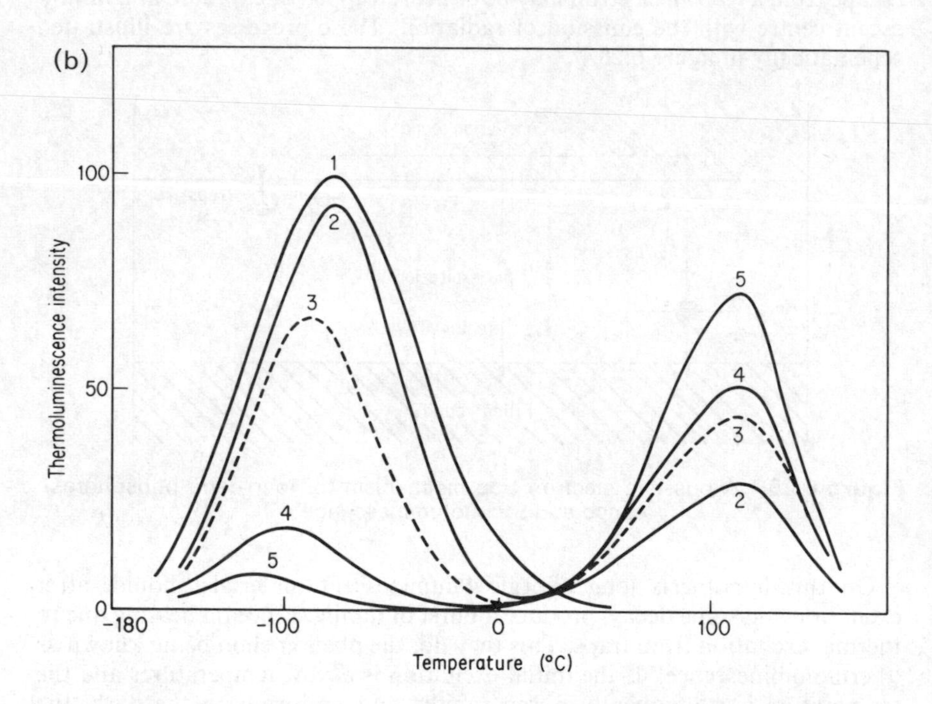

Figure 6.29 Thermoluminescence glow curves. Heating rate about 2.5 C s^{-1}.
(a) $SrSiO_3$, Eu activated. Width of main glow peak corresponds, at given heating rate, to approximately single trap depth of about 0.7 eV.
(b) ZnS, Cu activated, with Cl and Co coactivators. Cu concentration 10^{-4}. Cobalt concentrations:
1. Zero;
2. 3×10^{-6};
3. 6×10^{-6};
4. 10^{-5};
5. 1.5×10^{-5}.

(After W. Hoogenstraaten, *J. electrochem. Soc.*, **100**, 356 (1953).)

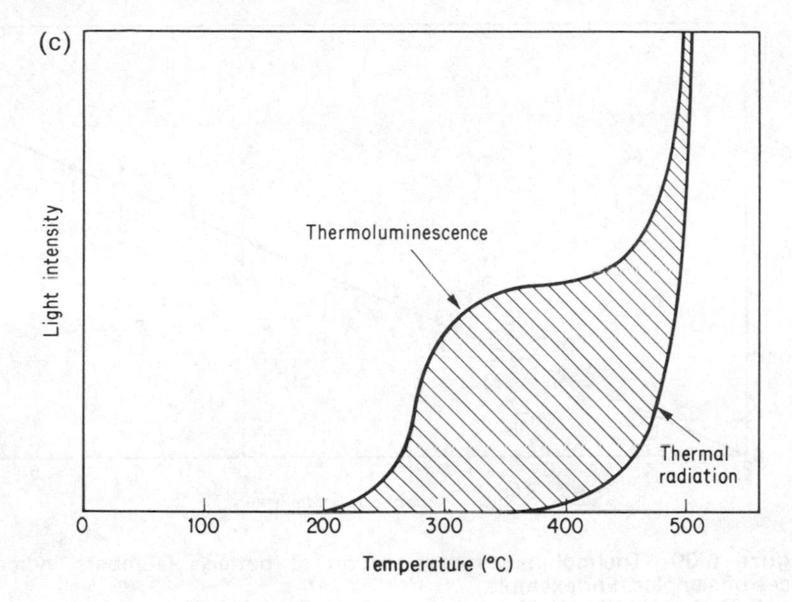

(c) Typical glow curve from ancient pottery. (From M. J. Aitkin, *Thermoluminescence of Geological materials*. D. J. McDougall (ed.), Academic Press (1968).)

traps are not filled to saturation, the integrated thermoluminescence output, when all the traps are emptied by heating, is a measure of the integrated excitation prior to heating. This effect is used in personal radiation dosemeters which are heated periodically to determine the total radiation dose which has been received since the previous heating. An even more extended application of the effect is thermoluminescence dating of geological and archaeological specimens. Pottery, for example, exposed to natural radioactivity for a long period after use as a cooking utensil exhibits thermoluminescence (figure 6.29(c)) and, if the natural radioactivity of the site is known, the elapsed time since the pot was last used may be determined. For time scales in the range 1000 to 5000 years the accuracy of the method is comparable with radio-carbon dating (see figure 6.30). For our present purpose the time scale involved confirms that transitions of electrons from traps to holes in the valence band have a truly negligible probability.

As might be expected, 'thermally stimulated conductivity' due to the release of electrons from traps accompanies thermoluminescence and indeed in non-luminescent materials trap distributions have been studied by conductivity measurements. With one or two notable exceptions (e.g. KCl–Tl) trap-controlled decay is limited to luminescent materials which are also photoconductive. Thus $ZnS–Mn^{2+}$ (see § 6.7.2) shows no photoconductivity and a decay like (6.50) after excitation by blue light but, when excited by short wavelength ultraviolet light to produce photoconductivity, the same exponential decay is followed by a long trap-controlled decay.

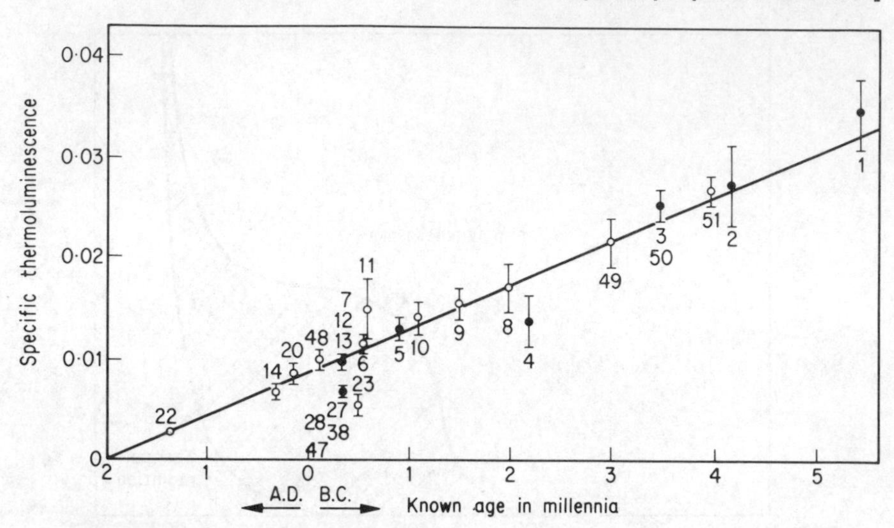

Figure 6.30 Thermoluminescence dating of pottery. Numbers indicate sources of samples. For example:

1, 2, 3, 4, 5, 9, 10 and 11 from Hasanlu, Iran
49, 50, 51 from Susa, Iran
22 from Southwest U.S.A.
23, 27, 28, 38, 47 and 48 from Athens, Greece.
6, 7, 12, 13, 14 and 20 from various parts of Italy.

(After E. K. Ralph and M. C. Han, *Thermoluminescence of Geological Materials*. D. J. McDougall (ed.), Academic Press (1968).)

Trapping centres, although particularly noticeable in luminescent materials, occur in many semiconductors and insulators. Their nature and origin is known in only a few materials. In N type germanium, for example, traps for holes are associated with the crystal surface. In ZnS phosphors the traps appear to be associated with co-activator impurities, as illustrated by figure 6.29(b).

6.7.5 Injection electroluminescence and injection lasers (d)

Electroluminescence is the excitation of luminescence by electric fields. We will consider only 'injection electroluminescence' in single crystal PN junctions and not the electroluminescence of the type observed in ZnS powders.

Junctions between P and N material in the same single crystal are discussed in Chapter 9. Here we shall make, where necessary, assertions about their electrical properties without comment.

When the P side of a PN junction is made positive, current is carried by a flow of electrons into the P material, where there are already free holes in equilibrium, and conversely holes flow into the N side. Recombination can then take place and optical emission occurs by one or more of the processes described in § 6.7.1. If, however, this were all, the study of luminescence due to the recombination of electrons 'injected' into P type material and *vice versa* would give no new information. In fact we shall show that in this situation laser action is possible and in the following discussion use GaAs PN junctions as our example.

GaAs is a direct gap semiconductor, its absorption edge rises very rapidly with photon energy and radiative recombination is highly probable. With the P side positive, junctions made with low P and N impurity concentrations show strong emission at low temperatures due to exciton annihilation. When high impurity doping concentrations ($>10^{18}$ cm^{-3}) are used, however, exciton formation is no longer possible as the conductivity of the material is too high (see § 6.5) and emission is primarily due to band-impurity transitions. Frequently the emission is due to conduction band-acceptor level transitions (process F). Emission is confined to a thin layer, a few μm thick, on the P side of the junction. The peak of this emission shifts to higher photon energies as the current through the junction is increased until, at a critical current, the emission peak sharpens considerably and laser action begins.

As we saw in § 6.2, laser action requires there to be more electrons in the upper state of a radiative transition than the lower one ($g_1 N_2 > g_2 N_1$). This condition applied to GaAs conduction band-acceptor transitions requires (a) more holes than electrons in the acceptor states, i.e. the Fermi level in the P type material below the acceptor energy level and (b) more than 50% occupation by injected electrons of the states at the bottom of the conduction band in the P material. As it is not possible at any current to inject a higher density of electrons into the P side than existed in equilibrium in the N side, the latter condition requires the Fermi level in the N material to lie above the bottom of the conduction band. This is why the impurity doping levels must be high. The shift with current of the peak emission to higher photon energies indicates that the density of electrons in the P material is rising: as the band is filled up, so the centre of gravity of the electron distribution moves to higher energies in the band.

Population inversion ($g_1 N_2 > g_2 N_1$ in the notation of (6.21)) is a necessary and sufficient condition for light amplification. Any amplifying device—transistor, laser or whatever—can be employed as an oscillator if a sufficient fraction of the output is fed back to the input in the right phase to be self-sustaining. Most lasers are in fact used as oscillators, not amplifiers, feedback being provided by two external mirrors aligned on a common axis or (more usually in PN junction lasers) by reflection from the crystal surfaces. If the reflectivity of the mirrors is $\mathscr{R}$ and the amplification obtained on passing through the crystal once is G_a, the condition $G_a \mathscr{R} > 1$ implies sufficient feedback for oscillation if all other energy loss processes are neglected. In practice there are additional losses, mainly due to diffraction, for which the gain must also compensate. The useful output from the oscillator is proportional to $(1 - \mathscr{R})$.

The external mirrors or crystal surfaces form a Fabry–Perot type cavity and, under oscillation conditions, light travelling round the cavity must reproduce itself in amplitude, phase and frequency. Any path which satisfies this condition determines a *mode* of the cavity. We have already indicated that the amplification must be sufficient for the wave to reproduce itself in amplitude. For axial modes, i.e. those in which, neglecting diffraction, the light always propagates normal to the mirror surfaces, the condition on the phase is satisfied if the length of the cavity, $\mathscr{L}$, is given by

$$\mathscr{L} = m(\lambda/2n_0) \tag{6.52}$$

where m is an integer and n_0 the refractive index. In words, $\mathscr{L}$ must be an integral number of half wavelengths long. The frequencies that satisfy (6.52) are spaced at intervals of $(cn_0/2\mathscr{L})$. The amplification will, of course, be a function of frequency with a sharp peak centred on the emission line of the radiative transition. Oscillation can only occur at mode frequencies such that the amplification is sufficient to compensate for the losses: at low amplification levels only one mode may oscillate but the number will increase as the amplification increases.

The features discussed above are illustrated in figure 6.31. Figure 6.31(a) shows the spontaneous emission of heavily doped GaAs PN junctions and the shift of the peak emission to higher energies with increasing current. Figure 6.31(b) shows the Fabry–Perot cavity modes affecting spontaneous emission. A little above the minimum threshold current for laser action (figure 6.31(c)) oscillation occurs on a single mode. At currents well above threshold more than one mode can oscillate simultaneously.

Some rough, order of magnitude, figures for the performance of GaAs junction lasers can be obtained from the following simple argument. Electrons penetrate the P material a distance (see Chapter 9) of order $(D\tau)^{1/2}$ before recombination. The diffusion constant for electrons in GaAs, D, $\sim 10^{-2}$ m^2 s^{-1} and τ, the mean time before recombination due to any cause, $\sim 10^{-9}$ s. Hence $(D\tau)^{1/2} \sim 3$ μm and laser action is confined to a sheet of this thickness near the junction, as observed. If the junction area is $\mathscr{A}$ the active volume is $\mathscr{A}(D\tau)^{1/2}$ and within this volume an electron density, N, large enough to invert the population must be maintained. The recombination rate per unit volume is $\sim N/\tau$ so the current density, J, must exceed J_c where

$$J_c \sim eN(D/\tau)^{1/2}$$

which, for $N \sim 10^{23}$ m^{-3}, gives $J_c \sim 50$ amps per mm^2. So high a current density causes considerable heating so that at least a good heat sink and often refrigeration is required. The value of τ determines the speed with which the emission responds to changes in junction current: with $\tau \sim 10^{-9}$ s the output can be modulated at frequencies up to about 1 GHz, which is a valuable feature of GaAs lasers.

Laser action can and has been obtained, at least at liquid He temperatures, in most III-V and II-VI semiconductors which show direct and allowed transitions at their absorption edges. PN junctions can be fabricated in only a few of these. Where injection electroluminescence is not possible, electron-beam and optical excitation have been used. Laser action has not been obtained in indirect gap materials such as Si, Ge and GaP. As phonon assistance is required, optical transition probabilities in these materials are two or three orders of magnitude lower and too low compared with other competing, non-radiative processes.

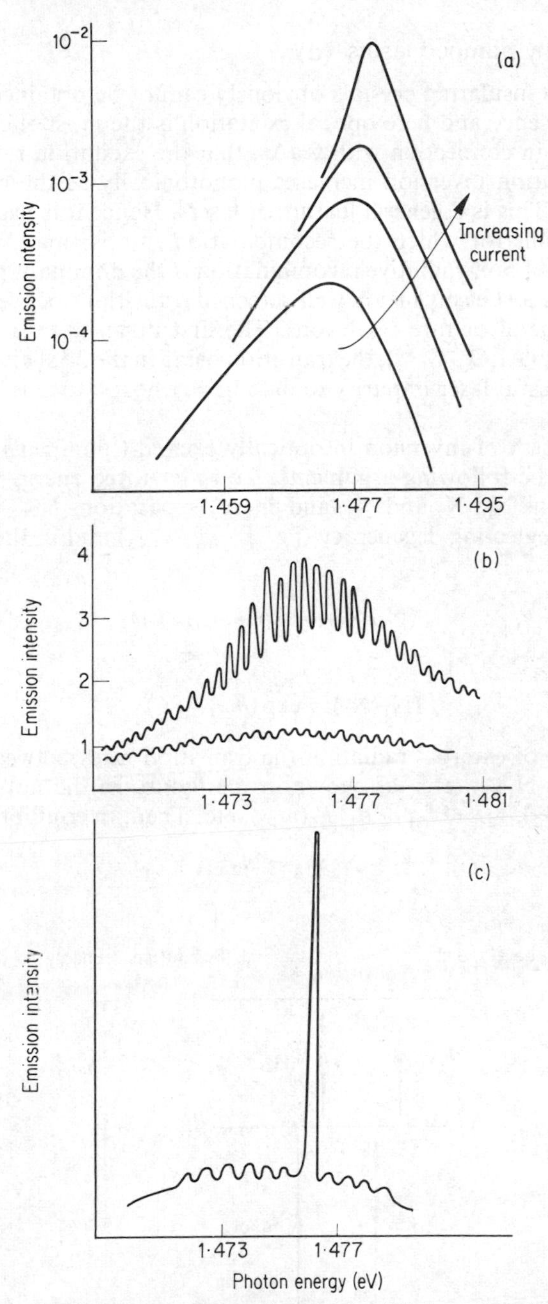

Figure 6.31. Emission spectrum of a GaAs PN junction in a Fabry–Perot cavity. (a) Low resolution; (b) high resolution; (c) at onset of laser oscillations. (After P. R. Thornton, *The Physics of Electroluminescent Devices*. E. & F. N. Spon Ltd. (1967).)

6.7.6 Optically pumped lasers (d)

Laser action in insulating crystals obviously cannot be obtained by injection electroluminescence and here optical excitation is the most effective method. We have seen, in connection with GaAs, that the excitation rate required to achieve population inversion increases monotonically as the recombination rate increases. This is a general feature of lasers. Hence it is usually desirable to use transitions for which the recombination rate is small, provided that emission and not non-radiative recombination is the dominant process. These conditions are most easily met in well-screened transitions between deep levels in transition metal or rare earth ions. The first material ever to show laser action was ruby (Al_2O_3–Cr^{3+}), the transition being in the 3d shell of chromium; the most successful laser impurity to date in any host lattice is Nd^{3+} (4f-shell transition).

The mechanism of inversion in optically excited ('pumped') lasers can be illustrated by the following arguments. Consider three energy levels, 0, 1, 2 with populations N_0, N_1 and N_2, and energy separations $\hbar\omega_{01}$ and $\hbar\omega_{12}$ (see figure 6.32). Neglecting degeneracy (i.e. $g_1 = g_2 = g_3$) and in thermal equilibrium we have

$$(N_0/N_1) = \exp(\hbar\omega_{01}/kT) \tag{6.53}$$

and

$$(N_1/N_2) = \exp(\hbar\omega_{12}/kT) \tag{6.54}$$

In the absence of external radiation the transition rates between these levels are designated $N_1\nu_{12}$, etc., as shown in the figure. In the notation of § 6.2, $\nu_{12} = B_{12}E_0(\omega_{12})$, $\nu_{21} = A_{21} + B_{12}E_0(\omega_{12})$, etc. Then, in equilibrium

$$N_1\nu_{12} = N_2\nu_{21}$$

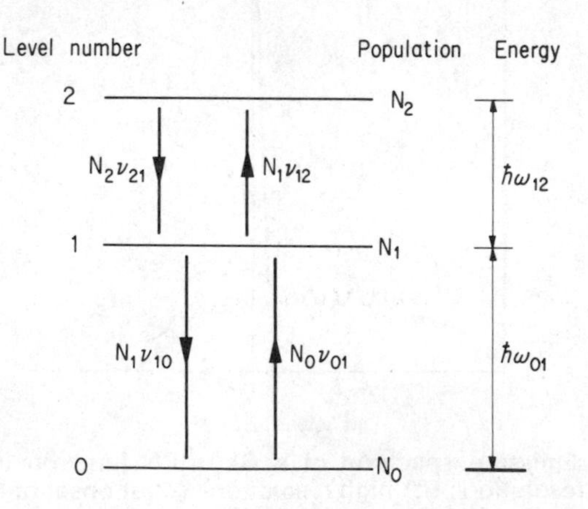

Figure 6.32 Three level system in thermal equilibrium

and

$$N_1 \nu_{10} = N_0 \nu_{01}$$

so that

$$(\nu_{21}/\nu_{12}) = \exp(\hbar\omega_{12}/kT) \tag{6.55}$$

and

$$(\nu_{10}/\nu_{01}) = \exp(\hbar\omega_{01}/kT) \tag{6.56}$$

The crystal is now illuminated (pumped) so strongly with radiation corresponding to the $0 \rightarrow 2$ transition (photon energy $\hbar(\omega_{01} + \omega_{12})$) that the populations of the 0 and 2 levels are virtually equalised. This does not affect the transition rates ν_{10}, ν_{12}, etc., as the pump photons have too high a frequency. A new steady state is now established such that

$$(dN_1/dt) = 0 = N_2(\nu_{21} + \nu_{01}) - N_1(\nu_{12} + \nu_{10})$$

putting $N_0 = N_2$. Using (6.55) and (6.56) we obtain

$$\frac{N_2}{N_1} = \frac{N_0}{N_1} = \frac{\exp(-\hbar\omega_{12}/kT) + (\nu_{10}/\nu_{21})}{1 + (\nu_{10}/\nu_{21})\exp(-\hbar\omega_{01}/kT)} \tag{6.57}$$

The above expression equals unity if $\nu_{10} = \nu_{21}$ and $\omega_{01} = \omega_{12}$. In general, however, these conditions are not satisfied and

$$N_2/N_1 = N_0/N_1 \lesseqgtr 1$$

If $N_2/N_1 > 1$, population inversion is achieved between the upper pair of levels. If N_2/N_1 is less than unity inversion occurs between the lower pair.

Thus population inversion can usually be obtained between one pair of levels in a three-level system by pumping between the outer levels. We have assumed equality of population between the outer levels although strictly this would require infinite pump power. The actual pump power needed to achieve adequate inversion will depend on the absolute values of ν_{10} and ν_{21}, rather than their ratio only. The smaller these values, the lower the pump power required.

It can be shown by the methods used above that greater inversion, for a given pump power, can be obtained in four or multi-level systems. Physically this can be seen by adding another level between 0 and 1 in the previous model. This will have a lower population than N_0 so that if pumping makes $N_1 > N_0$ the inversion between level 1 and the new level will be even greater. Cr^{3+} in ruby operates as a three-level system, Nd^{3+} in YAG, $CaWO_4$ and glass as a four-level system.

The operation of ruby-type lasers is described in many books and popular articles. Excitation is usually by the light emitted from gas discharge tubes and the cavity necessary to make an oscillator employs external mirrors. The laser material (e.g. YAG–Nd^{3+}) is usually in rod form, typically 5 to 10 cm long, though 1 m long glass-Nd^{3+} rods are used in very high power applications. Lasers of this type are often 'Q-switched'. In this mode of operation one mirror of the pair is rotated at high speed and oscillation is only possible when the two mirrors are nearly parallel. In the comparatively long period when the mirrors are not aligned a very substantial population inversion can be built up without oscillation occurring. When the mirrors become aligned oscillation begins and a substantial burst of radiation is emitted. Using this technique and subsequent amplifiers peak powers up to 10^{10} watts for a few nanoseconds can be obtained from glass-Nd^{3+} laser systems. If the beam is focused down, power densities up to nearly 10^{14} W cm^{-2} can be reached.

6.8 Non-linear optics

6.8.1 Photon—photon interactions (a)

In Chapter 3 we discussed phonon–phonon interactions which could take place through anharmonic contributions to the atomic displacements and we shall return to phonon–phonon interactions when we consider thermal conduction in Chapter 8. In this chapter we have so far considered only the interaction of photons with crystal excitations. We now go on to consider photon–photon interactions. Electromagnetic waves can be coupled in a solid in much the same way as lattice waves, i.e. through anharmonic effects, though at optical frequencies it is the anharmonic displacements of electrons bound to nuclei that are important since it is their motion, rather than the motion of the ions, which determines the high frequency dielectric constant. Anharmonic effects will be most noticeable if the electronic displacements are large, which in turn requires that the electric field of the incident radiation be large. We may therefore anticipate that photon–photon interactions will be most striking when the sources of radiation are powerful lasers. The effect of electric and magnetic fields on the optical properties of isolated atoms (the Stark and Zeeman effects) are well known. We are concerned here only with radiation fields and non-magnetic media so the effect of the magnetic components can be neglected. In solids the effects derived from the Stark effect are (a) the Pockels effect; a linear change in dielectric constant with electric field and (b) the Kerr effect; field-induced birefringence, proportional to the square of the electric field. In general we can write the polarisation of a dielectric as a series expansion in the electric field of the form

$$\mathbf{P} = \chi_1 \mathbf{F} + \chi_2 \mathbf{F}^2 + \chi_3 \mathbf{F}^3 \ldots \tag{6.58}$$

where the series converges quite rapidly, and in § 6.2 we neglected all but the first term. Photon interactions arise through the higher order terms. Thus if two beams of frequencies ω_1 and ω_2 are incident on the medium the total field $\mathbf{F}$ will be of the form

$$\mathbf{F}_1 \sin \omega_1 t + \mathbf{F}_2 \sin \omega_2 t$$

Due to the χ_2 term, the polarisation, and hence the emitted radiation, will then contain terms in

$$\chi_2 F_1^2 \cos 2\omega_1 t; \qquad \chi_2 F_2^2 \cos 2\omega_2 t; \qquad \chi_2 F_1 F_2 \cos(\omega_1 - \omega_2) t;$$
$$\chi_2 F_1 F_2 \cos(\omega_1 + \omega_2) t.$$

as well as a zero frequency (d.c.) term. These terms correspond respectively to two second harmonics ($2\omega_1$ and $2\omega_2$), the difference frequency and the sum frequency. The d.c. output (called optical rectification) and the harmonic terms can be considered as special degenerate cases ($\omega_1 = \omega_2$) of the difference and sum processes respectively. Note that the power of the radiation emitted at any frequency is proportional to the square of the polarisation and hence to χ_2^2.

To obtain strong photon–photon interactions we require (a) high incident radiation intensities, (b) materials of large χ_2 and (c) we may anticipate the need to satisfy the energy and wave vector conservation rules. We now show that for χ_2 to be non-zero, let alone large, the crystal must lack a centre of inversion symmetry. This immediately restricts the classes of crystal (see Chapter 1) that may be used. As in (3.59) we can write the potential energy of the bound electrons including anharmonic terms as

$$V(\boldsymbol{r}) = \Lambda r^2 - \Lambda' r^3 - \Lambda'' r^4 \ldots \tag{6.59}$$

where $\boldsymbol{r}$ is the displacement and each term corresponds to a term in (6.58). We conclude that if Λ' is not equal to zero

$$V(\boldsymbol{r}) \neq V(-\boldsymbol{r})$$

and the crystal lacks inversion symmetry in the direction $\boldsymbol{r}$. It should be noted, however, that Λ'' and χ_3 need not be zero (though they may be small) in isotropic media.

Since a crystal to be used for photon mixing must be transparent and large compared with the wavelength of any wave involved, the position of the photons is uncertain and the wave vector is a good quantum number which must be conserved in an interaction. It is convenient to use a general notation in which the energy and wave vector conservation rules are always written

$$\omega_3 = \omega_2 + \omega_1 \tag{6.60}$$

and

$$\boldsymbol{\beta}_3 = \boldsymbol{\beta}_2 + \boldsymbol{\beta}_1 \tag{6.61}$$

where the ω's and $\boldsymbol{\beta}$'s are the frequencies and wave vectors of the photons involved. We then suppose that only one of the various possible processes of harmonic, sum or difference-frequency generation can occur at any one time. Thus, if we are considering the sum process we shall assume that the input contains the frequencies ω_1 and ω_2, and the output contains ω_3; if the difference process, we assume the input contains ω_3 and ω_1, the output ω_2. Equations (6.60) and (6.61) then describe the conservation rules for all situations.

Equations (6.60) and (6.61) are not, of course, independent: ω/β is the phase velocity of each wave and hence, if absorption is neglected,

$$\frac{\omega_1}{\beta_1}=\frac{c}{\mathrm{n}_{o_1}}; \qquad \frac{\omega_2}{\beta_2}=\frac{c}{\mathrm{n}_{o_2}}; \qquad \frac{\omega_3}{\beta_3}=\frac{c}{\mathrm{n}_{o_3}} \tag{6.62}$$

where c is the velocity of light and n_{o_1}, n_{o_2} and n_{o_3} are the refractive indices at the three frequencies. In general $\mathrm{n}_{o_1} \neq \mathrm{n}_{o_2} \neq \mathrm{n}_{o_3}$. Satisfying (6.60), (6.61) and (6.62) simultaneously presents a problem which is central to non-linear optics and is referred to as *phase matching*.

6.8.2 Second harmonic generation (D)

If initially we consider only the special case of second harmonic generation

$$\omega_1=\omega_2=\omega \qquad \beta_1=\beta_2=\beta(\omega)$$
$$\omega_3=2\omega \qquad \beta_3=\beta(2\omega)$$

and the phase matching conditions reduce to

$$\omega/\beta(\omega)=2\omega/\beta(2\omega)$$

i.e. the phase velocities at the primary frequency and the second harmonic frequency must be equal. It is instructive to consider what happens when phase matching is not achieved. This behaviour is illustrated in figure 6.33. At the

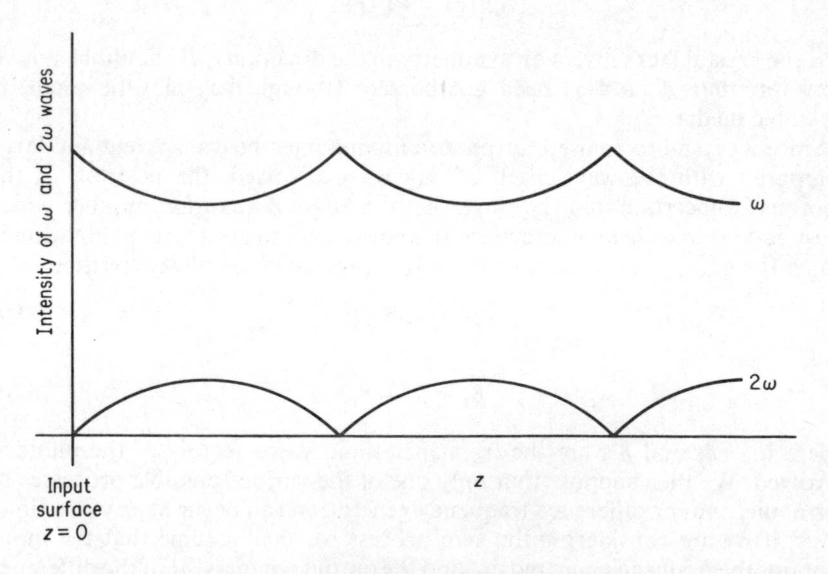

Figure 6.33 To illustrate schematically what happens in second harmonic generation in a block of non-linear dielectric when phase matching is not achieved. z is the distance into the block

input surface of the dielectric the primary wave at frequency ω generates 2ω-photons in phase with itself. It continues to do this throughout the crystal but if the two waves are not travelling with the same phase velocity there will come a time when the photons being generated by the primary wave are out of phase with the 2ω-photons that have travelled, at their velocity, from the input surface. Destructive interference then takes place and the intensity at 2ω falls, eventually to zero. The process then repeats itself. If, as we assume, there is no absorption in the dielectric the energy crossing any surface normal to the propagation direction must be constant. Hencc increasing or decreasing power at 2ω must be accompanied by decreasing or increasing power at the primary frequency, as shown in the figure.

The behaviour illustrated in figure 6.33 has been observed experimentally and correlated with the known refractive indices of the crystals at ω and 2ω. An example is shown in figure 6.34. When phase matching is achieved, however, the efficiency of second harmonic generation can be increased by as much as four orders of magnitude. Phase matching can be achieved experimentally in a variety of ways but by far the most popular is to use naturally birefringent crystals. The refractive index of birefringent crystals varies with polarisation for waves propagating down a symmetry axis. When propagation is at some angle to the symmetry axis the refractive index is double valued (see any textbook of optics). It is then possiblc to find a propagation direction in which the refractive index for the ordinary ray at ω, say, is equal to the refractive index for the extraordinary ray at 2ω. Equal phase velocities can then

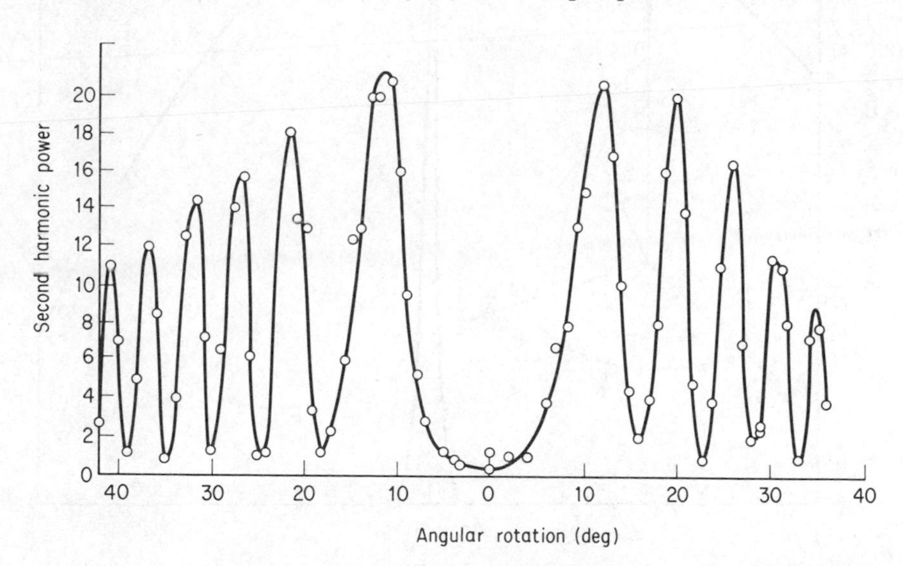

Figure 6.34 When phase matching is not achieved, intensity of second harmonic output is a periodic function of crystal thickness, from figure 6.33. The effective thickness may be varied by rotating the crystal in the beam. The experimental results refer to quartz plate, 0.077 cm thick, as the non-linear dielectric and a ruby laser. (After P. D. Maker *et al.*, *Phys. Rev. Lett.*, **8**, 21 (1962).)

be achieved. Using this technique, which is illustrated in figure 6.35, second harmonics have been generated with efficiencies as high as 30 %. The interaction volume for two waves, and hence the conversion efficiency, is limited by one of two effects: either diffraction or 'Poynting vector walk-off'. The latter arises because the phase and group velocities of extraordinary rays in birefringent crystals are not, in general, colinear. The phase velocities are fixed, in magnitude and direction, by the phase matching condition so that the wave energy, at the group velocity, 'walks out' of the crystal. This problem is often even more severe in the general case of three frequencies.

Phase matching can sometimes be achieved by adjusting the refractive index of a crystal by an external agency such as an applied electric or magnetic field. We shall show in the next chapter that the refractive index of semiconductors may be varied by varying the free-electron density and magnetic field. This approach is particularly successful in difference-frequency generation when the output is in the far infrared where free electron magnetoplasma effects (see § 7.2) can be large.

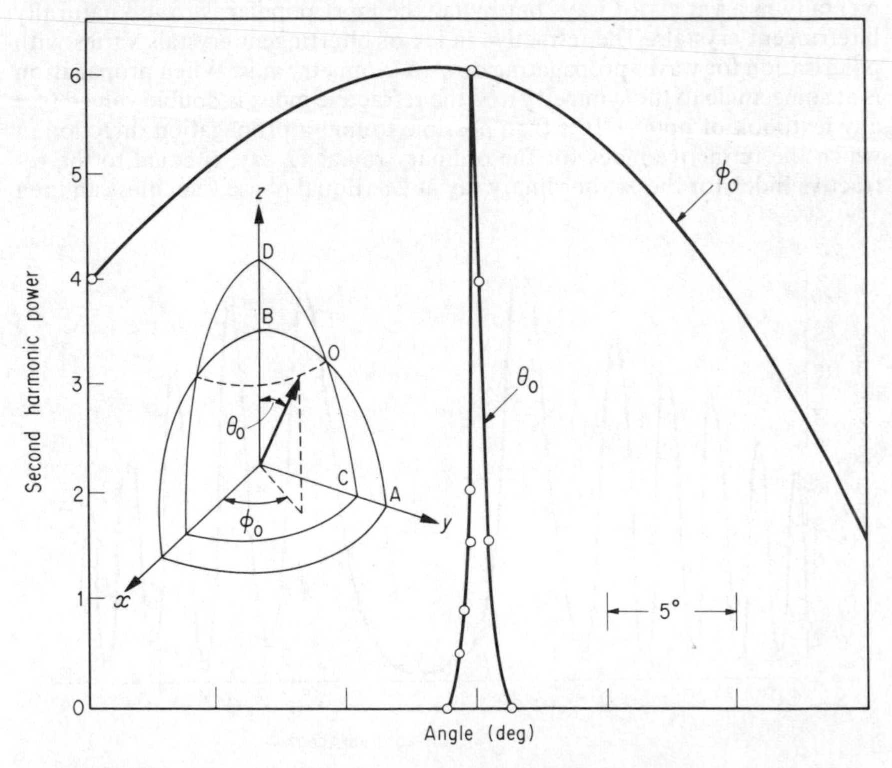

Figure 6.35. Phase matching in KH_2PO_4. Primary wave at frequency ω propagates as an ordinary wave, second harmonic at 2ω propagates as an extraordinary wave. Maximum second harmonic output is obtained with $\theta_0 = 52°$, $\phi_0 = 45°$ when pump is ruby laser. AOB is an arc on a constant index of refraction surface for ω, ordinary, COD similarly for extraordinary ray at 2ω. (After P. D. Maker *et al., Phys. Rev. Lett.* **8**, 21 (1962).)

6.8.3 Parametric amplification (D)

In connection with figure 6.33 we remarked that power was transferred continuously from the ω to the 2ω wave *and back again*. On the face of it, the latter process is subharmonic generation ($2\omega \rightarrow \omega$) but no mechanism for subharmonic generation is included in the expressions of § 6.8.1. In fact it is an example of *parametric amplification* of a signal at frequency ω by a pump at 2ω. We note again that second-harmonic generation is a special case of sum-frequency generation ($\omega_1 = \omega_2 = \omega$), referred to as degenerate. Similarly, amplification of a wave of frequency ω by a 2ω wave is degenerate parametric amplification, which occurs only when the phase difference between signal and pump waves is appropriate.

A swing is a familiar mechanical example of a degenerate parametric oscillator which we shall use to explain the mechanism of a parametric amplifier. We shall then go on to the more general case which corresponds essentially with the difference frequency generation process already discussed.

A swing is a damped harmonic oscillator described by

$$\frac{\mathrm{d}^2 r}{\mathrm{d}t^2} + 2\gamma \frac{\mathrm{d}r}{\mathrm{d}t} + \omega_0^2 r = 0 \tag{6.63}$$

We now suppose that some external agency (called the pump) varies a parameter of the motion (e.g. the effective length of the rope of a swing, the dielectric constant of a crystal) such that

$$\omega_0^2 = \omega_{00}^2(1 - \xi \sin 2\omega t) \tag{6.64}$$

where ξ is some constant less than unity so that $\omega_0 \sim \omega_{00} \sim \omega$, the frequency of the swing. We now insert (6.64) into (6.63) and seek a solution of the form

$$r = r_0 \exp(\gamma' t) \sin \omega t \tag{6.65}$$

to describe the displacement of the swing. If γ' is positive, the amplitude of the swing will grow with time; if negative, it will decay. Inserting (6.65) and solving the resulting equation for γ' we obtain, assuming $\xi < 1$ and $\gamma/\omega < 1$,

$$\gamma' \approx \xi\omega \sin^2 \omega t - \gamma$$

in which the first term is always positive and may exceed the losses represented by the damping coefficient γ. Hence a person who is swinging adjusts the position of his body to raise and lower his centre of gravity, thus altering the effective length of the swing rope. This movement is made at *twice* the natural frequency of the swing and in a suitable phase. The appropriate motion is illustrated in figure 6.36.

Light can be amplified by an analogous degenerate parametric process. The non-degenerate situation was first analysed in connection with microwave parametric amplifiers by Manley and Rowe in 1956, who produced two equations now known as the Manley–Rowe relations. We define Φ_1, Φ_2 and Φ_3 as the

Figure 6.36 A degenerate parametrically pumped oscillator. The occupant of a swing goes through one complete cycle of body movement during one half cycle of the swing

photon fluxes at frequencies ω_1, ω_2 and ω_3 crossing any plane perpendicular to the propagation direction in a non-absorbing, non-linear dielectric medium. In this notation the Manley–Rowe relations are

$$\Phi_1 + \Phi_3 = \text{constant} \tag{6.66}$$

$$\Phi_2 + \Phi_3 = \text{constant} \tag{6.67}$$

Consider, for example, the sum process (inputs ω_1 and ω_2, output ω_3). Whenever a photon at ω_3 is generated one photon at ω_1 (6.66) and one at ω_2 (6.67) vanishes. In the difference process (inputs at ω_3 and ω_1, output at ω_2 plus left-over power at ω_3 and ω_1) whenever an ω_2 photon is generated, Φ_3 decreases by one (6.67) and Φ_1 consequently *increases* by one. If we identify ω_1 with the signal and ω_3 with the pump, amplification has been obtained. In the jargon of parametric amplification ω_2 is called the 'idler' frequency as it does nothing, but must be generated.

Parametric amplification of light was first achieved in 1965 using the second harmonic of the ruby laser line as the pump. As with lasers, oscillators are more valuable than amplifiers and since that date the parametrically pumped oscillator has received most attention. A typical experimental arrangement will consist of a non-linear, birefringent crystal (e.g. $LiNbO_3$) mounted between two mirrors to provide positive feedback. Pump power is provided by a laser with the beam at a suitable angle with respect to the crystal and cavity axes. Oscillation is, in general, at two frequencies simultaneously (ω_1 and ω_2), the frequencies being such as to satisfy the phase matching conditions. The output can, as a result, be tuned by varying the crystal orientation, temperature, etc. Thus parametrically pumped light oscillators provide powerful and tunable sources of coherent radiation which seem likely to find wide application in spectroscopy in the future.

6.8.4 Three-photon mixing (D)

The third-order process, involving χ_3, can mix three photon frequencies. Since χ_3 does not vanish for any crystal symmetry, these processes may occur in any media, including liquids. Three-frequency mixing has been observed experimentally but phase matching of four different frequencies (3 in, 1 out) is extremely difficult and the effect is small except at very high powers. Among the more dramatic effects arising from the χ_3 term is self-focusing of laser beams. The beam, whose intensity falls off with radius, changes the refractive index of the medium in such a way as to bring itself to a focus inside the crystal. The consequences can be destructive. The relevant interaction term is of the form $\chi_3(\omega-\omega+\omega)$ where ω is the laser frequency. $(\omega-\omega)$ gives optical rectification and a steady potential which changes the dielectric constant seen by the remaining ω.

6.8.5 Two-photon absorption and the Raman effect (AD)

So far in this section we have considered processes involving only photons. We now consider processes involving two photons together with one or more excitations of the crystal. Interactions involving two photons and an excitation of the crystal are analogous to the one-photon, two-phonon, interactions discussed in § 6.6, coupling being provided *via* the anharmonic term χ_3 of (6.58). Although χ_3 is generally very small, in processes involving an excitation of the crystal two factors offset the smallness of the interaction term. For example, the energy of the crystal excitation may be substantially independent of wave vector over an appreciable range (cf. an optical phonon near $q=0$) which makes phase matching easy to achieve. Another possibility is that the excitation is highly localised and its wave vector indeterminate. In either case the need for ingenious phase matching techniques, essential when only photons are involved, can be avoided. The second factor is equally, if not more, important. When only photons are involved the observed quantities are proportional to the square of the amplitude and hence to χ_2^2 or χ_3^2. The density of excitations (e.g. phonons), on the other hand, is proportional to the amplitude of the wave and hence is linear in χ_3. Since χ_3 is usually of the order of χ_2^2, processes involving χ_3 and an excitation of the crystal are of the same magnitude as sum and difference frequency generation in anisotropic media. This result can also be deduced from the quantum mechanical approach to absorption outlined in § 6.2.3. Single photon absorption can be accounted for by first-order perturbation theory: the absorption coefficient is proportional to the square of a matrix element. The two-photon mixing discussed in § 6.8.1 and the processes discussed in this section appear in second-order perturbation theory and involve a matrix element to the fourth power. Three-photon mixing appears only in third-order perturbation theory.

The simultaneous absorption of two photons to generate an excitation of a crystal is, therefore, comparatively easy to observe with powerful lasers. An example is the promotion of electrons from the valence to conduction band of a semiconductor by two-photon absorption. Wave vector conservation presents no problem. Energy conservation requires

$$\hbar\omega_1 + \hbar\omega_2 \geqslant E_g$$

The probability of such an event is proportional to the probability of two photons being available and hence if $\omega_1 = \omega_2$ to the square of the incident intensity. The matrix element of a two-photon transition, unlike single photon absorption, is large only for initial and final states with the same wave function symmetry (cf. § 6.2.3). Hence single photon forbidden transitions are allowed and *vice versa* Two-photon spectroscopy, therefore, is capable of giving information not easily available from other experiments. The same statements are true of the Raman effect. As the Raman effect is generally easier to observe than two-photon absorption it is widely used as a spectroscopic technique for the study of many types of crystal excitation. We now consider the Raman effect in more detail.

In the Raman effect a single photon is absorbed and another of different frequency is emitted. The difference corresponds to an excitation of the crystal. As an example we consider Raman scattering by optical phonons. Raman scattering from other crystal excitations is, of course, also observed. Let ω_3 be the frequency of the incident radiation and ω_2 the frequency of the scattered radiation. If the third particle involved is an optical phonon of frequency $\omega(\boldsymbol{q})$ energy conservation requires

$$\omega_2 = \omega_3 \pm \omega(\boldsymbol{q}) \tag{6.68}$$

where the + sign implies that the phonon density decreases and the minus sign that it increases. Since $\omega(\boldsymbol{q})$ is practically invariant with $\boldsymbol{q}$ we can neglect wave vector conservation. The relative probabilities of the two possible processes are then proportional to $n(\omega)$ and $(n(\omega) + 1)$, respectively, where $n(\omega)$ is the Bose distribution function of the phonons. Consequently the so-called Stokes line ($\omega_2 < \omega_3$) is brighter than the anti-Stokes line ($\omega_2 > \omega_3$), particularly at low temperatures.

The Raman effect has been known and exploited for many years. Experimentally it is usual to observe the scattered radiation at right angles to the primary beam to reduce the amount of primary radiation reaching the detector. If wave-vector conservation is important the angle at which the scattered radiation is observed is, of course, significant. The ratio of scattered to primary radiation intensity is usually in the range 10^{-4} to 10^{-8}. Before lasers became commercially available, the most intense lines of a mercury discharge tube were widely used sources for Raman scattering experiments but these have been superseded by powerful continuous-wave lasers. An example of a Raman spectrum due to phonon scattering is shown in figure 6.37. The weak lines with large Raman shifts on the Stokes side are due to two-phonon Raman scattering (an interaction involving four particles in all).

When high intensity laser sources are used, Raman scattering may be accompanied by parametric amplification. Consider again the Manley–Rowe relations (6.66) and (6.67) and suppose we are concerned only with the Stokes

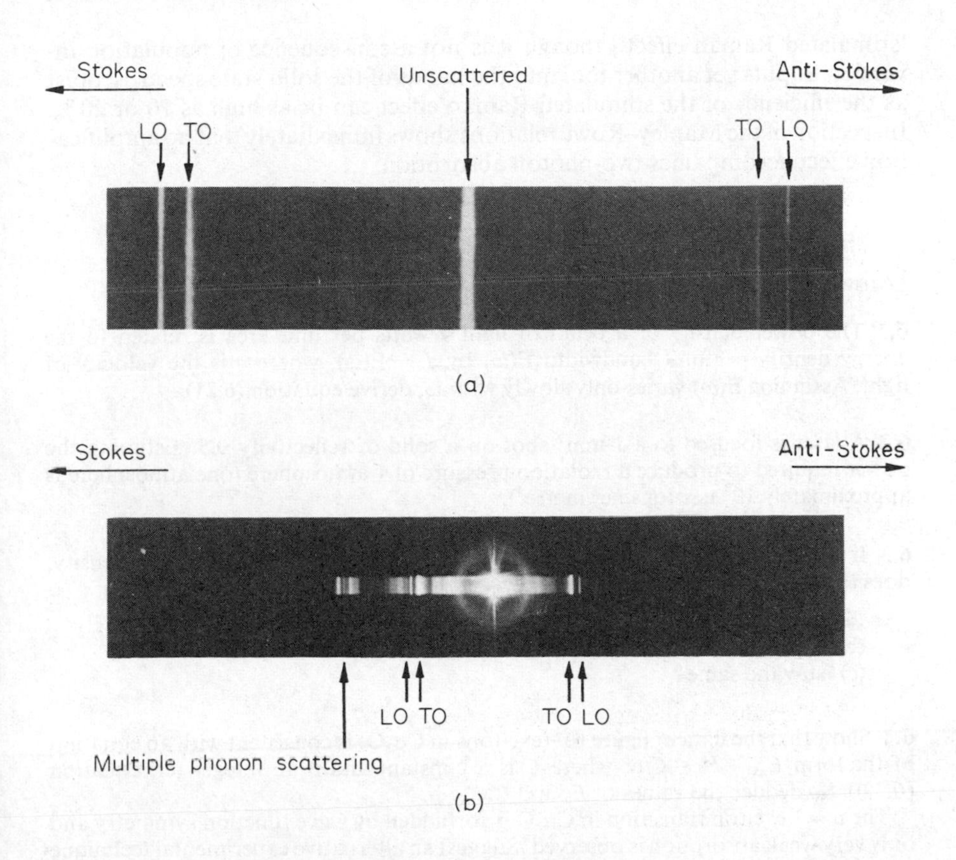

Figure 6.37 Spectrum of Raman-scattered light from gallium phosphide. Source: helium neon laser. 'Stokes' (bright) lines on left, anti-Stokes lines on right. Centre line: light scattered without change of frequency.
(a) High resolution, short exposure. First Raman line shifted by 370 cm^{-1} (T. O. Phonon), second Raman line shifted by 400 cm^{-1} (L. O. Phonon).
(b) Long exposure showing single phonon lines (marked by arrows) and, at larger energy displacements, light scattered by multiphonon Raman processes (mostly two phonon).
(Photographs kindly supplied by J. P. Russell, then at R.R.E. Malvern. See M. V. Hobden and J. P. Russell, *Phys. Let.*, **13**, 39 (1964).)

line (corresponding to the minus sign in (6.68)). To be consistent with our earlier notation we must identify ω_3 and Φ_3 with the incident radiation, ω_2 and Φ_2 with the scattered radiation, ω_1, Φ_1 with the phonons. Then phonon generation must, according to the Manley–Rowe relations, be accompanied by amplification at ω_2. Hence with external mirrors to form a laser type cavity around a Raman scattering crystal we should observe oscillations at the Raman shifted frequency ω_2, if the pump power at ω_3 is sufficiently high. This effect was first observed by accident but is now well established. It was named the 'stimulated Raman effect', though it is not a consequence of population inversion. It puts yet another tool into the hands of the solid state spectroscopist as the efficiency of the stimulated Raman effect can be as high as 10 or 20%. Inspection of the Manley–Rowe relations shows immediately that no amplification effect accompanies two-photon absorption.

PROBLEMS

(Answers, where appropriate, are given on page 481.)

6.1 The power density of a beam of light I watts per unit area is related to the energy density per unit bandwidth $\mathrm{E}(\omega)$ by $I = c\mathrm{E}(\omega)$, where c is the velocity of light. Assuming $\mathrm{E}(\omega)$ varies only slowly with ω, derive equation (6.21).

6.2 A laser is focused to a 1 mm^2 spot on a solid of reflectivity 0.5. Estimate the power required to produce a radiation pressure of 1 atmosphere (one atmosphere is approximately 10^5 newtons per metre^2).

6.3 If the line width of a laser is increased at constant population inversion density, does the gain at the optimum frequency

(a) increase,
(b) decrease,
(c) stay the same?

6.4 Show that the data of figure 6.7 (excitons in Cu_2O) is consistent with an equation of the form $E_{ex} = E_g - C/\mathrm{n}^2$ where C is a constant and n an integer (cf. equation (6.39)). So deduce the values of E_g and C.

The n = 1 exciton transition in Cu_2O is forbidden by wave function symmetry and only very weak absorption is observed. Suggest an alternative experimental technique which would permit observation of this line.

6.5 A certain semiconductor has simple, parabolic energy bands centred on $\boldsymbol{k} = 0$. Interband transitions are forbidden at $\boldsymbol{k} = 0$ but the transition probability increases linearly with $|k|$. Show that, in the absence of exciton effects, the absorption increases as $(\hbar\omega - E_g)^{3/2}$ for $\hbar\omega > E_g$.

6.6 If electrons and holes recombine by direct, interband transitions the decay in electron density after excitation has ceased is described by

$$N = 1/G(t + t_0)$$

(equation (6.51)). Suppose that, for $t < 0$ the crystal is uniformly excited by radiation which generates J_0 electron-hole pairs per unit volume per second. At $t = 0$ excitation ceases. Show that, for $t > 0$.

$$N = 1/G[t + (J_0 G)^{-1/2}]$$

and hence that the time taken for the emission to decay to 1/nth of its value at $t = 0$ is given by

$$(\mathrm{n} - 1)/(J_0 G)^{1/2}$$

6.7 The decay rate deduced in problem 6.6 is a function of the initial excitation rate. Under what other conditions might a decay rate which increases with J_0 be observed?

6.8 A rare earth luminescence impurity centre and its sensitiser are characterised by energies E_1 and E_2. Strong resonance coupling occurs when $|E_1 - E_2| = 0$. For what other values of $|E_1 - E_2|$ might strong coupling be expected?

6.9 A population inversion density $N_2 - N_1$ is maintained between two levels separated in energy by $\hbar\omega$. Excitation then ceases and the laser cavity is Q-switched very rapidly. Show that the maximum total energy in the output pulse is

$$\mathscr{V}\hbar\omega(N_2 - N_1)/2$$

where $\mathscr{V}$ is the volume of the medium.

6.10 The rate of growth in amplitude $\mathrm{F}(2\omega)$ of a second harmonic wave travelling in the z-direction in a lossless non-linear dielectric can be written

$$\frac{\mathrm{dF}(2\omega)}{\mathrm{d}z} = C\mathrm{F}^2(\omega)$$

(cf. equation (6.58)) where C is some constant. Show that, provided phase matching has been achieved, and neglecting diffraction, $\mathrm{F}(2\omega)$ increases with z as

$$\mathrm{F}(2\omega) = \mathrm{F}_0(\omega)\tanh[C\mathrm{F}_0(\omega)\,z]$$

where $\mathrm{F}_0(\omega)$ is the amplitude of the primary wave, frequency ω, at $z = 0$.

6.11 Electroluminescent GaAs PN junctions emit light of wavelength about 0.84 μm. Alloying with GaP shifts the emission to progressively shorter wavelengths. At a composition near GaP, 0.6 and GaAs 0.4 the efficiency drops sharply by a factor ~200. What deductions can be made about the band structure of GaP/GaAs alloys? (Refer to table 4.1 for data on GaAs and GaP.)

7

The optical and microwave properties of free carriers

7.1 Free-carrier absorption (a)

In the last chapter (§ 6.3) we indicated that 'free' charge carriers, i.e. electrons in a partially filled band, contributed to absorption in semiconductors and metals. Such electrons may, of course, be promoted to a higher band on absorbing a photon and if so the process is the same as that discussed in § 6.4. In this chapter we are concerned with transitions between initial and final electron states which lie in the same band minimum. When a band is nearly completely filled (e.g. a P type semiconductor) it is convenient to speak of hole transitions though in fact an electron is excited to occupy a vacant state at the top of the band.

If E_i, $\boldsymbol{k}_i$, E_f, $\boldsymbol{k}_f$ are the initial and final energies and wave vectors respectively of an electron confined to a single parabolic energy minimum we have

$$k_f - k_i = \frac{1}{\hbar}[(2m^*E_f)^{1/2} - (2m^*E_i)^{1/2}]$$

If only an electron and a photon are involved the conservation rules require that $E_f - E_i = \hbar\omega(\boldsymbol{\beta})$ and $\boldsymbol{k}_f - \boldsymbol{k}_i = \boldsymbol{\beta}$, which is not possible for any achievable value of m^* unless $\omega = \boldsymbol{\beta} = 0$. The conservation rules can therefore be satisfied only at zero frequency and, due to the uncertainty in the electron energy, up to frequencies of the order of the reciprocal of the electron mean free time (see § 2.7). Intraband electronic transitions require the co-operation of a third particle and are, in fact, normally phonon assisted. The absorption due to free carriers is therefore a function of the density of available final electron states, the density of suitable phonons and the square of the matrix element describing the electron–phonon–photon coupling, all summed over the initial electron distribution.

An alternative and much simpler description of free carrier absorption is given by a purely classical approach analogous to the Lorentzian model of § 6.2.1. We have seen (§ 4.6) that in the free electron approximation electrons and holes can be described as particles of a classical gas and the only concession we make to quantum mechanics is to ascribe to them an effective mass. A classical description is strictly only valid if $\hbar\omega \ll kT$, i.e. in the far infrared and microwave region of the spectrum at room temperature, but it is found that the quantum and classical pictures predict almost indistinguishable results well beyond this limit of validity. The classical description of free carrier absorption, which is based on early work by Drude (1904) and Zener (1933) has therefore been remarkably successful, considering its simplicity, and we will use it in what follows, indicating some limitations later.

7.1.1 The free-carrier contribution to the dielectric constant and conductivity (a)

The equation of motion of an electron (or hole) of effective mass m^* can be written

$$m^*(\mathrm{d}\boldsymbol{v}/\mathrm{d}t) + (m^*/\tau)\boldsymbol{v} = -e\mathbf{F}(t) \tag{7.1}$$

where $\boldsymbol{v}$ is the net drift velocity due to the field $\mathbf{F}(t)$ which will be a sinusoidal function of time. τ is a relaxation time describing the rate at which the velocity of the electron is randomised due to collisions. For the moment τ is just a phenomenological constant but we will discuss its relation to other properties of the solid in the next chapter.

The solution of (7.1) is

$$\boldsymbol{v} = \frac{e\mathbf{F}(t)\tau}{m^*(1 + i\omega\tau)} \tag{7.2}$$

Hence the conductivity, defined as the current density ($Ne\boldsymbol{v}$) per unit field is given by

$$\sigma = \frac{Ne^2\tau}{m^*}\left(\frac{1}{1 + i\omega\tau}\right)$$

where N is the electron density. The real and imaginary parts of σ contribute respectively to the conductivity and dielectric constant of the material as defined in § 6.1.3, so it follows that

$$2n_0 k_0 = \frac{\sigma}{\omega\varepsilon_0} = \frac{\sigma_0}{\omega\varepsilon_0}\left(\frac{1}{1+\omega^2\tau^2}\right) \tag{7.3}$$

and

$$n_0^2 - k_0^2 = \frac{\varepsilon_1}{\varepsilon_0} = \frac{\varepsilon_L}{\varepsilon_0} - \frac{\sigma_0}{\omega\varepsilon_0}\left(\frac{\omega\tau}{1+\omega^2\tau^2}\right) \tag{7.4}$$

where n_0 and k_0 are the optical constants of the material and $\sigma_0 = Ne^2\tau/m^*$. We have assumed there are no contributions to the absorption other than from free carriers and that the dielectric constant of the lattice due to all causes other than free carriers is included in ε_L. Note that the free-carrier and lattice contributions to the dielectric constant are of opposite sign. Physically this arises because the displacement current due to bound charges leads the field (i.e. is capacitative) while the free carriers lag behind the field and behave inductively.

Equations (7.3) and (7.4) give between them a full description (within the limits of the classical model) of the free-carrier contribution to the optical constants, n_0 and k_0 of a semiconductor or metal. For comparison with experimental data, however, it is convenient to find approximate solutions of (7.3) and (7.4) appropriate to limited ranges of frequency or conductivity and this we will now do.

7.1.2 Absorption in semiconductors (a)

If the conductivity is low, as in a semiconductor, we can assume that

$$(\sigma_0/\omega\varepsilon_1) \ll 1 \tag{7.5}$$

and hence $2n_0k_0 \ll (n_0^2 - k_0^2)$ so that $n_0 \gg k_0$ and $n_0 \approx (\varepsilon_1/\varepsilon_0)^{1/2}$. In addition the free-carrier contribution (n_0^2—k_0^2) to the dielectric constant is negligible at any frequency. The dielectric constant is then that of the lattice and the absorption coefficient given by

$$K = \frac{2k_0\omega}{c} = \left(\frac{\mu_0}{\varepsilon_L}\right)^{1/2}\left(\frac{\sigma_0}{1+\omega^2\tau^2}\right) \tag{7.6}$$

where μ_0 is the permeability of free space. The absorption is therefore independent of frequency at low frequencies but falls as ω^2 at high frequencies. The value of τ at room temperature has been found for many solids (including metals) to lie between 5×10^{-14} and 5×10^{-13} s. Hence $\omega\tau$ is of the order of unity around 10^{12} Hz ($\lambda = 300$ μm). On cooling to helium temperatures τ increases by as much as three orders of magnitude in pure materials, so that $\omega\tau$ can be made greater than unity even at microwave frequencies.

Equation (7.6) is found to give an excellent description of the experimentally observed absorption of many semiconductors throughout the infrared. When the absorption due to mobile carriers is underestimated by (7.6) the additional absorption is usually due to direct transitions to higher bands. This occurs, for example, in P type germanium where absorption proportional to the hole density is observed due to transitions between the light hole, heavy hole and spin-orbit split bands (§ 4.5). Even when this does not occur (7.6) is too good. It should not be valid if $\hbar\omega > kT$, where kT is the average thermal energy of the carriers. In fact it is often obeyed well beyond this limit. For example, N type InSb and N type Si show a $1/\omega^2$ dependence from their respective absorption edges up to $\omega\tau \sim 1$ and the absolute magnitude of the absorption is also predicted quite accurately, account being taken of the effective mass, which is known from other data.

An interesting feature of free-carrier absorption in semiconductors is the comparative ease with which it may be varied electronically. The carrier density can be increased by 'injection' from PN junctions (see Chapter 9), by illumination in the fundamental absorption region and by other methods. Hence it is possible to modulate the infrared absorption of semiconductors, a process which has found application in a number of semiconductor studies.

7.1.3 Absorption in metals (a)

If the conductivity is high, as in a metal, the situation is more complicated. Inspection of (7.4) shows that at sufficiently high frequencies the free-carrier contribution to the dielectric constant, even in metals, must become small and (7.6) will then apply. For most metals this condition requires visible or ultraviolet frequencies and interband absorption may begin first. If, however, it does not, the resulting very large value of $\omega\tau$ ensures that σ, and hence the free-carrier absorption, is small. The alkali metals (Li, K, Na, etc.) are transparent in the ultraviolet for this reason and Cu and Au owe their characteristic colours to an absorption minimum in the visible, bounded by interband absorption extending into the ultraviolet and free-carrier absorption extending into the infrared. We discuss these features more fully below.

At the low-frequency end of the spectrum—the far infrared and microwave region—the reverse of (7.5), namely

$$\sigma_0 \gg \omega\varepsilon_1 \tag{7.7}$$

will apply to a metal of sufficiently high conductivity. If this condition applies $2n_0 k_0 \gg (n_0^2 - k_0^2)$ in (7.3) and (7.4) and hence

$$n_0^2 \approx k_0^2 \approx \sigma/2\omega\varepsilon_0$$

If, additionally, the frequency is low so that $\omega\tau \ll 1$ we find for the absorption coefficient

$$K = (2\sigma_0\,\omega\mu_0)^{1/2} \tag{7.8}$$

an expression which is in good agreement with experiment. The quantity 2/K is called the skin depth, and is familiar in electrical engineering. An electromagnetic wave entering a metal is attenuated in power in a distance ~1/K so that a wave carried by a conductor may be considered to flow only in a surface layer whose thickness is about the skin depth. The factor 2 appears because the skin depth is defined in terms of amplitude rather than power. Using the same approximations ($\sigma_0 \gg \omega\varepsilon_1$ and $\omega\tau \ll 1$) the reflectivity of metals in the far infrared can be found from (6.4) and is

$$\mathscr{R} = 1 - 2\left(\frac{2\omega\varepsilon_0}{\sigma_0}\right)^{1/2} \tag{7.9}$$

which is called the Hagen–Rubens relationship and was shown experimentally by them as long ago as 1903 to give a good description of metals at wavelengths greater than about 30 μm.

As the frequency increases and $\omega\tau$ becomes greater than unity, k_0 starts to decrease and n_0^2 becomes negative. No simple analytic approximations are possible until $\omega\tau \gg 1$ so comparison with experiment requires numerical computation. As an example, figures 7.1 and 7.2 show the conductivity and

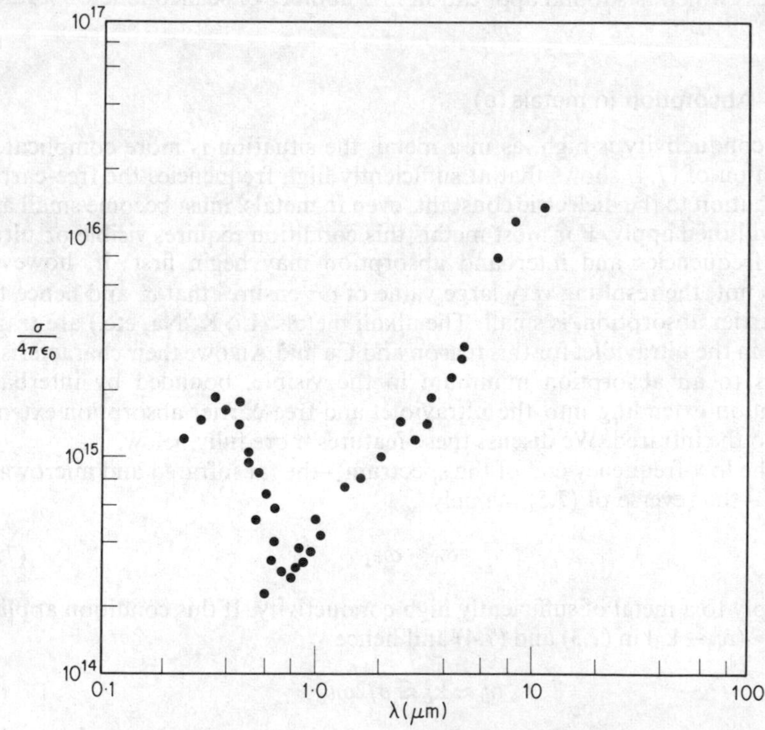

Figure 7.1 The conductivity of gold deduced from reflectivity and transmission measurements. Note the approximately ω^{-2} dependence in the infrared, as expected from equation (7.3). The rise in apparent conductivity in the visible is due to interband transitions. (After Givens, see Figure 7.2.)

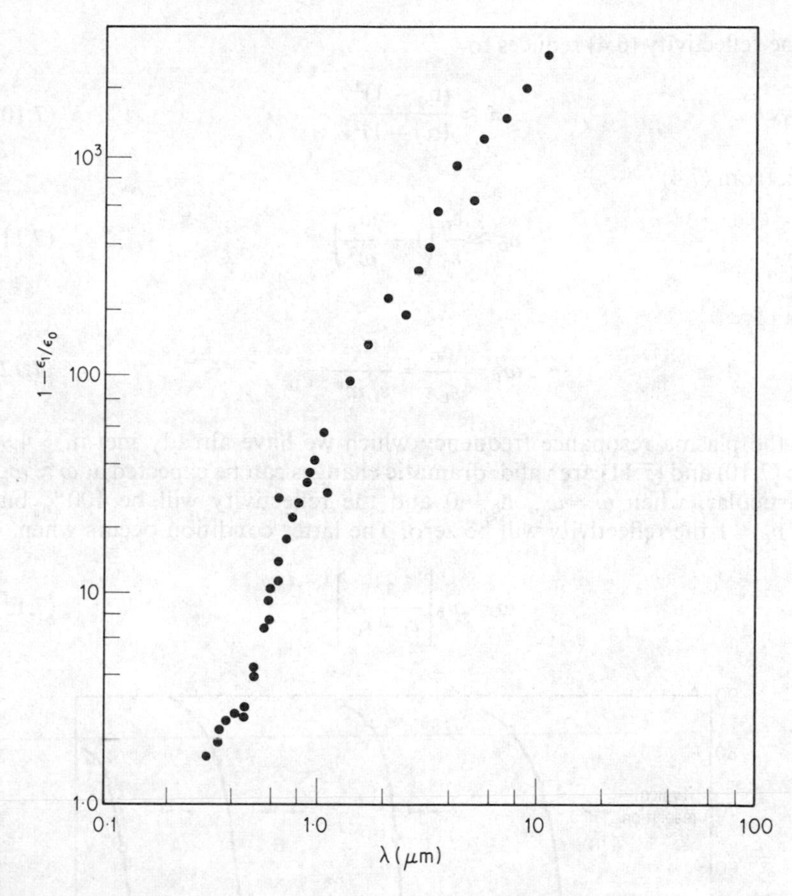

Figure 7.2 The variation of $(1-\varepsilon_1/\varepsilon_0)$ for gold with wavelength. The dielectric constant is dominated by the free electron contribution at long wavelengths and consequently is negative and varies as ω^{-2} for $\omega\tau > 1$. (After M. P. Givens, *Solid State Physics*, Vol. 6, F. Sietz and Turnbull (ed.), Academic Press.)

dielectric constant of gold in the infrared. Notice that the dielectric constant is large and negative at long wavelengths, due to the free-carrier term, but falls as the frequency increases. The rise in σ below 0.5 μm is due to the onset of fundamental interband transitions, as discussed above.

7.1.4 Plasma resonance (a)

When scattering is small, i.e. when $\omega\tau \gg 1$, simplification of (7.3) and (7.4) is again possible. Inspection of these equations shows that $2n_0k_0$ falls more rapidly with increasing ω than does $(n_0^2 - k_0^2)$ whence k_0 tends to zero while n_0^2 remains finite, though negative. Hence when $\omega\tau \gg 1$, k_0 can be neglected and the reflectivity (6.4) reduces to

$$\mathscr{R} \approx \frac{(n_0 - 1)^2}{(n_0 + 1)^2} \tag{7.10}$$

where, from (7.4)

$$n_0^2 \approx \frac{\varepsilon_L}{\varepsilon_0}\left(1 - \frac{\omega_p^2}{\omega^2}\right) \tag{7.11}$$

and

$$\omega_p^2 = \frac{\sigma_0}{\varepsilon_L \tau} = \frac{Ne^2}{\varepsilon_L m^*} \tag{7.12}$$

ω_p is the plasma resonance frequency which we have already met in § 4.9. When (7.10) and (7.11) are valid, dramatic changes can be expected at $\omega \approx \omega_p$. In particular, when $\omega = \omega_p$, $n_0 = 0$ and the reflectivity will be 100% but when $n_0 = 1$ the reflectivity will be zero. The latter condition occurs when

$$\omega = \omega_p \left[\frac{\varepsilon_L}{\varepsilon_L - \varepsilon_0}\right]^{1/2} \tag{7.13}$$

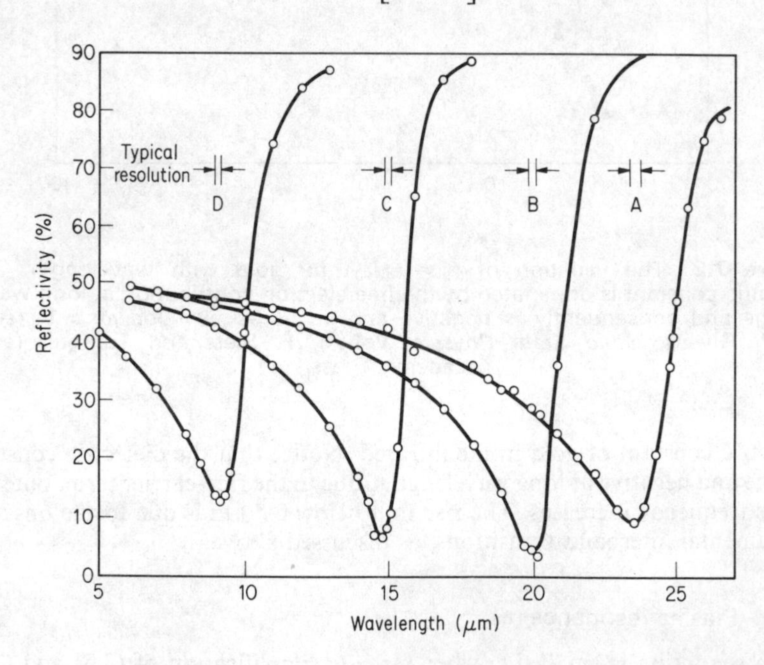

Figure 7.3 Reflectivity at 81 K and normal incidence of variously doped samples of P type PbTe, showing plasma resonance. Hole concentrations: A, 3.5×10^{18} cm^{-3}; B, 5.7×10^{18} cm^{-3}; C, 1.5×10^{19} cm^{-3}; D, 4.8×10^{19} cm^{-3}. (After J. R. Dixon and H. R. Riedl, *Phys. Rev.*, **138**, 873 (1965).)

which is not far removed from ω_p if $\varepsilon_L/\varepsilon_0$ is large. As ω increases further $n_0^2 \rightarrow \varepsilon_L/\varepsilon_0$ and $k_0 \rightarrow 0$ so the material behaves as a normal dielectric.

In practice the reflectivity will not change from unity to zero at $\omega \approx \omega_p$ as k_0, though small, is always finite. Practical examples illustrating this behaviour are given in figures 7.3 and 7.4, which show the reflectivity of heavily doped P type PbTe and N type InSb at various electron concentrations and temperatures respectively. The shift in frequency with temperature in InSb is primarily due to a change in the electron effective mass with temperature as the conduction band minimum of InSb is far from parabolic in $\boldsymbol{k}$ space. For the examples chosen, $\omega = \omega_p$ in the infrared while for many metals it occurs in the ultraviolet. As already stated, the alkali metals are transparent in the ultraviolet and their reflectivity edges have been tabulated in § 4.9.

The name plasma resonance arises by analogy with ionised gases, in which the same phenomenon occurs. Indeed within the free-electron approximation electrons or holes in a solid are an electron gas embedded in a material of dielectric constant ε_L. At the plasma resonance frequency $\varepsilon_1 = 0$ and hence, by the argument used in § 6.6.1, longitudinal electric oscillations are possible. These take the form of electron density fluctuations at frequency ω_p, as discussed in § 4.9.

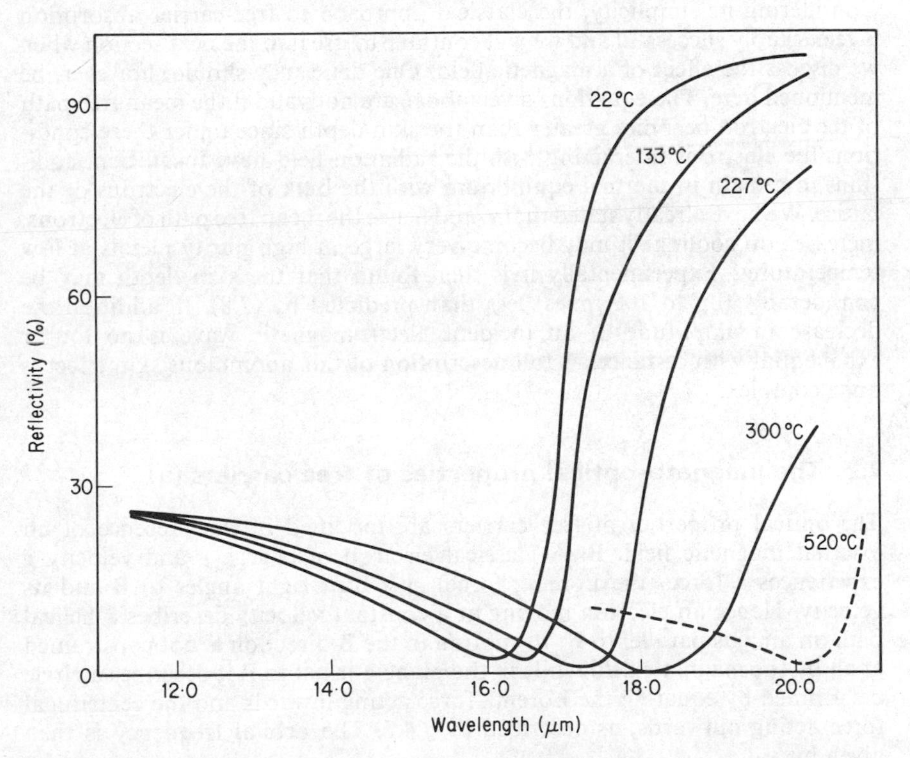

Figure 7.4 Plasma resonance in N type InSb at various temperatures. Electron concentration 2.9×10^{18} cm^{-3}. (After R. B. Hawkins, Thesis for Ph.D.)

Plasma resonance provides a method of determining m^* if N is known, or *vice versa.* It is frequently used to determine electron densities in ionised gases.

It should be noted that plasma resonance, unlike resonance absorption, has a finite frequency width even when no energy loss mechanism is involved. From (7.13), the change from zero to 100% reflection takes place in a frequency interval

$$\Delta\omega = \omega_p \left[1 - \left(\frac{\varepsilon_L}{\varepsilon_L - \varepsilon_0}\right)^{1/2}\right] \qquad (\tau = \infty)$$

As a consequence plasma resonance effects due to electrons in a completely filled energy band can be observed even though such electrons cannot contribute to the d.c. conductivity of the material (Chapter 4). The smooth fall in absorption on the high energy side of the fundamental band observed in most semiconductors and insulators (see § 6.3) is in fact the plasma edge of the valence-band electrons.

7.1.5 The anomalous skin effect (A)

Considering its simplicity, the classical approach to free-carrier absorption is remarkably successful and we will continue to use it in the next section when we discuss the effect of a magnetic field. One deficiency should, however, be mentioned here. The equations given above are not valid if the mean free path of the electron becomes greater than the skin depth since under these conditions the electrons interacting with the radiation field have insufficient collisions to remain in thermal equilibrium with the bulk of the electrons or the lattice. We have already stated that τ, and hence the mean free path of electrons, increases on cooling: it may become very large in high purity metals at low temperatures. Experimentally it is then found that the skin depth may be considerably (up to 100 times) less than predicted by (7.8). In addition the decrease in amplitude of an incident electromagnetic wave is no longer exponential with distance. A full description of this anomolous skin effect is quite complex.

7.2 The magneto-optical properties of free carriers (a)

The optical properties of free carriers are modified in the presence of an external magnetic field, **B**. A classical electron of charge e and velocity $\boldsymbol{v}$ experiences a force (the Lorentz force) $e\boldsymbol{v} \times \mathbf{B}$ at right angles to **B** and its velocity. Hence an electron moving at a constant velocity describes a helical path on an axis parallel to **B**: its motion in the **B**-direction is not constrained at all by the magnetic field while in the plane normal to **B** it describes a circle determined by equating the Lorentz force acting inwards and the centrifugal force acting outwards, as described in § 4.7. The orbital frequency is then given by

$$\omega_c = e\mathbf{B}/m^* \qquad (7.14)$$

where ω_c is the cyclotron frequency by analogy with the orbits of charged particles in cyclotron accelerators. When we discussed the effect of a magnetic field on the motion of free electrons in § 4.7 we showed that it could be described quantum mechanically as pulling together groups of conduction band states to form Landau levels. The energy spacing between successive Landau levels is $\hbar\omega_c$ and transitions between adjacent states are equivalent to the excitation of a classical oscillator of resonant frequency ω_c (see § 6.2). It is essential to use the Landau model if $\hbar\omega_c > kT$ but, as we saw in the previous section, the classical approach is simpler and valid if $\hbar\omega < kT$. Similarly magnetic field effects can be described classically if $\hbar\omega_c < kT$.

We consider two orientations in which an external magnetic field may be applied to a solid sample: either parallel with the radiation propagation direction or at right angles to it. We shall assume that the conductivity of the sample in the absence of a magnetic field is isotropic. With this assumption the form of the Lorentz force indicates that the magnetic field will affect the optical properties only if $\mathbf{F}(t)$ has a component at right angles to $\mathbf{B}$. This will apply for any polarisation when $\mathbf{B}$ is parallel with the propagation vector, $\boldsymbol{\beta}$. This is called the Faraday orientation. When $\mathbf{B}$ is perpendicular to the propagation direction the magnetic field will affect only the polarisation component having $\mathbf{F}(t) \perp \mathbf{B}$. If the radiation is plane polarised in this direction the arrangement $\mathbf{F}(t) \perp \mathbf{B} \perp \boldsymbol{\beta}$ is called the Voigt orientation.

7.2.1 The Faraday orientation (a)

In the Faraday orientation the circular motion of the electrons (or holes) in a plane normal to the propagation direction suggests immediately that it will be convenient to describe the plane polarised wave as the sum of two circularly polarised waves whose electric vectors rotate in opposite directions. One circularly polarised component will rotate in the same direction as the electrons: if $\omega = \omega_c$ the electrons will experience (classically) a steady outward force, increasing their orbit radius and energy. Energy is therefore absorbed from this circularly polarised component while the other is little affected.

To put the above argument on a more quantitative basis we suppose that the propagation and magnetic field direction is z. The equation of motion for the electrons in the x- and y-directions are then:

$$\frac{\mathrm{d}v_x}{\mathrm{d}t} + \frac{v_x}{\tau} = -\frac{e}{m^*}\mathrm{F}_x(t) - \frac{ev_y\,\mathrm{B}}{m^*} \tag{7.15}$$

and

$$\frac{\mathrm{d}v_y}{\mathrm{d}t} + \frac{v_y}{\tau} = -\frac{e}{m^*}\mathrm{F}_y(t) + \frac{ev_x\,\mathrm{B}}{m^*} \tag{7.16}$$

which may be compared with (7.1). The components of the two circularly polarised waves are:

$$\mathrm{F}_x(t) = \mathrm{F}_0 \cos\omega t; \qquad \mathrm{F}_y(t) = \mathrm{F}_0 \sin\omega t$$

and

$$\mathrm{F}_x(t) = \mathrm{F}_0 \cos\omega t; \qquad \mathrm{F}_y(t) = -\mathrm{F}_0 \sin\omega t$$

where F_0 is a constant. Substituting these into (7.15) and (7.16), and proceeding as in §7.1, we obtain:

$$2n_0 k_0 = \frac{\sigma}{\omega\varepsilon_0} = \frac{\varepsilon_L}{\varepsilon_0}\left(\frac{\omega_p^2\tau}{\omega}\right)\left(\frac{1}{1+\tau^2(\omega \pm \omega_c)^2}\right) \tag{7.17}$$

and

$$n_0^2 - k_0^2 = \frac{\varepsilon_1}{\varepsilon_0} = \frac{\varepsilon_L}{\varepsilon_0}\left[1 - \left(\frac{\omega_p^2\tau^2}{\omega}\right)\left(\frac{\omega \pm \omega_c}{1+\tau^2(\omega \pm \omega_c)^2}\right)\right] \tag{7.18}$$

where use has been made of (7.12). Equations (7.17) and (7.18) reduce to (7.3) and (7.4) if ω_c is put equal to zero. The $\pm$ signs before ω_c refer to the two circularly polarised components of the plane polarised incident wave respectively, the minus sign corresponding to the component for which the electric vector rotates in the same sense as the electrons. As before, it is convenient to examine (7.17) and (7.18) under certain simplifying conditions. The phenomena we shall describe and the conditions under which they are observed are given in the Table below. In all cases the Faraday orientation is assumed. For the first three phenomena the simultaneous presence of electrons and holes either enhances (Faraday rotation) or duplicates at a second frequency (cyclotron and magnetoplasma resonance) the phenomenon described. Under the conditions for helicon and Alfvén wave propagation, however, the simultaneous presence of equal electron and hole densities leads to qualitative differences in behaviour. Note that either $\omega\tau \gg 1$ or $\omega_c\tau \gg 1$ throughout the table. These conditions ensure that k_0 is small and consequently $(n_0^2 - k_0^2) \approx n_0^2$ except near plasma resonance.

Cyclotron resonance	$\omega\tau \gg 1$	$\omega \approx \omega_c$	$\omega \gg \omega_p$
Faraday rotation	$\omega\tau \gg 1$	$\omega \gg \omega_c$	$\omega \gg \omega_p$
Magnetoplasma resonance	$\omega\tau \gg 1$	$\omega \gg \omega_c$	$\omega \approx \omega_p$
Helicon wave propagation	$\omega_c\tau \gg 1$	$\omega \ll \omega_c$	$\omega \ll \omega_\rho^2/\omega_c$
Alfvén wave propagation	as helicons but equal electron and hole densities and $\omega\tau \gg 1$		

7.2.2 Cyclotron resonance (a)

For low conductivity material (a semiconductor) $\omega_p \ll \omega$ and the free-carrier contribution to the dielectric constant can be ignored. The absorption is simply proportional to σ (cf. (7.6)). Inspection of (7.17) shows immediately that the absorption of the appropriate circularly polarised component is a maximum when $\omega = \omega_c$, though the absorption band will be sharp only if $\omega\tau \gg 1$. Furthermore, if $\omega\tau > 1$ there is a significant net absorption of the two circularly polarised components taken together, resulting in observable absorption of an unpolarised primary electromagnetic wave.

The resonance condition $\omega = \omega_c = e\mathbf{B}/m^*$ provides, when applicable, an unambiguous method of obtaining the carrier effective mass, as discussed in

§ 4.7. To meet the requirement $\omega\tau \gg 1$ one needs high purity samples and either low temperatures to increase τ if ω is in the microwave range, or frequencies in the infrared if $\tau \sim 3 \times 10^{-13}$ s, a typical value at room temperature. At high frequencies, however, high magnetic fields are needed to achieve $\omega = \omega_c$. If $m^* = m$ the relationship $\omega_c = eB/m^*$ is numerically equivalent to 28 GHz per tesla. Magnetic fields in excess of 10 tesla are difficult to obtain, even with superconducting solenoids, so unless $m^* \sim 0.1\, m$ or less the observation of cyclotron resonance in the infrared presents serious experimental problems. Infrared cyclotron resonance has been observed in materials like N type InSb ($m^* \sim 0.013\, m$) both in reflection and transmission. The use of lasers has facilitated this work, the magnetic field rather than the frequency being tuned for resonance (see figure 7.9).

7.2.3 Faraday rotation (a)

The free-carrier contribution to the dielectric constant, though small for a semiconductor, does imply that there are different effective refractive indices and hence phase velocities for the right- and left-hand circularly polarised components of the wave. A plane polarised wave therefore emerges with its plane of polarisation rotated by an angle θ given by:

$$\theta = \frac{\mathscr{L}\omega}{2c}(\mathrm{n}_{0+} - \mathrm{n}_{0-}) \qquad (7.19)$$

where c is the velocity of light, $\mathscr{L}$ is the thickness of the sample and n_{0+} and n_{0-} the refractive indices appropriate to the two circularly polarised waves. This effect is called Faraday rotation. For semiconductors in the infrared it is comparatively easy to obtain the conditions given in the table, namely

$$\omega\tau \gg 1,\ \omega > \omega_p,\ \omega > \omega_c \qquad (7.20)$$

and inserting these conditions as approximations into (7.17) and (7.18) we obtain from (7.19)

$$\theta = -\mathscr{L}\left(\frac{\varepsilon_L}{\varepsilon_0}\right)^{1/2}\left(\frac{\omega_p^2\omega_c}{2c\omega^2}\right) \qquad (7.21)$$

for the Faraday rotation. θ increases linearly with magnetic field and carrier density through ω_c and ω_p^2 respectively. Through the product $\omega_p^2\omega_c$ it is inversely proportional to m^{*2}. The decrease in θ with frequency, ω, is the same, and arises from the same cause, as the equivalent decrease in absorption in a semiconductor (7.6).

Because of its dependence on m^{*2}, Faraday rotation represents quite a powerful method of obtaining the value of m^* for semiconductors. The rotation can be quite large if m^* is small: with InSb containing about 10^{17} electrons per cm^3, a field of 1 tesla gives a rotation of about 250°/mm at 15 μm wavelength. The experimental apparatus is also relatively simple: a laser or monochromator, polariser, analyser, sample and solenoid together with a

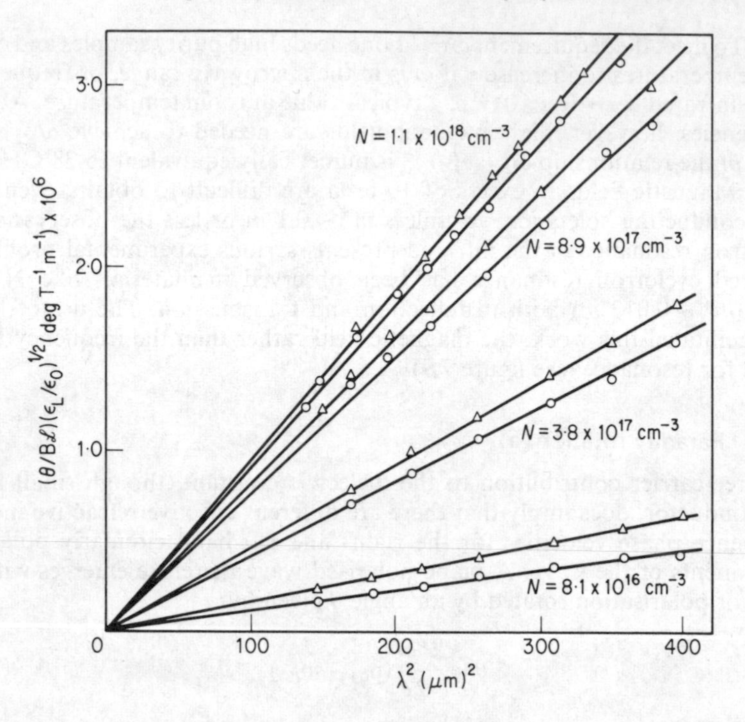

Figure 7.5 Faraday rotation due to free electrons in InAs: ○, 296 K; △, 77 K. Carrier concentrations shown on diagram. Straight lines confirm ω^{-2} dependence predicted by equation (7.21). (After C. J. Summers and S. D. Smith, *Proc. Phys. Soc.*, **92**, 215 (1967).)

detector. Figure 7.5 shows the Faraday rotation of InAs samples. It will be seen that (7.21) is well obeyed, although the effective mass is a function of temperature (cf. InSb, § 7.1.4, and see § 8.8.1).

7.2.4 Magnetoplasma resonance (A)

In § 7.1.4 we described how high conductivity materials showed a characteristic plasma resonance if $\omega \approx \omega_p$ when $\omega\tau \gg 1$. The presence of a magnetic field, in both the Faraday and Voigt orientations, splits the plasma resonance so that there are two frequencies at which $n_0 \approx 1$ and $n_0 \approx 0$. Using (7.18) which applies to the Faraday orientation, and inserting the conditions tabulated, we find that $n_0 \approx 0$ when

$$\omega = \tfrac{1}{2}[(\omega_c^2 + 4\omega_p^2)^{1/2} \pm \omega_c]$$

If $\omega_c < \omega_p$, (see table) we can expand the square root term and to a first approximation

$$\omega \approx \omega_p \pm \omega_c/2 \qquad (7.22)$$

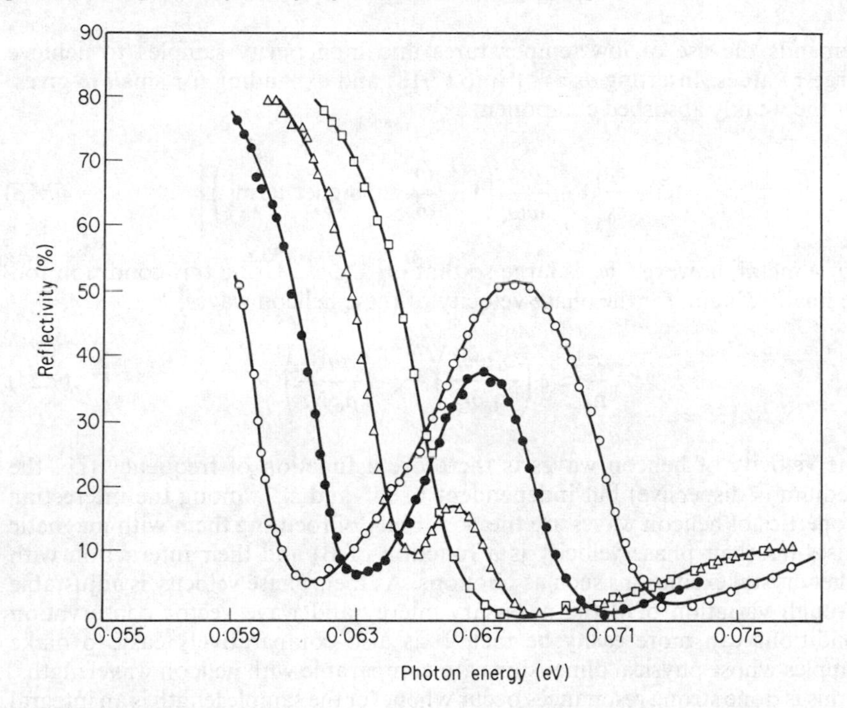

Figure 7.6 Magnetoplasma resonance in N type InSb. ○, 3.85 tesla, ●, 2.5 tesla, △, 1.2 tesla, □, zero field. (After B. Lax and G. B. Wright, *Phys. Rev.*, 4, 16 (1960).)

so the two plasma edges are separated by the cyclotron frequency. An experimental example—again InSb—is shown in figure 7.6. Thus magneto-plasma measurements, where they can be made, provide another way of obtaining ω_c and hence the effective mass of the carriers.

7.2.5 Helicon waves (AD)

Cyclotron resonance of the type discussed in § 7.2.2 cannot be observed in metals because the condition $\omega_c > \omega_p$ would require enormous magnetic fields. A form of cyclotron resonance in metals is possible if the magnetic field is transverse to the propagation direction (the Voigt orientation) and we discuss this below. There remains, however, one interesting set of conditions that can be met in metals in the Faraday orientation. These conditions lead to the propagation, with small attenuation, of circularly polarised electromagnetic waves called helicon waves. Helicon waves with frequencies in the optical range are rare but, having set up (7.17) and (7.18), it seems appropriate to examine all their interesting consequences here.

We suppose that $\omega < \omega_c$ but $\omega_c\tau \gg 1$. The latter condition (§ 7.2.1) ensures that the absorption is small for one circularly polarised component but demands the use of low temperatures and high purity samples to achieve large τ values. Inserting $\omega_c\tau \gg 1$ into (7.18) and expanding for small ω gives, for the weakly absorbed component,

$$n_0^2 = \frac{\varepsilon_L}{\varepsilon_0}\left[1 + \frac{\omega_p^2}{\omega\omega_c}\left(1 + \frac{\omega}{\omega_c} \dots \text{higher terms}\right)\right] \tag{7.23}$$

For a metal, however, ω_p is large so that $\omega_p^2 \gg \omega\omega_c$. Using this condition too we finally obtain, for the phase velocity of these helicon waves,

$$\frac{c}{n_0} \approx c\left(\frac{\varepsilon_0\,\omega\omega_c}{\varepsilon_L\,\omega_p^2}\right)^{1/2} = \left(\frac{\omega B}{\mu_0 Ne}\right)^{1/2} \tag{7.24}$$

The velocity of helicon waves is therefore a function of frequency (i.e. the medium is dispersive) but independent of m^* and ε_L. Among the interesting properties of helicon waves are the possibility of focusing them with magnetic lenses (as their phase velocity is a function of **B**) and their interaction with other crystal excitations such as phonons. As their phase velocity is adjustable through variation of **B** the necessary energy and wave vector conservation conditions can more easily be met. It is also comparatively easy to make samples whose physical dimensions are comparable with helicon wavelengths. If this is done strong resonances occur whenever the sample length is an integral number of half wavelengths in the propagation direction. An example of such geometrical resonances is shown in figure 7.7. Note that the resonance mode

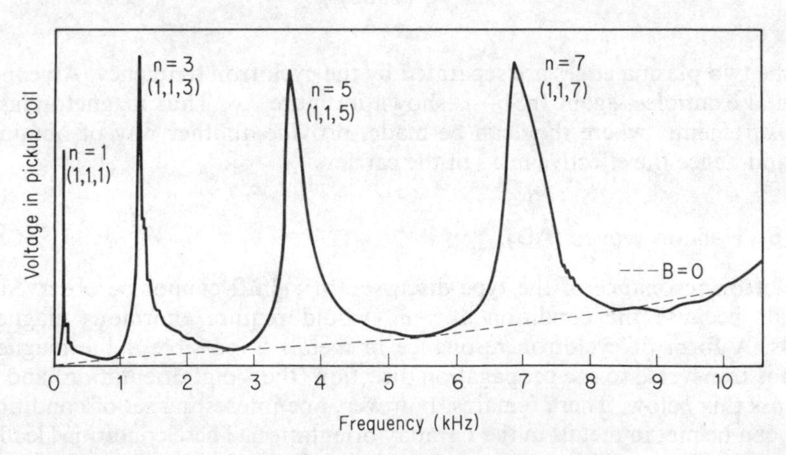

Figure 7.7 Geometrical helicon wave resonances in sodium at 4 K. Magnetic field 2.5 tesla. A resonance occurs whenever a dimension of the slab equals an integral number of half wavelengths of the helicon wave. Even harmonics are missing due to particular geometry of helicon detection pick up coil. Equation (7.24) plus resonance condition $n\lambda/2$ = length, where n is an integer, implies n proportional to $\omega^{1/2}$, as observed. (After R. Bowers, *Symposium on Plasma effects in Solids,* Dunod, Paris (1964).)

number increases as $\omega^{1/2}$, as expected from (7.24). Very sharp resonances can be obtained and metallic helicon wave magnetometers are available commercially. With pure metals and reasonable magnetic fields geometrical resonances occur in conveniently sized samples for frequencies between 10 Hz and 1000 Hz. Helicon phase velocities lie between 0.1 and 1.0 ms^{-1}. In semiconductors, on the other hand, microwave frequencies and higher magnetic fields are necessary to observe helicon waves and typical phase velocities are around 10^7 ms^{-1}.

The essential similarities between ionised gases and solid conductors has already been pointed out. It will be clear from the derivation of (7.17) and (7.18) that these similarities extend to magneto-optics. Faraday rotation measurements, particularly, have found application in the study of laboratory plasmas. Faraday rotation is also observed in the transmission of stellar radiation through inter-stellar gas. Cyclotron resonance was first observed in gases and cyclotron resonance absorption occurs in the ionosphere at about 1.5 MHz, corresponding to a magnetic field $\sim 2 \times 10^{-5}$ tesla at the height of the ionosphere. Plasma resonance is also observed in the ionosphere at frequencies around 5 MHz, but the exact value depends on height and solar activity. Helicon waves propagate freely above the ionosphere at frequencies below ω_c as τ is large due to the low particle density. Typical frequencies are around a few kHz. An initial electrical disturbance on one side of the earth may be detected on the other side after it has propagated as a helicon mode along a magnetic flux line. If the wave is made audible by a loudspeaker a characteristic whistle of steadily decreasing frequency is heard due to the higher phase velocity of the high frequency components. Because of this, helicon waves are called 'Whistler modes' in plasma physics.

7.2.6 Alfvén waves (AD)

Throughout the previous sections we have tacitly assumed that only one type of mobile carrier is present. If both electrons and holes are simultaneously present the two contributions simply add. It should be noted, however, that in a magnetic field holes orbit in the opposite direction to electrons so that the $\pm$ sign on ω_c is reversed. This introduces no new phenomena except under the conditions appropriate to a helicon wave. If then holes and electrons are simultaneously present n_0^2 becomes the sum of two series of terms like (7.23) but with successive terms of opposite sign. Furthermore, if the electron and hole densities are equal, the first term of each expansion cancels with the other, independent of τ or m^*, though the second terms of each series add. This leads to a totally different situation in which plane polarised waves may be propagated with low attenuation at a phase velocity

$$\frac{c}{\mathrm{n}_0} \approx \frac{\mathrm{B}}{[\mu_0 N(m_e^* + m_h^*)]^{1/2}} \tag{7.25}$$

where m_e^* and m_h^* are the effective masses of electrons and holes, respectively, and N their density. This is called an Alfvén wave, after its discoverer. Notice that it is of the same form as the vibrations of a string of mass density $N(m_e^* + m_h^*)$ and tension B^2/μ_0. The electrons and holes, whose charges compensated each other, may be considered as clamped to a magnetic flux line (by the condition $\omega_c \tau \gg 1$) which provides the tension in the string.

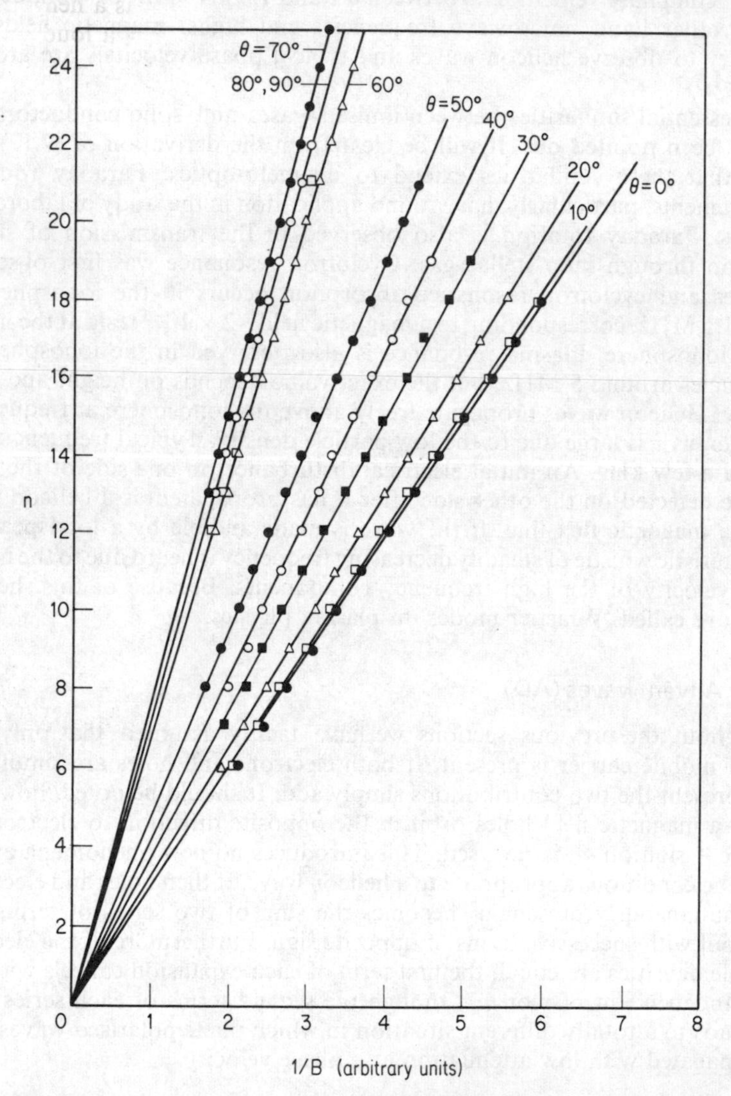

Figure 7.8 Alfvén wave resonances in bismuth at 4 K. From equation (7.25) and resonance condition $n\lambda/2 =$ sample length, n is proportional to 1/B. θ is angle between magnetic field and trigonal axis of the crystal. Variation in slope with θ implies that $(m_e + m_h)$ is anisotropic, i.e. energy is a function of $\boldsymbol{k}$ direction in bismuth (see Chapter 8). (Diagram after B. W. Faughnan, *J. Phys. Soc. Japan,* **20**, 574 (1965).)

The conditions necessary for Alfvén wave propagation are difficult to achieve in solids, though common in gases. Alfvén wave propagation has been observed in semi-metals such as bismuth (see figure 7.8) where geometrical resonances are observed but, unlike helicons, the mode number from (7.25) is proportional to B^{-1}. An interesting feature of these experiments is the observation that the condition $\omega < \omega_c$ is a necessary but not sufficient condition for low absorption. Electron and hole motion along the magnetic field is not constrained by the magnetic field so they are free to move at a velocity appropriate to their energy near the Fermi level. The cyclotron frequency is then shifted by the Doppler effect and the condition for low attenuation becomes

$$\omega < \omega_c - \beta v_F$$

where v_F is the Fermi velocity and $\boldsymbol{\beta}$ the wave vector of the Alfvén wave. In semi-metals there is quite a sharp drop in Alfvén wave transmission when this inequality is not met, which allows v_F to be determined. Similar remarks apply to helicon wave propagation.

7.2.7 The Voigt orientation (A)

When the magnetic field is transverse to both the direction of propagation of the electromagnetic wave and its electric vector—the Voigt orientation—the optical constants become quite complicated functions of ω, ω_p and ω_c. Expressions equivalent to (7.17) and (7.18) can be obtained by the same procedures as used in the Faraday orientation. A sharp resonance absorption is predicted when

$$\omega^2 = \omega_p^2 + \omega_c^2$$

provided $\omega\tau \gg 1$. The dependence on ω_c^2, and hence on B^2, is characteristic of the Voigt orientation and leads to a particularly sharp resonance condition. An experimental example of infrared cyclotron resonance in both Faraday and Voigt orientations is shown in figures 7.9(a) and (b).

The refractive index in the Voigt orientation, assuming $\omega\tau \gg 1$, is given by

$$n_0^2 = \frac{\varepsilon_L}{\varepsilon_0}\left[1 - \frac{\omega_p^2(\omega^2 - \omega_p^2)}{\omega^2(\omega^2 - \omega_p^2 - \omega_c^2)}\right] \tag{7.26}$$

so that singularities in reflectivity can occur at a number of different frequencies, depending on the conditions. Equation (7.26) only applies, however, to a wave for which $\mathbf{F}(t) \perp \mathbf{B} \perp \boldsymbol{\beta}$. For the orthogonal polarisation $\mathbf{B}$ is effectively zero so that the refractive index reduces to its usual value (7.11) of

$$n_0^2 = \frac{\varepsilon_L}{\varepsilon_0}\left[1 - \frac{\omega_p^2}{\omega^2}\right]$$

and, in the Voigt orientation, a semiconductor becomes birefringent.

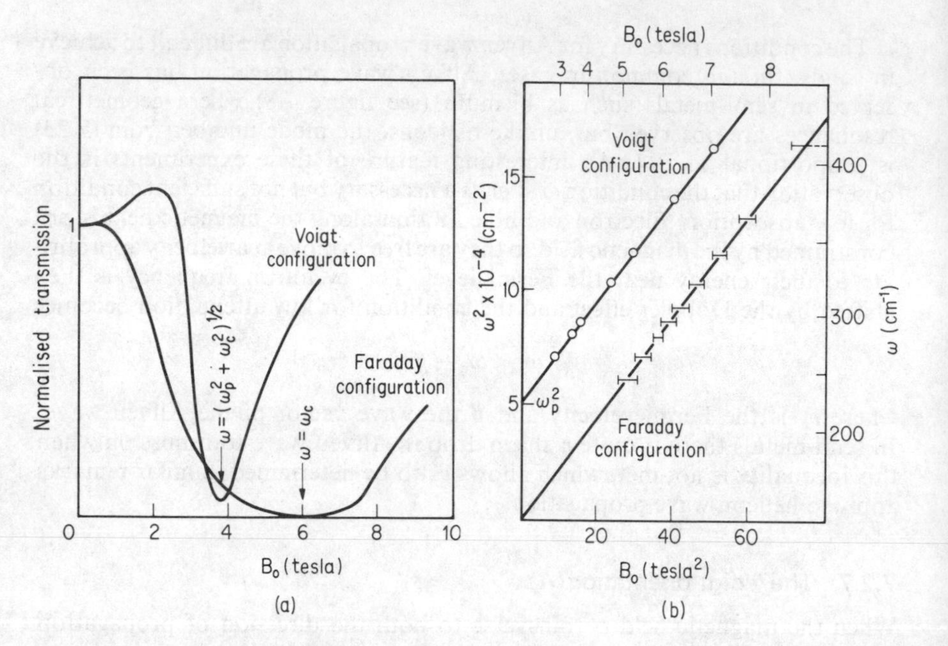

Figure 7.9 (a) Transmission of InSb crystal at $\omega = 324$ cm^{-1} as a function of magnetic field strength in the Faraday and Voigt configurations (b) Cyclotron resonance results of the type displayed in figure 7.9(a) plotted to show $\omega = \omega_c$ proportional to magnetic field in Faraday configuration and $\omega^2 = \omega_p^2 + \omega_c^2$ in the Voigt configuration. (After S. Iwasa, Y. Sawada and E. Burstein, *J. Phys. Soc. Japan,* **21**, (suppl) 742 (1966).)

The most important application of the Voigt orientation arises in metals. As already explained, conventional cyclotron resonance is not possible in metals as ω_p is greater than any achievable cyclotron frequency. In the Voigt orientation, however, and under anomalous skin effect conditions (§ 7.1.5) a form of cyclotron resonance, called Azbel–Kaner resonance after those who first predicted it, becomes possible. In the Voigt orientation the electron cyclotron orbits lie in a plane normal to the surface upon which the radiation is incident. If the mean free path of the electrons is much greater than the skin depth (anomalous skin effect condition) and $\omega_c \tau \gg 1$ the orbit radius is greater than the skin depth. Hence an orbiting electron may pass through the skin depth region for part of each revolution and during this time may be accelerated by the electric field of the radiation. Resonance will occur if the radiation frequency equals ω_c or any integral multiple of ω_c, as a 'kick' from the radiation field every second, third, etc., orbit still implies resonant excitation.

Azbel–Kaner resonance has been observed at low temperatures in a number of metals and semi-metals. As an example, we show in figure 7.10 results for zinc. That resonance occurs whenever $\omega = n\omega_c$ $(n = 1, 2, 3 \ldots)$ is clearly demonstrated in these results.

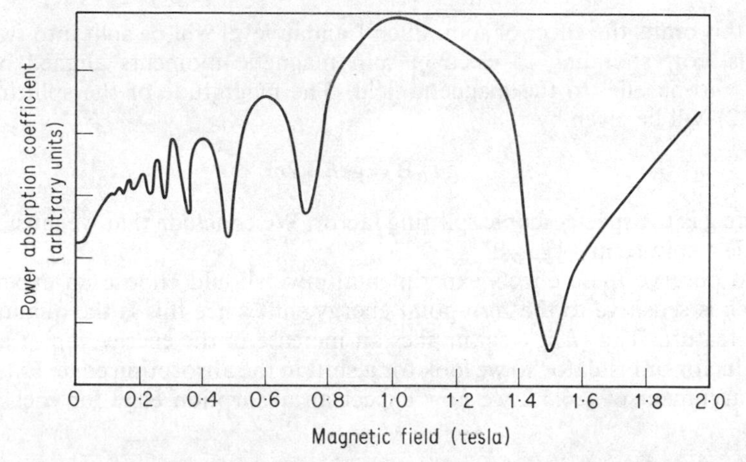

Figure 7.10 Azbel–Kaner cyclotron resonance in zinc at 1.3 K. Magnetic field in sample plane directed along sixfold axis. Frequency 72 GHz. $\omega\tau \sim 20$, the sample having a resistance ratio between room and liquid helium temperature of about 9000 (see Chapter 8). (After J. K. Galt *et al., Phys. Rev. Lett.*, **2**, 292 (1959).)

7.2.8 Quantum magneto-optic effects (A)

Notwithstanding its success, the classical picture of free-carrier absorption is at variance with the basic postulate of energy quantisation. In particular we have assumed that an electron in a magnetic field can gain energy continuously from the incident radiation, or in other words that any cyclotron orbit radius is allowed. In fact the angular momentum of the electron is quantised. When this is included, we get the Landau-level picture of § 4.7, the electron energy levels being those of an harmonic oscillator, namely

$$\hbar\omega_c(n + \tfrac{1}{2})$$

where $n = 0, 1, 2, 3$, etc. (4.47). The only allowed transitions are those in which n changes by ± 1 so absorption occurs at $\omega = \omega_c$, as it does classically. A qualitatively new feature is the appearance of the zero-point energy, $\frac{1}{2}\hbar\omega_c$, which implies a shift in the energy of the conduction and valence band edges with magnetic field. There is no quantisation along the direction of the magnetic field, z. Hence k_z remains a good quantum number and must be conserved in any interaction. The Landau levels have an energy width $\hbar/\tau$ so the levels are only well defined if $\omega_c\tau \gg 1$. Additionally the electron distribution function is only seriously perturbed if the spacing of the levels is greater than kT, i.e. $\hbar\omega_c \gg kT$. When these two conditions apply the quantum model must be used.

For a simple band (E proportional to $|k|^2$) the above argument gives the E, k relationship

$$E = \hbar\omega_c(n + \tfrac{1}{2}) + \frac{\hbar^2 k_z^2}{2m^*} \tag{7.27}$$

but this omits the effect of spin. Each Landau level will be split into two sub-levels, corresponding to electron spin magnetic moments aligned parallel and anti-parallel to the magnetic field. The magnitude of the splitting (see § 10.2) will be given by

$$g\mu_B \mathrm{B} = ge\hbar\mathrm{B}/2m$$

where g is the spectroscopic splitting factor. We conclude that we must add to (7.27) a spin term $\pm\frac{1}{2}g\mu_B\mathrm{B}$.

To observe these effects experimentally we should choose an experiment which is sensitive to the zero-point energy shift since this is the qualitatively new feature. The $\frac{1}{2}\hbar\omega_c$ term implies an increase in the energy gap of a semiconductor or insulator so we look for a shift in the absorption edge. Extending the argument of § 6.5.2 we now expect an absorption edge for each of the

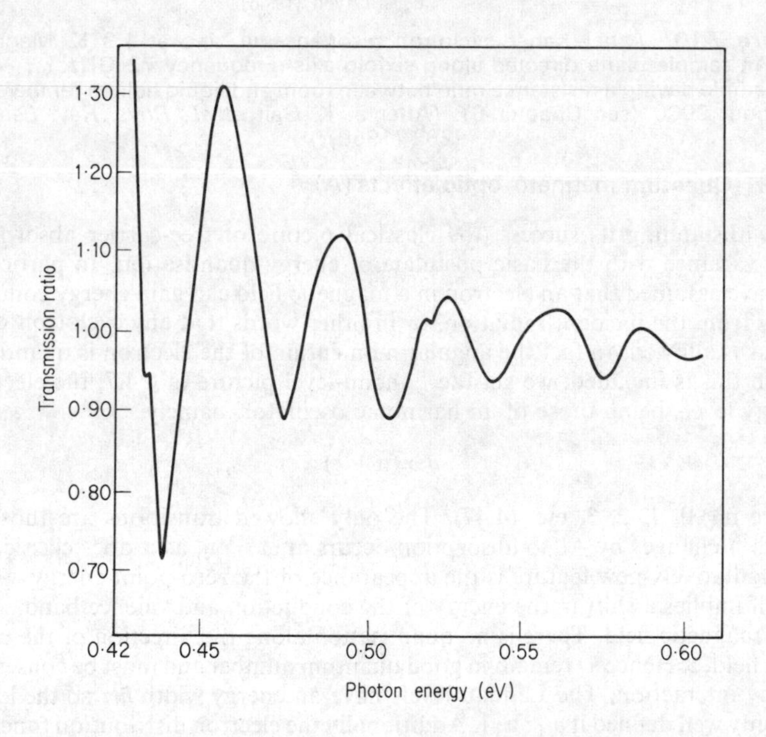

Figure 7.11 Ratio of intensities transmitted through InAs with and without a magnetic field of 9.75 tesla applied. Sample thickness 3 μm (epitaxial layer, see Chapter 5), electric vector $\perp\mathbf{B}$, temperature 77 K. Oscillatory behaviour as a function of photon energy due to transitions between valence and conduction band Landau levels. Second-order periodicity (e.g. kink at 0.52 eV) due to existence of light and heavy hole bands in InAs with consequently different values of ω_c. (After C. R. Pidgeon, D. L. Mitchell and R. N. Brown. *Phys. Rev.* **154**, 737 (1967).)

Landau sub-bands in the valence and conduction band. In a simple band structure this has the form (cf. (6.44))

$$\mathrm{K} = \text{constant} \sum (\hbar\omega - E_n)^{1/2} \tag{7.28}$$

where

$$E_n = E_g + \hbar\omega_c'(n + \tfrac{1}{2}) \pm \tfrac{1}{2}\mu_B B(g_h - g_e) \tag{7.29}$$

In equation (7.29) ω_c' has been written for

$$\omega_c' = eB\left(\frac{m_h^* + m_e^*}{m_h^* m_e^*}\right)$$

g_h, g_e are the splitting factors for the valence and conduction bands respectively. For free electrons $g \approx 2$, but may be very different in a solid (see Chapter 10).

Equations (7.28) and (7.29) give a good description of data obtained on germanium, InSb and other low effective-mass materials. The use of a low effective-mass material makes the condition $\hbar\omega_c \gg kT$, easier to achieve at temperatures sufficiently high for exciton effects to be small. As an example, figure 7.11 shows the ratio of the intensities of radiation transmitted through InAs near its absorption edge, with and without a magnetic field applied.

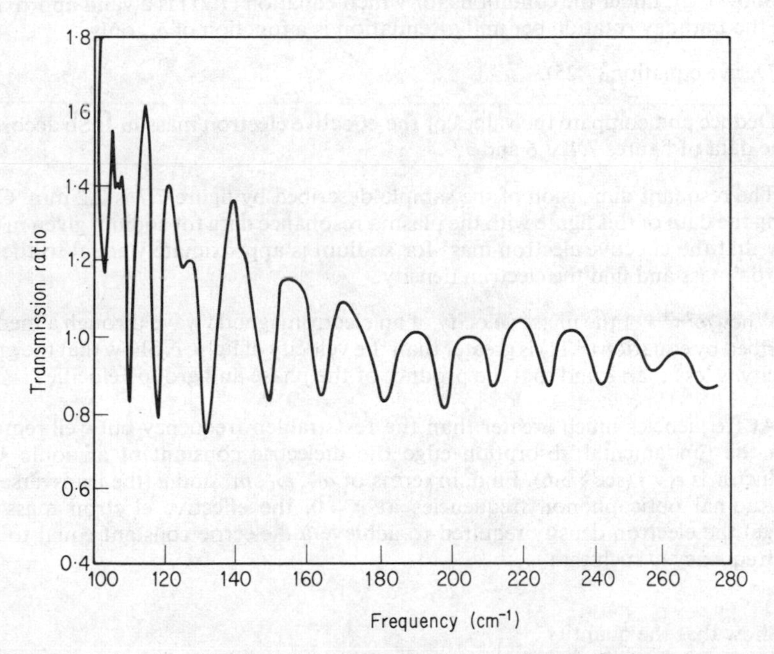

Figure 7.12 Ratio of intensities transmitted through arsenic doped germanium with and without a magnetic field of 2.18 tesla; temperature 4.2 K. Oscillatory behaviour as a function of photon energy is due to transitions between donor ground state and successive conduction band Landau levels. (After W. S. Boyle, *J. phys. Chem. Solids*, **8**, 321 (1959).)

The peaks correspond to transitions from spin-split Landau levels in the valence band to similar levels in the conduction band. In general, selection rules (§ 6.2.3) limit the number of possible transitions. Thus for a simple, parabolic, pair of bands centred at $\boldsymbol{k} = 0$ transitions between states of unequal n (7.29) are relatively forbidden. In InAs the most likely transitions are those in which n changes by 0 or -2.

Landau levels also show up in a striking way in the absorption spectra of impurities. Figure 7.12 shows the absorption of arsenic donor impurities in germanium for photon energies in excess of the ionisation energy (cf. figure 5.33). Each peak in the absorption corresponds to a transition to a progressively higher Landau level or into impurity bound states associated with each Landau level. The reduction in sharpness of the lines with increasing photon energy is due to a decrease in τ with increasing energy in the band. We shall see in the next chapter that τ decreases as the density of states increases, and hence with increasing E.

PROBLEMS

(Answers, where appropriate, are given on page 481.)

7.1 Show that, under the conditions for which equation (7.21) is a valid approximation, the Faraday rotation per unit attenuation is a function of ω_c only.

7.2 Derive equation (7.25).

7.3 Deduce and compare the values of the effective electron mass in InSb according to the data of figures 7.4, 7.6 and 7.9.

7.4 The resonant dimension of the sample described by figure 7.7 is 2.2 mm. Combining the data of this figure with the plasma resonance data for sodium given in § 4.7 show that the effective electron mass for sodium is approximately equal to the free electron mass and find the electron density.

7.5 When $\omega^2\tau^2 \gg 1$ the phase velocity of an electromagnetic wave through a medium described by equation (7.11) is greater than the velocity of light, c. Show that the group velocity is less than c and that the product of the phase and group velocities is c^2.

7.6 At frequencies much greater than the reststrahlen frequency but well removed from the fundamental absorption edge the dielectric constant of an ionic semiconductor is $\varepsilon(\infty)$ (see § 3.6). Find, in terms of ω_T, ω_L, m^* and e (the transverse and longitudinal optic phonon frequencies at $q = 0$, the effective electron mass and charge) the electron density required to achieve a dielectric constant equal to $\varepsilon(\infty)$ at a frequency ω such that

$$\omega_T \gg \omega \gg 1/\tau$$

7.7 Show that the quantity

$$[\omega_L^2 - \omega_T^2]^{1/2}$$

which appears in the answer to problem 7.6, can be written in the form $Ne^2/\varepsilon_0 M$ and can therefore be described as a lattice plasma frequency. Is there any analogue of ω_T in a free electron plasma?

8

Transport

8.1 Introduction

8.1.1 Conduction (a)

The definitive property of a conductor is that it conducts, both heat and electricity. Charge transport is, of course, by electrons; energy can be transported by electrons, phonons, excitons, photons and other particles. We therefore start with electrical conduction as this is easier.

We have seen (§ 7.1.1) that if we do not worry unduly about the interpretation of the relaxation time, τ, the d.c. conductivity due to N electrons of mass m^* per unit volume can be written

$$\sigma = \frac{Ne^2\tau}{m^*} = Ne\mu \tag{8.1}$$

where we have taken the opportunity to define the mobility, μ, as the average drift velocity per unit field. If two types of charge carrier are present simultaneously, as in an intrinsic semiconductor, the conductivity will be the sum of two terms like (8.1). It might reasonably be concluded that d.c. conductivity measurements could give no information not available from free-carrier absorption data and if τ and m^* are independent of E and $\boldsymbol{k}$ this is indeed true. We have seen for instance how N, m^* and τ may be deduced from plasma resonance, cyclotron resonance, Faraday rotation, etc. In many solids, however, m^* and τ appear to be functions of E and $\boldsymbol{k}$. Though this variation obviously affects the optical properties, it is often easier to take these complexities into account without the additional complication of a finite frequency.

In this chapter we first consider how the conductivity of metals and semiconductors varies with temperature. This variation, to a first approximation, is independent of any anisotropy in the band structure so that we neglect the possibility of any variation of τ or m^* with $\boldsymbol{k}$. Hence the Fermi surface is initially assumed to be spherical. The temperature dependence of σ can, however, be affected if τ or m^* are functions of energy, E. Consequently we shall retain these possibilities. To account for the temperature dependence of σ we shall have to probe more deeply into the interpretation of τ and the scattering processes (e.g. electron–phonon interactions) which determine its value. Towards the end of the chapter we will consider those phenomena which are sensitive to anisotropy of the band structure and discuss the dependence of m^*, in particular, on $\boldsymbol{k}$. Finally we describe the phenomenon of superconductivity.

8.1.2 Hall effect, magnetoresistance and drift mobility (a)

Apart from conductivity measurements, experimental information can also be obtained on the temperature coefficients of the parameters in (8.1) by a study of the phenomena summarised below.

The Hall Effect. Consider a rectangular sample and suppose that a current of density J_z flows in the x-direction. If a magnetic field is now applied in the z-direction an electric field is generated in the direction at right angles to both **B** and **J**, namely the y-direction. This 'Hall field' is simply due to the Lorentz force, $(e\boldsymbol{v} \times \mathbf{B})$ and the Hall coefficient is defined by

$$R_{\mathrm{H}} = \frac{\mathrm{F}_y}{\mathrm{B}_z \mathrm{J}_x} \tag{8.2}$$

We shall show that, for small magnetic fields, R_{H} is related to the electron (or hole) density by

$$R_{\mathrm{H}} = -\mathrm{r}_0/Ne \tag{8.3}$$

(negative for electrons) and hence

$$\mu = \sigma R_{\mathrm{H}}/\mathrm{r}_0 \tag{8.4}$$

where r_0 is a number of the order of unity. Note that R_{H} depends linearly on the electronic charge and is therefore of opposite sign for hole flow and electron flow. Physically this arises as follows: (a) holes and electrons flow in opposite directions when carrying current in the x-direction; (b) the Lorentz force depends on both the charge and the velocity so if both are reversed in sign (e.g. when changing from N type to P type) the force remains in the same, positive, y-direction. This behaviour is illustrated in figure 8.1. Consequently,

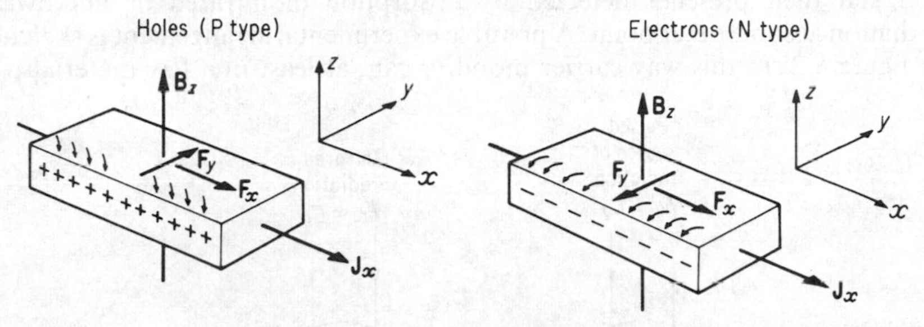

Figure 8.1 The Hall effect

if electrons and holes are both present simultaneously their charges on the $+y$ surface partially cancel and the Hall voltage is reduced. R_H then becomes

$$R_H(\text{two carriers}) = \frac{P\mu_h^2 r_{0_h} - N\mu_e^2 r_{0_e}}{e(P\mu_h + N\mu_e)^2} \tag{8.5}$$

where P, μ_h, N, μ_e are the hole and electron densities and mobilities, respectively. r_{0_h} and r_{0_e} are again numbers of order unity, but not necessarily equal. *Magnetoresistance*. With a magnetic field arranged for Hall measurements there is often a change in sample resistance, as measured in the direction of the current, J_x. We might be tempted to deduce the magnitude of this effect by using equation (7.17) and simply putting $\omega = 0$. We would then obtain

$$\frac{\Delta\sigma}{\sigma} = \omega_c^2\tau^2 \text{ where } \omega_c = eB/m^* \tag{8.6}$$

but we would be wrong. We shall show that it is necessary to return to the more basic equations (7.15) and (7.16). When discussing optical properties, as in Chapter 7, we supposed the sample to be large compared with the wavelength. At zero frequency this would require an infinite sample. The assumption of large dimensions makes the Hall field vanish. In fact the Hall field opposes any sideways motion of the carriers due to the magnetic field and ideally just balances the Lorentz force. The carriers should then move undeflected in the x-direction and there should be no magnetoresistance effect, but there usually is. We shall find that magnetoresistance arises because the velocities of the carriers are not all the same and the Hall voltage can only stop the motion of the *average* carrier, not the whole distribution. Even more mysterious is the appearance of a longitudinal magnetoresistance effect, when **J** and **B** are colinear. The Lorentz force, $e\boldsymbol{v} \times \mathbf{B}$, should then be zero. In fact, longitudinal magnetoresistance is associated with non-spherical Fermi surfaces, which we are ignoring for the present.

Drift Mobility. For those semiconductors in which PN junctions can be fabricated, it is possible to measure the time taken by carriers 'injected' (see Chapter 9) at one PN junction to drift in an electric field to another, similar, junction at which they may be detected. Alternatively, excess carriers may be generated by photons of energy greater than E_g at one point in a semiconductor rod and their presence detected by absorption of infrared or microwave radiation at some other point. A possible experimental arrangement is sketched in figure 8.2. In this way carrier mobility can, at least in a few materials, be

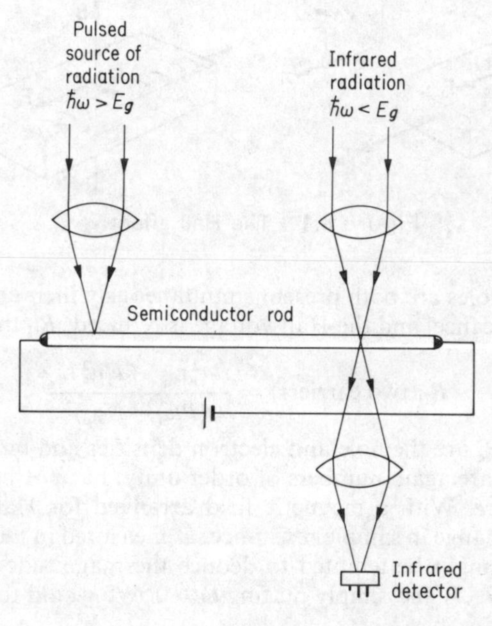

Figure 8.2 A possible method of measuring drift mobility (see text)

measured directly. The mobility measured in this way is not necessarily the same as that deduced from a combination of Hall and conductivity data. In particular it may be shown that the $R_H\sigma$ product is unchanged if electrons are trapped and released periodically, while the drift mobility is obviously reduced.

8.1.3 Temperature dependence of conductivity (a)

Before plunging into theory, we briefly summarise the experimental position on electrical conductivity as a function of temperature.

The variation of carrier density and mobility with temperature has been measured for a very large number of solids, chiefly using Hall and conductivity data. In metals, which for a discussion of transport properties we define as having a Fermi energy, E_F, much greater than kT, the electron concentration is independent of temperature. Ignoring superconductivity, the conductivity of a typical pure metal is essentially constant at very low temperatures, falls rapidly between about 10 K and 50 K when $\sigma \propto T^{-5}$ and more slowly at higher temperatures, so that $\sigma \propto T^{-1}$ at room temperature and above. As we shall see, the nearly constant resistivity at very low temperatures is dominated by residual impurities or other defects. Above 10 or 20 K the mobility is determined by the scattering of electrons by phonons. These results are illustrated in figure 8.3. To show up the variation in resistivity with temperature the residual resistivity of each sample has been subtracted.

The behaviour of semiconductors is more complicated. At low temperatures, and indeed up to room temperature and above in most semiconductors, the carrier concentration is determined by the number of ionised donors or acceptors, as explained in § 5.3.1. At the very lowest temperatures the mobility is sometimes determined by neutral impurity scattering, which is independent of temperature, but as soon as a few impurities are thermally ionised, scattering by ionised impurities becomes the dominant process and the mobility increases

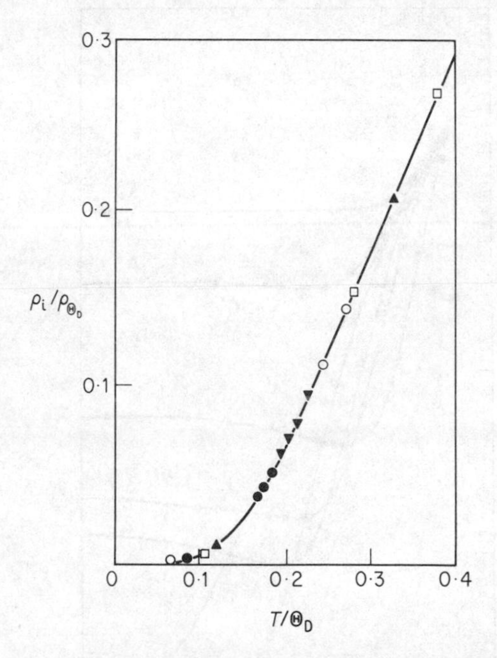

Figure 8.3 The resistivity of metals at low temperatures tends to a constant, residual, value determined by impurities. If this is subtracted the resistivity due to electron scattering by lattice vibrations increase monotonically with temperature. According to the theory of § 8.5.1 the ratio of the lattice resistivity at temperature T to the resistivity at a characteristic temperature, Θ_D, called the Debye temperature, should be the same function of T/Θ_D for all metals. This function is the curve shown and given in equation (8.35). Samples are

▲ Au $\Theta_D = 175$ K ▼ Al $\Theta_D = 395$ K
□ Na $\Theta_D = 202$ K ○ Ni $\Theta_D = 472$ K
● Cu $\Theta_D = 333$ K

with increasing temperature. Above about 100–200 K, depending on purity and other factors, scattering by phonons becomes the dominant process. The mobility now decreases with temperature, often as T^{-n} where $\eta \sim 2$. When intrinsic conduction becomes important the conductivity rises very rapidly with temperature due to the rapid increase in total (electron plus hole) carrier density. In this temperature region it is not usually possible to deduce the carrier mobilities with any confidence. Some representative data on charge carrier concentration and mobility are shown in figure 8.4.

With the experimental background thus established, we shall discuss the factors affecting carrier mobility more thoroughly in the next two sections. We shall speak of electrons but most remarks will be equally applicable to holes.

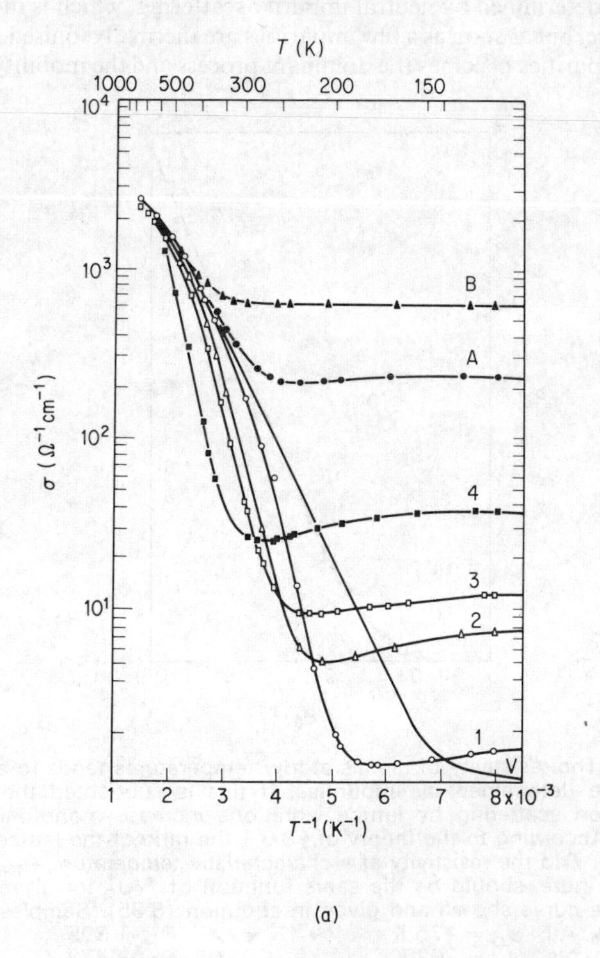

Figure 8.4 (a) For caption see page 293.

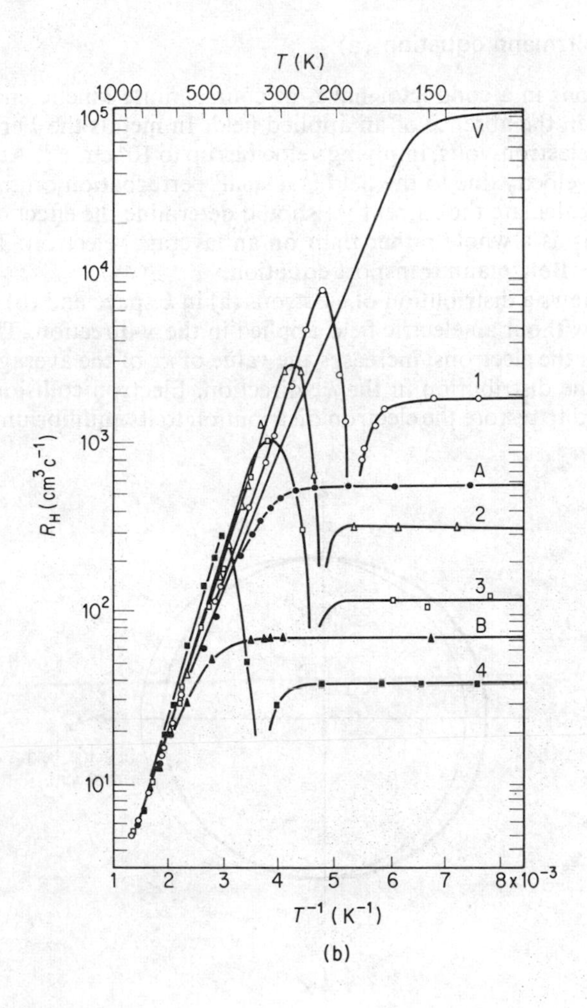

Figure 8.4 The temperature dependence of (a) the conductivity, σ, and (b) the Hall coefficient, R_H, of the semiconductor indium antimonide. Samples A, B and V are N type; 1 to 4 are P type. Note that the carrier density varies exponentially with temperature and is practically the same for all samples in the intrinsic range at high temperatures but is constant in the low temperature exhaustion region. The Hall coefficient of P type samples changes sign as the intrinsic region is approached because the electron mobility exceeds the hole mobility. The product σR_H, which is proportional to mobility, generally decreases with increasing temperature. (After O. Madelung and D. Meyerhofer, *The Physics of III/V Compounds*, John Wiley and Sons.)

8.2 Charge transport in an electric field

8.2.1 The Boltzmann equation (a)

The free electrons in a conductor have, of course, finite kinetic energies and velocities even in the absence of an applied field. In metals the Fermi energy may be several electron-volts, implying velocities up to 10^8 cm s^{-1}. At moderate fields the drift velocity due to the field is a small perturbation on the random motion and to calculate the current we should determine the effect of the field on the electrons as a whole rather than on an 'average' electron. This is the approach of the Boltzmann transport equation.

Figure 8.5 shows a distribution of electrons (a) in $\boldsymbol{k}$ space and (b) in energy, each with and without an electric field applied in the x-direction. The electric field accelerates the electrons, increases the value of k_x of the average electron and displaces the distribution in the k_x-direction. Electron collisions, on the other hand, tend to restore the electron distribution to its equilibrium position.

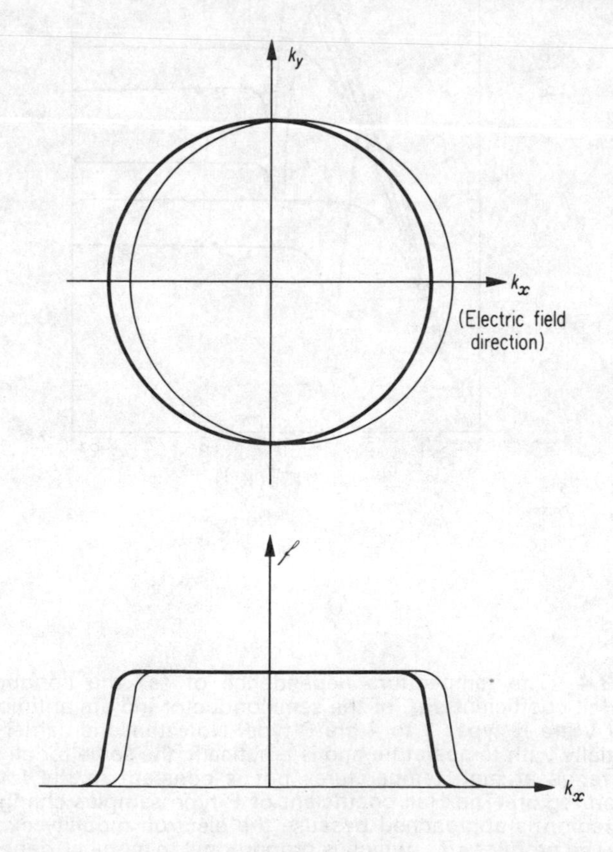

Figure 8.5 Electron distribution in a metal in $\boldsymbol{k}$-space, showing the effect of applying a small electric field

If the electric field has a steady value (d.c.) a steady state will result when the effect of the field and the collisions just counterbalance one another. The distribution will then be displaced as shown in figure 8.5. The current flow will be determined by the (relatively few) electrons on each side of the distribution whose velocity in the positive x-direction has been increased and decreased respectively. We shall now put this argument on to a more quantitative basis, using f to denote the electron distribution function.

The Boltzmann equation describes the rate of change of the distribution of electrons in $\boldsymbol{k}$ space. In the absence of temperature gradients and magnetic fields the rate of change of the distribution function is given by

$$\frac{\mathrm{D}f(\boldsymbol{k})}{\mathrm{D}t} = \left[\frac{\partial f}{\partial t}\right]_{\text{Field}} + \left[\frac{\partial f}{\partial t}\right]_{\text{Collisions}} \tag{8.7}$$

and when a steady state is established, $\mathrm{D}f(\boldsymbol{k})/\mathrm{D}t$ is equal to zero.

For a small field in the x-direction

$$\left[\frac{\partial f}{\partial t}\right]_{\text{Field}} = \left(\frac{\mathrm{d}f}{\mathrm{d}k_x}\right)\left(\frac{\mathrm{d}k_x}{\mathrm{d}t}\right) = \left(\frac{\mathrm{d}f}{\mathrm{d}k_x}\right)\frac{e\mathrm{F}}{\hbar} \tag{8.8}$$

as the field determines the rate of change of momentum (see (4.31)).

If the perturbation due to the field is small, we suppose that the distribution function returns to its unperturbed form exponentially with time so that

$$\left[\frac{\partial f}{\partial t}\right]_{\text{Collisions}} = \frac{f_0 - f}{\tau_{\mathrm{D}}} \tag{8.9}$$

where f is the perturbed and f_0 the unperturbed distribution, and τ_{D} the relaxation time of the distribution. Equation (8.9) assumes that τ_{D} is isotropic. There is no *a priori* reason for believing that τ_{D} is the same as the τ of (8.1), although we shall find that they are closely related.

A solution of (8.7), (8.8) and (8.9) with $\mathrm{D}f/\mathrm{D}t = 0$, using (4.28), is

$$f = f_0 + e v_x \mathrm{F} \tau_{\mathrm{D}} (\mathrm{d}f_0/\mathrm{d}E) \tag{8.10}$$

where v_x is the electron velocity in the x-direction and the second term of (8.10) represents the perturbation due to the field.

We now assert that the current density, J_x, is the electronic charge times the velocity, summed over all carriers in $\boldsymbol{k}$ space in the Brillouin zone (B.Z.).

$$\mathrm{J}_x = e \int_{\mathrm{B.Z.}} v_x f(\boldsymbol{k}) \frac{\mathscr{V}}{(2\pi)^3} \, \mathrm{d}\boldsymbol{k} \tag{8.11}$$

where $\mathscr{V}/(2\pi)^3$ is the density of states per unit volume in $\boldsymbol{k}$ space and $\mathrm{d}\boldsymbol{k} = \mathrm{d}k_x \mathrm{d}k_y \mathrm{d}k_z$ is the volume element (cf. (2.16)). We may write

$$v_x = v_k \cos\theta \quad \text{where } v_k^2 = 2E/m^*$$

and, using polar co-ordinates, $\mathrm{d}\boldsymbol{k} = \sin\theta \mathrm{d}\theta \mathrm{d}\phi k^2 \mathrm{d}k$.

Substituting (8.10) for $f(\boldsymbol{k})$ but dropping the f_0 term as it cannot contribute to the current, we obtain

$$\mathrm{J}_x = e^2 \mathrm{F} \int_{\mathrm{B.Z.}} \int_0^{\pi} \int_0^{2\pi} v_k^2 \cos^2\theta \tau_{\mathrm{D}} \left(\frac{\mathrm{d}f_0}{\mathrm{d}E}\right) k^2 \mathrm{d}k \sin\theta \mathrm{d}\theta \mathrm{d}\phi (\mathscr{V}/(2\pi)^3)$$

Integrating with respect to θ and ϕ assuming τ_{D} to be isotropic and assuming a parabolic band, namely $E = \hbar^2 k^2/2m^*$, J_x becomes

$$\mathrm{J}_x = \frac{4\pi e^2 \mathrm{F}}{3m^*} \int E^{3/2} \tau_{\mathrm{D}} \left(\frac{\mathrm{d}f_0}{\mathrm{d}E}\right) \frac{2^{1/2} m^{*3/2}}{\hbar^3} \mathrm{d}E (\mathscr{V}/(2\pi)^3) \tag{8.12}$$

The electron density is now written in the same form, namely

$$N = \int_{\mathrm{B.Z.}} f_0(\boldsymbol{k}) \mathrm{d}\boldsymbol{k} (\mathscr{V}/(2\pi)^3)$$

$$= \int 2\pi f_0(\boldsymbol{k}) \left(\frac{2^{1/2} m^{*3/2} E^{1/2}}{\hbar^3}\right) \mathrm{d}E (\mathscr{V}/(2\pi)^3) \tag{8.13}$$

so the mobility, μ, is given by

$$\mu = \frac{\mathrm{J}_x}{Ne\mathrm{F}} = \frac{2e \int E^{3/2} \tau_{\mathrm{D}} (\mathrm{d}f_0/\mathrm{d}E) \mathrm{d}E}{3m^* \int E^{1/2} f_0 \mathrm{d}E} \tag{8.14}$$

To make further progress we need to insert an expression for $f_0(\boldsymbol{k})$. This is the Fermi distribution function, though for a semiconductor it reduces to the Boltzmann distribution. Hence we now consider metals and semiconductors separately.

8.2.2 Solution of the Boltzmann equation for metals (a)

For a metal the Fermi energy is very large compared with kT and hence f_0 is either unity or zero at all energies except for a short transition region at the Fermi energy. Consequently the derivative $(\mathrm{d}f_0/\mathrm{d}E)$ is zero everywhere except for a narrow energy interval near E_F, as shown in figure 8.6. Equation (8.14) then simplifies enormously to give

$$\mu = \frac{e\tau_F}{m^*} \tag{8.15}$$

where τ_F is to be interpreted as the effective value of τ_{D} at the Fermi surface. This is the same as the simple result from the equation of motion expressed in (8.1) provided we interpret τ as the scattering time at the Fermi level. This is, in fact, physically reasonable. Thus, although the N in (8.1) represents the total density of free electrons, only those with energies near the Fermi level

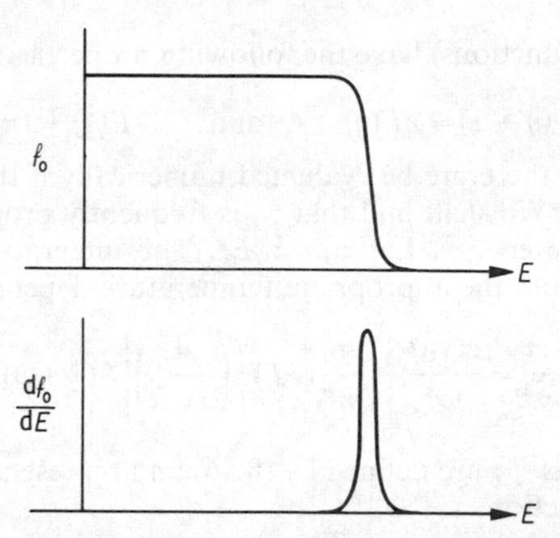

Figure 8.6 The Fermi distribution function and its derivative

have any vacant energy states nearby into which they can be scattered. Hence they alone can gain energy from the applied electric field and they alone determine the value of τ for the whole distribution. Even if τ_D was a function of energy in the band, it would not matter as scattering takes place at essentially one energy only. By a similar argument we can conclude that if the mass is dependent on energy and consequently E is not proportional to k^2 there is still no energy (and hence temperature) dependent term in the final result. In semiconductors this is not true but (8.15) goes a long way towards explaining the success of the equation of motion approach used in Chapter 7.

8.2.3 Solution of the Boltzmann equation for semiconductors (a)

For a semiconductor in which the Fermi level is below the bottom of the conduction band the Boltzmann distribution (cf. (4.67))

$$f_0 \propto \exp(-E/kT)$$

is more appropriate. Inserting this into (8.14) gives

$$\mu = \frac{2e\int E^{3/2}\tau_D \exp(-E/kT)\,dE}{3m^* kT\int E^{1/2}\exp(-E/kT)\,dE} \tag{8.16}$$

which for brevity we write as $\mu = e\langle\tau_D\rangle/m^*$. As the Brillouin zone extends to energies large compared with the value of E at which there is any appreciable electron density in a semiconductor, we can replace the integration over the zone in (8.16) by integration to infinity. Evaluation of (8.16) is made easy by the fact that integrals of the form

$$\int_0^\infty g^{(\eta-1)}\exp(-g)\,dg = \Gamma(\eta)$$

(known as gamma functions) have the following properties:

$$\Gamma(\eta+1)=\eta\Gamma(\eta) \qquad \text{and} \qquad \Gamma(\tfrac{1}{2})=(\pi)^{1/2}$$

Equation (8.16) can therefore be evaluated numerically if the energy dependence if τ_D is known. We shall find that τ_D is frequently proportional to some simple power of the energy. Let $\tau_D=\tau_0 E^p$. The integration then produces a numerical factor and the appropriate temperature dependence, as follows:

$$\mu=\frac{e\tau}{m^*}=\frac{e\langle\tau_D\rangle}{m^*}=\frac{e\tau_0}{m^*}(kT)^p\left(\frac{4}{3\sqrt{\pi}}\right)\left[\Gamma(\tfrac{5}{2}+p)\right] \tag{8.17}$$

The angular brackets $\langle\ \rangle$ are defined by (8.16) and represent the average over the distribution function.

As an example, suppose $p=-\frac{1}{2}$. Then

$$\mu=\frac{e\langle\tau_D\rangle}{m^*}=\frac{4e\tau_0}{3m^*(\pi kT)^{1/2}}=\frac{4}{3\sqrt{\pi}}\left(\frac{e\tau_D}{m^*}\right)$$

Thus the Boltzmann equation again produces the same result as the simple theory, except for a numerical factor of order unity. If the factor is as small as $4/3\sqrt{\pi}$ it is comparable with experimental errors due to inhomogeneity of practical samples, etc., but it is not always quite as small as that. We are, however, reassured that we can use the kinetic equation of motion approach provided we use the appropriate averaging procedure for τ. Thus the Hall effect and transverse magnetoresistance can be derived (for a spherical energy surface) from (7.15) and (7.16) by putting $J_y=0$, where y is the direction of the Hall field. At low magnetic fields, such that $\omega_c\tau \ll 1$, we obtain

$$R_H=\frac{1}{Ne}\left(\frac{\langle\tau_D{}^2\rangle}{\langle\tau_D\rangle^2}\right)=\frac{r_0}{Ne} \tag{8.18}$$

and

$$\frac{\Delta\sigma}{\sigma}=\omega_c^2\left[\frac{\langle\tau_D^3\rangle\langle\tau_D\rangle-\langle\tau_D^2\rangle^2}{\langle\tau_D^2\rangle}\right] \tag{8.19}$$

which should be compared with (8.3) and (8.6) respectively.

8.2.4 The transverse magnetoresistance of a semiconductor (a)

For a metal, as we have seen, τ_D is just a constant and the r_0 of (8.3) is equal to unity. Otherwise r_0 is a small numerical factor, of order unity, whose value can be found using (8.18) if the energy dependence of τ_D is known. The magnetoresistance is more complicated and involves the small difference between large quantities. If the scattering time is a constant, as in a metal, the transverse magnetoresistance should vanish. This accords with our argument in § 8.1.2. When, however, the electrons are distributed in energy and τ_D (or m^* for that matter) is energy dependent, the Hall field cannot stop all motion in the Hall field direction. There is a flow against the Hall field of electrons with more than average velocities and an equal and opposite flow with the Hall field of electrons with below average velocities. We can consider that these two opposite but equal currents are caused by the magnetic field but may interact again with the magnetic field. This second interaction produces motion of charge in the original primary current direction, and subtracts from that current. Thus a magnetoresistance effect occurs, proportional to B^2 because of the double interaction. This is the physical interpretation of (8.19).

8.2.5 Electron scattering (AD)

The Boltzmann equation and its solution shows us that the time constant, τ, defined by (8.1) is numerically equal to τ_D at the Fermi surface in metals and differs from τ_D by only a small numerical factor in semiconductors. Small though the latter difference may be, it does lead to a finite transverse magnetoresistance effect in semiconductors without recourse to a non-spherical energy surface. We now wish to find a value for τ_D (and hence τ) in terms of phonon density, etc., and hence the temperature.

The momentum of electrons is randomised and consequently the distribution function restored to equilibrium in $\boldsymbol{k}$ space by electron scattering 'collisions'. The rate of restoration, and hence τ_D, will be determined by (a) the frequency of collisions and (b) how efficacious such events are in randomising the motion of an electron. Clearly a collision which simply deflects an electron through a small angle produces a smaller change in momentum and hence current than total reversal of the electron's motion. If a collision changes the direction of motion by an angle θ without changing the speed the change in momentum is proportional to $(1 - \cos\theta)$ and we should average over all possible values of θ. Hence

$$\tau_D = \frac{\tau_c}{1 - \langle \cos\theta \rangle} \tag{8.20}$$

where τ_c is the average time between collisions. Later we snall also need to know the efficiency with which collisions enable electrons to lose *energy* gained from the field. It is immediately apparent that this process must be less efficient than the randomisation of momentum for an electron never loses all its energy in a collision. Indeed we shall find that, in solids, most collisions are nearly elastic so that the energy relaxation time, τ_E, is usually considerably greater than τ_D.

In later sections we shall be concerned with calculating the value, and particularly the temperature dependence, of τ_c. We note now, however, that it is sometimes more convenient to speak of a mean free path, l_f, between collisions than of a mean free time τ_c (see § 2.7). We then define a scattering cross section, σ_d. If N_c is the density of scattering centres we have

$$\tau_c = 1/N_c \sigma_d v \tag{8.21}$$

where $v = l_f/\tau_c$ and hence $l_f = 1/N_c\sigma_d$.

Any disturbance of the perfect periodicity of the lattice (phonons, the crystal surface, impurities, dislocations, etc.) will act as scattering centres. It is evident that the crystal surface, or boundary scattering, will be important only when l_f is greater than the crystal dimensions. In practice phonon and impurity scattering are the most important and we consider these in detail in the following sections. An order of magnitude estimate of impurity scattering in metals can be obtained immediately from (8.21). For a metal v is essentially the same for all electrons and equal to the velocity at the Fermi surface—say 10^8 cm s^{-1}. Using (8.1), putting $\sigma_d \sim 10^{-16}$ cm^2 (of atomic dimensions) and $\tau_D \approx \tau_C \approx \tau$ we obtain for the resistivity, ρ

$$\rho = 1/\sigma = \frac{m^* N_c \sigma_d v}{Ne^2} \tag{8.22}$$

which is ~0.3×10^{-6} ohm cm for $N_c/N \sim 10^{-2}$ (1 atomic percent of impurity). Experimentally the resistivity of metals is usually determined by impurities at low temperatures when phonon scattering is small. Typical values of ρ lie between 0.1 and 0.5×10^{-6} ohm cm per atomic percent for homovalent impurities and rise to about 3×10^{-6} ohm cm when the impurity differs in valency by 2 from the host element.

The very large difference in the density of electrons in the conduction bands of metals and semiconductors leads to a substantial difference in the frequency of phonon scattering in each case. We have already seen (§ 8.1.2) that the temperature dependence of the mobility, which we hope to predict, is very different in metals and semiconductors. In metals the electrons occupy a substantial fraction of the Brillouin zone and since they can only be scattered from one point on the Fermi surface to another (§ 8.2.2), any substantial change in momentum requires a substantial change in wave vector. The same is not true in semiconductors in which the electrons occupy a very much smaller volume in $\boldsymbol{k}$ space. Not only is a large change in wave vector not required to change $\boldsymbol{k}$ to (say) $-\boldsymbol{k}$ but the E, $\boldsymbol{k}$ conservation rules do not allow large changes in $\boldsymbol{k}$. Because of these differences we shall consider the factors determining τ_c in semiconductors and metals separately, beginning with semiconductors.

8.3 Scattering processes in semiconductors

8.3.1 Electron scattering by phonons in semiconductors (AD)

We have seen that electrons in semiconductors may be scattered by phonons and crystal imperfections. At ordinary temperatures phonon scattering is usually dominant. As the energies and wave vectors of electrons and phonons are well defined the conservation rules apply. If the initial and final electron energies and wave vectors are written E_i, E_f, $\boldsymbol{k}_i$ and $\boldsymbol{k}_f$ respectively we have

$$E_i - E_f = \pm\hbar\omega(\boldsymbol{q}) = \frac{\hbar^2}{2m^*}(\boldsymbol{k}_i^2 - \boldsymbol{k}_f^2) \tag{8.23}$$

and

$$\boldsymbol{k}_i - \boldsymbol{k}_f = \pm\boldsymbol{q} \tag{8.24}$$

where $\hbar\omega(\boldsymbol{q})$ and $\boldsymbol{q}$ are the energy and wave vector of the phonon, which may be emitted or absorbed. If, as we have assumed in (8.23), electrons are constrained to remain in one energy band minimum (8.23) and (8.24) limit the possible phonons involved to those of small q near the centre of the zone. The probability of an electron making a transition from the i state to the f state is determined by the probability of an empty f state being available and the probability of a phonon of energy $\hbar\omega(\boldsymbol{q})$ being created or destroyed. In a semiconductor the electron density is low so we neglect the possibility of the final state being occupied. The total transition probability is then proportional to the two factors:

(a) The density of final states into which an electron may be scattered (§ 6.2.3). For a band which is spherical in $\boldsymbol{k}$ space and parabolic in E this will be (from (2.43))

$$S(E_f) = \frac{(2m^*)^{3/2}(E_i \pm \hbar\omega(q))^{1/2}}{4\pi^2\hbar^2} \tag{8.25}$$

(b) The sum of the probabilities of the electron absorbing or emitting a phonon of energy $\hbar\omega(\boldsymbol{q})$, which in turn are proportional to $n(\boldsymbol{q})$ and $(n(\boldsymbol{q})+1)$ respectively, where $n(\boldsymbol{q})$ is the Bose distribution function (see §§3.5 and 6.2.2).

The total scattering probability, and hence $1/\tau_c$, is then proportional to the product of the two above factors, summed over all phonons and averaged over the electron distribution function

$$\frac{1}{\langle\tau_c\rangle} \propto \frac{(2m^*)^{3/2}}{4\pi^2\hbar^3}\sum_{\boldsymbol{q}}\left\langle G(\boldsymbol{q})\left[\frac{(E+\hbar\omega(q))^{1/2}}{\exp(\hbar\omega(\boldsymbol{q})/kT)-1} + \frac{(E-\hbar\omega(\boldsymbol{q}))^{1/2}\exp(\hbar\omega(q)/kT)}{\exp(\hbar\omega(\boldsymbol{q})/kT)-1}\right]\right\rangle \tag{8.26}$$

where the first term in the summation corresponds to the electron absorbing a phonon and the second to emission. If $E < \hbar\omega(\boldsymbol{q})$ the latter term is set equal to zero. The factor $G(\boldsymbol{q})$ contains the matrix element (which may be a function of $\boldsymbol{q}$) and its magnitude depends on the mechanism of coupling between electrons and phonons, which is discussed in the next section.

8.3.2 Electron–phonon coupling in semiconductors (AD)

There are three important coupling mechanisms in semiconductors. These may be referred to as electromagnetic, piezoelectric and deformation potential coupling.

Electromagnetic coupling arises from the electric fields generated by the motion of charged ions. The electric dipole moments generated by optical modes can be very large in ionic crystals (cf. § 3.4). In metals the electrons may move to smooth out, to a large extent, the fields so generated but in low conductivity ionic crystals the coupling between electrons and optical phonons, particularly the longitudinal modes, can be very strong. In extreme cases the coupling may be so strong that the electron and optical phonon must be considered to form a composite particle called a 'polaron'. We have discussed this case in § 4.11. The mobility is then very low (e.g. $<1\ \text{cm}^2\ \text{V}^{-1}\ \text{s}^{-1}$).

Piezoelectric scattering, like piezoelectricity, is limited to crystals which lack inversion symmetry. Semiconductors in these crystal classes are usually also ionic or partially ionic so that piezoelectric scattering is often overshadowed by optical mode scattering. However, the piezoelectric effect can couple electrons to acoustic phonons of low frequency and, as we shall show later, this has device applications. The coupling arises because acoustic modes generate regions of compression and rarefaction in a crystal (§ 3.3.1) and in piezoelectric crystals these lead to electric fields.

Compression and rarefaction of crystals also cause changes in the energy of energy band minima. This provides a further mechanism for electron–phonon coupling called deformation potential coupling. Compared with the mechanisms considered previously deformation potential coupling is relatively weak but it can occur in non-ionic crystals with inversion symmetry (e.g. Ge and Si) and is then dominant by elimination. If the minimum of the energy band is at the centre of the Brillouin zone the electrons are coupled only to longitudinal phonons since, by symmetry, such a minimum is unaffected by shear forces. Transverse vibrations can produce energy changes away from the zone centre so that conduction band electrons in Si and Ge (see table 4.1) are coupled, *via* an appropriate deformation potential constant, to transverse phonons.

A knowledge of the electron–phonon coupling mechanism allows the $\boldsymbol{q}$ dependence of $G(\boldsymbol{q})$ to be determined. The mobility as a function of temperature can then be found using (8.17) and (8.26). For comparison with experiment two approximate solutions are of value, as follows.

(i) *Optic phonon scattering*. The energy, $\hbar\omega_0$, of optic phonons of small $\boldsymbol{q}$ is practically independent of $\boldsymbol{q}$ and below room temperature $\hbar\omega_0 \gg kT$ and E. Compared with the resulting exponential the variation of $G(\boldsymbol{q})$ with $\boldsymbol{q}$ is relatively slow for any coupling mechanism so the mobility is given approximately by

$$\mu_{\text{opt}} \propto (m^*)^{-5/2}\left[\exp(\hbar\omega_0/kT)-1\right] \tag{8.27}$$

(ii) *Acoustic phonon scattering due to deformation potential coupling*. The energies of acoustic phonons of small $\boldsymbol{q}$ are proportional to $\boldsymbol{q}$ and $\hbar\omega(\boldsymbol{q}) \ll kT$ or E. For deformation potential coupling $G(\boldsymbol{q}) \propto \boldsymbol{q}$ so that when the exponentials are expanded the summation in (8.26) becomes independent of $\boldsymbol{q}$ and

$$\mu_{\text{ac}} \propto (m^*)^{-5/2}\, E^{-1/2}\, (kT)^{-1} \tag{8.28}$$

where the $E^{-1/2}$ dependence leads to a factor $4/3\sqrt{\pi}$ between τ and τ_D in (8.17) and the value of r_0 in (8.3) and (8.18) becomes $3\pi/8$.

Experimentally, the mobility of electrons and holes in many semiconductors can be fitted to a power law, $T^{-\eta}$, over the (relatively limited) temperature range in which phonon scattering is dominant (§ 8.1.3). Frequently $1.5 < \eta < 2.5$. Using (8.17), (8.28) predicts a $T^{-1.5}$ dependence of mobility on temperature for acoustic mode scattering. Any coupling to optical modes will tend to increase the rate of variation with temperature and an exponential term like (8.27) plus a power law can usually be fitted to a higher power law over a limited range. In addition it is possible that the effective mass, which appears to a high power, may be energy and hence temperature dependent (see § 8.8.1). Some representative values of the index η are given in the table below. It is almost always possible to account for such values by a suitable admixture of acoustic and optical mode scattering and in a few cases the choice is confirmed by other experiments (see, for example, § 8.7.2).

$\mu \propto T^{-\eta}$. Values of η.

Germanium	electrons	1.7
	holes	2.3
Silicon	electrons	2.4
	holes	2.6
InSb	electrons	1.7
	holes	2.1
PbS	electrons	2.5
	holes	2.5

Table 8.1

8.3.3 Scattering by impurities in semiconductors (A)

The fall in scattering rate by phonons as the temperature decreases implies that either the mobility becomes very large or some other scattering mechanism becomes relatively more important. When two or more scattering mechanisms are important simultaneously it is usual simply to add the scattering probabilities. Thus

$$1/\tau_{\text{eff.}} = 1/\tau_1 + 1/\tau_2 + \ldots$$

although strictly this procedure is only valid if the τ's have the same dependence on wave vector. In most semiconductors ionised-impurity scattering dominates below 100 K and may, in heavily doped materials, be important at room temperature. The mobility of holes in germanium, measured by the drift method (§ 8.1.2) is shown as a function of impurity concentration in figure 8.7.

The scattering of electrons by ionised impurities is similar to the scattering of charged particles by nuclei, usually referred to as Rutherford scattering.

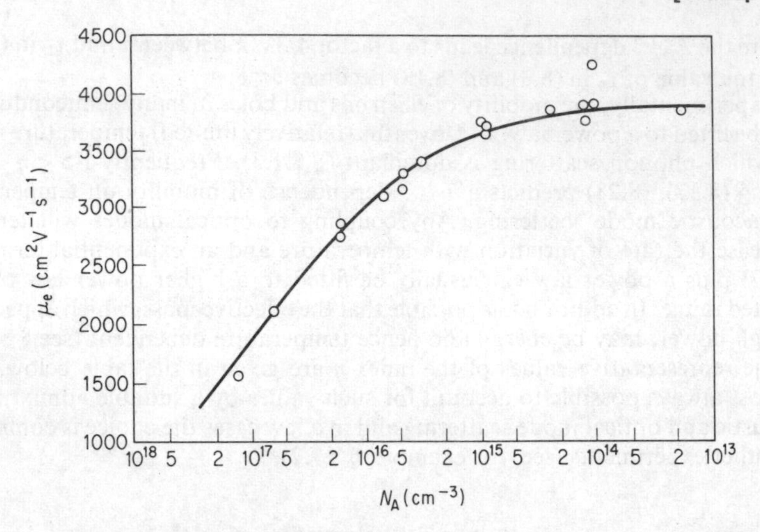

Figure 8.7 The drift mobility of injected electrons in P type germanium at 300 K as a function of impurity concentration. Points experimental, curve calculated using Conwell–Weisskopf formula for ionised impurity scattering (equation (8.29)). (After M. B. Prince, *Phys. Rev.*, **92**, 681 (1953).)

Similar behaviour occurs in gaseous plasmas, electrons being scattered by positive ions whose mass is relatively so large that they may be considered stationary. An electron of charge e and velocity $\sqrt{(2E/m^*)}$ passing a fixed centre of charge Ze will be deflected by the Coulomb field of the ion. Most electrons are scattered through small angles. To find the rate at which the momentum of the electrons is randomised one must average $(1-\cos\theta)$ over all scattering angles and this calculation is set out in a number of textbooks (such as R. A. Smith's *Semiconductors*) so will not be reproduced here. Two expressions for impurity scattering are often used. In the first, due to Conwell and Weisskopf, the field around an ionised impurity is assumed to have a simple Coulomb dependence, though reduced by the dielectric constant of the medium (see § 5.6.5). It is then further assumed that the field ceases to be effective at a radius equal to half the mean inter-impurity spacing. The latter assumption clearly becomes increasingly unsatisfactory as the impurity concentration increases. An alternative approach, associated with Mott, Brooks, Herring and others, takes into account the screening action of the free electrons and employs an expression for the Coulomb potential of the form

$$\frac{Ze}{4\pi\varepsilon r}\exp(-r/w)$$

where w is a screening length (§ 4.9). The results of the two approaches for the scattering time, τ_I, due to impurities are

$$\frac{1}{\tau_I} = \frac{Z^2 e^4 N_I}{16\pi(2m^*)^{1/2}\ E^{3/2}\ \varepsilon^2}\log \Lambda_0 \qquad (8.29)$$

where N_I is the impurity density and Λ_0 is given by

$$\Lambda_0 = 1 + \left(\frac{4\pi\varepsilon E}{Ze^2 N_I^{1/3}}\right)^2 \qquad \text{(Conwell–Weisskopf)}$$

or

$$\Lambda_0 \approx \frac{24m^*\varepsilon E^2}{Ze^2 \hbar^2 N_I} \qquad \text{(Brooks–Herring)}$$

As they differ only in the logarithmic term the difference is small, and, as shown in figure 8.7, quite good agreement is obtained with experiment. The $E^{-3/2}$ dependence implies (8.17) that ionised impurity scattering decreases with increasing temperature as $T^{-3/2}$. Physically this arises because the higher the electron velocity, the smaller the angle through which it is scattered. It should also be noticed that this relatively strong dependence of τ_I on electron energy leads to a non-trivial numerical ratio between τ_I and $\langle\tau_I\rangle$. Equation (8.17) gives

$$\langle\tau_I\rangle = \left(\frac{8}{\sqrt{\pi}}\right)\tau_I$$

8.3.4 Other scattering mechanisms in semiconductors (A)

Phonons and ionised impurities are the most important sources of electron scattering in semiconductors. We now briefly enumerate a number of other processes which can be observed under appropriate conditions.

(a) At very low temperatures, when most free electrons are 'frozen out' (i.e. captured) at impurities, scattering by neutral impurities may be important. A number of expressions have been derived for neutral impurity scattering, of which the most widely adopted is due to Erginsoy and is

$$\frac{1}{\tau_N} = \frac{20\hbar a_I N_N}{m^*} \tag{8.30}$$

where a_I is the Bohr radius of the ground state of the impurities and N_N is their density. In practice neutral-impurity scattering is rarely important even when most free electrons have been frozen out as there are usually charged donor and acceptor impurity centres present whose charges compensate one another (§ 5.3.1).

(b) Dislocations (§ 5.4.1), both edge and screw, may act as scattering centres and can lead to anisotropic scattering.

(c) When electrons occupy a number of equivalent minima in the conduction band, as for example in silicon, there are electron states available into which electrons may be scattered by phonons of large wave vector and consequently large energy. This 'intervalley scattering', as it is called, produces an additional term with a temperature dependence similar to optical-mode scattering (8.27). Scattering between *non-equivalent* minima is also possible if the electrons are sufficiently energetic, as we shall see in § 8.7.

(d) Electron–electron scattering and electron–hole scattering are similar to ionised impurity scattering though particles of comparable mass are involved. Electron–electron scattering can be effective in redistributing momentum amongst the electrons and hence can influence other mechanisms of momentum loss. Electron–hole scattering obviously requires the simultaneous presence of both types of particle, which in an applied field will be drifting in opposite directions. The electrons are scattered into a frame of reference which is stationary with respect to the holes, and *vice versa*. If the ratio of the electron and hole mobilities is large, as in InSb, this effect could in principle reverse the direction of motion of one species, though this has not been observed experimentally.

8.4 Acoustic amplification in piezoelectric semiconductors (d)

The amplification of sound is usually done electronically: the sound wave is converted by a suitable transducer (e.g. a microphone if in the audio range) to an electric wave, amplified and reconverted to sound. An interesting application of electron–phonon interaction in piezoelectric semiconductors is direct amplification of acoustic phonons by drifting electrons. As we shall see, the technique is most suitable for frequencies greater than 10 MHz. It is possible (but admittedly unlikely) that the best way of amplifying electrical signals in the future may be their conversion to sound by a suitable transducer, amplification acoustically and reconversion to electrical signals.

To obtain acoustic amplification in a bar of piezoelectric semiconductor the sound wave must be spatially phase coherent: i.e. across any plane normal to the propagation direction the waves must be in phase. If conventional piezoelectric transducers (e.g. quartz) are used to launch and extract the acoustic wave they must be flat and parallel to a small fraction of an acoustic wavelength. Such transducers can be made for use up to about 10 GHz. At this frequency the wavelength of sound, λ_s, is of order 0.1 μm. Additionally the rod of piezoelectric semiconductor which forms the amplifying medium must be acoustically uniform. These represent the chief technological limitations of acoustic amplifiers but they have been used successfully from a few megahertz to microwave frequencies.

A qualitative picture of the origin of the amplification effect can be obtained as follows. A sound wave, i.e. a stream of phonons, launched into a rod of N type semiconductor will be coupled to the free electrons present. If the semiconductor is also a piezoelectric material the coupling for low frequency acoustic phonons can be strong. Energy and momentum will be transferred from the phonon stream to the electrons. In consequence the electrons will be driven down the bar and the sound wave attenuated (in addition to any lattice attenuation). If, however, we now apply an electric field to the semiconductor so that the electrons move at the velocity of the sound wave no net momentum transfer will take place between electrons and phonons and the attenuation will be zero. Finally if the electron velocity exceeds the phonon velocity the energy exchange moves in favour of the phonons and the acoustic

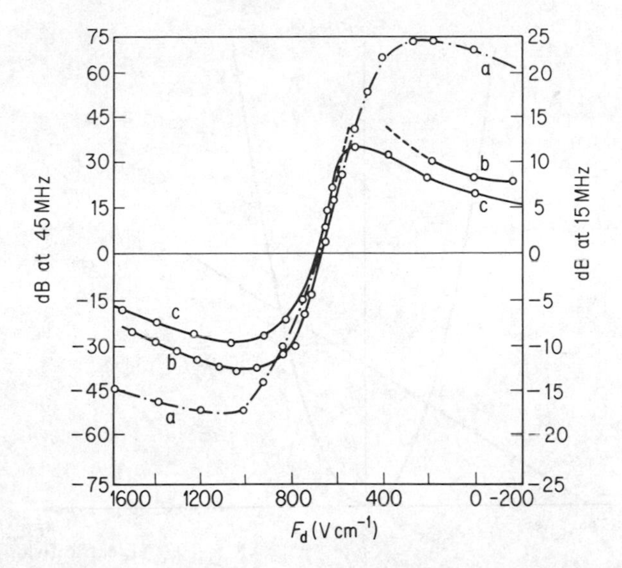

Figure 8.8 Experimentally observed acoustic attenuation in CdS as a function of electric field. Conductivity varied by illumination (photoconductivity).
Curve (a) Acoustic frequency = 15 MHz $\sigma/\omega\varepsilon = 1.2$
Curve (b) " " = 45 MHz $\sigma/\omega\varepsilon = 0.24$
Curve (c) " " = 45 MHz $\sigma/\omega\varepsilon = 0.21$
(After Hutson, McFee and White, *Phys. Rev. Lett.*, **7**, 237 (1961).)

wave is amplified. This behaviour is somewhat analogous to photon amplification in a microwave travelling wave tube.

Coherent phonon amplification was first observed in CdS in 1961 and figure 8.8 shows some early results. It has since been studied in a number of piezoelectric materials for the theoretical amplification achievable at small signal levels is very substantial (of the order of 1000 dB cm^{-1} at 1 GHz). The properties of acoustic amplifiers can be accounted for quite adequately by a purely classical approach as $\hbar\omega \ll kT$. We give the results of this analysis below. In view, however, of the approach used elsewhere in this book, it is of interest to show that the major features arise quite naturally from the E, $\boldsymbol{k}$ conservation rules and energy level population inversion. To do this we plot in figure 8.9 $E/\hbar$ against k_z for the electrons in a simple parabolic band where z is the direction of the applied field. If an electron is to emit an acoustic phonon in the z-direction it must make a transition between two points such as B and A where the conservation rules require

$$\omega_B - \omega_A = \omega(q)$$

and

$$k_B - k_A = -q$$

where $\omega(q)/q = s$, the velocity of sound. We assume that the acoustic frequency is so low (less than 10^{11} Hz) that phonon dispersion can be neglected. Hence

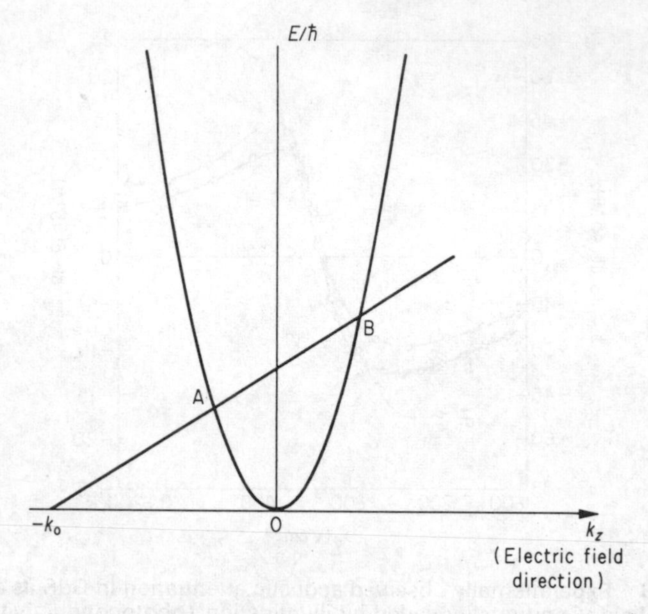

Figure 8.9 The conduction band minimum of a semiconductor and construction line for the derivation of equation (8.31). Slope of line AB equals velocity of sound in the crystal, ω/q

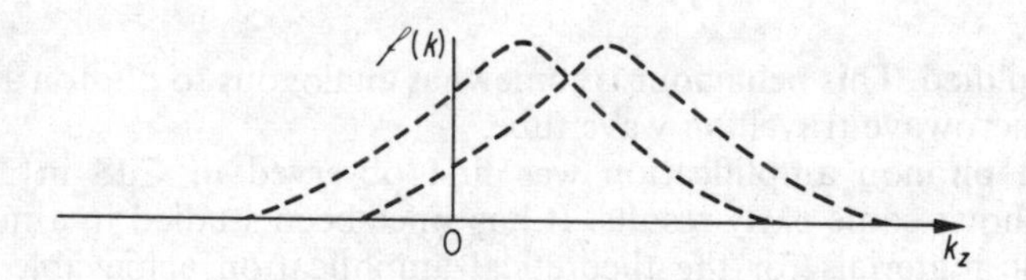

Figure 8.10 Electron distribution function in semiconductor at two electric fields

the slope of the line AB is s and as the equation of the electron parabola is $\omega = \hbar k_z^2/2m^*$ the values of the wave vector at A and B are given by

$$k_{A,B} = \frac{m^* s \pm (m^{*2} s^2 + 2m^* s \hbar k_0)^{1/2}}{\hbar} \tag{8.31}$$

where $-k_0$ is the intercept of the line AB on the k_z-axis. To obtain phonon amplification the net stimulated emission must exceed the stimulated absorption; i.e. the electron occupation probability at B must exceed the occupation probability at A (see §§ 6.2 and 6.8). This is achieved by the application of the electric field. At zero electric field the electron distribution is centred on $k_z = 0$. When the field is applied the distribution is displaced in the k_z-direction.

When the centre of the distribution is half-way between k_A and k_B the occupation probabilities will be equal and the net attenuation zero; beyond this point there will be amplification (figure 8.10). From (8.31) the midpoint between k_A and k_B is

$$k_c = m^* s/\hbar$$

The electron group velocity is then

$$\left(\frac{d\omega}{dk}\right)_{k=k_c} = \frac{\hbar k_c}{m^*} = s \tag{8.32}$$

so we have shown that amplification occurs when the mean velocity of the electron distribution exceeds the velocity of sound, whatever the value of $\omega(q)$. The actual magnitude of the amplification will depend on the difference in electron population at A and B. For small displacements of the distribution function beyond k_c this is obviously proportional to the velocity difference between electrons and sound waves $(v - s)$ and to the total electron density (and hence the conductivity σ). We have, however, totally ignored space charge effects. Not only will the free electrons smooth out the piezoelectric field over distances of the order of the screening length but also their mutual repulsion will oppose the tendency of the wave to 'bunch' them in the potential minima that occur each wavelength. The latter effect reduces the coupling at short wavelengths and limits piezoelectric scattering to relatively low-frequency phonons. This is why piezoelectric scattering by thermal phonons is not very important in determining mobility, notwithstanding the strength of the coupling. These features are most easily described by a classical theory.

Acoustic amplification was first discussed theoretically by Hutson and White in 1962. Using current and charge continuity equations, Poisson's equation and the piezoelectric equations relating field, stress and strain they showed that the attentuation coefficient of sound by free electrons was given by

$$\alpha = \frac{e_p^2}{2\varepsilon^2 C_c}\left[\frac{-\sigma(v-s)}{(v-s)^2 + s^2\left(\dfrac{\sigma}{\omega\varepsilon}\right)^2(1+q^2w^2)^2}\right] \tag{8.33}$$

where e_p is the piezoelectric constant, s the sound velocity, C_c is an elastic constant, v is the electron drift velocity, w is the screening length, and the other symbols have their usual meanings. Note that α is negative, corresponding to amplification, if $v > s$. Equation (8.33) can be fitted to the experimental data of figure 8.8. Its chief deficiency is that it applies only to small wave amplitudes such that $eF_s\lambda_s$, where F_s is the piezoelectric field due to the sound wave, is much less than kT. For large amplitudes the gain naturally saturates but the non-linear theory becomes very complex. When $eF_s\lambda_s \gg kT$ the electrons are effectively trapped in the trough of each wave and cannot move faster than the sound velocity at any field. The current then becomes independent of field, a condition which is very easily achieved experimentally.

As with lasers (§§ 6.7.5 and 6.7.6), the achievement of amplification permits the construction of oscillators provided the crystal is fabricated as a Fabry–Perot cavity and the round trip gain exceeds the losses. The oscillating modes are defined by the condition that the cavity length be an integral number of half wavelengths. The availability of d.c. driven acoustic oscillators is obviously necessary for the development of 'acousto-electronics'.

The technological development of acoustic amplifiers has been held back by the need for highly homogeneous crystals and the lack of simple low-loss transducers for input and output. These problems can be partially overcome by the use of surface (Rayleigh) waves rather than bulk waves. Surface waves on solids are equivalent to surface tension waves on liquids. Both longitudinal and transverse modes are possible, but unlike bulk waves, they travel with the same phase velocity. It has recently been shown that surface waves can be launched and detected with low loss on piezoelectric crystals by the use of interdigital arrays of metal strips, spaced $\lambda_s/2$ apart and prepared by evaporation or photolithography on the crystal surface. To make an amplifier the piezoelectric crystal carrying the wave need not be a semiconductor provided a semiconductor crystal or film is in close proximity with the surface used. Coupling between piezoelectric crystal and the electrons in the semiconductor is by the electric field generated in the piezoelectric crystal, which extends into the semiconductor. Amplification can be obtained as before. At the time of writing, surface-wave devices seem likely to grow in importance.

We note finally that acoustic amplification is in principle possible through deformation-potential coupling when this is large enough. Equation (8.33) shows that, so long as the space charge effects ($q^2 w^2$) are small, the power gain per unit length increases as ω^2. This is a characteristic feature of travelling wave amplifiers (e.g. travelling wave electron tubes, optical parametric amplifiers as discussed in Chapter 6, etc.) for it is the amplitude gain per wavelength which is the fundamental 'unit'. Furthermore, the matrix element for deformation potential coupling increases with q. At sufficiently high frequencies, then, a deformation-potential coupled amplifier with appreciable gain is possible. Acoustic amplification has in fact been observed in germanium at gigahertz frequencies due to deformation potential coupling between electrons and phonons.

8.5 Electron scattering in metals

8.5.1 Electron scattering by phonons in metals (A)

In a metal, coupling between electrons and phonons is primarily electromagnetic due to the large electron, and hence positive ion, density. Electric fields extending over distances greater than a screening length are smoothed out and there is no piezoelectric coupling. Deformation-potential coupling can occur. In any event for elemental metals the dominant phonons are generally from the longitudinal acoustic branch, though some coupling to transverse modes can occur.

The procedure for calculating the scattering probability due to phonons as a function of temperature is in principle the same as for semiconductors (§ 8.3.1), namely to find the number of available states into which an electron may be scattered and the number of suitable phonons available for scattering. For electrons on a spherical Fermi surface, however, the density of available states is practically independent of the energy of the electron compared with the Fermi energy but it is a strong function of the occupation probability. This arises because (a) the phonon energies are negligible compared with the Fermi energy, so a scattered electron changes its energy very little and (b) the occupation probability of a state near the Fermi surface is (by definition) comparable with unity and not negligibly small, as in a semiconductor. We therefore conclude that the expression for the rate of change of the distribution function due to collisions, for use in (8.7), will be of the form

$$\left[\frac{\partial f(\boldsymbol{k})}{\partial t}\right]_{\text{collisions}} \propto \sum_{\boldsymbol{q}} G(\boldsymbol{q})\{(1-f_0(\boldsymbol{k}))[f_0(\boldsymbol{k}+\boldsymbol{q})(n(\boldsymbol{q})+1)+f_0(\boldsymbol{k}-\boldsymbol{q})(n(\boldsymbol{q}))]$$

$$-f_0(\boldsymbol{k})[(1-f_0(\boldsymbol{k}+\boldsymbol{q}))(n(\boldsymbol{q}))+(1-f_0(\boldsymbol{k}-\boldsymbol{q}))(n(\boldsymbol{q})+1)]\} \qquad (8.34)$$

where the first two terms refer to electrons scattered into the state $\boldsymbol{k}$ and the last two out of state $\boldsymbol{k}$, either by the absorption $n(\boldsymbol{q})$ or emission $(n(\boldsymbol{q})+1)$ of a phonon of wave vector $\boldsymbol{q}$. $n(\boldsymbol{q})$ is given by the Bose distribution, f_0 by the Fermi distribution and the E, $\boldsymbol{k}$ conservation rules apply. Application of the conservation rules, together with the limitation that scattering is confined to the Fermi surface, places a limitation on the largest wave vector phonon that may take part, namely $q_{\max} \leqslant 2k_F$, where k_F is the wave vector at the Fermi level. The summation in (8.34) may then be replaced by an integral with an upper limit of $2k_F$. We now insert the Debye expression for the density of phonon states (§ 3.6) and average over the scattering angles $(1-\cos\theta)$ to find the relaxation time for use in the Boltzmann equation. The manipulation, integration and resulting expressions are rather formidable. Reference may be made to F. J. Blatt's article in *Solid State Physics*, vol. 4 or J. M. Ziman's *Electrons and Phonons*. No complete analytic solution is possible but an approximate solution is

$$\sigma \propto \left[\left(\frac{T}{\Theta_D}\right)^5 \int_0^{\Theta_D/T} \frac{x^5\,dx}{[\exp(x)-1][1-\exp(x)]}\right]^{-1} \qquad (8.35)$$

which is known as the Bloch–Grüneisen relation. Θ_D is the Debye temperature (§ 3.6). The temperature dependence predicted by (8.35) is in good agreement with experimental results on many metals, as shown by figure 8.3, though the values of Θ_D deduced are not always the same as the Debye temperatures deduced from specific heat measurements.

At high and at low temperations (8.35) can be approximated by

$$\sigma \propto \left(\frac{\Theta_D}{T}\right) \qquad T \gg \Theta_D \tag{8.36}$$

and

$$\sigma \propto (\Theta_D/T)^5 \qquad T \ll \Theta_D \tag{8.37}$$

Physically, the reason for these expressions is as follows. At high temperatures $kT \gg \hbar\omega(\boldsymbol{q})$ as $\hbar\omega(\boldsymbol{q})$ cannot, on the Debye model, exceed $k\Theta_D$. The electron scattering rate is then simply proportional to the density of phonons which, when $T > \Theta_D$, is proportional to the absolute temperature. At low temperatures the T^{-5} law arises from two factors. First, the effective phonon density multiplied by the coupling factor varies as T^3 at low temperatures (cf. § 3.6.) Secondly at low temperatures only phonons of small wave vector are available, where q_{max} is proportional to T. Scattering is then limited to small angles and the vector $\boldsymbol{q}$ joins two nearby points on the Fermi surface, which is assumed to be spherical. Then

$$k_i \approx k_f \approx k_F$$

and the angle θ between $\boldsymbol{k}_i$ and $\boldsymbol{k}_f$ is approximately q/k_F, as shown in figure 8.11. The change in momentum is

$$|\boldsymbol{k}_i - \boldsymbol{k}_f| = k_i(1 - \cos\theta) \approx \tfrac{1}{2}k_F\theta^2 \approx \tfrac{1}{2}q^2/k_F$$

for small angles. Hence the efficiency of randomisation of the momentum falls off as T^2 at low temperatures when phonons with $q \approx k_F$ are no longer available and this, together with a T^3 effective phonon density dependence, leads to a T^5 relationship.

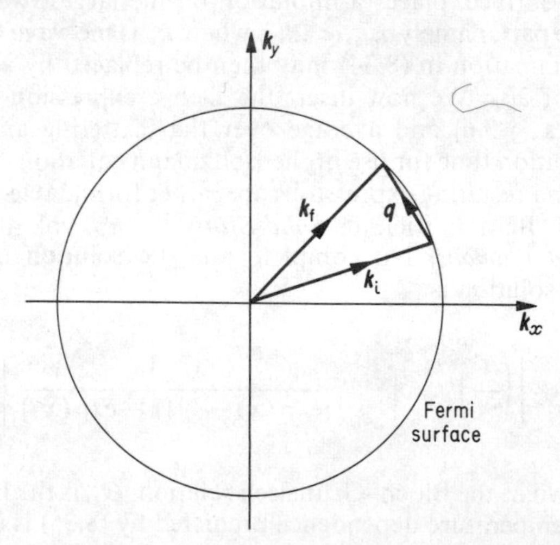

Figure 8.11 Spherical Fermi surface in $\boldsymbol{k}$-space illustrating the small change in electron wave vector on scattering by a phonon of small $\boldsymbol{q}$. Wave vector conservation requires $\boldsymbol{k}_f - \boldsymbol{k}_i = \boldsymbol{q}$

8.5.2 Impurity scattering in metals (A)

As in semiconductors, the electron mobility in metals at low temperatures is limited by impurities. Due to screening by free electrons the effect of the impurity charge is much less in a metal than in a semiconductor. The Brooks–Herring expression (8.29), which includes screening, is the most appropriate. It is found experimentally that, up to a few atomic percent, the increase in resistivity of monovalent metals on addition of impurity is given by

$$\Delta\rho = c_0(a' + b'Z^2) \tag{8.38}$$

where c_0 is the concentration, Z is the difference in valence between solute and solvent and a′ and b′ are constants. The values of a′ and b′, are comparable and of order 10^{-6} ohm cm per atomic percent (see § 8.2.5). The dependence on c_0 and Z^2 follows immediately from (8.29). The constant a′ can be considered a size effect and depends on the relative ionic radii of impurity and host metal.

Equation (8.38) cannot of course apply at high impurity concentrations or to alloys. If two metals form a continuous range of solid solutions and no separate phases occur the resistivity is evidently proportional to $c_0(1 - c_0)$ as atoms of either species are interchangeable. This (admittedly oversimplified) expression is often obeyed in practice, and is illustrated in figure 8.12.

Vacant lattice sites and interstitial ions behave in much the same way as impurities, in metals as in semiconductors. Dislocations may also scatter electrons. All these effects are important only at low temperatures, when

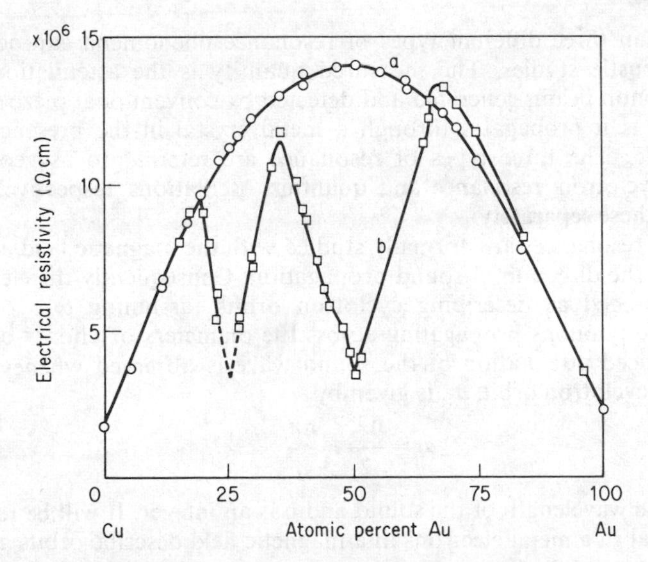

Figure 8.12 Resistivity of Cu–Au alloys. Curve (a) unannealed, curve (b) annealed at 200°C. Curve (a) follows relation $c_0(1 - c_0)$, see text. Curve (b) shows precipitation of separate phases ($AuCu_3$ and AuCu). (After C. H. Johansson and J. O. Linde, *Ann. Physik,* **25**, 1 (1936).)

phonon scattering is small. All are independent of temperature also, for, unlike semiconductors, the electron energy appearing in (8.29) is the Fermi energy and is independent of temperature. Hence a description in terms of a scattering cross section, as in (8.21) is appropriate. In most real metals the 'residual resistivity' range is reached at temperatures above the boiling point of helium (4.2 K) so a measure of ρ (room temperature)/ρ (helium) is sometimes quoted for commercial samples of high purity metals. A ratio in excess of about 100 implies an impurity concentration below spectroscopic detection limits and the resistivity ratio is then the best remaining criterion of quality.

8.5.3 Magneto–acoustic effects in metals (A)

We have seen that the coupling between electrons and phonons impedes electron flow and provides the major source of electrical resistance in both semiconductors and metals. Conversely the presence of electrons can impede the flow of phonons, giving rise to acoustic absorption. We have seen (§ 8.4) how this coupling can be exploited in semiconductors to provide phonon amplification. No equivalent device is possible in metals as the high electric fields needed would lead to intolerably large currents. The study of coherent phonon scattering in metals can, however, give valuable information about the metal provided we have a way of picking out the electron–phonon interaction from other phonon scattering processes. This may be achieved by the application of a magnetic field and is the basis of magneto–acoustic studies in metals.

No less than three different types of resonance phenomena can occur in magneto–acoustic studies. The measured quantity is the attenuation of a coherent phonon beam, generated and detected by conventional piezoelectric transducers, as it propagates through a metal crystal in the presence of a magnetic field. The three types of resonance are referred to as geometric resonance, cyclotron resonance and quantum oscillations respectively. We now discuss these separately.

Geometric resonances are normally studied with the magnetic field applied transverse to the direction of sound propagation. Consequently the electrons may be envisaged as describing cyclotron orbits (assuming $\omega_c\tau > 1$, see § 7.2) and the phonons propagating across the diameters or chords of such orbits. Enhanced attenuation of the sound wave is observed whenever the radius of the cyclotron orbit, a_c, is given by

$$a_c = \frac{n\lambda_s}{2} = \frac{n\pi}{q} \tag{8.39}$$

where λ_s is the wavelength of the sound and n is an integer. It will be recalled from § 4.7 that in a metal electrons in a magnetic field describe orbits around the Fermi surface at the Fermi velocity v_F so that

$$a_c = v_F/\omega_c = \frac{\hbar k_F}{m^*\omega_c} \tag{8.40}$$

where $\omega_c = eB/m^*$. The resonance condition is therefore a measure of the electron wave vector, k_F, at the Fermi surface. Resonance can be considered to arise because the maximum scattering of coherent phonons occurs when the electron and phonon velocities are equal. This is essentially the phase-matching condition discussed in connection with non-linear optics (§ 6.8), and is a consequence of wave-vector conservation requirements. The electron velocity at the Fermi surface is very large compared with the velocity of sound so that only for electrons travelling nearly at right angles to the sound propagation can the component of v_F along the propagation direction be equal to the sound velocity. Maximum attenuation of a coherent phonon beam will then occur when electrons in cyclotron orbits cross and recross the phonon wave at points of the same phase, for then the scattering will be coherent. The maximum number of electrons in these positions correspond to orbit diameters. Note that this argument makes no assumptions about the shape of the orbit and is therefore applicable to non-spherical energy surfaces. The measured quantity, from (8.40), is the diameter of the cyclotron orbit, whatever its shape, in the direction of the sound wave. Consequently it is possible, in principle at least, to use crystals of various orientations and to map out the caliper dimensions of a Fermi surface by this means. We show an example of geometrical magneto–acoustic resonances in figure 8.13.

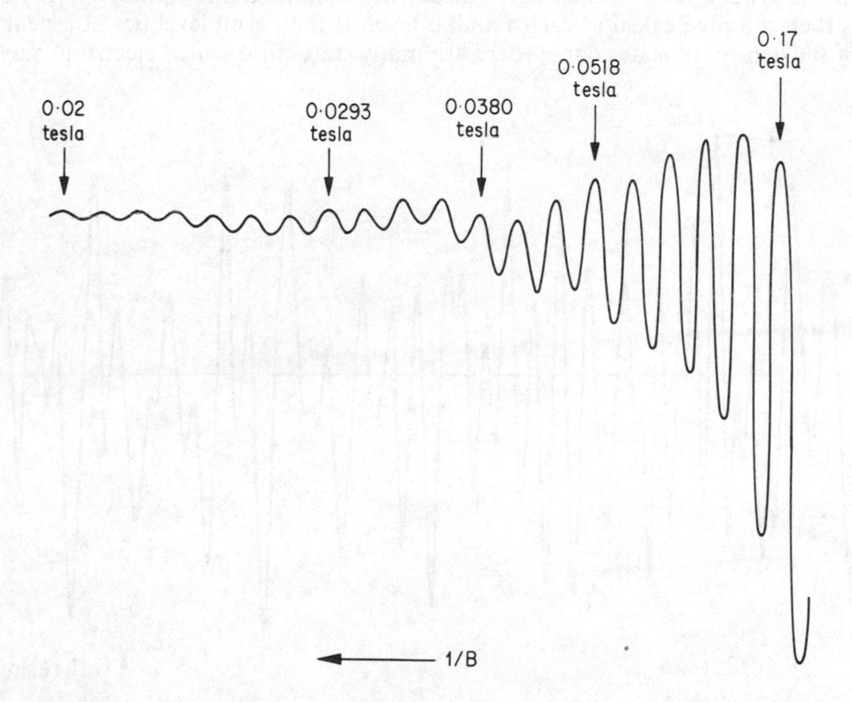

Figure 8.13 Geometrical magneto–acoustic resonances in cadmium at 1.2 K, periodic in the reciprocal of the magnetic field; acoustic frequency 170 MHz (from Dr. L. Mackinnon, unpublished)

To obtain sharp geometrical resonances, it is clear that the wavelength of the phonons must be well defined. This condition may be written as $ql_f(\boldsymbol{q}) \gg 1$ where $l_f(\boldsymbol{q})$ is the phonon mean free path. In most metals at low temperatures $ql_f(\boldsymbol{q}) \gg 1$ can be achieved using acoustic frequencies around 50 MHz or greater.

Acoustic cyclotron resonance can be observed most easily with the magnetic field orientated parallel with the sound propagation (Faraday orientation) but naturally requires the condition $\omega_c \tau \gg 1$. As the electron motion is not quantised along the field direction electrons are moving in this direction at the Fermi velocity. The resonance is then Doppler shifted and occurs when

$$\omega(q) = \omega_c - qv_F \tag{8.41}$$

where $\omega(q)$ is the phonon frequency and q the phonon wave vector. The Doppler shift arises from the need to conserve wave vector in the direction of the magnetic field (cf. § 7.2.6).

The third type of acoustic resonance phenomenon—quantum oscillations—requires two conditions, $\hbar\omega_c \gg kT$ and $\hbar\omega_c \gg \hbar\omega(q)$ to be satisfied. As we have seen in §§ 4.7 and 7.2.8, the former condition implies splitting of the conduction band into Landau levels separated in energy by more than the normal thermal spread around the Fermi energy. The density of states in the conduction band is then sharply peaked at each Landau level. If the Fermi level lies at a peak of the density of states curve there are many states into which electrons may

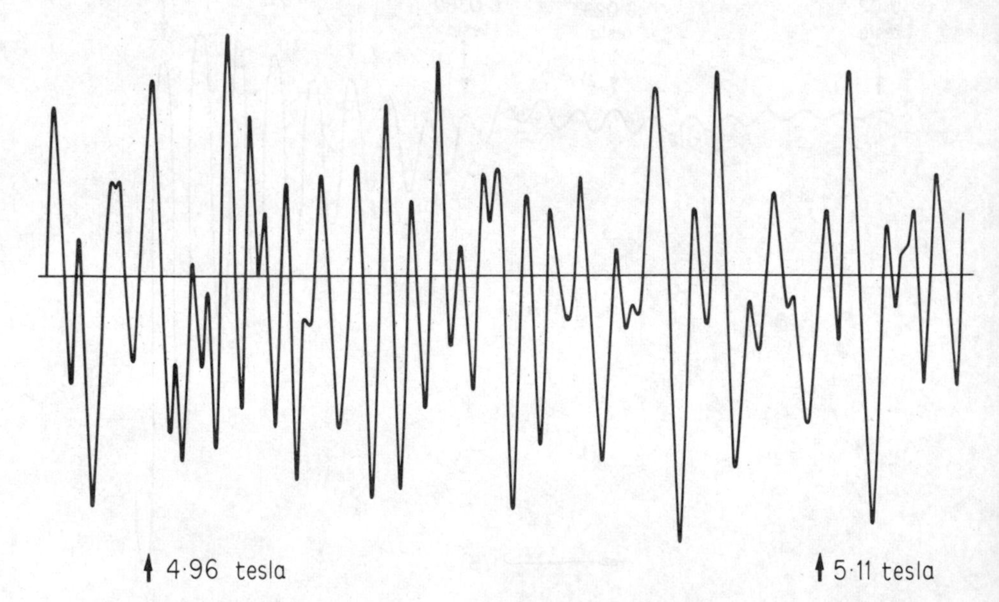

Figure 8.14 Quantum magneto–acoustic oscillations in zinc at 1.2 K, acoustic frequency 170 MHz, acoustic propagation parallel to [0001], magnetic field 40° from [0001] towards [1120] (from Dr. L. Mackinnon, unpublished)

be scattered by interaction with a phonon so that the sound wave is strongly absorbed; otherwise the absorption is reduced. As the magnetic field is varied so the sound attenuation oscillates between maximum and minimum values, being periodic in $1/B$ in precisely the same way as the other properties discussed in § 4.7 (see also § 8.8.4). The protagonists of magneto–acoustics claim that phonon absorption can be the most sensitive method of observing 'de Haas–van Alphen' type quantum oscillations. An example of quantum magneto–acoustic oscillations is shown in figure 8.14.

8.6 Heat transport

Heat flow in solids is primarily due to electrons and phonons, which carry energy. Excitons may also transport energy if they are created at one point and recombine with the emission of phonons at another. Similarly light emitted at one luminescent centre may be reabsorbed elsewhere, with the eventual emission of phonons, The latter two processes are rare, and are important only under rather special conditions. We do not consider them further. It is convenient to consider the electronic contribution to thermal conduction first, as it is related to electrical conduction, and then to consider phonon transport.

8.6.1 The electron contribution to heat transport (a)

As free electrons can move and carry energy it is natural that they contribute to heat conduction. It is also not surprising that when the electron density is large, as in a metal, the electronic contribution to the thermal conductivity is dominant. We can then expect a linear relationship between thermal and electrical conductivity. This relationship is called the Wiedemann–Franz law and is well established experimentally at room temperature, although substantial deviations occur at low temperatures.

Before considering a more formal derivation of the Wiedemann–Franz law an approximate treatment may be helpful. Electrons 'flow down' a temperature gradient because they have more energy, and hence higher velocities, at the hot end than at the cold end. The incremental temperature above ambient of an electron will be about $l_f(\mathrm{d}T/\mathrm{d}x)$ where l_f is the electron mean free path and $(\mathrm{d}T/\mathrm{d}x)$ is the temperature gradient. If the electrons are moving at a velocity v in the x-direction the rate of heat flow per unit area will be

$$\mathscr{U} \sim vC_V l_f(\mathrm{d}T/\mathrm{d}x)$$

where C_V is the specific heat of the electron gas. The specific heat of a classical gas, from kinetic theory, is about Nk.

Putting $l_f \approx v\tau$ (§ 8.2.5) the thermal conductivity due to the electrons is about

$$\mathscr{K}_e = \mathscr{U}/(\mathrm{d}T/\mathrm{d}x) \sim v^2\tau Nk \sim (2kT/m^*)\tau Nk$$

The electrical conductivity is given by (8.1). Hence

$$\frac{\mathscr{K}_e}{\sigma T} = L \approx 2\,(k^2/e^2) \tag{8.42}$$

which differs from the expression for the Wiedemann–Franz law obtained below by a numerical factor of order unity. By coincidence (8.42) is numerically correct for a semiconductor in which acoustic mode scattering is dominant.

A more formal treatment must use the Boltzmann transport equation. Furthermore, we must take into account the fact that the electrons carry charge as well as energy so that their flow constitutes a current. If, as is usual, thermal conductivity is measured with the ends of the sample electrically open circuit, the current will be zero but an electric field will build up to oppose the electron motion (cf. the Hall effect). In general we can write, in one dimension

$$\left.\begin{aligned} J_x &= \frac{\chi_1}{T}\left(\frac{dT}{dx}\right) + \chi_2 F_x \\ \mathscr{U}_x &= \frac{\chi_3}{T}\left(\frac{\partial T}{dx}\right) + \chi_4 F_x \end{aligned}\right\} \tag{8.43}$$

where χ_1, χ_2, χ_3 and χ_4 are constants. J_x is the electric and $\mathscr{U}_x$ the thermal current density. By symmetry arguments one can show that $\chi_1 = -\chi_4$. Equations (8.43) describe, in general terms, the three thermo-electric effects (i.e. the Seebeck, Peltier and Thomson effects). Since thermal conductivity is measured with $J = 0$ we can write, from (8.43)

$$\mathscr{K}_e = \frac{\mathscr{U}}{(dT/dx)} = \frac{1}{T}(\chi_3 + \chi_1^2/\chi_2) \tag{8.44}$$

To find χ_1, χ_2 and χ_3 we have to solve the Boltzmann equation in the presence of a temperature gradient. We then find J_x and $\mathscr{U}_x$ for $F_x = 0$ or $(dT/dx) = 0$. This procedure gives three equations for the χ's.

In the presence of a temperature gradient and an electric field in the x-direction the force term of the Boltzmann equation becomes

$$\left[\frac{d f}{dt}\right]_{\text{Field plus gradient}} = \frac{eF_x}{\hbar}\left(\frac{d f}{dk}\right) + v_x\left(\frac{d f}{dx}\right) \tag{8.45}$$

which should be compared with (8.8). Equation (8.9) remains unchanged, by assumption. The solution for $f(\boldsymbol{k}, x)$ is

$$f = f_0 + v_x\tau\left[eF_x\left(\frac{d f_0}{dE}\right) + \left(\frac{\partial f_0}{\partial T}\right)\left(\frac{\partial T}{\partial x}\right)\right] \tag{8.46}$$

We already have an expression for J_x in terms of f, namely (8.11). The equivalent expression for $\mathscr{U}_x$ is the same except that the electrons are carrying energy rather than charge so we replace e by E. Hence

$$\mathscr{U}_x = \int_{\text{B.Z.}} E v_x f(\boldsymbol{k}, x) \, \mathrm{d}\boldsymbol{k} \left(\frac{\mathscr{V}}{(2\pi)^3} \right) \tag{8.47}$$

With this equation, (8.11) for the current and (8.46) for $f(\boldsymbol{k},x)$ we can find χ_1, χ_2 and χ_3 from equations (8.43). Thus to find χ_2 we put $(\mathrm{d}T/\mathrm{d}x) = 0$ everywhere and solve. χ_2 is, of course, the electrical conductivity, σ. To find χ_1 and χ_3 we put $F_x = 0$ everywhere. We can then obtain the relationship between $\mathscr{K}_e$ and σ by inserting these solutions into (8.44) and applying the conditions to $(\mathrm{d}f_0/\mathrm{d}E)$ appropriate to a metal and a semiconductor (following §§ 8.2.2 and 8.2.3). We finally emerge with the Wiedemann–Franz law as

$$\mathscr{K}_e = \mathrm{L}\sigma T \tag{8.48}$$

where

$$\mathrm{L} = (\pi^2/3)(k^2/e^2) \tag{8.49a}$$

for a metal and

$$\mathrm{L} = (p + \tfrac{5}{2})(k^2/e^2) \tag{8.49b}$$

for a semiconductor in which $\tau_D = \tau_0 E^p$.

The Wiedemann–Franz law is in good agreement with experiment at room temperature and again in the impurity-scattering region at very low temperatures. For metals, its validity is independent of the band structure. The law arises essentially from the assumption that the relaxation time, τ_D, of the distribution function in $\boldsymbol{k}$ space applies to both thermal and electrical conduction. It may not be valid if either charge or energy transport is not simply controlled by τ_D. We have seen (cf. (8.37)) that there is a temperature region, when $T \ll \Theta_D$ but phonon scattering of the electrons is still dominant, in which the electrical conductivity of a metal varies as T^{-5} while τ_D is varying as T^{-3}. This arises because the charge transport is determined by the randomisation of electron *momentum* and the momentum change per scattering event varies as T^{-2}. Energy transport, on the other hand, is much more determined by the rate at which the distribution of electron *energies* is restored to equilibrium. This argument suggests that we should use τ_D for thermal conductivity at all temperatures and expect marked deviations from the Wiedemann–Franz law when $T \ll \Theta_D$ but electron scattering is still primarily by phonons.

Hence we predict for the electron contribution to the thermal conductivity:

(1) $\mathscr{K}_e \propto T$ — at very low temperatures; impurity scattering; τ_D constant. Wiedemann-Franz law obeyed.

(2) $\mathscr{K}_e \propto T^{-2}$ — for $T \ll \Theta_D$ phonon scattering; $\tau_D \propto T^{-3}$. Wiedemann-Franz law not obeyed.

(3) $\mathscr{K}_e = \text{constant}$ — for $T \gg \Theta_D$; phonon scattering; $\tau_D \propto T^{-1}$. Wiedemann-Franz law obeyed.

and these predictions imply that $\mathscr{K}_e$ passes through a maximum. Experimental results on metals, in which the electronic contribution to the thermal conductivity is large, tend to confirm these predictions though in the intermediate temperature range $\mathscr{K}_e \propto T^{-x}$ where $2 < x < 2.5$. This discrepancy is well within the general reliability of the argument given above. There are more serious discrepancies when the absolute magnitude of $\mathscr{K}_e$ is considered but we shall not explore this further. Some representative experimental results are shown in figure 8.15.

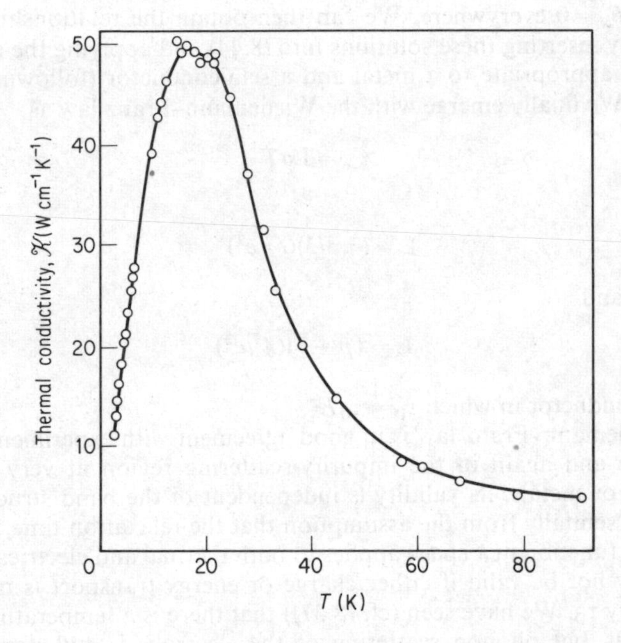

Figure 8.15 The thermal conductivity of copper as a function of temperature. (After R. Berman and D. K. C. MacDonald, *Proc. roy. Soc.*, **A211**, 122 (1952).)

8.6.2 The phonon contribution to heat transport; phonon–phonon scattering (a)

In insulators heat is carried primarily by the diffusion of phonons from the high temperature region, where their density is highest, to the low temperature region, where there are few. Phonon scattering turns out to be quite a complicated process and what follows is an oversimplified picture. Thermal conductivity is in any case very much an 'averaged' type of process with few clear cut features.

We note first that phonons can be scattered by other phonons, coupling arising through anharmonic terms in the atomic displacements as discussed in § 3.7. This is exactly analogous to the photon–photon interaction discussed in § 6.8. In terms of atomic displacements u the potential energy (3.59) is

$$V = \Lambda u^2 + \Lambda' u^3 + \Lambda'' u^4$$

The term in u^3 permits the combination of two phonons to form one, or the break up of one phonon to produce two. Similarly the term in u^4 allows four phonon interactions. The series is rapidly convergent and Λ' and Λ'' are small: how small can be appreciated by recalling that it is the anharmonic terms that lead to thermal expansion (cf. (3.61)) which is usually a small effect. Phonon–phonon scattering does, however, provide an upper limit to the mean free path and mean free time of phonons. It is the origin of the finite line-widths observed in optical absorption (§ 6.6) and neutron scattering by phonons (§ 3.7).

We now apply the energy and wave vector conservation rules. These produce additional constraints on the possible interactions, analogous to the need for phase matching in non-linear optics. These do not need to concern us as much as the realisation that, if the wave vector conservation rule

$$\boldsymbol{q}_1 + \boldsymbol{q}_2 = \boldsymbol{q}_3 \tag{8.50}$$

applies, there is no net change in phonon momentum on scattering. As with electron–electron scattering (§ 8.3.4), the momentum is simply redistributed. We conclude that normal processes, as described by (8.50), provide no resistance to heat flow in crystals.

This means that however low the probability of umklapp processes (§ 2.6) we need them to explain why the thermal conductivity of pure crystals is not infinite. If we write, for an umklapp process;

$$\boldsymbol{q}_1 + \boldsymbol{q}_2 = \boldsymbol{q}_3 + \mathbf{Q} \tag{8.51}$$

where $\mathbf{Q}$ represents a reciprocal lattice vector, the possibility arises that $\boldsymbol{q}_3$ is negative, or at least has a component against the direction of the stream. The momentum of the phonon flow, then, can be absorbed by the crystal as a whole through $\mathbf{Q}$. We conclude that umklapp processes are responsible for thermal resistance. A further conclusion we draw from this, together with the energy conservation rule, is that the phonons involved in an umklapp process must all have wave vectors whose magnitudes are about one-third to one-half of $\mathbf{Q}$; i.e. quite large. Such phonons will have substantial energies (~one-third to one-half of $k\Theta_D$). Hence the thermal conductivity due to phonons will be inversely proportional to the (relatively small) number of high energy phonons that can be scattered by an umklapp process. That is, from (2.39)

$$\mathscr{K}_u \sim \exp(\hbar\omega(\boldsymbol{q})/kT) - 1 \tag{8.52}$$

where $\hbar\omega(\boldsymbol{q}) \approx k\Theta_D/\eta$ and η lies between 2 and 3. We consequently expect $\mathscr{K}_u$ to vary as $1/T$ at high temperatures when $T > \Theta_D$ but to increase exponentially on cooling below about $\Theta_D/3$. Experimentally this equation is quite well obeyed, but it is only applicable if phonon transport is the dominant process in heat conduction and the scattering of phonons is primarily by other phonons. Furthermore, even though the resistance to energy transport by phonons arises from umklapp scattering, the normal processes, described by (8.50), are responsible for restoring the phonon distribution as a whole to equilibrium. Thus only high energy phonons can 'umklapp' but normal processes transmit the consequences of umklapp scattering to the low-energy phonons. In this respect the normal phonon–phonon scattering is analogous to electron–electron scattering (§ 8.3.4) and plays an important, if subsidiary, role in thermal resistance.

8.6.3 Boundary scattering of phonons (AD)

If (8.52) were obeyed to very low temperatures the thermal conductivity, even of insulators, would become very large indeed. Not surprisingly, it is limited by other scattering processes. Of these we first consider boundary

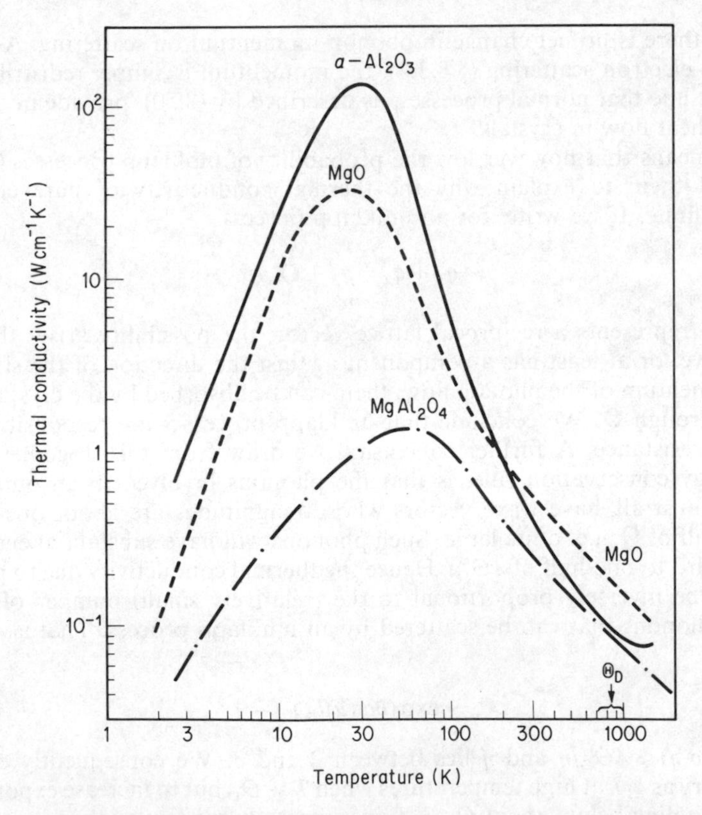

Figure 8.16 The thermal conductivity of α-Al_2O_3, MgO and $MgAl_2O_4$ as a function of temperature. (After G. A. Slack, *Phys. Rev.*, **126**, 427 (1962).)

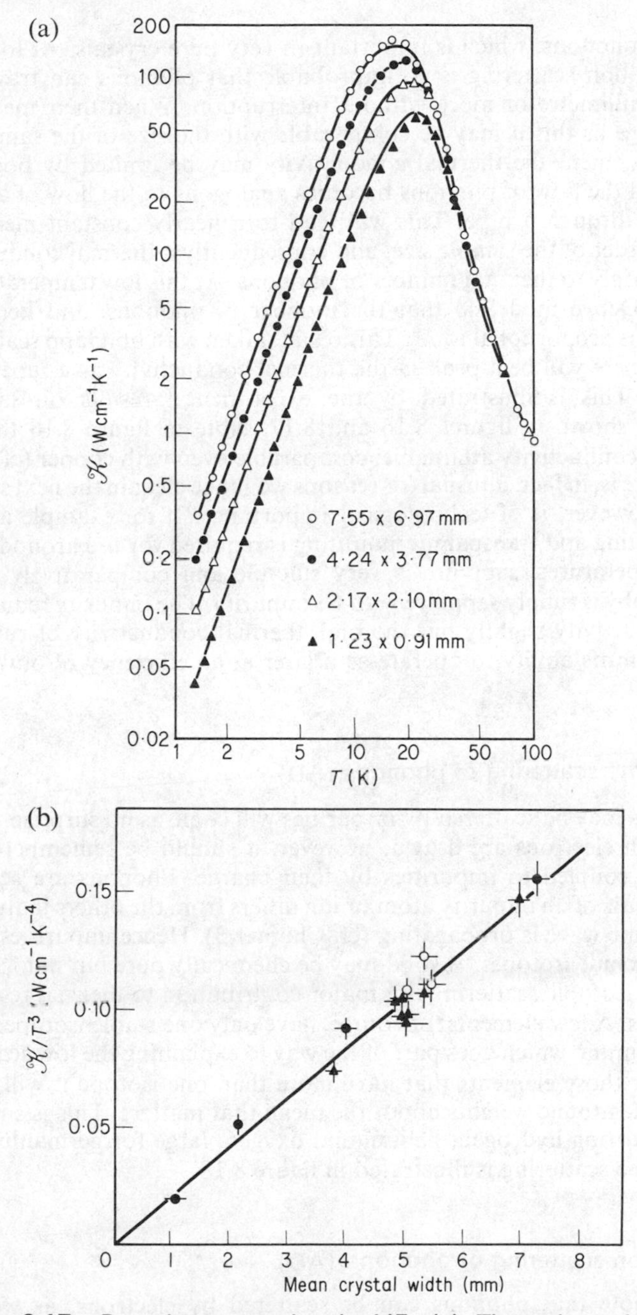

Figure 8.17 Boundary scattering in LiF. (a) Thermal conductivity of LiF as function of temperature for various sample sizes. Low temperature region varies as T^3. (b) Thermal conductivity over (temperature)^3 as a function of sample width. (After R. Berman, *Cryogenics*, **5**, 297 (1965) quoting Thatcher, Ph.D. thesis, Cornell University (1965).)

scattering of phonons, which is important in very pure crystals. At low temperatures phonon scattering is so improbable that phonons can travel distances of a millimetre or more without interruption. When their mean free path is as large as this it may be comparable with the size of the sample. In finite samples, then, the thermal conductivity may be limited by boundary scattering and the flow of phonons becomes analogous to the flow of a gas at low pressure through a pipe. This will lead to a nearly constant mean free path, of the order of the sample size, and consequently a thermal conductivity proportional only to the total number of phonons. At this low temperature we can use the Debye model so that the number of phonons, and hence the conductivity, is proportional to T^3. This result, taken with umklapp scattering, implies that there will be a peak in the thermal conductivity as a function of temperature. This is illustrated by the experimental results on sapphire (Al_2O_3), etc., shown in figures 8.16 and 8.17. Note in figure 8.16 the very high thermal conductivity attainable, comparable even with copper (cf. figure 8.15). Sapphire is, in fact, unusual for reasons we make clear in the next section. The result, however, is of technological importance. If for example an electrically insulating and transparent mounting is required for use around liquid nitrogen temperatures, sapphire is very suitable and comparatively cheap. Notice that ruby is simply sapphire with Cr impurity. The impurity reduces the thermal conductivity slightly but the high thermal conductivity of ruby is a major factor in its ability to operate as a laser at an efficiency of only a few per cent.

8.6.4 Impurity scattering of phonons (AD)

That phonons may be scattered by impurities will come as no surprise. Before analogies with electrons are drawn, however, if should be remembered that electrons are coupled to impurities by their charge. Phonons are scattered because the *mass* of an impurity atom or ion differs from the others in the chain along which the wave is propagating (cf. Chapter 3). Hence impurities in this connection include isotopes. A solid may be chemically pure but not isotopically pure and isotopic scattering is a major contribution to thermal resistance in many solids. A few elements, of course, have only one stable isotope. These include aluminium, which goes part of the way to explaining the low scattering in Al_2O_3. For those elements that have more than one isotope it will be the variance of the atomic weights about the mean that matters. This is small for naturally occurring hydrogen, helium and oxygen, large for germanium and silicon. Isotope scattering is illustrated in figure 8.18.

8.6.5 Electron scattering of phonons (AD)

Finally, we note that phonons can be scattered by electrons, as we have discussed in connection with magneto–acoustics. Hence electrons may reduce the phonon contribution to thermal conduction while contributing themselves. We are now in danger of a circular argument which we should clarify.

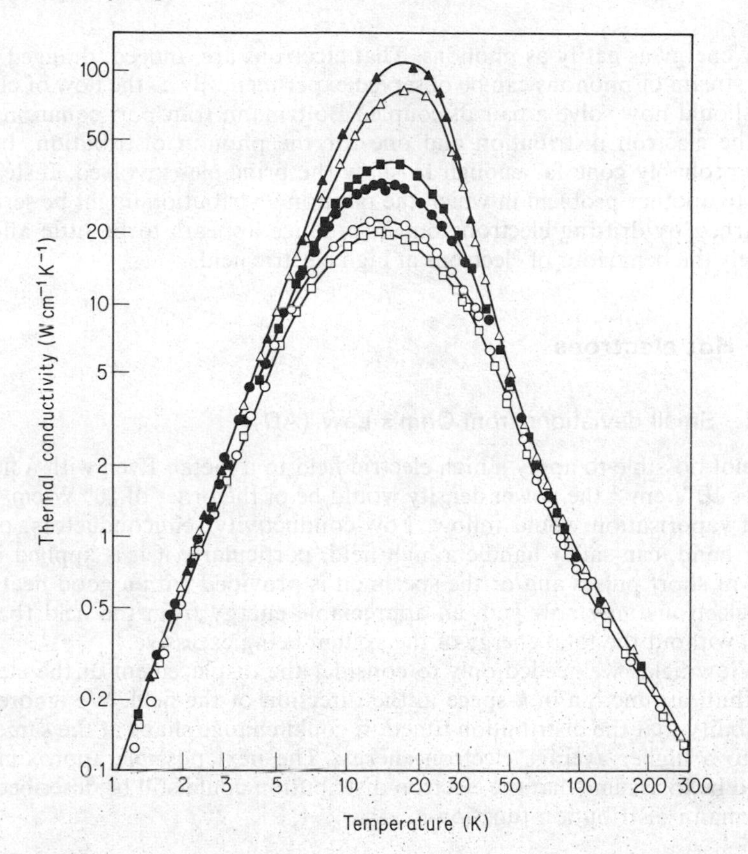

Figure 8.18 Thermal conductivity of LiF with various concentrations of ^{6}Li showing isotope scattering.

Sample	^{6}LiF, Percent
△	0.01
▲	0.02
■	4.6
●	25.0
○	50.1
□	90.4

(After R. Berman and J. C. F. Brock, *Proc. roy. Soc.*, **A289**, 46 (1965).)

When discussing electrical conduction we tacitly assumed that the phonons were a bottomless sink for electron energy and that the phonon distribution remained undisturbed. This was a good assumption because even if the distribution were disturbed this would produce only a change in the scattering, which is itself a small perturbation on the electrons. At least phonons carried down stream with the electrons do not carry charge and contribute to the current. The converse situation is much less reassuring. If phonons pass momentum to electrons and the electron distribution is affected, not only is the scattering changed but also the electrons carried down the phonon stream can carry energy as easily as phonons. That electrons are, indeed, dragged along by a stream of phonons can be observed experimentally as the flow of charge. We should now solve a pair of coupled Boltzmann transport equations, one for the electron distribution and one for the phonon distribution, but we have probably gone far enough to show the principles involved. Instead we turn to another problem in which the phonon distribution might be seriously disturbed by drifting electrons but in practice appears to be little affected, namely the behaviour of electrons at high electric fields.

8.7 Hot electrons

8.7.1 Small deviations from Ohm's Law (AD)

It is not possible to apply a high electric field to a metal. Even with a field as low as 10 V cm^{-1} the power density would be of the order of 10^8 W cm^{-3} and rapid vaporisation would follow. Low-conductivity semiconductors, on the other hand, can safely handle a high field, particularly if it is applied in the form of short pulses and/or the specimen is provided with a good heat sink. The electrons may now gain an appreciable energy from the field (become 'hot') without the total energy of the system being excessive.

At low fields we needed only to consider the displacement of the electron distribution function in ***k***-space in the direction of the field. We ignored the possibility that the distribution function could change shape at the same time due to a higher average electron energy. The next possible approximation would be to assume that the electron distribution could still be described by a Boltzmann distribution function

$$f_0 \propto \exp(-E/kT_e)$$

but that T_e, the electron temperature, was greater than the lattice temperature. This approach leads to the term 'hot electrons'. In fact the approximation implied is rarely justified but widely used.

To find out how hot electrons can get, we should set up a Boltzmann-type energy equation and equate the rate of change of 'temperature' of the electrons due to the field to the equivalent term due to collisions. Again we have a problem for we will wish to use the result far beyond the limit of a small perturbation on the temperature. We might as well, therefore, adopt a phenomenological approach and write an energy balance equation for the electrons, equivalent to the equation of motion used so successfully in Chapter 7. We suppose

$$\frac{d(\Delta E)}{dt} = e\mu F^2 - \frac{\Delta E}{\tau_E} \tag{8.53}$$

where $d(\Delta E)/dt = 0$ when a steady state is established. Here ΔE is the excess electron energy due to the field F and τ_E is some energy relaxation time. As the mobility, μ, is energy dependent in semiconductors, we can write

$$\mu = \mu_0 \left(\frac{E_0 + \Delta E}{E_0} \right)^p \tag{8.54}$$

where E_0 and μ_0 refer to the electron energy and mobility at zero field.
Using (8.53) and expanding (8.54) for low fields gives

$$\mu \approx \mu_0(1 + \mathrm{F}^2/\mathrm{F}_0^2) \tag{8.55}$$

where

$$1/\mathrm{F}_0^2 = ep\mu_0 \tau_E / E_0$$

The form of (8.55) might have been anticipated simply on the grounds that the mobility change could not depend on the sign of F and the neglected higher order terms must also be in even powers of F. Notice that for acoustic mode scattering, when $p = -\frac{1}{2}$, F_0^2 is negative. For impurity scattering, on the other hand, $p = +\frac{3}{2}$ and F_0^2 is positive. This is in accord with experimental observation.

8.7.2 Scattering of hot electrons (AD)

It is useful to obtain a rough theoretical estimate of the magnitude of τ_E in terms of the collision time, τ_c. Their ratio is simply a measure of the degree to which collisions are elastic. For acoustic phonon scattering the change in electron velocity cannot be much greater than s, the velocity of sound. Hence for acoustic phonons

$$\tau_E \sim \left(\frac{kT_e}{m^* s^2} \right) \tau_c \tag{8.56}$$

and scattering is nearly elastic. This is not true of optical-mode scattering where the energy loss per collision is $\sim k\Theta_D$, where Θ_D is the Debye temperature, and large. Hence τ_E is very sensitive to the phonon scattering mechanism and measurement of τ_E can indicate the relative strength of coupling of electrons to acoustic and optic branch phonons. Suitable data can best be obtained by a differential experimental technique which measures the *change* in mobility (the F^2 term of (8.55)) rather than the mobility itself. One such technique is to measure the change in absorption at microwave frequencies ($\omega\tau \ll 1$) when an external electric field is applied. Experimental confirmation of (8.55) by this method is shown in figure 8.19. It is also possible to find τ_E more directly from microwave absorption data if $\omega\tau_E \sim 1$ while $\omega\tau \ll 1$. In germanium, for example, it is found that holes are much more strongly coupled to optical modes than electrons, in agreement with the observations on the variation of mobility with temperature (§ 8.3.2). Polar crystals with strong electron optic phonon coupling, generally show small hot electron effects.

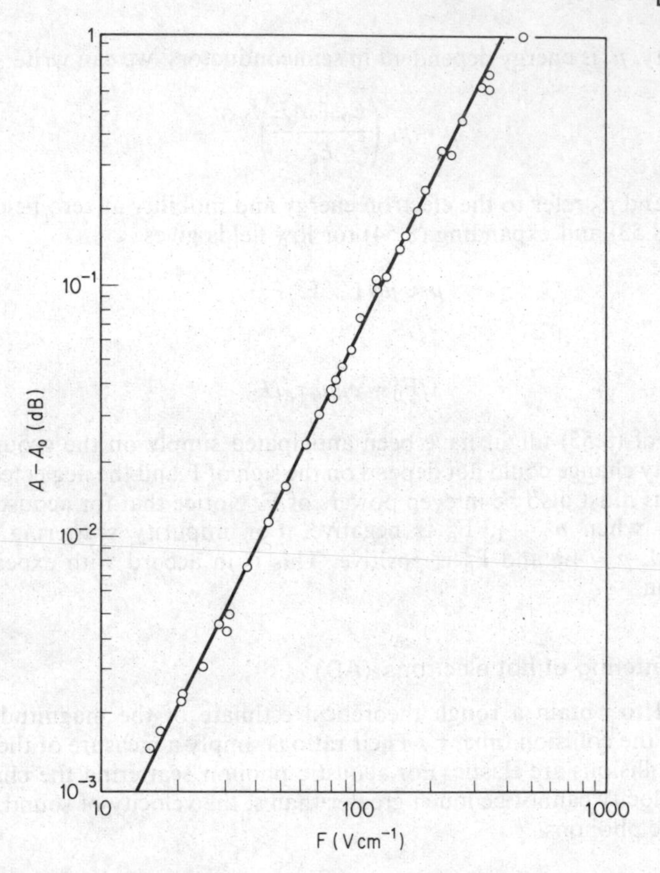

Figure 8.19 The reduction in microwave absorption by free carriers on application of an electric field to germanium, which causes carrier heating. Curve drawn to demonstrate validity of equation (8.55). $A_0 = 12$ dB. (After J. B. Arthur, A. F. Gibson and J. W. Granville, *J. Electron.*, **2**, 145 (1956).)

Impurity scattering is at the other extreme and is highly elastic simply because the ratio of the mass of the impurity ion to that of the electron is so large ($\sim 10^4$ or more). We conclude that at low lattice temperatures impurity scattering may be the dominant mechanism for randomising momentum, but acoustic phonon scattering will still predominantly determine the rate of energy loss. The number of phonons available will be small at low temperature so that τ_E will be large, particularly in materials of low effective mass (8.56).

The large value of τ_E possible in InSb ($m^* = 0.013\ m$) at low temperatures is exploited commercially in an 'electronic bolometer' for use in the wavelength range around 1 mm. We have seen (Chapter 7) that free-carrier absorption is high in this wavelength region. Absorption of radiation leads to appreciable carrier heating if τ_E is large because the average excess energy per electron is

$$\Delta E = W\tau_E/N$$

where W is the rate of power absorption from the radiation. At liquid helium temperature, τ_E in InSb is of the order of 10^{-7} s, and this also determines the speed with which the device can respond to changes in radiation intensity. The radiation is detected by monitoring the sample resistance, which changes by an amount determined by (8.54). Notice that the value of $1/F_0^2$ is increased by operation at low temperatures (small E_0) and by the use of high purity samples (large μ_0). In practice a gain in overall performance can also be obtained by the application of a magnetic field, which raises the resistance of the InSb sample to a more convenient value for use with conventional electronics. A superconducting solenoid (§ 8.9.8) is convenient since a low temperature is required in any case.

8.7.3 Large deviations from Ohm's Law (AD)

Since impurity scattering decreases with increasing electron energy, phonon scattering, and particularly optical-mode scattering, dominates at sufficiently high fields and high electron temperatures whatever the lattice temperature. The mobility then falls with field and in a number of semiconductors the electron (hole) drift velocity, μF, tends to an approximately constant value independent of F. This is illustrated by the results for N type germanium in figure 8.20. Equation (8.53), however, is unreliable if $\Delta E \gtrsim E_0$. In any case new effects occur. One such effect is impact ionisation: any electrons with energy greater than the energy gap, E_g, may ionise a lattice atom to produce an additional hole-electron pair. Since the hole so generated will be accelerated in the opposite direction and may also cause ionisation, the ionisation process is self-sustaining, as in a gas discharge. Before this breakdown effect occurs

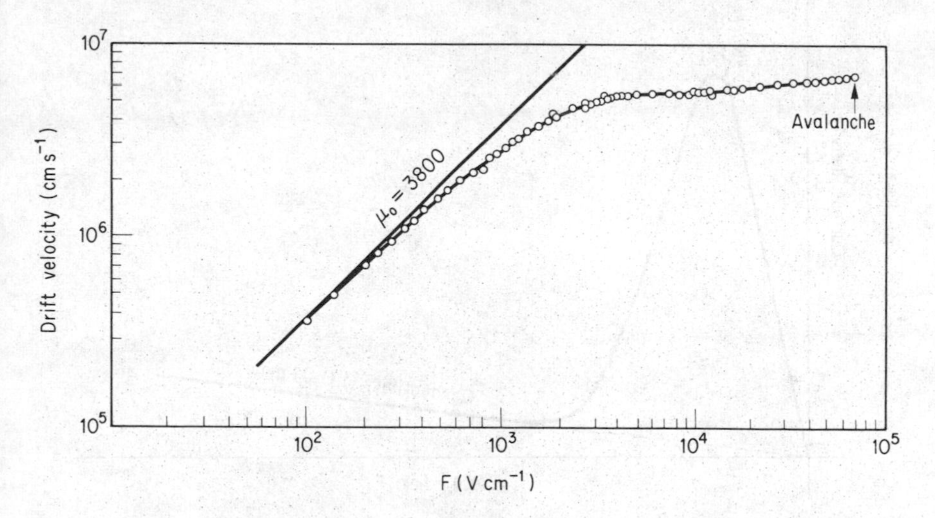

Figure 8.20 Drift velocity of electrons in N type germanium as a function of electric field. (After J. B. Gunn, *J. Electron.*, **2**, 87 (1956).)

(which limits the observations of figure 8.20) the electron energy may become large enough to populate higher energy minima of the conduction band. In such minima the effective mass of the electron may be different. This is the basis of the 'transferred electron amplifier' or Gunn-effect device which was discussed theoretically by Watkins, Ridley, Hilsum and others but first observed experimentally by J. B. Gunn in GaAs.

The lowest conduction band minimum of GaAs (cf. figure 4.19) is at the centre of the zone ($\boldsymbol{k} = (0,0,0)$) and it is characterised by an effective mass, m^*, of about 0.07 m. There are also, however, 'silicon-like' minima at the zone edges in the (1,0,0) directions and 'germanium-like' minima in the (1,1,1) directions. The latter are the highest in energy and need not concern us here. The (1,0,0) minima are about 0.36 eV above the (0,0,0) minimum and are characterised by an effective mass of about 0.3 m, i.e. nearly a factor of 5 higher than that of the (0,0,0) minimum. As a result of this mass ratio the ratio of the density of states is about 10 ($\propto m^{*3/2}$) and the mobility ratio for phonon scattering around 50 ($\propto m^{*5/2}$).

At low electric fields and electron energies all free electrons occupy the (0,0,0) minimum. At very high fields and electron energies in excess of 0.36 eV most of the electrons will occupy the (1,0,0) minima because of the higher density of available states. Furthermore, the higher effective mass implies that the electrons have a much reduced mobility while in the (1,0,0) states.

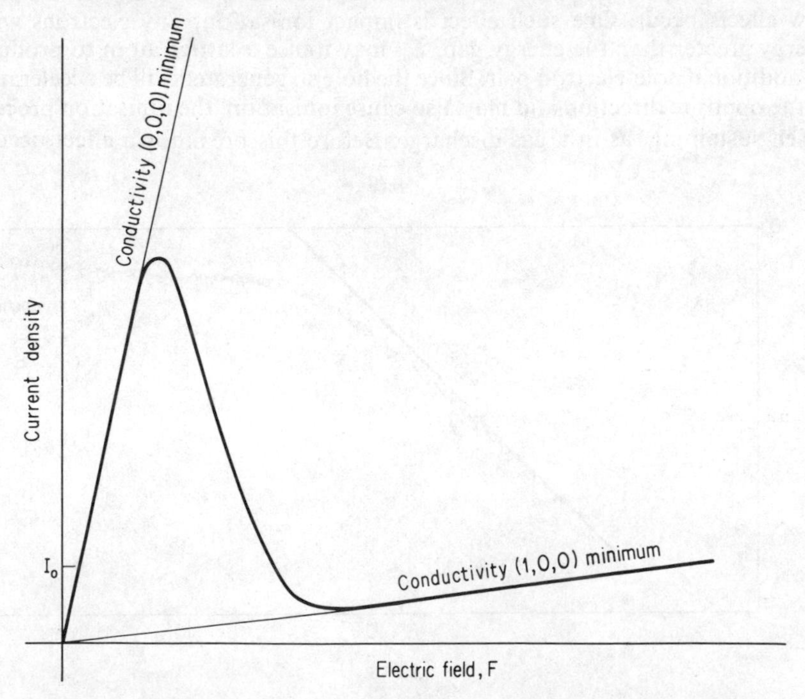

Figure 8.21 Schematic diagram showing possible current–field curve due to intervalley electron transfer in GaAs

Hence it is possible that the conductivity, and consequently the current, might actually fall with increasing field leading to a region of negative differential mobility. This is illustrated schematically in figure 8.21. A negative resistance region can be observed under suitable circuit conditions in GaAs, the onset being at about 3 kV cm^{-1} and the minimum occurring between 5 and 8 kV cm^{-1}. These figures are, of course, highly sensitive to the phonon scattering mechanisms which determine the electron energy and the inter-minima scattering rate which is phonon assisted to conserve wave vector.

It is well known that negative resistance devices can be used for amplification (hence 'transferred electron amplifier') but are inherently unstable. In particular if the current in the sample is fixed at some value such as I_0 in figure 8.21, and is everywhere constant, the electric field in the sample can have one of two values. Depending on the external voltage and the length of the sample there will be one or more regions of high electric field separated by low field regions. The high field regions are called 'electric domains'. But the change in field at the edge of a domain cannot be infinitely abrupt. Poisson's equation relates the gradient of the field and the space charge density, $e\Delta N$, in one dimension x by

$$\frac{\mathrm{dF}}{\mathrm{d}x} = \frac{e\Delta N}{\varepsilon}$$

and $e\Delta N$ is consequently negative (excess electrons) on one side of the domain and positive (deficit of electrons) on the other. This space charge difference leads to domain movement at a velocity approximately equal to the electron drift velocity in the low field outside the domain. Hence we must envisage domains as moving rapidly ($\sim 10^7$ cm s^{-1} in GaAs) towards the anode. As a domain disappears at one end of the crystal a new one must be formed at the other.

The movement of domains, their destruction and creation, leads to fluctuations in the current in the external circuit. It was this oscillatory component of current, superimposed on a d.c. component, that Gunn first observed. The oscillation frequency is determined by the sample length and electron drift velocity. With short samples, say, 0.01 cm, oscillation frequencies are in the gigahertz range. Domains and domain movement can more easily be studied by the use of fine probes placed at strategic points along the sample. The field profile of the domain can then be obtained directly as the domain moves past the probe. Theory, based on Poisson's equation, together with the condition of current continuity, predicts a truncated triangular shape for a domain with a sharp rear (electron excess) edge and a more gently sloping leading edge. Experiment confirms this general picture (see figure 8.22).

Space-charge considerations determine not only the shape but also the length of a domain which consequently depends on the electron density in the undisturbed material. We conclude that, for a given electron density, controlled by impurity addition, there is a sample length below which domain formation is not possible as there simply is not room to contain a whole domain. We can now have a new mode of operation, the 'limited space charge accumulation'

(LSA) mode, in which the velocity of domains no longer determines the frequency of oscillation. LSA–Gunn devices can be mounted in resonant microwave cavities to produce coherent, and to some extent tunable, microwave oscillators. LSA-mode operation is strictly still possible with samples long enough to contain a domain provided the cavity frequency is high enough to reverse the field before the domain has had time to grow. The maximum permissible length is then a function of the cavity quality factor, *Q*, and other factors, but is about one domain length.

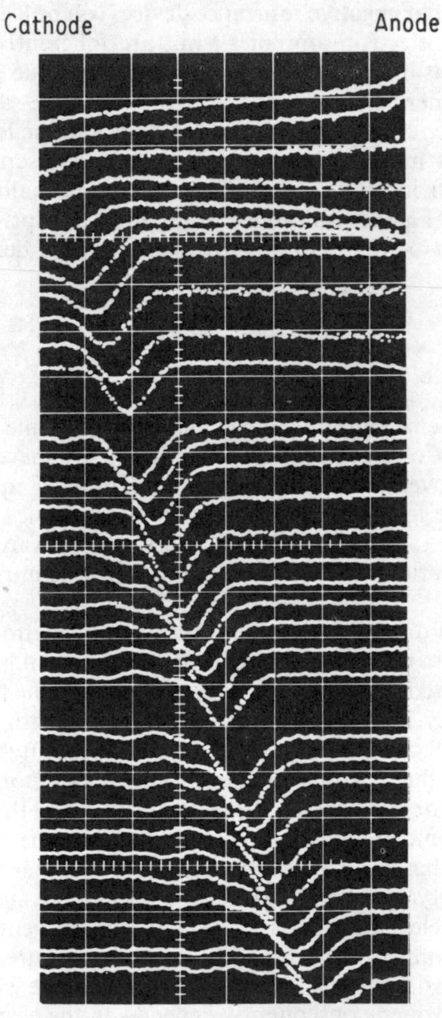

Figure 8.22 The motion of a high field domain through a sample of GaAs. Each trace shows the time derivative of potential as a function of position. Successive traces are separated by time intervals of 6.6×10^{-11} s showing movement of domain. Horizontal scale 26.5 μm per division. (After J. B. Gunn, *Proceedings of Symposium on Plasma Effects in Solids,* Dunod, Paris (1964).)

The Gunn effect has also been observed in InP, whose band structure is similar to that of GaAs, and in CdTe. In CdTe the satellite minima are in the (1, 1, 1) direction and about 0.5 eV above the central minimum, leading to a high onset field (~12 kV cm^{-1}). There is no doubt about the considerable technological importance of Gunn GaAs oscillators (which are available commercially as fully packaged units) as microwave sources. Though output powers are low (milliwatts) and efficiencies of the order of 10%, the device is robust and comparatively cheap. Low cost can go a long way towards offsetting low power as it is possible to envisage extensive arrays of oscillators suitably phased to give electronically steerable beams of high total power.

The Gunn effect relies on the special features of the band structure of GaAs. In the next section we discuss the influence of band structure on general transport properties.

8.8 The effect of band structure on electron transport (a)

The convenience of being able to assume that electrons occupy an energy minimum in the conduction band which is related to electron wave vector by

$$E = \hbar^2 \boldsymbol{k}^2/2m^* \tag{8.57}$$

will be clear from all that has gone before. In general, however, we must allow that (8.57) can only be a first approximation. It is frequently convenient, if not strictly justified, to continue to use (8.57) but to assume that the effective mass is a function of either the magnitude of $\boldsymbol{k}$ or its direction or both. There are a number of phenomena which are particularly sensitive to anisotropy of the energy bands in $\boldsymbol{k}$ space and which consequently yield information on the anisotropy. The most important have already been described in Chapter 4 but in this section we discuss those that are particularly associated with transport measurements.

8.8.1 Energy dependence of effective mass (AD)

If the effective mass is a function of the magnitude, but not the direction, of k, we can always write the energy in terms of even powers of the wave vector as in (4.56)

$$E = A'k^2 + B'k^4 + \dots \tag{8.58}$$

and speak of m^* as being energy, rather than wave-vector, dependent. There are quite a large number of semiconductors for which the term in k^4 is non-trivial for electron energies around kT at room temperature. A particular example is InSb. We have already seen (§ 7.1.4) that the effective electron mass in this material, as deduced from plasma resonance, appears to be a function of temperature. To a large extent this apparent increase in mass with temperature is due to the occupation by the electrons of higher energy and higher effective mass states as kT increases. The form of (8.58) is similar to that of the expansions used in describing non-linear interactions between photons (§ 6.8) and between phonons (§§ 3.7 and 8.6.2). Since free electrons absorb photons we might anticipate, from our discussion of non-linear optics, that electrons in InSb could 'mix' three input frequencies (not two, since there is no term in k^3, the conduction band having inversion symmetry). Non-linear mixing of three infrared laser frequencies has, in fact, been observed in N type InSb and, in principle at least, the value of B' in (8.58) can be deduced from these data.

By far the most satisfactory data on effective masses in semiconductors come, as we have seen (§ 7.2.2), from cyclotron resonance. If the effective mass is a function of energy only we find that the mass as a function of energy may be deduced from cyclotron resonance as a function of magnetic field. To see how this comes about recall that, as an electron describes an orbit in real space, it

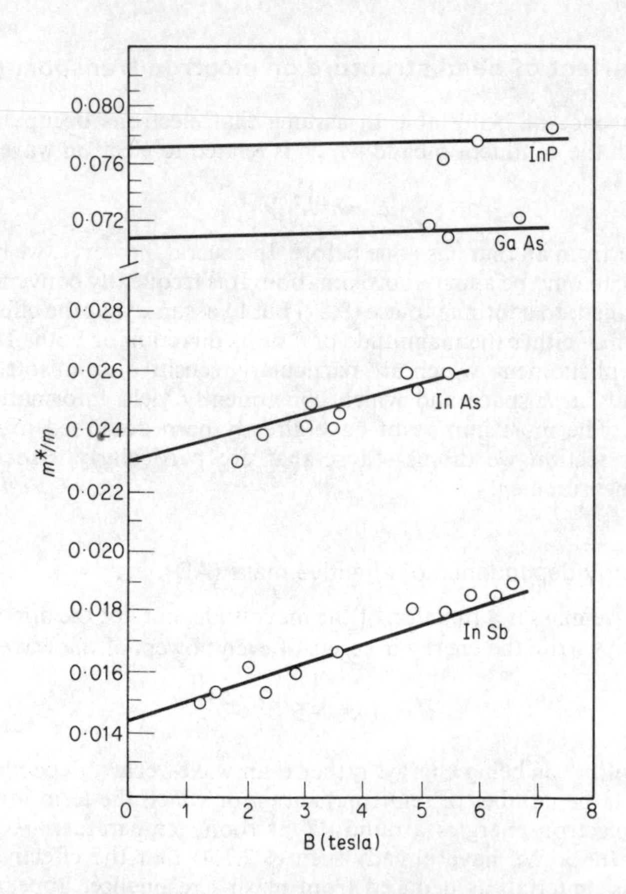

Figure 8.23 Variation of effective mass ratio m^*/m with magnetic field at low temperatures deduced from cyclotron resonance data on various semiconductors. (After S. D. Smith, article in *Optical Properties of Solids*. S. Nudelman and S. S. Mitra (eds.), Plenum Press.)

also describes an orbit in $\boldsymbol{k}$ space (§ 4.7). If $\omega_c\tau \gg 1$, as we may assume if the resonance is to be sharp, the electron is orbiting at a constant energy $\hbar\omega_c$ above the bottom of the band (constant because scattering is small). As far as the orbit is concerned the electron is moving with an effective mass appropriate to an energy $\hbar\omega_c$. Hence the effective mass as a function of energy can be deduced by varying the magnetic field. Not surprisingly, the effective mass increases with E for most semiconductors. The increase is particularly large in InSb and some representative experimental results are shown in figure 8.23.

An energy-dependent effective mass can be incorporated into the treatment of the Boltzmann equation as given in § 8.2 in much the same way as an energy-dependent relaxation time, though with an appreciable increase in algebraic complexity. For metals we have seen that an energy-dependent scattering time has no effect on the transport properties, and neither does an energy-dependent mass. On the other hand, the mobility of a semiconductor in the phonon scattering range depends on $m^{*5/2}$ (8.28) so that a relatively small dependence of mass on energy can lead to a significant change in the variation of mobility with temperature. If, as is usual, m^* increases with E the rate of variation of mobility with temperature in the phonon scattering range (§ 8.3.2) is increased.

8.8.2 Tensor conductivity (AD)

We now consider how the conductivity, and (in the next section) the magneto-resistance, is affected if E is a function of the direction of $\boldsymbol{k}$ so that a surface of constant energy in $\boldsymbol{k}$ space is not spherical.

Suppose that a surface of constant energy in the x, y-plane is an ellipse as shown in figure 8.24. The ellipse may be specified by the values m_x^* and m_y^* of the effective mass along the major and minor axes of the ellipse. We now apply an electric field, F, in an arbitrary direction, as shown. The field may be resolved into two components, F_x and F_y, along the axes of the ellipse. The components of the acceleration $(dv_x/\mathrm{d}t)$ and $(\mathrm{d}v_y/\mathrm{d}t)$, are then

$$(\mathrm{d}v_x/\mathrm{d}t) = e\mathrm{F}_x/m_x^* \quad \text{and} \quad (\mathrm{d}v_y/\mathrm{d}t) = e\mathrm{F}_y/m_y^*$$

If the masses are unequal the resultant acceleration will not be in the same direction as the field but turned in the direction of the lower effective mass, m_y^* (see figure 8.24). Nor is this all. The components of velocity will be

$$v_x = (\mathrm{d}v_x/\mathrm{d}t)\tau_x \quad \text{and} \quad v_y = (\mathrm{d}v_y/\mathrm{d}t)\tau_y$$

so that if the scattering time is also anisotropic ($\tau_x \neq \tau_y$) the velocity will not be collinear with the acceleration. In fact the angle between field and resultant velocity, and hence current, is determined by the ratio τ/m^* and the orientation of the field with respect to the axes of the ellipse. It is evident that the angle between field and current is zero if the field is aligned with a symmetry axis of the ellipse, i.e. the x- or y-axes.

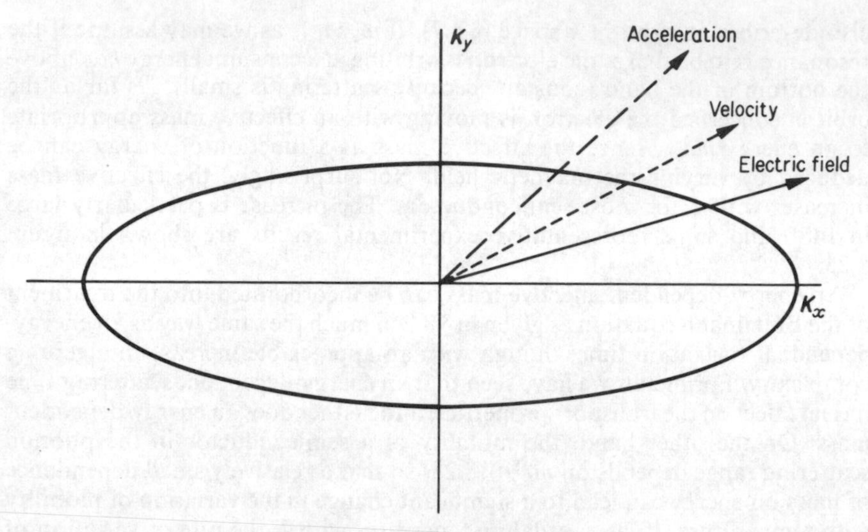

Figure 8.24 Diagram illustrating effect of an anisotropic energy surface on the acceleration and resultant velocity of electrons in an applied electric field

The above properties are described in terms of tensors. If we have a laboratory co-ordinate system specified by X, Y and Z, not necessarily aligned with the major axes of our ellipsoid, x, y and z, we write, in tensor notation,

$$J_i = \sigma_{ij} F_j \qquad \text{where} \qquad i, j = X, Y, Z \tag{8.59}$$

σ_{ij} is the tensor conductivity, **J** and **F** both being vectors. Equation (8.59) is a compact notation for

$$\begin{bmatrix} J_x \\ J_y \\ J_z \end{bmatrix} = \begin{bmatrix} \sigma_{xx} & \sigma_{yx} & \sigma_{zx} \\ \sigma_{xy} & \sigma_{yy} & \sigma_{zy} \\ \sigma_{xz} & \sigma_{yz} & \sigma_{zz} \end{bmatrix} \cdot \begin{bmatrix} F_x \\ F_y \\ F_z \end{bmatrix} \tag{8.60}$$

The values of the nine components of the tensor $\boldsymbol{\sigma}$ give the resultant current components in the X, Y, Z co-ordinate frame for a given set of field components in the same frame.

As an example, consider again an ellipsoidal energy surface set at an angle θ with respect to the X-axis in the X,Y-plane. If an electric field is applied along the X-direction the velocity components along the ellipsoidal axes x and y are

$$v_x = \frac{eF\tau_x \cos\theta}{m_x^*} \qquad \text{and} \qquad v_y = \frac{eF\tau_y \sin\theta}{m_y^*}$$

and the resultant velocity along the X-direction is

$$eF\left(\frac{\tau_y \sin^2\theta}{m_y^*} + \frac{\tau_x \cos^2\theta}{m_x^*}\right)$$

so the component σ_{xx} is given by

$$\sigma_{xx} = Ne^2\left[\frac{\tau_y \sin^2\theta}{m_y^*} + \frac{\tau_x \cos^2\theta}{m_x^*}\right]$$

In a similar way one can show that for this orientation

$$\sigma_{yy} = Ne^2\left[\frac{\tau_y \cos^2\theta}{m_y^*} + \frac{\tau_x \sin^2\theta}{m_x^*}\right]$$

$$\sigma_{xy} = -\sigma_{yx} = Ne^2\left[(\cos\theta \sin\theta)\left(\frac{\tau_x}{m_x^*} - \frac{\tau_y}{m_y^*}\right)\right]$$

$$\sigma_{zz} = \frac{Ne^2 \tau_z}{m_z^*}$$

and

$$\sigma_{xz} = \sigma_{yz} = \sigma_{zx} = \sigma_{zy} = 0$$

It will be noticed that all the terms except σ_{xx}, σ_{yy} and σ_{zz} vanish if the ellipsoid is aligned with the laboratory axes.

A single ellipsoidal energy surface is comparatively rare in solids. Tellurium has a complex multi-valley conduction band but behaves as if

$$\sigma_{xx} = \sigma_{yy} = 1.95\,\sigma_{zz}$$

and consequently has an anisotropic conductivity. It should be evident that if a field is applied at an angle θ to the z-axis in the x,y-plane the conductivity will be given by

$$\frac{1}{\sigma} = \frac{\sin^2\theta}{\sigma_{xx}} + \frac{\cos^2\theta}{\sigma_{zz}}$$

Many solids of interest have cubic symmetry and on these grounds alone cannot have anisotropic conductivities. This does not, however, mean that their energy surfaces are spherical. Thus the conduction band of silicon, as shown in figure 4.17, consists of six equivalent spheroids in the (1,0,0) directions of the crystal. To solve for the conductivity we write down the contribution from each spheroid and add. The sum is isotropic, even though the contribution of each spheroidal valley is not and the resultant conductivity is

$$\sigma = \frac{Ne^2}{3}\left(\frac{\tau_x}{m_x^*} + \frac{2\tau_y}{m_y^*}\right)$$

where $(\tau_y/m_y^*) = (\tau_z/m_z^*)$. Similar remarks apply to the more complicated shape of the valence band of silicon or other cubic crystals like germanium or copper. Hence conductivity data give no information on anisotropic surfaces in cubic crystals. This is not true of magnetoresistance.

8.8.3 Magnetoresistance of non-spherical energy surfaces (A)

We have seen that if the Fermi surface is spherical there should be no magneto-resistance effect in metals for any orientation of magnetic field. If, on the other hand, τ or m^* is a function of energy in a semiconductor there can be a magneto-resistance effect for transverse fields but not if the magnetic field and primary current are colinear. The transverse magnetoresistance arises from the interaction of the magnetic field with two equal and opposite currents flowing transverse to both field and primary current, as explained in § 8.2.4. Essentially the same situation arises if the Fermi surface is not spherical. We have seen in the preceding section that unless all fields are directed along major symmetry axes of the band structure transverse currents will flow. If the crystal has cubic symmetry the transverse currents must sum to zero but they can still interact with the magnetic field to produce a finite magnetoresistance.

The conductivity tensor $\boldsymbol{\sigma}$ is said to be second rank. If a magnetic field is also applied we can permute the orientation of this with respect to the crystal axes and the primary current direction. The Hall coefficient is then a third-rank tensor and we may write

$$\mathbf{F}_i = R_{\mathrm{H}ijk}\, \mathbf{J}_j \mathbf{B}_k$$

where F_i is the Hall field. R_H has 27 components. Magnetoresistance involves the square of the magnetic field and consequently the magnetoresistance coefficient is a fourth-rank tensor with 81 components. Due to the symmetry of crystals, however, the number of independent and non-zero components is usually much less. For example, it should be clear that band anisotropy will

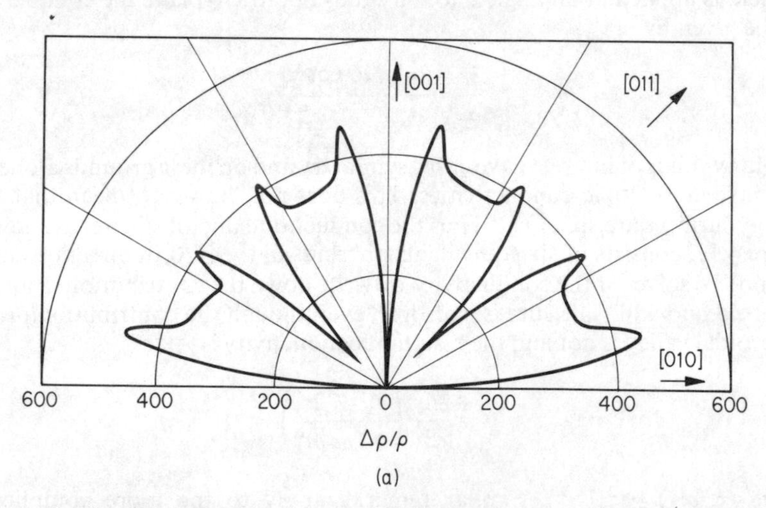

Figure 8.25 (a) Transverse magnetoresistance of a high purity copper sample (resistance ratio 4000) at 4.2 K. Note zeros in (1,0,0) directions. (From J. R. Klauder and J. E. Kunzler, *Proceedings of the International Conference on the Fermi Surface*. John Wiley (1960).)

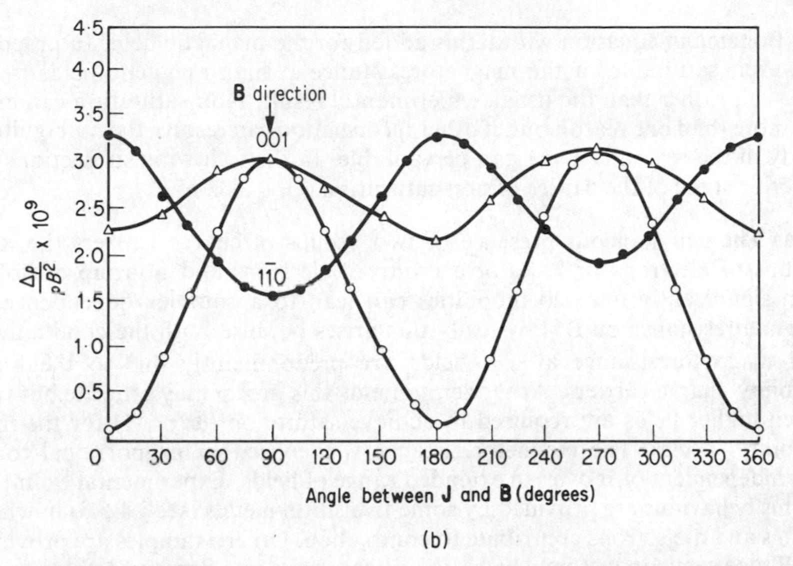

Figure 8.25 (b) Transverse and longitudinal magnetoresistance of silicon at 77 K. Direction of magnetic field, as shown in diagram. Direction of current, 1 : ○ (1,0,0) △ (1,1,0) ● (1,1,1). (After G. L. Pearson and C. Herring, *Physica* **20**, 975 (1954).)

produce no magnetoresistance effect if both **B** and **J** are aligned with a major symmetry axis of the Fermi surface. Hence we can use this property to obtain information on the symmetry of the Fermi surface. Thus we conclude that N type silicon, which has equivalent spheroidal valleys arranged along (1,0,0) axes, will show zero longitudinal magnetoresistance only when **B** and **J** are also orientated along a (1,0,0) direction. For a metal the same symmetry arguments apply to transverse magnetoresistance as for longitudinal magnetoresistance. The Fermi surface of copper (figure 4.12), like that of silicon, has $\pi/2$ rotational symmetry around any (1,0,0) axis and consequently copper has nearly zero magnetoresistance if **B** and **J** are orientated along these directions. Experimental results for Si and Cu are given in figure 8.25. It will be seen that these confirm the symmetry of the Fermi surface in each case. It is worth noting that historically the multi-valley model of N type germanium was first suggested on the basis of magnetoresistance data and later confirmed by cyclotron resonance measurements.

8.8.4 Field dependence of magnetoresistance (A)

Sometimes further information on band structure can be obtained from the dependence of magnetoresistance on the magnitude of the magnetic field. For small B and in cubic crystals the fractional change in resistance, $\Delta\rho/\rho$, must be proportional to B^2 on symmetry grounds: reversal of the magnetic field should have no effect. The classical equation of motion approach ((7.15) and (7.16)) predicts this and also that the magnetoresistance will saturate at high magnetic fields when $\omega_c \tau \gg 1$. These predictions are confirmed by using the Boltzmann equation with terms added for the magnetic field. In practice, however, saturation of the magnetoresistance at high magnetic fields is the unusual, rather than the usual, experimental result. Non-saturation can arise for more than one reason but, if other information can resolve the ambiguities, study in the region $\omega_c \tau \gg 1$ can be valuable. In the following subsections we describe some of the causes of non-saturation.

(a) The simultaneous presence of two groups of charge carriers (i.e. two groups of electrons or holes or a group of electrons and a group of holes) with significantly unequal mobilities can lead to a complex dependence of magnetoresistance on **B**. Physically this arises because both the conductivity and magnetoresistance at low fields are predominantly due to the high-mobility charge carriers. At moderate fields this group may saturate but very much higher fields are required to achieve saturation ($\omega_c \tau \gg 1$) for the low-mobility species. The magnetoresistance is then neither proportional to B^2 nor independent of B over an extended range of fields. Experimental examples of this behaviour are provided by some transition metals (see § 4.5.3) in which both s and d-electrons contribute to conduction. Other examples are provided by P type semiconductors which, like silicon, have two degenerate hole bands at the centre of the Brillouin zone and hence simultaneous conduction by low- and high-mass holes (see figure 4.18).

(b) We have seen that for metals a magnetoresistance effect exists only because the Fermi surface is not spherical. Not only the magnitude of the magnetoresistance but also the degree of saturation then depends on the orientation of the magnetic field. In the simpler odd-valent metals (e.g. Cu, Ag, Al) the magnetoresistance is observed to saturate for most orientations of the magnetic field. Non-saturation, when it occurs, usually indicates a multiply-connected Fermi surface. Thus if, as in Cu (figure 4.12), the Fermi surface touches the Brillouin zone boundary there can exist paths for electrons on the Fermi surface that progess indefinitely in $\boldsymbol{k}$ space without ever completing a closed cyclotron-type orbit. For magnetic field orientations such that these 'open orbits' (see figure 4.25) are possible the condition $\omega_c \tau \gg 1$ has no significance for at least some of the electrons because the cyclotron frequency is indefinite.

(c) If very high magnetic fields are not available it may be necessary to cool the sample to increase the scattering time sufficiently to reach the classical saturation condition $\omega_c \tau \gg 1$. Depending on the sample purity the range of magnetic fields satisfying the conditions $\omega_c \tau > 1$ and $\hbar\omega_c < kT$ may then be quite small. Classical theory is no longer valid when $\hbar\omega_c > kT$. We have seen (§§ 4.7, 7.2.8 and 8.5.3) that the density of states within a band becomes strongly peaked at the Landau levels when $\hbar\omega_c > kT$. Quantisation leads to a number of effects resulting in non-saturation of the magnetoresistance. In metals the magnetoresistance becomes oscillatory in 1/B as in the de Haas–van Alphen effect, and for the same reason. In a semiconductor, with the Fermi level in the forbidden energy gap, there is no de Haas–van Alphen effect but the changing density of states affects the probability of scattering

(8.25). Generally τ decreases monotonically with B but resonance effects can occur: thus when $\hbar\omega_c$ is equal to the longitudinal optic phonon energy (or some integral multiple of it) optical phonon scattering is enhanced by the high density of electron states at the right energy difference. An oscillatory magnetoresistance is then observed.

(d) The break-up of otherwise quasi-continuous bands into Landau levels can be considered as a modification of the band structure of the solid by the magnetic field. In particular there is a change in the effective energy gap of a semiconductor (cf. § 7.2.8). This can lead to a change in carrier density in an intrinsic semiconductor and a new term in the magnetoresistance. A more interesting effect occurs if the semiconductor, like N type silicon, has a number of equivalent conduction band minima occupied by electrons. The shift in energy, $\frac{1}{2}\hbar\omega_c$, of each minimum will depend on the appropriate effective mass and hence on the orientation of the magnetic field with respect to the major axes of each ellipsoid. For example, if a magnetic field is applied in a (1,0,0) direction in silicon, two minima appear to have a low mass (ω_c large) and four present a high mass. If $\frac{1}{2}\hbar\omega_c \gg kT$ all electrons collect in the lowest Landau level of the four lowest valleys. The result is an apparent change in effective mass with field.

(e) Magnetic fields large enough for $\hbar\omega_c$ to be greater than E_g, where E_g is a semiconductor energy gap, have not been achieved experimentally. In some metals, however, (e.g. Mg) energy gaps exist between branches of the conduction band at the zone edges which are as small as 0.01 eV. A phenomenon known as 'magnetic breakdown' due to electron tunnelling may then occur. We discuss the tunnelling of electrons between bands due to the application of high electric fields in the next chapter. In magnetic breakdown the driving force is the Lorentz force resulting from the motion of the electron on the Fermi surface and the magnetic field. Breakdown occurs when

$$E_F\,\hbar\omega_c > E_g^2$$

where E_F is the Fermi energy. When magnetic breakdown occurs new electron orbits are possible and a closed orbit at low fields may become open, or *vice versa*.

Magnetic breakdown has been studied primarily in even-valent metals (e.g. Mg, Zn, Sn). In contrast to the situation for odd-valent metals, non-saturation of the magnetoresistance in these metals is the usual experimental result although saturation can be observed for some orientations of the magnetic field. Non-saturation is associated with open orbits (which may close on a new path on breakdown) but also occurs with closed orbits in these metals due to the simultaneous presence of hole and electron orbits. We have seen in (a) above that a multiplicity of carriers can cause deviations from simple behaviour. Additionally, if electrons and holes are present in equal numbers compensation effects occur, analogous to those discussed in connection with Alfven waves in § 7.2.6. Altogether, the even-valent metals are very complicated.

8.8.5 Piezoresistance (AD)

The detection of anisotropy in the Fermi surface of an otherwise symmetric (e.g. cubic) conductor requires that a preferred direction in space is specified by some external agency. Fundamentally, this is the role of the magnetic field in magnetoresistance and cyclotron resonance. A similar role can be played by uniaxial stress or by an electric field and we close this section by a brief account of the effect of stress.

There is a change in resistance, referred to as piezoresistance, of any conductor on the application of hydrostatic pressure. This is essentially due to the change in volume on compression which in turn changes the band structure. In particular the minima in the conduction band and maxima in the valence band will shift in energy. This is, in fact, the origin of deformation-potential coupling (§ 8.3.2). In general the shift of different minima will be different so that in a material like GaAs, for example, it is possible to alter the relative energies of the conduction band minima at the zone centre and in the (1,0,0) and (1,1,1) directions quite considerably. This is illustrated in the following table.

Minimum	Relative energy (zero pressure) (eV)	Pressure coefficient ($\text{eV kg}^{-1}\text{ cm}^2 \times 10^{10}$)
(0,0,0)	0	12.5
(1,0,0)	0.36	−2.0
(1,1,1)	0.5	5.0

The use of hydrostatic compression provided one of the most convincing pieces of evidence to show that the origin of the Gunn effect was electron transfer between the (0,0,0) and (1,0,0) conduction band minima (§ 8.7.3). By the application of pressure the (1,0,0) valleys are lowered to and eventually below the central minimum. At the same time the electric field for onset of oscillations decreases and finally the effect vanishes.

Thus uniform hydrostatic compression can provide information about band structure by changing it. A preferred direction, however, is only obtained if the stress is uniaxial. The resultant strain in various directions then depends on Poisson's ratio: the relation between stress and strain is a second-rank tensor. Because of the tensor nature of conductivity, piezoresistance is a fourth-rank tensor like magnetoresistance and may be used to study symmetry properties of the Fermi surface in the same way. Uniaxial stress produces particularly large changes in resistance in N type silicon and germanium due to relative changes in energy of the otherwise equivalent conduction band minima, much as in a magnetic field (§ 8.8.4(d)). This effect is exploited commercially to make strain gauges. In compression, germanium will withstand a considerable length change (up to 1 %) without fracture so that it is possible to lower one valley with respect to the others by an energy greater than kT at 100 K and below. All the conduction electrons then occupy the one lowest valley and it is possible to study the anisotropy of (m^*/τ) by simple conductivity measurements. Used with cyclotron resonance data, such measurements permit the anisotropy of τ to be deduced.

8.9 Superconductivity

8.9.1 Introduction (a)

So far in this chapter we have considered electron transport in metals and semiconductors on the basis of the single electron approximation used in band theory (Chapter 4), where the residual interaction between the electrons is the screened Coulomb interaction. This gives only small effects (§ 4.9). In these circumstances the resistivity, ρ, of metals normally decreases as T decreases and at low temperatures tends to a constant value determined by impurities (§ 8.1.3). For a large number of elements and alloys, and a few semiconductors, however, there is a transition at a critical temperature, T_c, below which ρ is identically equal to zero (figure 8.26(a)). This phenomenon is called superconductivity. The complete disappearance of all resistance to the flow of a direct current means that a current induced in a superconducting ring will flow indefinitely—experimentally the decay time is certainly greater than 10^5 years. A number of other properties are also changed drastically at and below T_c. The optical properties in the visible and near infrared, however, do not change so we can deduce that the condition $\rho = 0$ does not persist to indefinitely high frequencies.

It appears that this phenomenon must involve some interaction between the conduction electrons and indeed it was eventually satisfactorily explained on this basis by Bardeen, Cooper and Schrieffer (B.C.S., see § 8.9.5). The observed values of T_c are all low—in the elements they range from 0.01 K in W to 9.3 K in Nb. The highest T_c known at present is 20.2 K for the alloy $Nb_3\,(Al_{0.8}\,Ge_{0.2})$. This means that the effective interaction energy per electron taking part in conduction can only be ~10^{-4} to 10^{-3} eV. Thus we expect that

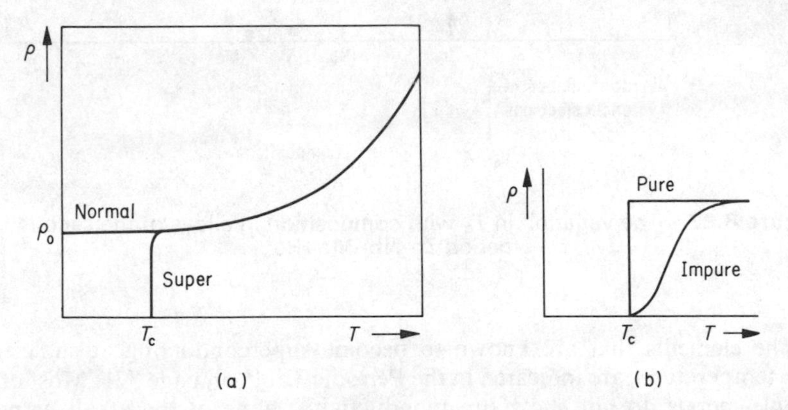

Figure 8.26 Typical variation in resistivity, ρ, of a metal at low temperatures. The resistivity of a superconductor disappears abruptly at $T = T$, which usually lies in the temperature region where $\rho = \rho_o$, the residual resistivity due to impurities. The width of the transition region at T_c is usually only about $T_c/100$ in elemental superconductors but is broadened by the addition of impurities, as shown in (b) on a much expanded temperature scale

the band theory of electrons will still provide a basic description and the interaction can be introduced as a perturbation. The transition at $T = T_c$ is observed to take place very abruptly in pure materials—the width of the transition region may be less than 10^{-3} K. Therefore it is certainly a co-operative effect, involving many electrons acting together. Impurities may broaden the transition (cf. figure 8.26(b)).

The value of T_c is found to be different in different allotropic forms of the same element. This is not surprising since the electronic band structure varies with the crystal structure. Also, as we shall see, phonons play an important part in the B.C.S. theory and the energy-wave vector relationship of these, too, depends on crystal structure. Because of its sensitivity to crystal structure, T_c may be changed by pressure: in a few materials superconductivity is only observed at high pressures. Changes in the number of electrons may also change T_c substantially. The variation of T_c with electron number along the alloy series in the second long period between Zr and Re is indicated in figure 8.27.

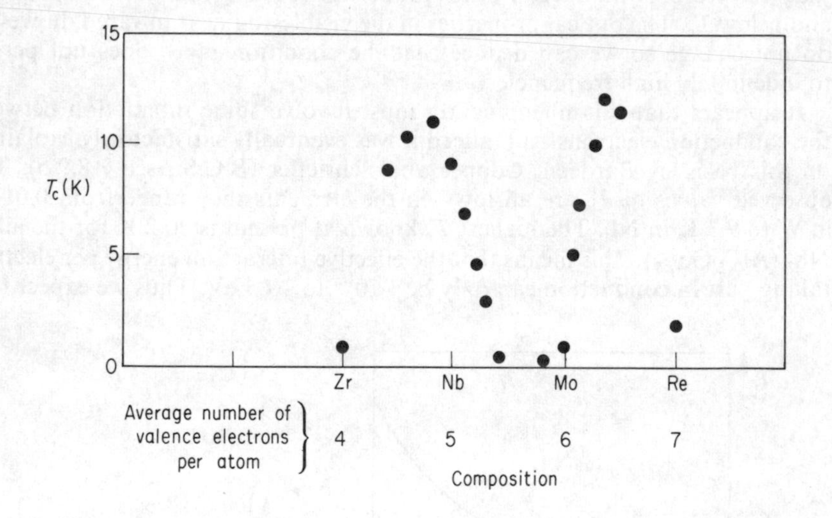

Figure 8.27 The variation in T_c with composition in alloys of the second long period Zr–Nb–Mo–Re

The elements that are known to become superconducting at sufficiently low temperatures are indicated in the Periodic Table on page 371. Most of the simple metals do not show superconductivity; none of the alkali or noble metals and few of the elements in Group II are superconducting. This is explained on the B.C.S. Theory by the smallness of the electron–phonon interaction in these cases. The magnetic transition elements and the rare earths are not superconducting and it is found that even very small concentrations of magnetic impurities depress the value of T_c in many metals.

8.9.2 Magnetic effects (a)

When classified according to their behaviour in an external magnetic field superconducting materials appear to fall into one of two groups, referred to as Type I and Type II. We discuss the origin of these two groups in the next section. The properties described in this, introductory, section on magnetic effects are those of macroscopic type I superconducting samples of simple geometry, though type II superconductors show related properties. Most of the elemental superconductors, when in a pure state, show type I behaviour.

It is hardly surprising that substances with zero resistance to current flow have unusual magnetic properties, since a magnetic field will readily induce persistent currents in them. This follows to some extent from the ordinary formulation of electromagnetic theory in Maxwell's equations. An electric field **F** would accelerate 'superconducting' electrons according to

$$m(\mathrm{d}\boldsymbol{v}/\mathrm{d}t) = e\mathbf{F}$$

and the 'supercurrent' density $\mathbf{J}_s = N_s e\boldsymbol{v}$ obeys the equation

$$\frac{\partial \mathbf{J}_s}{\partial t} = \left(\frac{e^2 N_s}{m}\right)\mathbf{F} \tag{8.61}$$

where N_s is the density of 'superconducting' electrons. Such indefinite acceleration is clearly impossible and hence **F** is usually zero inside a superconductor. From the Maxwell equation

$$\boldsymbol{\nabla} \times \mathbf{F} = -\partial \mathbf{B}/\partial t \tag{8.62}$$

it follows that **B** will be unchanged in time. In particular if an external magnetic field is applied to a sample in the superconducting state we would expect B to remain equal to zero in the specimen and magnetic flux to be excluded. This is found experimentally (though see § 8.2.3) and moreover it is found that B is excluded even when the sample is cooled through T_c in the *presence* of a field. The latter result, known as the Meissner effect after its discoverer, is not predicted by (8.62). To take account of it we require an additional condition, namely

$$\mathbf{B} = 0 \tag{8.63}$$

The Meissner effect is due to the generation of currents in the surface of the specimen. These must flow in a layer of finite thickness and hence (8.63) cannot be expected to hold near the surface. In 1935 F. and H. London proposed a phenomenological 'explanation' of the Meissner effect by combining (8.61) and (8.62) and integrating over time to give an additional equation

$$\left(\frac{N_s e^2}{m}\right)\mathbf{B} = -\boldsymbol{\nabla} \times \mathbf{J}_s \tag{8.64}$$

(Since $\mathbf{B} = \nabla \times \mathbf{A}$ where $\mathbf{A}$ is the vector potential this implies a local relation between $\mathbf{A}$ and $\mathbf{J}_s$). Equation (8.64) is not a consequence of Maxwell's equations but simply an extra equation justified by the fact that it describes an observable property of a superconductor.

With the usual Maxwell equation

$$\nabla \times \mathbf{B} = \mu_0 \, \mathbf{J}_s$$

where μ_0 is the permeability of free space and the displacement current $\partial \mathbf{D}/\partial t$ has been neglected, (8.64) gives

$$\nabla^2 \mathbf{B} = \mathbf{B}/\lambda^2 \qquad \text{where} \qquad \lambda^2 = m/N_s e^2 \mu_0 \tag{8.65}$$

and if N_s is put equal to the total density of conduction electrons, N, $\lambda \sim 10^{-5}$ cm. λ is called the penetration depth since B will fall off as exp $(-x/\lambda)$ with distance x away from the surface. Such a penetration is in fact found experimentally, although the experimental values of λ are somewhat larger than predicted by (8.65). Furthermore λ is found to depend on impurity content and also varies with B so that the above discussion is inadequate. Nevertheless it gives a simple description of an important aspect of superconductivity.

The penetration depth is also found to depend strongly on temperature and to become much larger as T approaches T_c. The observations can be fitted extremely well by a simple expression of the form

$$\left[\frac{\lambda(T)}{\lambda(0)}\right]^2 = \left[1 - \left(\frac{T}{T_c}\right)^4\right]^{-1} = \frac{N}{N_s(T)} \tag{8.66}$$

The second equality in (8.66), giving the temperature dependence of the density of 'superconducting' electrons, is deduced using (8.65) and assuming that $N_s = N$ at $T = 0$. The idea that a superconductor contains two types of electron, superconducting and normal, whose relative density varies with temperature, has been used in a 'two-fluid' model of superconductivity similar to that used to describe some aspects of superfluid helium. Such a model provides a very convenient description and it can, on the basis of simple assumptions, predict the form of (8.66). However, the model must not be taken too literally since electrons cannot be labelled as being of one type or the other. Nevertheless N_s is a measure of the amount of superconductivity and may be regarded as a parameter which characterises the extent of the 'order' in the superconducting state. It is somewhat analogous to the magnetisation of a ferromagnet (cf. § 10.8).

If the magnetic field B_0 applied externally exceeds a critical value B_c it is found that the superconductivity is destroyed. A current sufficient to generate B_c at the surface will have the same effect. In most elemental semiconductors B_c is found to decrease with temperature. Empirically it obeys a relation of the form

$$\frac{B_c(T)}{B_c(0)} = 1 - \left(\frac{T}{T_c}\right)^2 \tag{8.67}$$

to a few percent. B_c falls to zero at $T = T_c$. This equation is illustrated by the phase diagram of figure 8.28.

The magnetically induced transition from the superconducting to the normal state arises because of the extra energy of a superconductor in a magnetic field. Since the transition is reversible, it can be discussed thermodynamically.

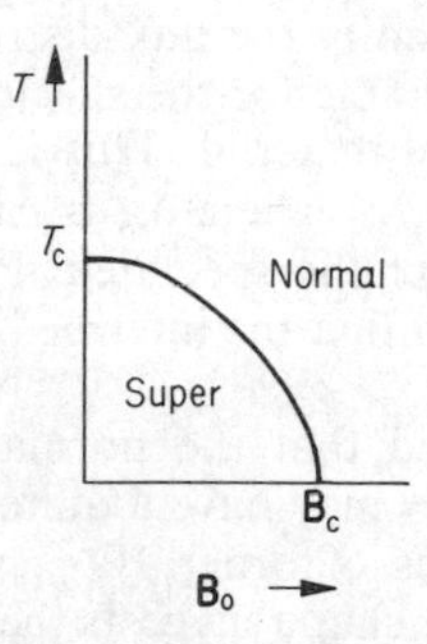

Figure 8.28 Phase diagram of a type I superconductor in an external magnetic field, B_0. The magnitude of the field required to destroy the superconductivity decreases with increasing temperature (equation (8.67)) to reach zero at $T = T_c$

The field energy density, $B_0^2/2\mu_0$, is absent over the volume of the specimen since $B = 0$. Thus when this magnetic energy equals the difference in free energy between the normal (n) and superconducting (s) states it will stabilise the normal state

$$\mathscr{F}_n - \mathscr{F}_s = B_c^2/2\mu_0 \tag{8.68}$$

The change in entropy $S = -\partial\mathscr{F}/\partial T$ gives rise to a latent heat. By the Clausius–Clapeyron equation

$$L_H = \mathscr{V}T(S_n - S_s) = -\frac{\mathscr{V}T}{\mu_0} B_c \left(\frac{\partial B_c}{\partial T}\right) \tag{8.69}$$

where L_H is the latent heat and $\mathscr{V}$ the volume of the specimen. At $T = T_c$, where $B_c = 0$, there is no latent heat but the specific heat $(= \mathscr{V}T(\partial S/\partial T))$ does have a discontinuity

$$C_{V_s} - C_{V_n} = -\frac{\mathscr{V}T_c}{\mu_0}\left(\frac{\partial B_c}{\partial T}\right)^2_{T=T_c} \tag{8.70}$$

which is characteristic of a second-order transition (cf. figure 3.25). Such specific heat anomalies are observed in superconductors.

8.9.3 The intermediate and mixed states—Type I and Type II superconductors (a)

The behaviour of superconductors described in the previous section only applies to specimens with a simple geometry, when the magnetic field is applied along the axis of a thin wire or in the plane of a flat plate. For short specimens the currents on the end surfaces give rise to magnetic fields called demagnetising fields (cf. § 10.6.1) which modify the flux distribution. As B_0 approaches B_c it is now energetically favourable for the specimen to divide into domains of superconducting and normal material. This is called the *intermediate state.* It sets in when $B_0 > B_c(1 - \delta_m)$ where δ_m is called the demagnetising factor and depends on the shape of the specimen. For B_0 perpendicular to a flat plate δ_m approaches unity so that the intermediate state is found in quite low fields.

Experimentally it is found that the normal material takes the form of laminae or filaments but they may have a quite complicated shape (see figure 8.29). They have dimensions of order 10^{-2} cm. The fact that the domain patterns take a characteristic shape suggests that there is a contribution to the energy of the system from the interfaces between the superconducting and normal regions, analogous to the energy which gives rise to surface tension at a liquid–vapour interface and controls the shape of droplets. (Compare also the domain wall energy in ferromagnets (§ 10.11.3).)

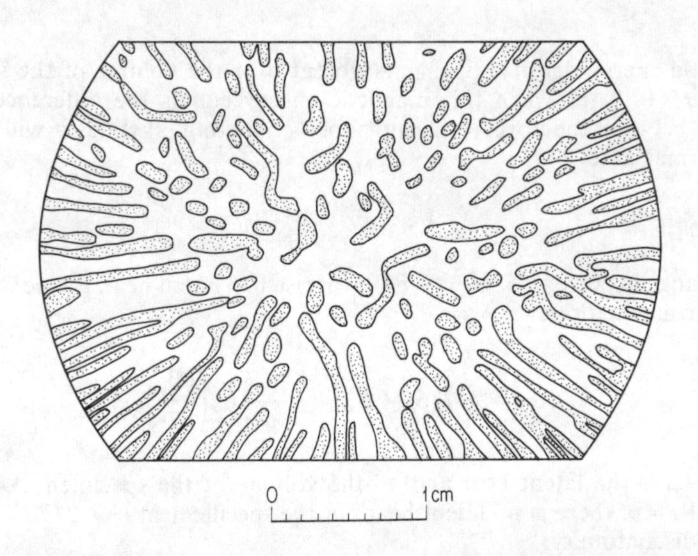

Figure 8.29 Domains in a type I superconductor (tin) in the intermediate state: shaded regions are normal, the remainder superconducting. Maps of this type are obtained using magnetoresistance in a bismuth probe to measure the magnetic field. Bismuth shows an unusually large magnetoresistance effect due to its complex band structure and very low average electron effective mass, which makes ω_c large. (After A. G. Meshovsky, *J. exptl theoret. Phys.,* **19**, 1 1949).)

London first recognised the importance of this surface energy. For the situation so far described it is positive, that is to say it costs energy to make the interfaces, and the domain structure will not occur in appropriate geometries. Such materials are called type I and include most of the elemental superconductors. Other materials, notably alloys or elements in which the electrons have a very small mean free path (e.g. Nb and V) show a more complex behaviour in an applied field. These are called type II superconductors and can be understood on the basis of a negative surface energy (i.e. there is a gain in energy in forming an interface). In these materials there are two critical fields. For $B_0 < B_{c1}$, B is completely excluded as before but for $B_{c1} < B_0 < B_{c2}$ incomplete exclusion occurs—see figures 8.30 and 8.31. This is called the

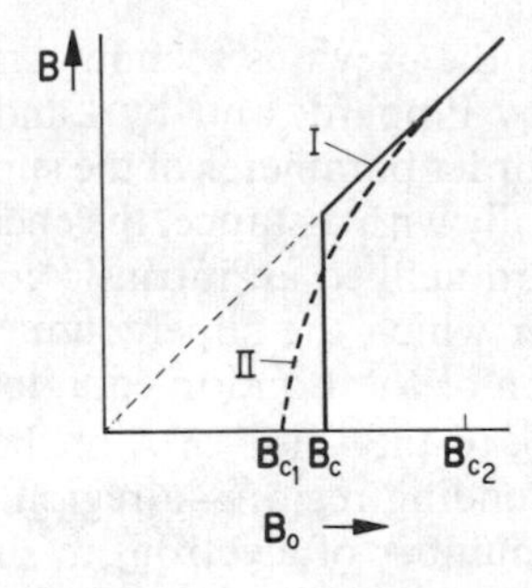

Figure 8.30 The average value of B inside type I and type II superconducting samples of simple geometry as a function of applied field, B_0

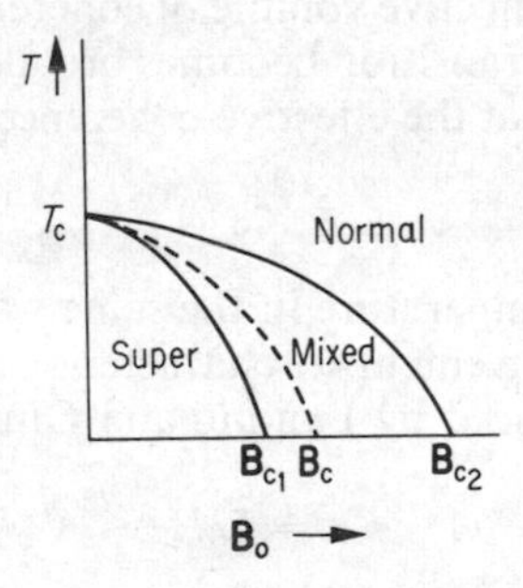

Figure 8.31 Phase diagram of a type II superconductor in a field. Compare with figure 8.28

mixed state. For $B_0 > B_{c2}$ all bulk superconductivity is destroyed. Both B_{c1} and B_{c2} vary like (8.67). Since $B \neq 0$ in some regions of a specimen in the mixed state the free energy difference (8.68) is less than $B_0^2/2\mu_0$ in this case, and the thermodynamic critical field defined by that equation lies between B_{c1} and B_{c2} (figure 8.31).

In the mixed state the normal material takes the form of threads running parallel to B_0, surrounded by current loops. These vortices have a very small cross section and have some remarkable properties to which we shall return in § 8.9.7. In particular the amount of flux along a thread is found to be quantised.

In special geometries, superconductivity can exist to fields $B_0 > B_{c2}$. One case of special interest occurs when B is applied parallel to the surface. A thin sheath of superconducting material then forms on the surface which persists up to $B = B_{c3} \sim 1.7\ B_{c2}$.

8.9.4 Coherence (A)

The properties described in the previous section can be related to the concept of coherence introduced by Pippard, and by Landau and Ginzburg. These authors showed that if the order parameter of the superconductivity, measured by N_s of § 8.9.2, varied rapidly with distance, the energy of the superconductor would be increased. Pippard defined an intrinsic coherence length, ξ_0, as the characteristic distance over which the superconducting order would persist. This leads to a modification of the London equations. In particular $\mathbf{J}_s$ is not simply related, as in (8.64), to the value of **A** at the same point in space but depends on **A** in the surrounding region—a region of volume approximately ξ_0^3. It is because the large number of electrons in such a volume act together in superconductivity that the transition is extremely sharp; otherwise statistical fluctuations would cause broadening.

Electrons cannot, however, behave in a coherent manner over distances large compared with their mean free path, l_f. The presence of impurities decreases l_f and this can reduce the effective volume of coherence. We have already noted that the superconducting transition becomes broader if impurities are present. When $l_f \ll \xi_0$ it is found that the effective coherence length, ξ, varies like

$$\xi \propto (\xi_0 l_f)^{1/2} \tag{8.71}$$

ξ and ξ_0 also vary with temperature in the same way as the penetration depth λ given in (8.66). It is convenient to characterise the relative size of λ and ξ by a parameter κ, introduced by Landau and Ginzburg, where

$$\kappa = \zeta\lambda/\xi \tag{8.72}$$

and ζ is a constant close to unity. In pure materials showing type I behaviour $\xi \approx \xi_0$ and κ is small. For example it is about 0.03 in Al and 0.15 in In. However, for alloys with small l_f, i.e. typical type II materials, κ is much larger.

This difference in κ is related to the difference in surface energy discussed in the previous section. Near the boundary of a superconducting region the magnetic energy density $B^2/2\mu_0$ rises in a characteristic distance λ. But the energy associated with the superconducting electrons, $\mathscr{F}_n - \mathscr{F}_s$, decreases in a distance ξ. At the critical field these energies balance in the bulk of the

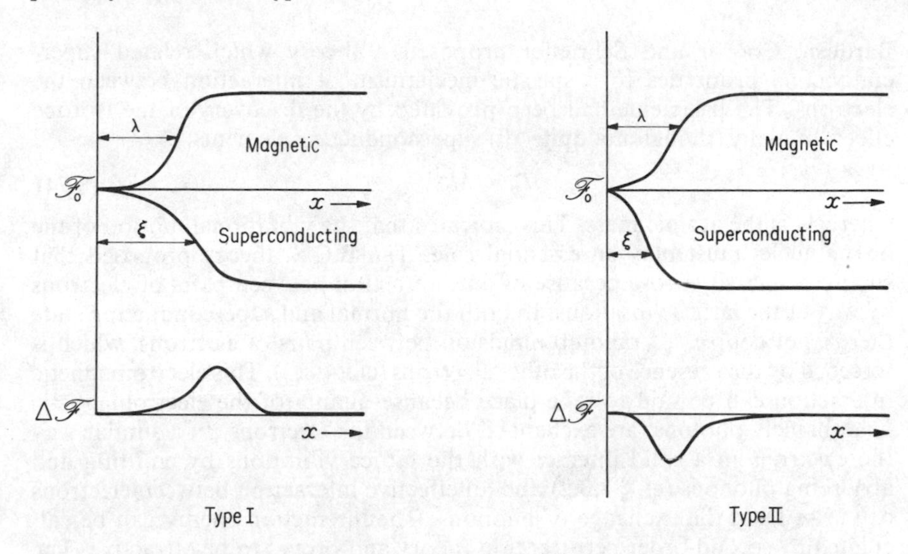

Figure 8.32 Boundary energies $\Delta\mathscr{F}$ are positive ($\xi > \lambda$) in type I superconductors and negative in ($\xi < \lambda$) type II superconductors. λ is the magnetic field penetration depth and ξ the electron coherence length

specimen, according to (8.68), but because of the difference between ξ and λ there is now a surface energy. This is positive if ξ is bigger than λ and negative if ξ is smaller (see figure 8.32). The actual criterion which determines whether the system will show type I or type II behaviour is found from the theory to be $\kappa \lessgtr 1/\sqrt{2}$.

The upper critical field, B_{c2}, which can be reached by the formation of such boundaries can also be obtained from the theory. It is the field at which the magnetic energy in a distance λ is equal to the superconducting energy in a distance ξ. This is given by

$$B_{c2} = B_c \kappa \sqrt{2} \tag{8.73}$$

Hence B_{c2} is larger than the thermodynamic B_c in type II materials. In type I superconductors $B_{c2} < B_c$ but it still has experimental significance. It represents the lowest field in which the normal state can persist in these materials in a metastable fashion. When B falls below this value the surface energy becomes negative so that boundaries can form spontaneously without nucleation.

8.9.5 The Bardeen–Cooper–Schrieffer theory of superconductivity (AD)

The phenomenological theories outlined in the previous sections can provide a satisfactory account of many of the phenomena associated with superconductivity but by definition they give no indication of the microscopic origin of the effect. This remained an unsolved problem until 1957 when

Bardeen, Cooper and Schrieffer proposed a theory which related superconducting properties to a specific mechanism of interaction between the electrons. The basic clue had been provided by the discovery of the isotope effect. In many (though not quite all) superconducting elements

$$T_c \propto M^{-1/2} \tag{8.74}$$

where M is the atomic mass. This indicates that the vibrational motion of the heavy nuclei must play an essential role. The B.C.S. theory proposed that superconductivity arose because of an interaction between pairs of electrons by way of the lattice vibrations. In both the normal and superconducting state there is, of course, a Coulomb repulsion between pairs of electrons, which is screened by the presence of the other electrons (cf. (4.81)). This electromagnetic interaction can be said to take place because quanta of the electromagnetic field, namely photons, are exchanged between the electrons. In a similar way the electrons in a solid interact with the lattice vibrations by emitting and absorbing phonons (cf. § 8.3.2) and an effective interaction between electrons can arise from the exchange of phonons. The interaction energy can be calculated by second-order perturbation theory and proves to be attractive. The phonons may be said to be 'virtual' since by the uncertainty principle energy conservation is not required in the transitional state if an emitted phonon is reabsorbed by a second electron in a very short time. The existence of a phonon implies the motion of ions and the attractive effect may be visualised by imagining one electron of the pair surrounded by a distribution of positive charge due to the ion movement. The essential requirement for superconductivity, according to B.C.S., is that the attraction due to phonon exchange should exceed the screened Coulomb repulsion.

The effect of the attraction is to bind the electrons together into pairs. By a simple statistical argument we can show that pairs with zero total momentum are the most likely. Consider two electrons in states $\boldsymbol{k}_i$ and $\boldsymbol{k}_j$ with energies equal to the Fermi energy so that $k_i = \kappa_j = k_F$. The total momentum of the pair is $\boldsymbol{k}_i + \boldsymbol{k}_j = \boldsymbol{K}$. For a particular value of $\boldsymbol{K}$, of magnitude $K = 2k_F \sin\theta$, only the electron states on the ring marked on figure 8.33 can form such pairs. But for $\boldsymbol{K} = 0$, i.e. $\boldsymbol{k}_i = -\boldsymbol{k}_j$, all the electron states on the Fermi sphere can be paired. Thus pair states with $\boldsymbol{K} = 0$ are much more likely to occur than states with finite $\boldsymbol{K}$. Strictly, because of the finite phonon energy, $\hbar\omega(\boldsymbol{q})$, only electron states in a shell of width Δk where $\hbar^2 k_F \Delta k/m = \hbar\omega(\boldsymbol{q})$ can take part so that a finite shell of $\boldsymbol{k}$ states is involved, but this does not invalidate the conclusion. Finally, as a consequence of the exclusion principle, the pairs are formed by electrons of opposite spin: just as in a molecule (cf. figure 1.11) electron pairs of opposite spin have the lower energy.

In a superconductor at $T = 0$ all the electrons in states near the Fermi surface are bound into pairs and are in the ground state of the system. A finite energy 2ε† is needed to break such a pair and give normal, single electrons. Hence there are no states of the system with energies between the ground

† This symbol is used universally for the pair binding energy and we shall follow this convention, notwithstanding the use of ε elsewhere to indicate the dielectric constant. For the remainder of this chapter, ε is reserved for this energy.

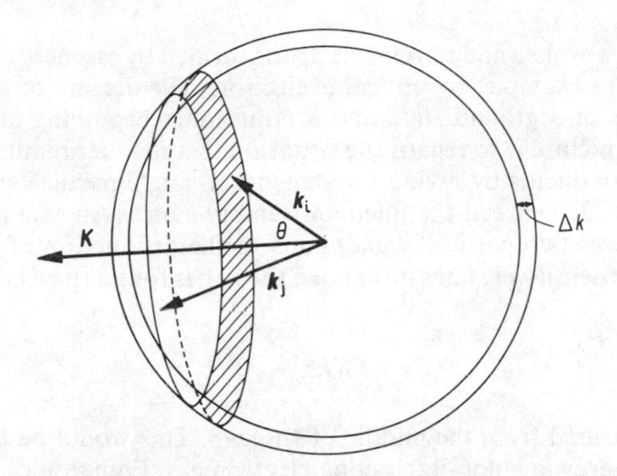

Figure 8.33 A shell of states of width Δk on a spherical Fermi surface showing a typical pair of states giving a total wave vector $\boldsymbol{K}$ where $|\boldsymbol{K}| = 2k_F \sin\theta$. When $\boldsymbol{K} = 0$ all states can form pairs

state and 2ε. This implies that there is an energy gap. Many of the properties of superconductors may be accounted for simply by the existence of this gap and we discuss some of these in the next section. The B.C.S. theory gives, for half the energy gap,

$$\varepsilon = 2\hbar\omega_L \exp\left[-1/S(E_F)G\right] \tag{8.75}$$

where $\hbar\omega_L$ is some average phonon energy and from (3.6) accounts for the isotope effect (8.74). $S(E_F)$ is the density of electron states at the Fermi energy and G is a measure of the strength of the electron–phonon interaction. A large density of states and a large value of G favour the occurrence of superconductivity.

Electrons in pairs may be regarded as the 'super' electrons of the two-fluid model. The most striking evidence for pairing arises from the fact that the quantum of flux discussed below has a value appropriate to that for particles of charge $2e$. Since the ground state of the system containing pairs is a single state of the whole system there is no possibility of scattering since there are no states to scatter into, unless the electrons gain sufficient energy to cross the gap. Hence at $T < T_c$ there is no resistance to current unless this becomes too large. If electrons at the Fermi surface drift with velocity $\boldsymbol{v}$ they gain an extra energy

$$(\hbar\boldsymbol{k} + m\boldsymbol{v})^2/2m - \hbar^2\boldsymbol{k}^2/2m \approx \hbar\boldsymbol{k}\cdot\boldsymbol{v} \tag{8.76}$$

When this energy is greater than the gap, i.e. when $\boldsymbol{v}$ exceeds a critical velocity

$$v_c \sim \varepsilon/\hbar k_F \tag{8.77}$$

the superconductivity will disappear because scattering begins.

A pair may be broken and normal electrons formed by an energy 2ε provided thermally or, for example, by optical excitation. The density of states of the system consists of a ground state and a continuum beginning at energy ε.† An alternative picture is to regard the situation in a manner reminiscent of an intrinsic semiconductor by giving the density of single particle states. There is now a gap of 2ε between the filled and empty states. An excitation across this gap produces two particles, analogous to the production of an electron and a hole. The density of states just above the gap is found to be large, varying as

$$S(E) \sim E/(E^2 - \varepsilon^2)^{1/2} \tag{8.78}$$

where E is measured from the middle of the gap. This would be the position of the Fermi energy in a non-interacting electron gas. Equation (8.78) and the occupation of the states at finite $T < T_c$ is illustrated in figure 8.34. The analogy

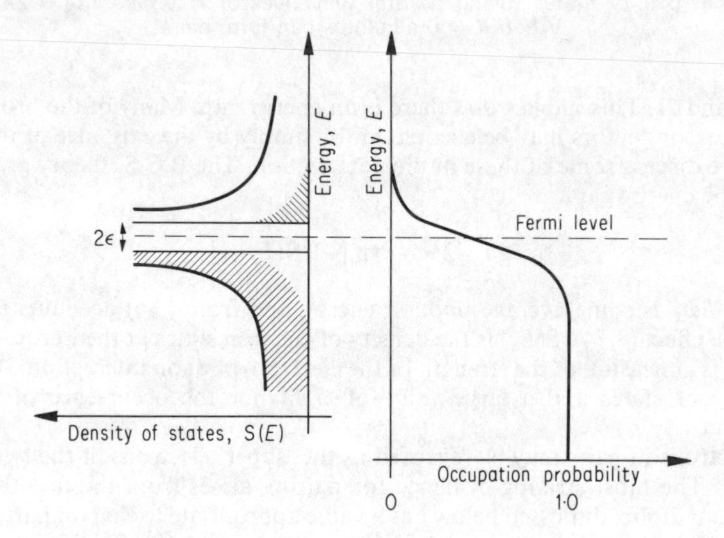

Figure 8.34 The density of states in a superconductor of energy gap 2ε (equation (8.78)) and their occupation according to the Fermi function

with the band structure of semiconductors must not, however, be pushed too far. In a superconductor all the states belong to the same electron band and the gap arises at the Fermi energy in this band because of the interelectronic interaction. Thus as T increases the number of normal electrons increases, as in a semiconductor. But the smearing of the Fermi distribution also affects the effective strength of the pair bonding so that the energy gap $2\varepsilon(T)$ falls to zero as T approaches T_c. In fact $\varepsilon(T)$ is an order parameter and behaves like

† But note that an energy 2ε is required to excite electrons into this continuum since they must be excited in pairs.

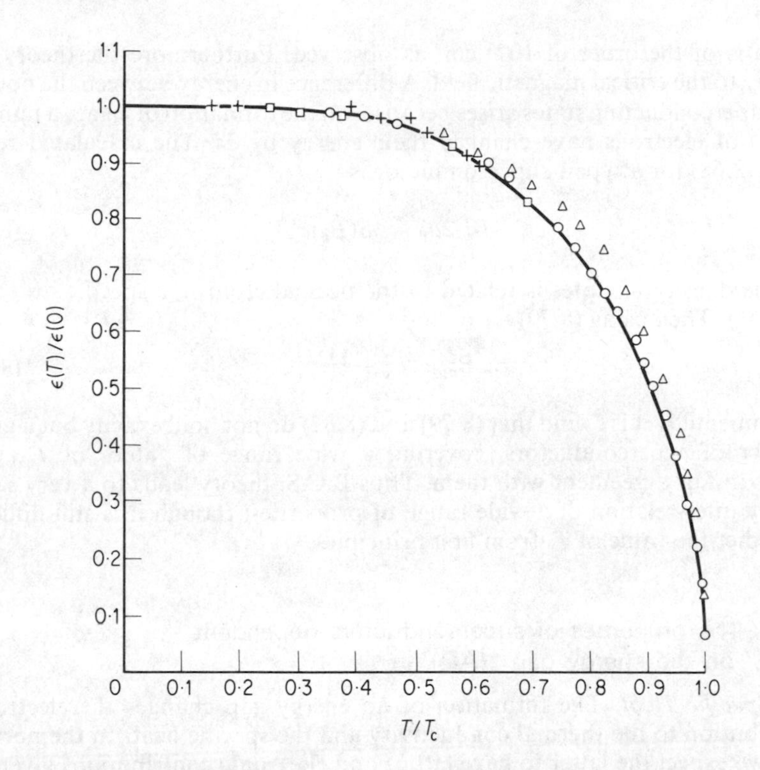

Figure 8.35 Reduced energy gap, $\varepsilon(T)/\varepsilon(0)$, as a function of reduced temperature, T/T_c, deduced from tunnelling experiments (§ 8.9.6). Curve B.C.S. Theory. Experimental points △, tin, ○, tantalum, +, lead, □, niobium. Approximate values of $2\varepsilon(0)/kT_c$ deduced from the experimental data are 3.5, 3.6, 4.3 and 3.8 respectively (cf. equation (8.79)). (After P. Townsend and J. Sutton, *Phys. Rev.*, **128**, 591 (1962).)

analogous quantities in other order–disorder transitions (cf. figure 8.35 with figures 3.24 and 10.17).

The width of the energy gap can be related, using B.C.S. theory, to a number of superconductor parameters. Thus the transition temperature is related to the pair binding energy and so to the energy gap at $T = 0$. Calculation gives

$$2\varepsilon(0) = 3.5kT_c \tag{8.79}$$

The coherence length, ξ_0, can also be related approximately to the energy gap by using the uncertainty principle. The range in electron momentum, $\Delta \boldsymbol{p}$, associated with the energy gap is given by $\hbar k_F \Delta \boldsymbol{p}/m \sim \varepsilon$. The coherence length on this criterion is then

$$\xi_0 \sim \hbar/\Delta \boldsymbol{p} \sim \hbar^2 k_F/m\varepsilon \tag{8.80}$$

which is of the order of 10^{-4} cm, as observed. Furthermore the theory can relate ε to the critical magnetic field. A difference in energy between the normal and superconducting states arises because, on the formation of a gap, a number $\varepsilon S(E_F)$ of electrons have changed their energy by 2ε. The calculated result using (8.68) for a type I superconductor is

$$B_c^2/2\mu_0 = \tfrac{1}{2}S(E_F)\varepsilon^2 \tag{8.81}$$

But the density of states is related to the normal electronic specific heat, γT, by (4.71). Then using (8.79)

$$B_c^2/\mu_0 T_c^2 \sim \tfrac{1}{2}\gamma \tag{8.82}$$

Experimentally it is found that (8.79) and (8.82) do not hold exactly but a great number of superconductors, covering a wide range of values of T_c, give results in fair agreement with them. Thus B.C.S. theory leads to a very satisfactory inter-relation of a wide range of properties, though it is still difficult to predict the value of T_c from first principles.

8.9.6 The properties of superconductors dependent on the energy gap (AD)

The Specific Heat. The formation of an energy gap changes the electronic contribution to the thermal conductivity and the specific heat. In the normal state we expect the latter to have lattice and electronic contributions given by

$$C_{V_n} = AT^3 + \gamma T \tag{8.83}$$

while in the superconducting state only the normal electrons contribute to the specific heat as the paired electrons in the ground state have zero energy. Hence the specific heat is given approximately by an expression of the form

$$C_{V_s} = AT^3 + \Gamma \exp(-\varepsilon/kT) \tag{8.84}$$

where the second term is the electronic contribution, proportional to the density of normal electrons. Γ is a constant and ε is approximately equal to $\varepsilon(0)$. Some typical experimental results are shown in figure 8.36.

Thermal Conductivity. In normal metals, as discussed in § 8.6, the heat current is predominantly carried by the conduction electrons and at low temperatures the electronic contribution to the thermal conductivity $\mathcal{K}_{e_n}$, is given by the Wiedemann–Franz Law. In a superconductor, however, the electron pairs have zero energy so they cannot contribute to energy transport and hence to the heat current (but being charged they can still contribute to the electric current). Hence the electronic contribution to the heat current depends on the number of normal electrons and $\mathcal{K}_{e_s}/\mathcal{K}_{e_n} \sim \exp(-\varepsilon/kT)$ like the electronic specific heat in (8.84). This is illustrated in figure 8.37. When $T \ll T_c$, $\mathcal{K}_{e_s} \to 0$ and the only thermal current will be carried by the phonons (as in an insulator!

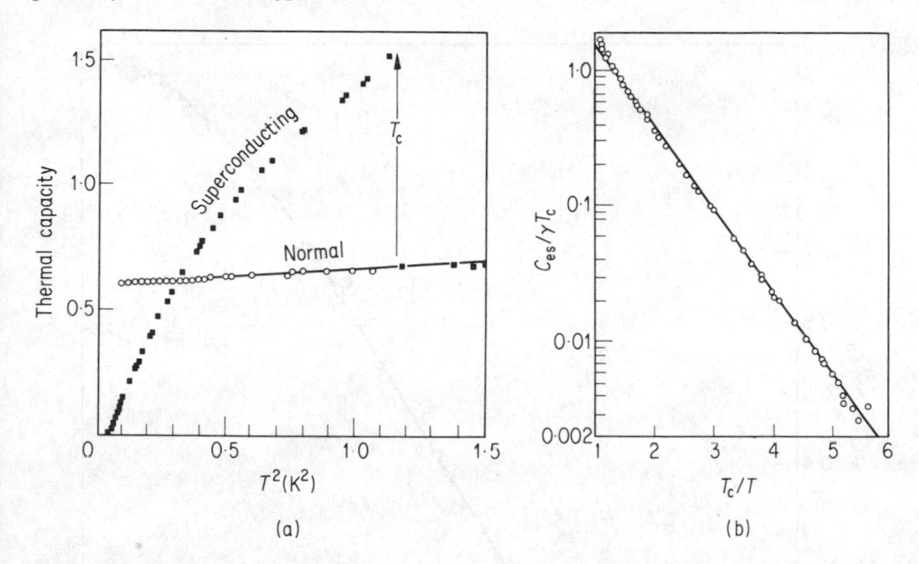

Figure 8.36 (a) The thermal capacity of gallium in the superconducting and normal (0.02 tesla field applied below T_c) states as a function of T^2. Normal state follows (8.83) over temperature range shown. (b) Electronic contribution, C_{es} to specific heat plotted to show exponential dependence on $1/T$ (equation (8.84)). Since $\varepsilon \sim \varepsilon(0) \sim (3.5/2)kT_c$ from (8.79), we expect C_{es} proportional to $\exp(-1.75\ T_c/T)$. In fact curve shown is $\exp(-1.39\ T_c/T)$. (After N. E. Phillips, *Phys. Rev.*, **134**, 385 (1964).)

cf. § 8.6.2). Under suitable conditions $\mathscr{K}_{e_n}/\mathscr{K}_{e_s}$ may be very large ($\sim 10^3$) and this property can be used to make a heat switch, the heat flow being controlled by a magnetic field ($B > B_c$, or zero).

The phonon contribution to thermal conduction will actually increase in the superconducting state since the scattering of phonons by electrons (§ 8.6.5) is reduced by the formation of pairs. In fact B.C.S. theory predicts a reduction in the attenuation coefficient, α, of acoustic waves in the superconducting phase and a ratio of phonon mean free path in the normal and superconducting states of

$$\frac{\alpha_s}{\alpha_n} = \frac{2}{1 + \exp(+\varepsilon/kT)} \qquad (8.85)$$

which varies as $\exp(-\varepsilon/kT)$ at low temperatures (cf. figure 8.38).

In extreme cases when $\mathscr{K}_{e_n}$ is made small by the introduction of impurities the increase in the phonon contribution to the thermal conductivity below T_c may outweigh the reduction in the electronic contribution so that the total conductivity increases in the superconducting state. To achieve this condition an impurity of similar mass but different valence, which will reduce $\mathscr{K}_{e_n}$ without greatly affecting phonon transport, should be used. An example is Bi in Pb.

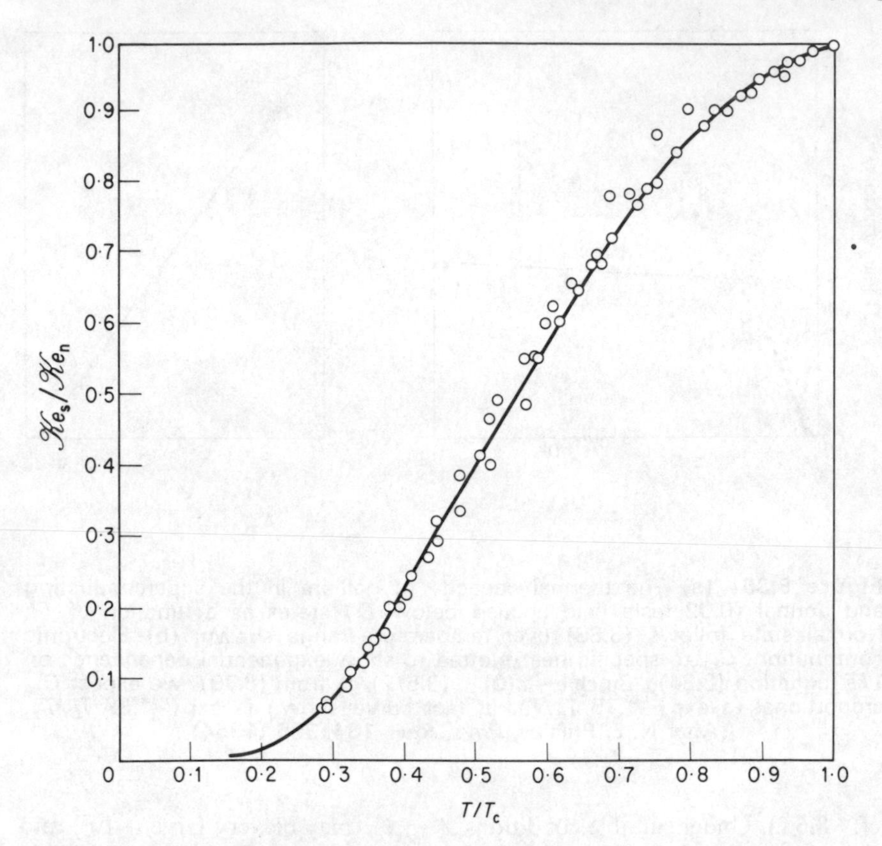

Figure 8.37 Ratio of the electronic contributions to the thermal conductivity of aluminium in the normal and superconducting states. Curve B.C.S. theory. (After C. B. Satterthwaite, *Phys. Rev.*, **125**, 873 (1962).)

Optical Properties. Since ε is a function of temperature it is not easy to extract its variation from the measurement of thermal properties. Optical absorption should give more direct information. The analogy with semiconductors suggests that superconductors should show an absorption edge (§ 6.3) at a frequency $\omega_g = 2\varepsilon/\hbar$ and that for photon energies much greater than 2ε the optical properties will be the same in the normal and superconducting states. For values of T_c ranging from 0.1 K to 10.0 K, (8.79) shows that the absorption edge will occur in the far infrared, at wavelengths between 5 cm and 0.05 cm. Transmission measurements at the low temperatures required have proved difficult in this region but the absorption can be deduced from reflectivity data. Some experimental results, which strikingly confirm the existence of an energy gap, are shown in figure 8.39.

Tunnelling. An experimentally easier technique which gives direct information on the energy gap is electron tunnelling through an insulating film between superconducting metals. An electron may tunnel at constant energy through—

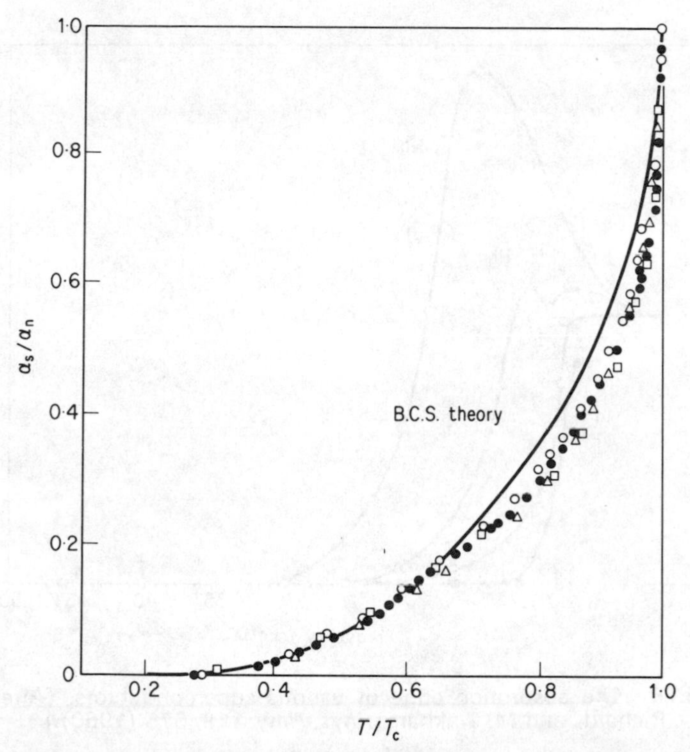

Figure 8.38 Ratio of attenuation coefficients for acoustic waves in superconducting and normal states. ○, tin at 33.5 MHz, ●, tin at 54.0 MHz, □, indium at 28.5 MHz, △, indium at 35.2 MHz. (After R. W. Morse and H. V. Bohm, *Phys. Rev.*, **108**, 1094 (1957).)

i.e. it has a finite probability of being found on the other side of—an insulating film whose thickness is of the order of or less than a mean free path (see also § 9.2.5). Tunnelling can, however, only occur if there are empty states available for the electron to occupy. Figure 8.40(a) shows the available states when a superconductor and a normal metal are separated by a thin insulating film and figure 8.41(a) when two superconductors are so separated. As shown in Chapter 9, the Fermi levels must be at the same level in equilibrium (if this were not so, charge would flow to equalise the levels) and, from the B.C.S. model, the Fermi level must be at the midpoint of the energy gap in a superconductor. At $T=0$ there are therefore no filled and empty states at the same energy on each side to permit tunnelling. When, however, a voltage V is applied across the insulator the electron potential energy is raised on the negatively charged side and the Fermi levels are separated by an energy eV. If, as in figure 8.40(b), the normal metal is made negative a tunnel current can flow for $V \geqslant \varepsilon/e$. When $T>0$ a few electrons are in states above the Fermi level on both sides so that some current flows even for $V<\varepsilon/e$ but a rapid increase can still be observed for $V>\varepsilon/e$ (figure 8.40(c)). The variation in

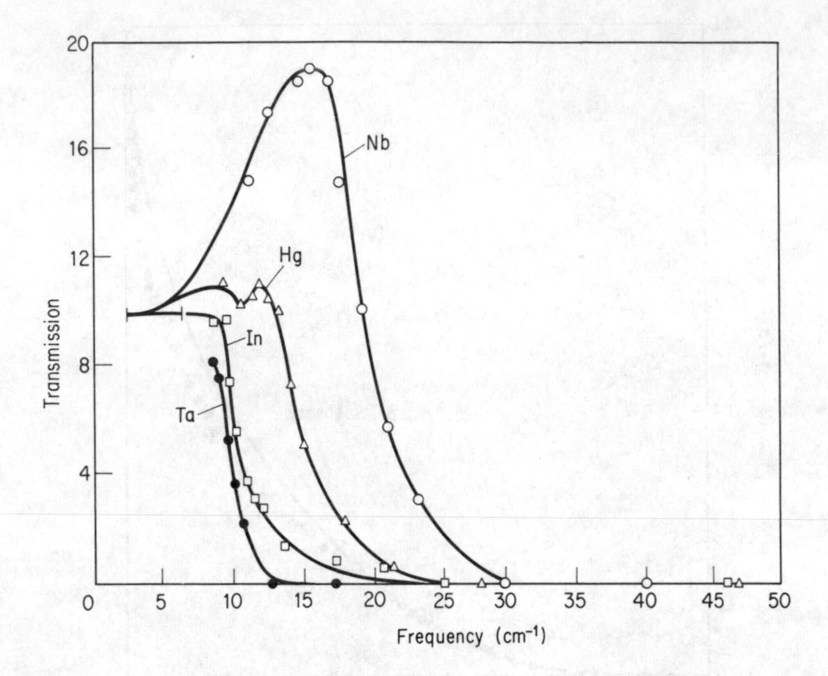

Figure 8.39 The absorption edge of various superconductors. (After P. L. Richards and M. Tinkham, *Phys. Rev.*, **119**, 575 (1960).)

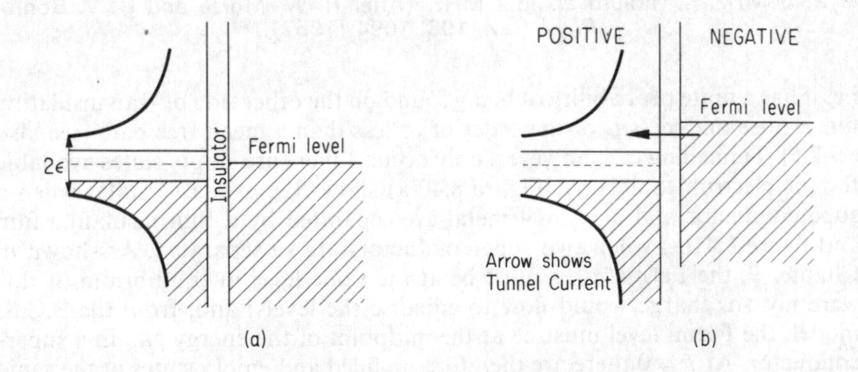

Figure 8.40 (a) A superconductor and a normal metal separated by an insulating film. States occupied at $T = 0$ shown shaded

(b) When a voltage is applied a field appears across the insulator and, in the polarity shown, the potential energy of electrons in the normal metal is increased relative to the superconductor. Tunnelling of electrons is possible when the Fermi level in the metal is above the top of the energy gap in the superconductor

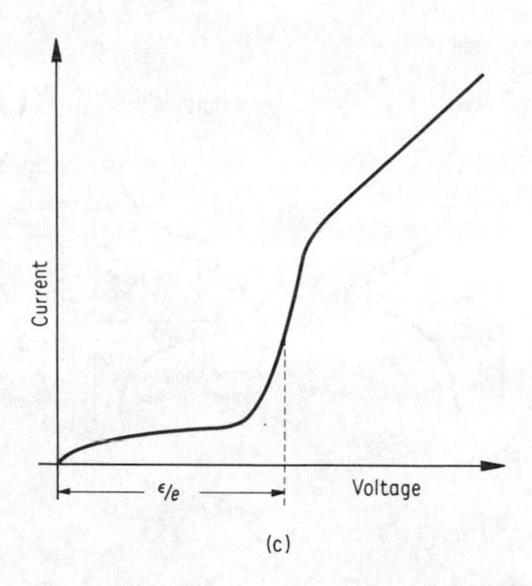

Figure 8.40 (c) Experimentally observed current voltage characteristic of structure illustrated in (a) and (b). Finite current at $V < \varepsilon/e$ due to finite temperature of sample

current with voltage above ε/e depends on the density of states distribution and the tunnelling probability (see § 9.2.5) but need not concern us here.

With the arrangement shown in figure 8.41 the resultant current–voltage curve is more complex, as shown in figure 8.41(c). The small peak arises only because T is finite: at $T = 0$ inspection of figure 8.41(b) shows that tunnelling can only begin when $V = (\varepsilon_1 + \varepsilon_2)/e$. However, for $T > 0$ tunnelling of normal electrons occurs below this voltage and is a maximum when the two density of states maxima are coincident in energy. Hence the peak occurs at $V = (\varepsilon_1 - \varepsilon_2)/e$ and the results permit both ε_1 and ε_2 to be determined separately.

Tunnelling provides the simplest and most direct method of determining energy gaps in superconductors and has been widely used. The results are generally in good agreement with other methods and with (8.79). Similar remarks apply to the variation of energy gap with temperature (see figure 8.35). Variations in ε with crystal orientation are observed, indicating an anisotropic energy gap. This is expected, in view of the occurrence of non-spherical Fermi surfaces, but not directly predictable from B.C.S. theory which assumes a spherical Fermi surface.

In recent years the study of tunnelling between two superconductors has been given a new impetus by the prediction of Josephson of the properties of very thin tunnel junctions (insulator thickness 10^{-7} cm or less). This arises from the quantum mechanical coherence in the wave function of the superconducting state, i.e. tunnelling of electron pairs takes place. We discuss these properties in the next section.

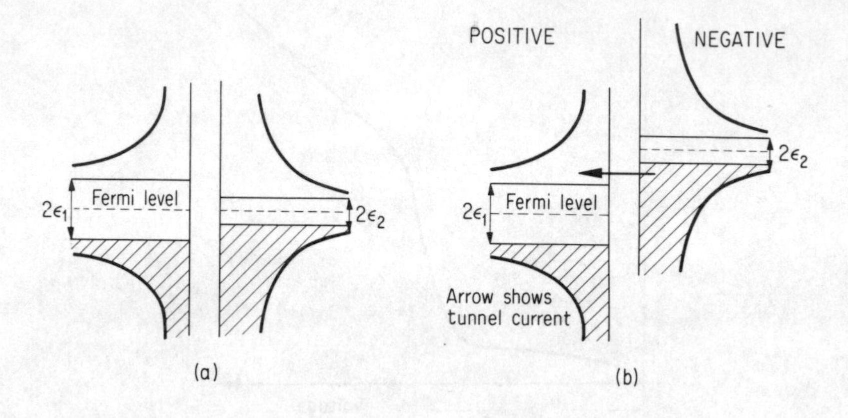

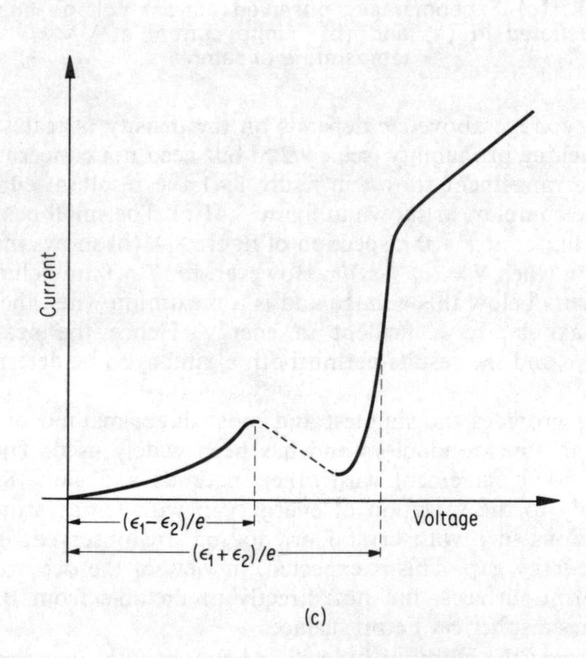

Figure 8.41 (a) As figure 8.40 (a) but for two superconductors. (b) As figure 8.40 (b) but for two superconductors. (c) Experimentally observed current voltage characteristic of structure illustrated in (a) and (b). Current peak at $V = (\varepsilon_1 - \varepsilon_2)/e$ due to finite temperature: at $T = 0$ current would not begin until $V = (\varepsilon_1 + \varepsilon_2)/e$

8.9.7 Quantum effects (AD)

Flux Quantisation and Vortices. In the superconducting ground state, there are, according to the B.C.S. theory, electron pairs. The pairs behave as single entities. If a current is set up these pairs must have a wave function corresponding to a certain momentum $\boldsymbol{p}$ and hence have the form

$$\Psi \exp(\mathrm{i}\boldsymbol{p}\cdot\boldsymbol{r}/\hbar) \tag{8.86}$$

In the presence of a magnetic field such pairs will follow a circular path and the wave function is modified. In order that it shall be single valued and return to the same value after a complete revolution it is necessary for

$$\int \boldsymbol{p}\mathrm{d}\boldsymbol{r} = \mathrm{n}h \tag{8.87}$$

where the integral is around the path, and n is an integer. This is simply the quantisation condition in the Bohr–Sommerfeld theory of atoms.

In the case of a particle of charge q in a magnetic field B, $\boldsymbol{p}$ must be replaced by $\boldsymbol{p} + \mathrm{q}\mathbf{A}$ where $\mathbf{A}$ is the vector potential. $\boldsymbol{p}$ is related to the supercurrent, j ($=\mathrm{q}\boldsymbol{p}/m$) and if the external current is zero, we can take the integral in the bulk superconducting region away from surface currents so that $\mathrm{j} = 0$. Condition (8.87) then becomes

$$\mathrm{q}\int \mathbf{A}\mathrm{d}\boldsymbol{r} = \mathrm{q}\int \mathbf{B}\mathrm{d}\boldsymbol{s} = \mathrm{n}h \tag{8.88}$$

using Stokes theorem and $\mathbf{B} = \boldsymbol{\nabla} \times \mathbf{A}$. Hence the flux through any region must be quantised in integral multiples of $\Phi_o = h/\mathrm{q}$.

We have seen (§ 8.9.3) that type II superconductors in the mixed state have threads of normal material running through them. These threads are surrounded by circular supercurrents. They are called *vortices* and enclose unit flux. The value of h/q turns out to be 2.07×10^{-15} weber corresponding to $\mathrm{q} = 2e$, thus confirming the proposition that it is electron pairs which carry the supercurrent. The vortices have a structure and a finite cross section. The flux falls off in a distance λ from the centre of the thread while the supercurrent is largely confined in a cylinder outside the coherence length ξ which defines the minimum normal region, i.e. it lies in the region $\xi < r < \lambda$ (cf. § 8.9.4 and figure 8.32).

The vortices interact with each other and form a regular array—in fact a triangular lattice of them has been observed by neutron diffraction. The pattern can also be made visible by decoration of the surface of a superconductor in the mixed state with ferromagnetic particles which cling to the ends of the flux lines. The density of vortices increases as B is increased from B_{c1} towards B_{c2}.

Vortices play an important role in the current carrying properties of type II materials in the mixed state. Suppose current of density $\mathbf{J}$ is passing along a wire subjected to a perpendicular magnetic field. Each vortex then experiences a Lorentz force of magnitude $\mathbf{J}\Phi$ per unit length which will try to drive the vortex in a direction perpendicular to both **J** and $\mathbf{B}_0$. In an imperfect specimen the vortices are found to be 'pinned' by impurities and imperfections (cf. the pinning of dislocations, § 5.4.1, and of magnetic domains, § 10.11.2). In this situation a supercurrent can flow in the presence of the magnetic field until the current becomes larger than a critical value at which the force on the vortices is sufficient to overcome the pinning. The vortices will then move but still be impeded by the impurities so that work must be done to sustain their motion. The necessary energy can only be provided by the e.m.f. driving the current and hence there appears to be some resistance in the wire.

Thus in impure type II materials the critical current in a transverse magnetic field can be much larger than in pure specimens. Furthermore, for much of the range $B_{c1} < B_0 < B_{c2}$, the critical current is substantially independent of B_0. These features, which are illustrated in figure 8.42, are important in the fabrication of superconducting solenoids (§ 8.9.8).

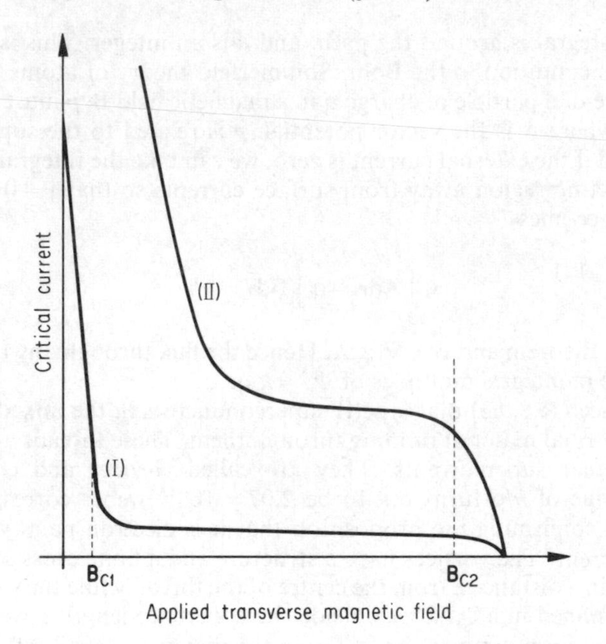

Figure 8.42 Critical current in a typical type II superconducting wire as a function of magnetic field applied at right angles to the current flow. (I) pure sample (II) impure sample

Josephson Junctions and Quantum Interference. In addition to the tunnelling properties discussed in § 8.9.6 some very remarkable effects occur when the insulating barrier between two like superconductors is very thin. If it is sufficiently thin (a few atomic spacings) there is coherence between the wave functions of electron pairs on each side of the barrier and the two superconducting sections are coupled so as to form a single entity. Electrons tunnelling through the barriers can do so as pairs. Hence the barrier region behaves as a superconductor but with a very low critical current. Such barriers in a superconducting circuit are called 'weak links'.

The behaviour of a weak link, then, depends on the phases of the pair wave functions like (8.86) on either side of the link. Specifically Josephson showed that, in the absence of an applied voltage, a supercurrent would flow through a sufficiently thin barrier and the magnitude of the current would be

$$\mathrm{j} = \mathrm{j}_c \sin \delta \tag{8.89}$$

where j_c is the critical current of the weak link and δ the phase difference across it. If now a voltage V is applied so that there is a potential drop $\Delta E = 2\,e\mathrm{V}$ across the barrier, there is an extra phase difference from the term $\exp(\mathrm{i}\Delta Et/\hbar)$ in the wave function. In this case, therefore

$$\mathrm{j} = \mathrm{j}_c \sin(\delta - 2e\mathrm{V}t/\hbar) \tag{8.90}$$

and the supercurrent oscillates with angular frequency $2\,e\mathrm{V}/\hbar$. (In addition there may be a tunnelling current due to 'normal' electrons, as above, but we ignore this here.) The emission of radiation of frequency $2\,e\mathrm{V}/\hbar$ from Josephson junctions has, in fact, been observed experimentally: for $\mathrm{V} = 10^{-4}$ V the frequency is about 50 GHz. This photon emission can be considered to arise from the need for the paired electrons to conserve energy—unless the energy ΔE is emitted, there are no pair states available into which the pairs may tunnel (cf. § 8.9.6 and figure 8.41).

An interesting and potentially important feature of (8.90) is that it relates voltage to frequency only through fundamental constants. Furthermore, frequencies in the microwave range can be measured with great precision and in turn related to atomic (e.g. Cs) standards. Hence Josephson tunnelling may permit improved standardisation of electrical units.

The properties of weak links can also be used to demonstrate electron pair interference and diffraction. Interference effects between two supercurrents can be observed in the experimental arrangement illustrated schematically in figure 8.43, which is analogous to Young's double slit experiment in optics. A current j is split into two (ideally equal) currents, passed through the weak links at A and B and recombined at C. We suppose that j is much less than the critical current everywhere except at the weak links so that significant phase differences occur only across these links (cf. (8.89)). The current at C is then

$$\tfrac{1}{2}\mathrm{j}(\sin\delta_\mathrm{A} + \sin\delta_\mathrm{B}) = \mathrm{j}\sin\tfrac{1}{2}(\delta_\mathrm{A} + \delta_\mathrm{B})\cos\tfrac{1}{2}(\delta_\mathrm{A} - \delta_\mathrm{B}) \tag{8.91}$$

where δ_A and δ_B are the phase differences across the links at A and B. If the links are identical, as we shall assume for simplicity, (8.91) is simply $\mathrm{j}\sin\delta_\mathrm{A}$. In the presence of a magnetic flux, however, a circulating current is induced in the ring so that the current through each link is not the same. δ_A is then not

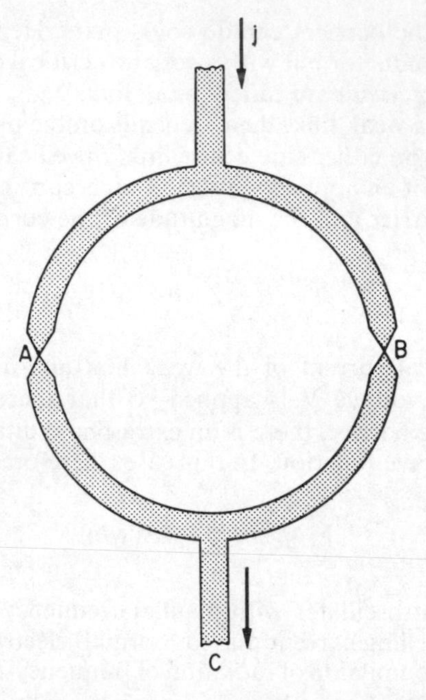

Figure 8.43 Schematic diagram illustrating superconducting circuit in which interference between supercurrents can be observed. A and B are weak links

equal to δ_B. The change in phase around the loop in the presence of a magnetic field of vector potential **A** into the paper is (cf. (8.88))

$$\frac{2e}{\hbar}\int \mathbf{A}\,d\mathbf{r} = \frac{2e}{\hbar}\int \mathbf{B}\,ds = \frac{2e}{\hbar}\Phi \qquad (8.92)$$

where Φ is the flux through the loop. Hence the current at C varies with field as

$$\cos\tfrac{1}{2}(\delta_A - \delta_B) = \cos(\pi\Phi/\Phi_0) \qquad (8.93)$$

where Φ_0 is the magnitude of the flux quantum. The variation in current observed in an experiment of this type is shown in figure 8.44. Note that the period of the oscillation corresponds to a very small field strength.

Devices constructed on the above principles have been christened 'squids' (Superconducting Quantum Interference Devices). They are capable, as figure 8.44 illustrates, of detecting an extremely small magnetic flux and have been used to detect 10^{-13} tesla. So small a field can be generated in a low inductance coil by a small current so that related structures can be used as Superconducting Low Inductance Galvanometers, known as 'slugs'. Since the current circuit can be entirely superconducting the voltage drop across a very small resistance can be observed—a voltage as small as 10^{-14} V has been detected.

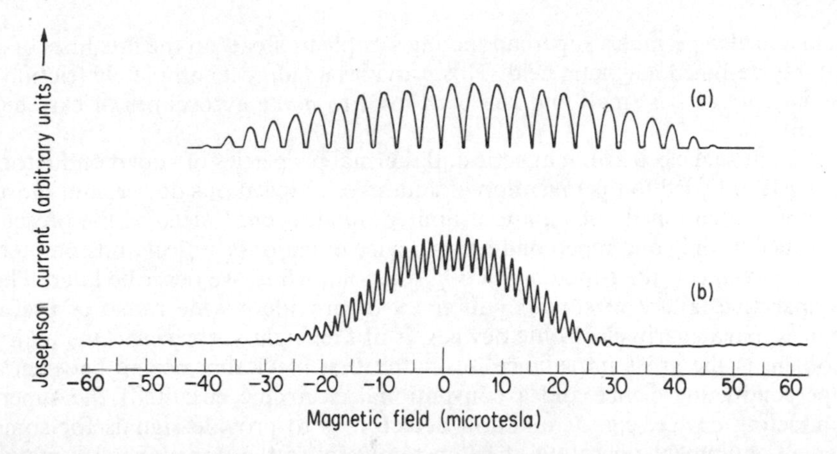

Figure 8.44 Supercurrent interference observed experimentally in a structure based on that of figure 8.43. In the structure used diffraction also occurs and the envelope is a diffraction peak. Structure (a) had a lower inductance than (b) and a steady magnetic field was applied in addition to the variable field displayed on the horizontal axis. Field periodicities for the two structures are (a) 3.95 μT and (b) 1.6 μT respectively. (After R. C. Jaklevic *et al., Phys. Rev.,* **140A**, 1628 (1965).)

8.9.8 The applications of superconductivity (D)

Conductors of zero resistance conjure up visions of lossless electrical power distribution resulting in economies that would more than offset the high capital and running costs implied by the use of sufficiently low temperatures. There are, however, technical advantages in the use of low frequency a.c. for power distribution and the resistance of superconductors is identically zero only at zero frequency. We have seen that flux penetration occurs in type II superconductors above B_{c1} and any change in current causes rearrangement of the flux distribution. This is opposed by pinning forces so that work must be done and as a result type II superconductors show significant hysteresis loss even at 50 Hz. Alternating current losses are much lower in type I superconductors due to the smallness of the penetration depth, λ, but the critical currents of this type are unfortunately too low to be of value in power transmission.

The resistance of type I superconductors is extremely small for all frequencies below the absorption edge at $2\varepsilon/\hbar$ (§ 8.9.6) and this can be exploited at 'laboratory' current densities. For example, microwave resonant cavities with a lead ($T_c = 7$ K) lining can be made with resonator quality factors (Q) as high as 10^9. Type I superconductors are excellent electrical screening materials and for some devices—including 'squids'—the exclusion of all stray external fields is essential for satisfactory operation. Conversely the flux exclusion property of a cylinder of type I material can be used to guide magnetic flux, provided it does not exceed B_c, round a curved path. Flux exclusion also permits a superconducting sample to 'float' on the flux lines of a suitably designed magnetic field. This provides an almost completely frictionless bearing as λ is small and has been used to make gyroscopes of extreme stability.

The unusual electrical, magnetic and thermal properties of superconductors have naturally led to the invention of a number of ingenious devices but most, like those mentioned above, are of limited application. Indeed at the present time there is only one superconducting device of major technical and commercial importance—the superconducting solenoid, which we describe later. The comparative failure of superconductivity to provide a wide range of useful devices, particularly electronic devices, is at first sight surprising. One major problem is the gross impedance mismatch that inevitably occurs between a superconducting device and a conventional electronic circuit. If the superconducting device, e.g. a radiation detector, is to provide signals for some display equipment operating at room temperature the impedance mismatch always degrades the signal to noise ratio, sometimes very seriously. This interface problem can, of course, be avoided if all operations are confined to the superconducting circuit but this is not a sufficient condition for success, as evidenced by the history of the cryotron. Enthusiasm for this device reached a peak about 1960 and has declined steadily ever since, though arrays of cryotrons may yet prove the most economical way of providing a really large ($>10^{10}$ bits) fast access data store.

The cryotron is a superconducting switch (superconducting to normal) which relies on the fact that the current through a type I superconductor of high T_c can generate a sufficiently high magnetic field to destroy the superconductivity of another material of lower T_c without itself switching. In one form of cryotron insulated crossed strips of tin ($T_c = 3.7$ K) and lead ($T_c = 7$ K) are prepared by evaporation to form the switched path and control elements respectively. The device is operated at a temperature just below the lower T_c. Since current flow in either direction can be controlled, the performance resembles that of unipolar transistors (§ 9.3) though, of course, even the 'off' impedance is very low. Cryotrons can perform all the logic and storage operations required in computing with essentially zero power dissipation except when switching. The time taken to switch an element is usually limited by its normal resistance and the inductance of the (evaporated) wiring but is comparable with competitive devices. The low power consumption and low unit production cost should permit a high packing density and the fabrication of large capacity machines. However, the high reproducibility required in manufacture and the inconvenience (e.g. in servicing) of equipment which will only operate at 3 K have proved formidable obstacles.

Unless a superconducting device offers some major advantage not available at room temperature its commercial future is doubtful. At the present time only the superconducting solenoid meets this criterion. Ideally, no power is required to simply maintain a steady field in a solenoid—in practice the power required in a conventional solenoid is quite substantial, and increases roughly as B^2. A magnetic field of, for example, 10 tesla in a reasonable working volume requires a power of the order of 1 MW and this is used only to heat the windings, which must be watercooled. Fields of this magnitude and greater are, of course, of importance for research in solid state, nuclear and plasma physics. High fields are also likely to prove valuable in some branches of electrical engineering, notably power generation. The only power required to maintain a 10 tesla field in a superconducting solenoid is that required for refrigeration, which is measured in kilowatts.

The maximum field attainable by a superconducting solenoid is, of course, limited by the critical field of the material and this is only large in type II superconductors (B_{c2}). We have seen (§ 8.9.7) that there is no simple relationship between critical current and critical field in a type II material when, as in a solenoid, they are at right angles to each other and B_0 lies in the range $B_{c1} < B_0 < B_{c2}$. The critical current is then primarily determined by the average pinning strength of the lattice defects and impurities. The only elemental superconductor of interest for solenoid fabrication is niobium ($T_c = 9.3$ K) and its properties well illustrate the effects of mechanically induced defects—the maximum field that can be reached in a small Nb coil varies between 0.6 and 0.2 tesla, depending on the degree to which the wire is annealed after drawing.

Taking into account mechanical properties and the ease of fabrication into wire or tape the most successful high field superconductors to date are those listed in the table below. The design of solenoids for optimum performance requires quite accurate information on the critical current as a function of field, particularly near B_{c2}, and this has to be found by experiment.

Material	Critical field, B_{c2}, at 4.2 K in tesla	Critical current density at 4.2 K and $B_{c2}/2$, in A cm^{-2}
Nb–25% Zr alloy	7	$\sim 10^5$
Nb–50% Ti alloy	12	$\sim 2 \times 10^5$
Nb_3Sn (compound)	22	$\sim 10^6$

Unless precautions are taken, superconducting solenoids are prone to 'go normal' at currents well below the critical currents listed above due to 'flux jumping'. When in operation there is a force on the vortex threads in the wire which is opposed by the pinning but occasionally the vortices will slip to be repinned a short time later. During flux movement resistance appears in the coil and power is dissipated. Since the pinning strength decreases with temperature the effect can be cumulative and disastrous. The energy stored in the field may be substantial (at 10 tesla the energy density, $B^2/2\mu_0$, is about 40 J cm^{-3}) and if the field collapses more rapidly than the resulting heat can be dissipated to the surrounding liquid helium the coil may melt. Additional heating arises from the mechanical relaxation of the coil and its mount when the field, and hence the electromechanical forces between turns, is reduced.

The problem of flux jumping can be surmounted in two ways: (a) it is found that the frequency of flux jumping at a given current density is much reduced by the use of thin wire, and (b) the effects of a flux jump, when it does occur, can be reduced by cladding the wire with copper and by interleaving copper foil between layers of wire. The two techniques can be used together.

Copper is not a superconductor but at 4.2 K its conductivity is an order of magnitude greater than, for example, Nb–Ti in the normal state. Copper cladding can therefore provide an alternative current path for a section of wire which has gone normal for a brief period and the resultant heating is reduced. For small (~0.5 tesla) Nb solenoids of the type used in conjunction with InSb electronic bolometers (§ 8.7.2) copper cladding alone is sufficient to stabilise the magnet. At higher fields copper interleaving is essential. The effect of interleaving is to reduce the rate of collapse of the magnetic field as eddy currents are induced in the sheets. This gives more time for heat dissipation to the He bath and additionally the presence of the copper improves the thermal contact. However, the use of this technique naturally increases the time required to establish the field after switching on and a run-up time of much more than, say, ten minutes is a serious inconvenience.

The use of copper is a palliative, not a cure. The frequency of flux jumping is much reduced by the use of wire of less than about 10^{-2} cm diameter though it is then necessary, at least in high field magnets, to use multistranded wire to carry sufficient current. Considerable technical effort has been devoted to the fabrication of such wire and multistranded Nb–Ti wire in a copper matrix has recently become commercially available. Copper interleaving can then be reduced, for the same solenoid performance, and run-up times reduced to a few seconds.

At the present time complete solenoids capable of operation up to 10 tesla are available commercially and higher fields can be attained. Except for the largest solenoids it is usual to provide a magnetically or thermally operated superconducting switch in parallel with the coil so that the current, once established, can be switched to circulate indefinitely (subject to flux jumping and maintenance of the bath temperature). The maximum steady field that can be obtained in any solenoid, superconducting or conventional, is set by the mechanical strength of the mounting. From the known yield strength of available materials this limit is around 25 to 30 tesla.

SUPERCONDUCTIVITY IN THE PERIODIC SYSTEM

(The number below the element is the approximate transition temperature in K)

IA	IIA	IIIA	IVA	VA	VIA	VIIA	VIII	VIII	VIII	IB	IIB	IIIB	IVB	VB	VIB	VIIB	0
1																	2
3	4 Be 0·03											5	6	7	8	9	10
11	12											13 Al 1·2	14	15	16	17	18
19	20	21	22 Ti 0·4	23 V 5·3	24	25	26	27	28	29	30 Zn 0·88	31 Ga 1·1	32	33	34	35	36
37	38	39	40 Zr 0·75	41 Nb 9·3	42 Mo 0·93	43 Tc 7·7	44 Ru 0·49	45	46	47	48 Cd 0·5	49 In 3·4	50 Sn 3·7	51	52	53	54
55	56	57 La † α4·8 β5·9	72 Hf ?	73 Ta 4·5	74 W 0·01	75 Re 1·7	76 Os 0·71	77 Ir 0·14	78	79	80 Hg † α 4·2 β 4·0	81 Tl 2·4	82 Pb 7·2	83	84	85	86
87	88	89	90 Th 1·4	91 Pa 1·4	92 U† α 0·6 β 1·8												

† α and β refer to alternative crystal structures of the same element

Periodic Table Periodic Table showing superconducting elements and their approximate transition temperatures

PROBLEMS

8.1 A semiconductor crystal 12 mm long, 5 mm wide and 1 mm thick has a magnetic field of flux density 1 tesla applied perpendicular to the largest faces. When a current of 20 mA flows lengthways through the specimen the voltage measured across its width is found to be 7.4 mV. What is the value of the Hall coefficient?

8.2 Assuming $r_{oe} = r_{oh} = 1$ show that the Hall coefficient of a P type semiconductor is zero when the excess density of acceptors over donors

$$N_A - N_D = n_i\left(\frac{b^2 - 1}{b}\right)$$

where n_i is the intrinsic density and $b = \mu_e/\mu_h$. So estimate, from the data of figure 8.4, the mobility ratio in InSb.

8.3 A semiconductor sample at a low temperature has a Hall coefficient of 3.7×10^{-2} $m^3\ C^{-1}$, a d.c. conductivity of 4.83×10^2 $ohm^{-1}\ m^{-1}$ and a lattice dielectric constant of 16. Cyclotron resonance is observed at 56 GHz and a magnetic field of 0.2 tesla. Estimate the absorption coefficient at 56 GHz in the absence of the magnetic field and the sharpness of the cyclotron resonance.

8.4 Show that, if the momentum scattering time of electrons in a semiconductor varies with electron energy as $E^{-1/2}$, the value of r_0 in equation (8.3) is $3\pi/8$.

8.5 Derive the expression given in § 8.8.2 for the conductivity of N type silicon, namely

$$\sigma = \frac{Ne^2}{3}\left(\frac{\tau_x}{m_x^*} + \frac{2\tau_y}{m_y^*}\right)$$

by adding the contributions of each valley when the electric field makes an angle θ with a (1,0,0) direction.

8.6 For what magnetic field orientation, if any, will the longitudinal magnetoresistance of N type germanium vanish?

8.7 Noting that microwave absorption (provided $\omega\tau \ll 1$) is proportional to the slope conductivity, deduce the magnitude of F_0 in equation (8.55) from the data of figure 8.19.

8.8 A semiconductor bar is illuminated at one end by a parallel beam of infrared radiation. Assume that only free carrier absorption is significant and that the photons transfer their momentum to the free carriers with an efficiency of 100%. Neglecting diffusion, show that under open circuit conditions a field is built up in the bar to oppose the motion and that, if reflection is neglected, the voltage developed across a bar of length $\mathscr{L}$ is given by

$$V = \frac{I[1 - \exp(-K\mathscr{L})]}{cNe}$$

where I is the incident power density, c the velocity of light, N the carrier density, e their charge and K the absorption coefficient. Then show, if reflection at both ends of the bar is included, that the above expression is multiplied by the factor

$$\frac{1 - \mathscr{R}}{1 + \mathscr{R}\exp(-K\mathscr{L})}$$

8.9 Suppose that the free carriers of problem 8.8 above were electrons in a simple parabolic band. Would the assumption of 100% efficiency then be justified?

8.10 The resistivity of copper at 273 K is 1.55×10^{-6} ohm cm. Estimate the thermal conductivity at the same temperature and compare your result with figure 8.15.

8.11 The Hall coefficient of copper at 273 K is 5.5×10^{-11} m^3 C^{-1}. Using the resistivity given in problem 8.10 above and figure 8.3 estimate the temperature at which $\omega_c\tau$ would equal unity in a magnetic field of 10 tesla.

8.12 The velocity of sound in cadmium is 2307 ms^{-1}. Assuming the Fermi surface to be spherical, deduce its radius from the data of figure 8.13.

8.13. Boundary scattering of electrons has been observed in thin wires of lithium at 4.2 K. From this data the mean free path of electrons in bulk lithium can be estimated to be 1.6×10^{-2} cm. At the same temperature the bulk resistivity is 7.0×10^{-10} ohm cm and the electron density from Hall data 3.7×10^{22} cm^{-3}.

Calculate the electron wave vector at the Fermi surface, assuming this to be spherical, and so the ratio of k_F to the shortest distance to the boundary of the Brillouin zone.

8.14 From § 8.5.2 we can deduce Matthiessen's empirical rule, namely that the resistivity of a metal is the sum of the residual resistivity ρ_r and the ideal (zero impurity) resistivity ρ_i. In practice the addition of impurities often affects the Debye temperature.

Copper containing 0.78 atomic per cent of As impurity has a residual resistivity of 4.83×10^{-6} ohm cm. The Debye temperature of this alloy is 313 K. The resistivity of pure copper at 50 K is 4.55×10^{-8} ohm cm.

Show that the resistivity of the alloy, according to the Bloch–Grüneisen formula, is 0.27% greater than that predicted by Matthiessen's rule.

8.15 Estimate the superconducting energy gaps, and hence the position of the far infrared absorption edges, of Ta, In, Hg and Nb at $T = 0$. Compare your estimates with figure 8.39. If the comparison suggests that a re-examination of the far infrared absorption of superconductors would be desirable, what techniques developed since 1960 could be applied in the hope of obtaining better data?

9

Semiconductor junction devices

The commercial interest in solid state physics has been engendered primarily by semiconductor devices. An important feature of semiconductors is the comparative ease with which the electrical characteristics can be adjusted by the chemical addition of impurities (§ 5.3.1). Given suitably doped materials the density of free electrons and/or holes can be further varied over quite wide limits *electronically*. We have seen that excess electrons and holes can be generated in semiconductors by the absorption of photons of sufficient energy (Chapter 6). We shall show in this chapter how excess charge carriers can be injected at PN junctions. Naturally, such excess electrons and holes recombine but in some semiconductors the recombination rate is sufficiently slow to allow us to maintain an excess, non-equilibrium, population with the expenditure of very little power. In this chapter we shall be concerned with non-equilibrium conditions and we begin by considering how a carrier density disturbance (excess or deficiency) moves in a semiconductor.

9.1 Bipolar transport

9.1.1 Recombination and generation (d)

Consider an N type bar of some semiconductor (e.g. silicon). In equilibrium the free electron and hole densities are determined by the position of the Fermi level (§ 4.8.2) but we can immediately write for the conductivity, σ, from (8.1)

$$\sigma = e[N_0 \mu_e + p_0 \mu_h] \tag{9.1}$$

where

$$N_0 p_0 = n_i^2$$

from (5.9). In this chapter a capital letter (N) indicates the density of the carriers in the numerical majority (electrons in this case) and a lower-case letter a minority carrier density. The subscript $_0$ is used to indicate thermal equilibrium values. n_i is the intrinsic density and μ_e, μ_h the mobilities (Chapter 8) of electrons and holes respectively.

Equation (9.1) describes an equilibrium, but not a static, situation. In thermal equilibrium electrons and holes are recombining and being generated all the time. Though these two rates are necessarily equal on average, statistical fluctuations will occur. It is convenient to define the rates of recombination and generation in terms of the average *lifetime* of a minority carrier. This lifetime will usually be determined by recombination *via* impurities (see § 5.6.5), the nature of which do not concern us at the moment.* Its value can be determined experimentally by measurement of the decay of photoconductivity after illumination by radiation with $\hbar\omega > E_g$. If the illumination produces only a small excess of holes, Δp (and necessarily an equal excess of electrons, ΔN), the recombination rate, r, will be given by

$$\mathrm{r} - \mathrm{g} = -\frac{\mathrm{d}\,\Delta p}{\mathrm{d}t} = -\frac{\mathrm{d}\,\Delta N}{\mathrm{d}t} = C_0(N_0 + \Delta p)(p_0 + \Delta p) - \mathrm{g}$$

where C_0 is some constant and g is the thermal generation rate. Since $\mathrm{d}\Delta p/\mathrm{d}t = 0$ when $\Delta p = 0$, $\mathrm{g} = C_0 N_0 p_0$. If Δp is small, $\Delta p \ll N_0$ and the decay is exponential,

$$\Delta p = \Delta N \propto \exp(-t/\tau_h)$$

where $1/\tau_h = C_0 N_0$ so that

$$\mathrm{g} = p_0/\tau_h = n_i^2/N_0 \tau_h \tag{9.2}$$

and g = r in thermal equilibrium. τ_h is the average lifetime of the minority carriers, in this case, holes. In semiconductors of technological importance the minority carrier lifetime is of the order of 10^{-6} s or greater. Hence a charge carrier suffers many millions of collisions (Chapter 8) before recombination.

* Deathnium, see § 6.7.1.

The carrier recombination rate at the surface of a crystal is often greater than in the bulk. It will prove convenient to describe recombination at a surface by a surface recombination velocity, S_v, the rate of recombination being $S_v \Delta p$ per unit area. A surface in this context is not limited to the physical surface of a crystal: it will be convenient to apply the same terminology to a junction between two portions of the same crystal. We shall find that the effective value of S_v at the boundary between P and N regions is very large, between N and N^+ (heavily doped N type) very small.

9.1.2 Charge transport (d)

Free electrons and holes drift when a field **F** is applied and diffuse in the presence of a concentration gradient. The electron and hole current densities in a given direction are then given by the combined effects of field and concentration gradient:

$$\mathbf{J}_e = e[N\mu_e \mathbf{F} + D_e \operatorname{grad} N]$$

and

$$\mathbf{J}_h = e[p\mu_h \mathbf{F} - D_h \operatorname{grad} p] \tag{9.3}$$

where D_e and D_h are the diffusion constants of electrons and holes, respectively, and we have arbitrarily assumed that the material is N type. In general $\mu_e \neq \mu_h$ due to differences in effective mass and scattering time. The alternation in sign in the second equation (9.3) arises because diffusion takes place in the same direction regardless of charge (i.e. down the concentration gradient) while field-induced drift and the resultant current reverse on changing from electrons to holes. We note in passing that since the current must be zero in equilibrium it follows that there exists a relationship between mobility, μ, and diffusion constant, D. To find this is left as a problem: the solution is the Einstein relation, referred to in § 5.2.3, namely

$$D_e = \frac{\mu_e kT}{e} \quad \text{and} \quad D_h = \frac{\mu_h kT}{e}$$

Particle flow is also constrained by the requirement of continuity, i.e. particle conservation. In any small element of volume the rate of change of particle density will be determined by generation minus recombination and the divergence of the current. Hence

$$\left.\begin{aligned} \frac{dN}{dt} &= g - r + \frac{\operatorname{div} \mathbf{J}_e}{e} \\ \text{and} \qquad \frac{dp}{dt} &= g - r - \frac{\operatorname{div} \mathbf{J}_h}{e} \end{aligned}\right\} \tag{9.4}$$

where the minus sign in the second equation arises from the reversal in sign of the charge. Equations (9.3) and (9.4) determine the flow of holes and electrons.

As in this chapter we shall be concerned with the motion of an excess number (or deficiency) of carriers above or below equilibrium numbers we can always write $N = N_0 + \Delta N$, $p = p_0 + \Delta p$, etc. The excess electron and hole density may lead to a space charge and consequently, from Poisson's equation, a field F_s given by

$$\operatorname{div} \mathbf{F}_s = \frac{e}{\varepsilon}(\Delta p - \Delta N) \tag{9.5}$$

where ε is the dielectric constant. The insertion of (9.5) into (9.3) and (9.4) leads to an intractable set of equations so that some approximation must be made. The most usual, and the most useful, approximation is that of quasi-neutrality, namely to assume that $\Delta p = \Delta N$ everywhere. This is justified on the grounds that, if it were not so, the space charge would generate a field large enough to restore neutrality. With this approximation we obtain, after some manipulation of (9.3) and (9.4), a single equation to describe the motion of excess electrons or holes in a semiconductor, namely

$$\frac{\mathrm{d}\,\Delta N}{\mathrm{d}t} = \frac{\mathrm{d}\,\Delta p}{\mathrm{d}t} = \mathrm{g} - \mathrm{r} + \mu^* \mathrm{F}.\operatorname{grad} \Delta p + D^* \operatorname{div}\operatorname{grad} \Delta p \tag{9.6}$$

which is known as Van Roosbroeck's continuity equation. We should note that the equation assumes μ to be independent of field, F, which is not valid at high electric fields (§ 8.7). The effective mobility μ^* and diffusion constant, D^*, are given in the general case (p not necessarily small) by

$$\mu^* = \frac{P - N}{(N/\mu_h) + (P/\mu_e)} \tag{9.7}$$

and

$$D^* = \frac{N + P}{(N/D_h) + (P/D_e)} \tag{9.8}$$

If the material is strongly N type or P type and the disturbance in carrier density (ΔN or Δp) is small, (9.7) and (9.8) immediately simplify. Thus

In N type material, $N > P = p$: $\mu^* = -\mu_h$, $D^* = D_h$
In P type material, $P > N = n$: $\mu^* = \mu_e$, $D^* = D_e$

We conclude that a density disturbance moves, as a whole, with a velocity determined solely by the minority carriers. In intrinsic material ($N = P$) the effective mobility of excess carriers is zero (see (9.7)). This state of affairs is a direct consequence of the assumption of space charge neutrality. When the material is strongly N or P type there are sufficient carriers to smooth out the space charge fields and consequently the effective mobility reduces to that of the minority carrier. For simplicity, we assume this below. The predictions of (9.6) can be tested experimentally in a variety of ways. One direct method is the drift mobility experiment described in § 8.1.2. Here the motion of a small pulse of excess carriers can be observed directly. The predicted variation in effective mobility and diffusion constant with carrier concentration has been confirmed.

9.1.3 Diffusion flow (d)

A particularly simple application of (9.6), which will be useful later, arises when the flow of carriers is restricted to one dimension and is entirely by diffusion. Consider an N type rod of length w terminated at $x = w$ by a boundary of recombination velocity S_v. Excess holes are generated at $x = 0$ by light or by some other means. Because of this excess there is a concentration gradient and diffusion in the x-direction. Putting $(g - r) = -\Delta p/\tau_h$ from § 9.1.1 and assuming steady state conditions with $F = 0$, (9.6) reduces to the diffusion equation

$$\frac{d^2 \Delta p}{dx^2} = \frac{\Delta p}{L_h^2} \tag{9.9}$$

where $L_h = (D_h \tau_h)^{1/2}$ and is called the *diffusion length* of holes. Equation (9.9) may be solved by inserting the boundary conditions

$$\Delta p = \Delta p(0) \quad \text{at} \quad x = 0$$

and

$$-D_h\left(\frac{d\,\Delta p}{dx}\right) = S_v \Delta p \quad \text{at} \quad x = w$$

The solution of (9.9) is then

$$\frac{\Delta p}{\Delta p(0)} = \cosh X - \sinh X\left[\frac{\sinh W + (S_v L_h/D_h)\cosh W}{\cosh W + (SL_h/D_h)\sinh W}\right] \tag{9.10}$$

where $X = x/L_h$ and $W = w/L_h$. Certain special cases of this solution to (9.9) are useful, as follows:

(a) $W \to$ infinity. Then

$$\Delta p = \Delta p(0)\exp(-x/L_h) \tag{9.11}$$

and the hole current density, from (9.3), is then given by

$$-J_h = eD_h\left(\frac{d\,\Delta p}{dx}\right) = e\,\Delta p(0)\left(\frac{D_h}{L_h}\right)\exp(-x/L_h) \tag{9.12}$$

(b) S_v large ($\gg D_h/L_h$). Then

$$J_h = -e\,\Delta p(0)\left(\frac{D_h}{L_h}\right)[\sinh X - \cosh X \coth W]$$

so that the ratio, β, of the hole currents at $x = 0$ and $x = w$, which we will want to know later, is

$$\beta = \frac{J_h(x = w)}{J_h(x = 0)} = \operatorname{sech} W$$

$$\approx 1 - \frac{W^2}{2} \qquad \text{for} \qquad w \ll L_h \tag{9.13}$$

(c) S_v small ($\ll D_h/L_h$) and $w \ll L_h$.

$$\Delta p = \Delta p(0)\left[1 - \frac{wx}{L_h^2}\right] \tag{9.14}$$

We note also that if $\Delta p(0)$, the excess density at $x = 0$, is a sinusoidal function of time, namely

$$\Delta p(0) = \Delta p'(0)[1 + \eta \exp(i\omega t)]$$

where η is some constant less than unity, the solutions given above are unaltered except that the diffusion length, L_h, is replaced by L_h' where

$$L_h' = L_h/(1 - i\omega t)^{1/2} \tag{9.15}$$

9.1.4 Numerical values (d)

We conclude this section with some numerical values for lightly doped material at room temperature (300 K) in the mixed practical–CGS units normally used in semiconductor device physics.

Germanium	*Silicon*
$n_i = 2.4 \times 10^{13}\ \text{cm}^{-3}$	$n_i = 1.5 \times 10^{10}\ \text{cm}^{-3}$
$\mu_e = 3900\ \text{cm}^2\ \text{v}^{-1}\ \text{s}^{-1}$	$\mu_e = 1350\ \text{cm}^2\ \text{v}^{-1}\ \text{s}^{-1}$
$\mu_h = 1900\ \text{cm}^2\ \text{v}^{-1}\ \text{s}^{-1}$	$\mu_h = 480\ \text{cm}^2\ \text{v}^{-1}\ \text{s}^{-1}$
$D_e = 100\ \text{cm}^2\ \text{s}^{-1}$	$D_e = 35\ \text{cm}^2\ \text{s}^{-1}$
$D_h = 49\ \text{cm}^2\ \text{s}^{-1}$	$D_h = 12\ \text{cm}^2\ \text{s}^{-1}$
τ_e, τ_h: typically in range 10^{-6} s to 10^{-3} s	τ_e, τ_h: typically in range 10^{-6} s to 10^{-4} s

9.2 PN junctions

9.2.1 The electrostatic properties of PN junctions

Though PN junctions are not made by simply putting P and N material together (as explained in § 5.5.4), the electrostatic conditions existing at a junction between P type and N type material can be determined by supposing that a piece of N and a piece of P material are brought into intimate contact. Electrons immediately flow into the P material, recombine with the free holes and produce a space charge (negative in the P material, positive in the N) which inhibits further flow. Assuming a one-dimensional model with an abrupt change from N to P at $x=0$ the electric potential, V, is given by Poisson's equation:

$$\frac{d^2V}{dx^2}=\frac{-e}{\varepsilon}(N_D+p-N) \qquad \text{for} \qquad 0<x<\ell_N \quad \text{(N side)}$$

and (9.16)

$$\frac{d^2V}{dx^2}=\frac{e}{\varepsilon}(N_A+n-P) \qquad \text{for} \qquad -\ell_P<x<0 \quad \text{(P side)}$$

where N_D and N_A are the densities of charged donor and acceptor centres, on each side respectively; p, N, n, P are free-carrier concentrations and ℓ_N, ℓ_P, are the thicknesses of the space charge regions in the N and P regions respectively. We assume in what follows that the ionisation energies of both donors and acceptors are small compared with kT so that $N_D \approx N_0$ and $N_A \approx P_0$, the thermal equilibrium carrier densities. It is not possible to obtain explicit analytic solutions of (9.16) since the carrier concentrations are functions of position, x. Numerical solutions have been found. To proceed analytically we assume that, within the space charge region, the electric field is so high as to exclude all mobile carriers. This will be a good approximation except at the edges of the space charge region provided the current through the junction is not too large. Electron and hole drift velocities are roughly independent of electric field at high fields (§ 8.7) and are about 10^7 cm s^{-1}. The space charge of the free carriers will therefore be important when $J \sim evN_D$ or evN_A where $v \sim 10^7$ cm s^{-1}; typically 10^2 A cm^{-2} or greater. Hence at moderate currents we can neglect the free carrier terms in (9.16) and assume that all the space charge resides at the donors and acceptors. We can now solve (9.16) subject to boundary conditions:

$$\frac{dV}{dx}=0 \qquad \text{at } x=\ell_N \text{ and } x=-\ell_P$$

$$\frac{dV}{dx} \text{ continuous at } x=0$$

and

$$V(x=\ell_N)-V(x=-\ell_P)=V_B$$

where V_B is called the barrier or potential step height. When the externally applied voltage is zero $V_B=V_{BO}$ where V_{BO} is the difference in potential between the Fermi levels on the P and N sides. Otherwise $V_B=V_{BO}\pm|V_J|$ where V_J is the voltage across the junction due to an external potential (+ sign if N side positive). Equation (9.16) can in principle be solved for any simple distribution of donors and acceptors as a function of x but the most common are (a) a step

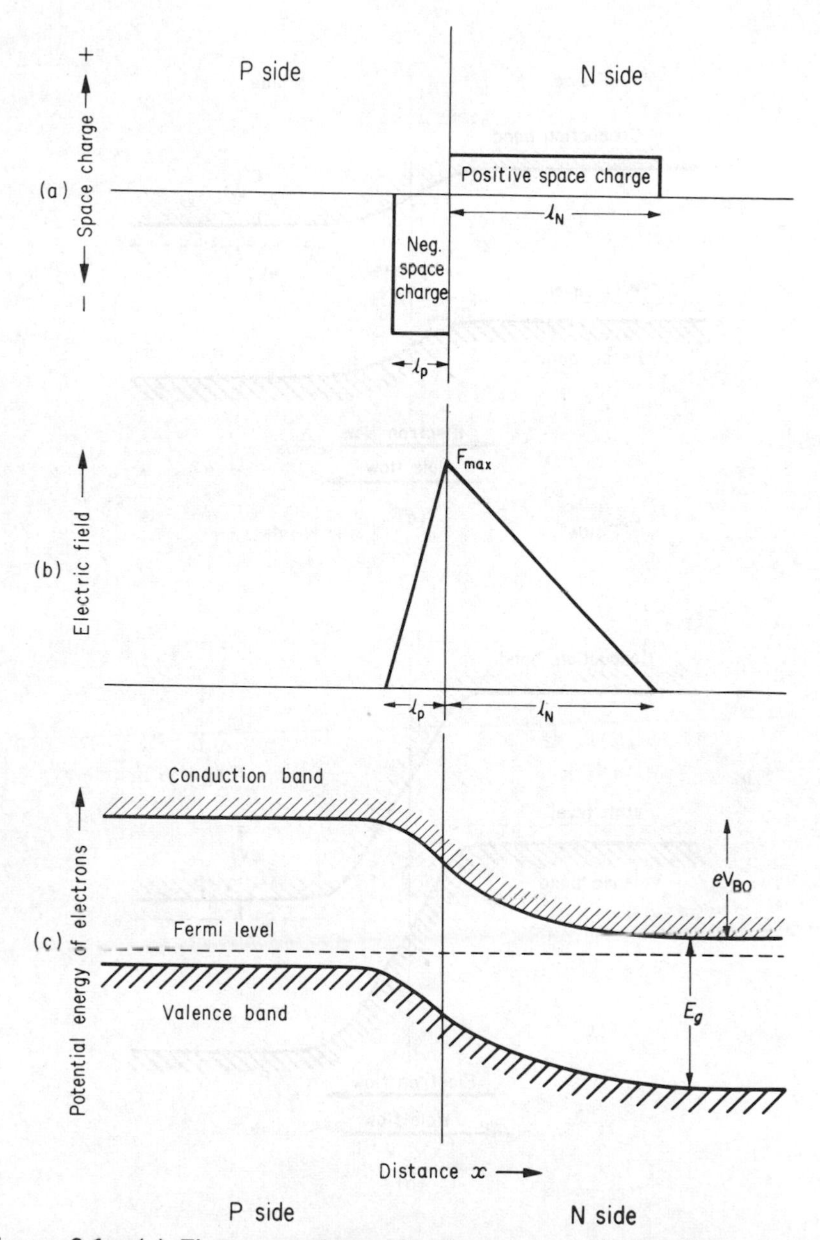

Figure 9.1 (a) The space charge distribution at a PN junction neglecting the effect of mobile carriers. In this approximation $\ell_N N_D = \ell_P N_A$
(b) The magnitude of the electric field at a PN junction (proportional to the integral of the space charge density with respect to distance)
(c) Potential energy of electrons at a PN junction (integral of field with respect to distance)
All curves assume that no external bias voltages are applied

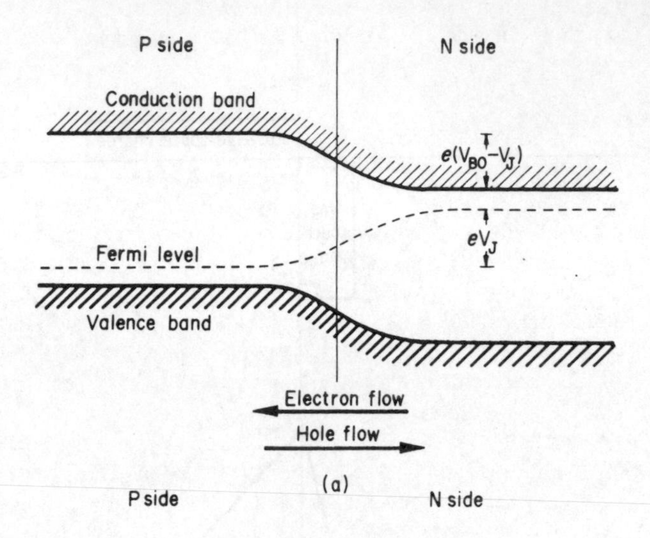

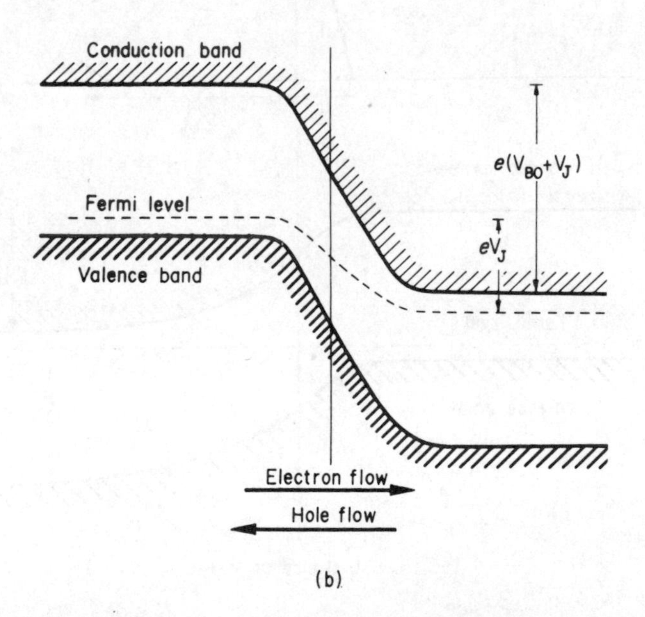

Figure 9.2 Electron potential energy as a function of position near a PN junction when (a) a forward bias of V_J volt is applied and (b) a reverse voltage of V_J volt is applied. Note the position of the Fermi levels and the directions of charge carrier flow

junction in which N_A and N_D are constants, independent of x, and change discontinuously at $x = 0$, and (b) a graded junction such that

$$\frac{d(N_D - N_A)}{dx} = a \qquad \text{for} \qquad -\ell_P < x < \ell_N$$

where a is some constant. The former applies to alloyed and the latter is often a good description of diffused junctions (see § 5.5.4). The distribution of space charge, electric field and potential energy of electrons ($= e$V) is shown for a step junction in figures 9.1(a), (b) and (c). Figure 9.2 shows how the barrier height varies when external bias is applied. Note that the edges of the space charge region, etc., are sharp as a direct result of the neglect of the free-carrier space charge: in reality they would be rounded off as in figure 9.1(c). The properties of PN junctions of most concern to us here are the barrier height, V_B, the maximum value of the electric field, F_{max}, and the total space charge width, ℓ $(= \ell_N + \ell_P)$. Expressions for these quantities, deduced from (9.16), are given in the table below. Since the space charge region is substantially devoid of mobile carriers (indeed, it is assumed to be totally devoid) the space charge region acts like a block of dielectric between semiconducting electrodes. Its capacitance per unit area, C_B, is also included in the table as $C_B = \varepsilon/\ell$.

Although the step in electrostatic potential, and hence in the energy bands, shown in figure 9.1(c) can be derived directly from the above space charge consideration, it is instructive to consider how one could start from the potential step and then work back to the space charge. We could have argued that a potential step large enough to ensure zero charge-carrier flow at zero applied voltage would be necessary. Considering the electrons only, their density far from the junction is n_0 on the P side and N_0 on the N side. For zero electron flow

$$(n_0/N_0) = \exp(-e V_{BO}/kT) \tag{9.17}$$

where V_{BO} is the potential step height and is equal to V_B when $V_J = 0$. Using (9.1) and writing $N_0 = N_D$ and $P_0 = N_A$ (i.e. assuming neither side near intrinsic) we obtain immediately

$$V_{BO} = \frac{kT}{e} \log_e \left(\frac{N_A N_D}{n_i^2} \right)$$

which is the expression given in the table, since the same argument applies to holes. This derivation also implies that, at zero bias, the Fermi levels in both N and P type material must be at the same potential, as shown in figure 9.1(c).

	Abrupt junction	Graded junction
Impurity concentration gradient	$\dfrac{d(N_D - N_A)}{dx} = \infty$ at $x=0$; $= 0$ at $x \neq 0$	$\dfrac{d(N_D - N_A)}{dx} = a$ for $-\ell_P < x < \ell_N$; $= 0$ otherwise
V_B	$\dfrac{kT}{e} \ln \left(\dfrac{N_A N_D}{n_i^2} \right) + V_J$	As abrupt junction
ℓ	$\left[\dfrac{2\varepsilon V_B (N_A + N_D)}{e N_A N_D} \right]^{1/2}$	$\left[\dfrac{3\varepsilon V_B}{2ea} \right]^{1/3}$
$F_{max}(x=0)$	$2V_B/\ell$	$3V_B/4\ell$
C_B	$\left[\dfrac{e\varepsilon (N_A N_D)}{2V_B (N_A + N_D)} \right]^{1/2}$	$\left[\dfrac{2ea\varepsilon^2}{3V_B} \right]^{1/3}$

Table 9.1

9.2.2 The applications of PN junction capacitance (d)

Before continuing with a discussion of conduction through PN junctions we note from the previous section and table that PN junctions can behave as voltage-variable capacitors. The d.c. conductivity of the junction obviously provides a parallel conduction path but this can be neglected at sufficiently high frequencies. Inspection of the table shows that the form of variation of capacitance with voltage is a function of the spatial distribution of donors and acceptors near the junction so that its measurement can give information about the distribution. Furthermore, to a limited extent impurity distributions can be 'tailored' to give a particular variation of capacitance with voltage.

PN junctions are used as variable capacitances in two main areas of application: microelectronics and parametric devices. We have seen (§ 5.5.4) that diffusion techniques permit the fabrication of numerous PN junction diodes and transistors on a single slice of semiconductor. Some of the junctions can be used as variable capacitors although the magnitude of the capacitance achievable is rather limited. More important, perhaps, is the use of PN junctions as voltage dependent (and consequently non-linear) capacitors for frequency mixing, harmonic generation and parametric amplification. As isolated components, these devices play the same role as the distributed non-linear dielectric discussed in § 6.8 on non-linear optics. Thus a convenient way of providing power in the gigahertz region and above is to use a high-power transistor oscillator at, say, 100 MHz and multiply up the frequency in a chain of PN junction diodes used as second harmonic generators. Quite high efficiencies can be obtained. Variable capacitance diodes can also provide the non-linear element of a tuned circuit and hence parametric amplifiers can be made following the principles discussed in § 6.8. An ideal—that is a purely reactive and lossless—variable capacitance generates no electrical noise, since noise is a function only of resistive elements. PN junction diodes, particularly if cooled to, say, liquid nitrogen temperature (77 K) provide extremely low noise parametric amplifiers for use at microwave frequencies and have found application in radio astronomy and satellite communication. It should perhaps be added that fairly sophisticated electronic techniques, beyond the scope of this book, are involved and the brevity of the above account is not indicative of the technical problems encountered.

9.2.3 The conductance of PN junctions (d)

When discussing current flow through PN junctions it is convenient to assume, at least initially, (a) that any electrical contacts to the P and N regions are many diffusion lengths (§ 9.1.3) from the junction and (b) that all the voltage applied externally appears across the junction space charge region. As a consequence of the latter assumption the electric field outside the junction region is zero and all carrier flow towards or away from the junction is by diffusion.

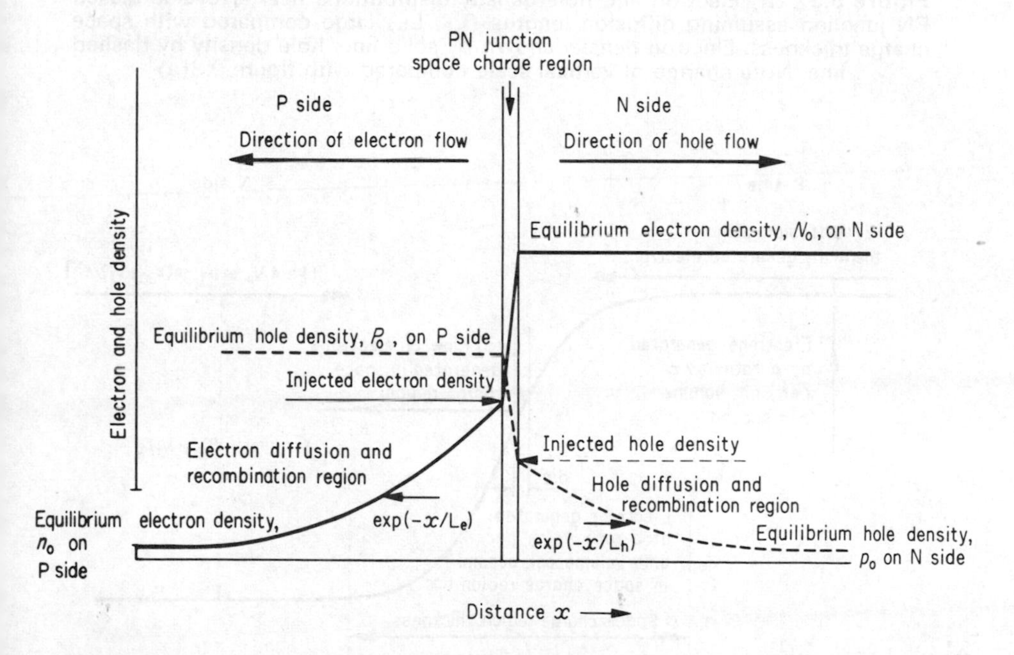

Figure 9.3 (a) Electron and hole density distributions near a forward biased PN junction assuming diffusion lengths (L_e, L_h) large compared with space charge thickness. Electron density shown by solid line, hole density by dashed line. Note that electrons are injected into the P side and holes into the N side, where they move primarily by diffusion. The magnitude of the bias voltage determines the injected densities at the edge of the space charge region. At these surfaces injected densities are approximately $P_0 \exp[e(V_J - V_{BO})/kT]$ and $N_0 \exp[e(V_J - V_{BO})/kT]$ respectively

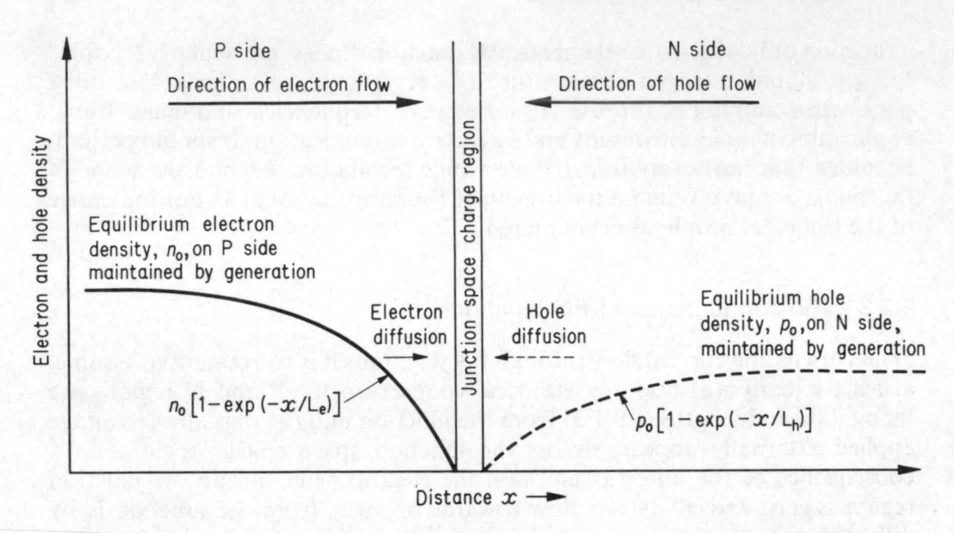

Figure 9.3 (b) Electron and hole density distributions near a reverse biased PN junction assuming diffusion lengths (L_e, L_h) large compared with space charge thickness. Electron density shown by solid line, hole density by dashed line. Note change of vertical scale compared with figure 9.3(a)

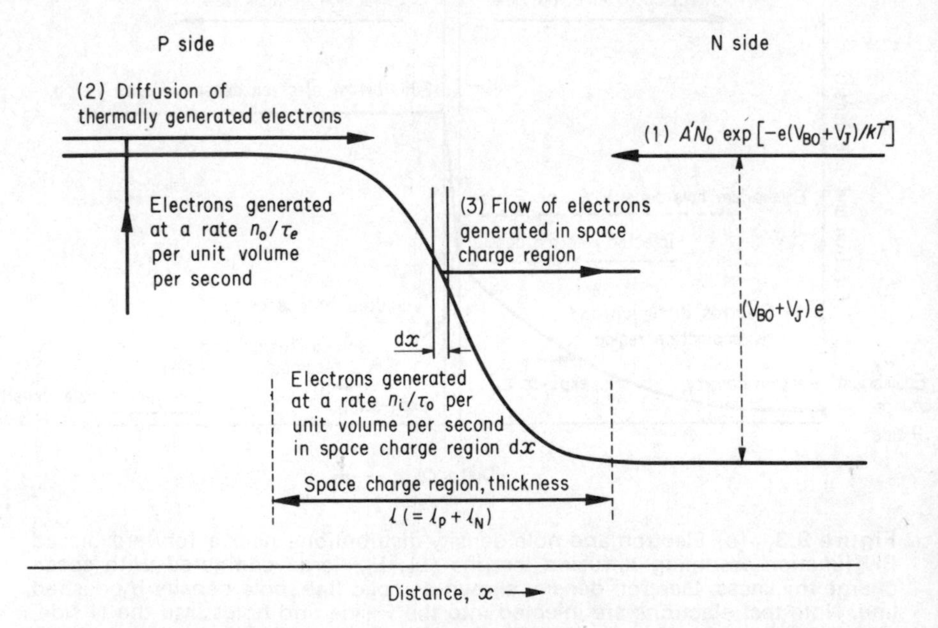

Figure 9.4 The sources contributing to the current flow in a PN junction. The numbers (1), (2) and (3) refer to equations (9.18), (9.19) and (9.20) in the text

The conductance of a PN junction is non-reciprocal. We shall find that its conductance is greatest when the N side is made negative with respect to the P side. This is called the forward direction and the reverse polarity the reverse direction. Forward bias has the effect of reducing the potential step (figure 9.2(a)) between the N and P sides and hence facilitates the flow of electrons into the P material and holes into the N side. We conclude that the current in the forward direction is carried by a flow of *injected minority carriers*. From our assumptions and (9.12), which applies to this situation, we conclude that current flow is due to diffusion and recombination. This is illustrated by figure 9.3(a).

When the polarity of the applied voltage is reversed V_B is increased. Electron flow into the P side and hole flow into the N side is reduced to a negligible value. Minority carriers, however, can still flow. Thus electrons generated (thermally, optically or otherwise) in the P material which diffuse to the junction are swept by the field into the N side. Similarly, minority holes generated in the N side may diffuse to the junction and into the P side. We conclude that the reverse current flow is due to generation and diffusion of minority carriers. The mechanism is illustrated in figure 9.3(b).

To put the above on to a quantitative basis, we consider the three sources of electron flow across a PN junction as indicated in figure 9.4. Hole flow is similar, with suitable alternation of subscripts.

(a) There is electron flow from N to P, proportional to the number of electrons on the N side with energies great enough to surmount the potential step This contribution is given by

$$J_{(N\to P)} = A' N_0 \exp[-e(V_{BO}+V_J)/kT] \qquad (9.18)$$

where A' is some constant to be determined.

(b) In the dark electrons are generated at a rate n_0/τ_e per unit volume on the P side (cf. (9.2)). If generated a distance x from the junction they have a probability

$$\exp(-x/L_e)$$

of reaching the edge of the junction by diffusion, where L_e is the electron diffusion length. These electrons contribute an electron current from P to N given by

$$J_{(P\to N)} = e\int_0^\infty \frac{n_0}{\tau_e}\exp(-x/L_e)\,dx \qquad (9.19)$$

(c) Electrons generated within the space charge region are swept by the electric field into the N side. Neglecting the few such electrons that climb up the barrier to enter the P side, this contributes a current from P to N of

$$J_{(P\to N)} = e\int_0^\ell \frac{n_i}{\tau_0}\,dx \qquad (9.20)$$

where τ_0 is the electron lifetime in the space charge region, which will in general not be equal to τ_e.

The net current from P to N, then, when V_J is positive (reverse bias) is given by the sum of (9.19) and (9.20) minus (9.18). The constant A' can be found from the condition that $J = 0$ when $V_J = 0$ and using (9.17) we obtain for the *electron* contribution to the reverse current:

$$J(\text{reverse}) = J_0[1 - \exp(-e V_J/kT)] \tag{9.21}$$

where

$$J_0(\text{electrons}) = e n_i \left[\frac{n_i D_e}{N_A L_e} + \frac{\ell}{\tau_0} \right] \tag{9.22}$$

An exactly similar expression holds for holes with appropriate changes of subscripts and the total current is the sum of the electron and hole contributions.

To a first approximation, J_0 may be taken as independent of voltage. If this is assumed, (9.21) shows that the reverse current saturates (i.e. becomes

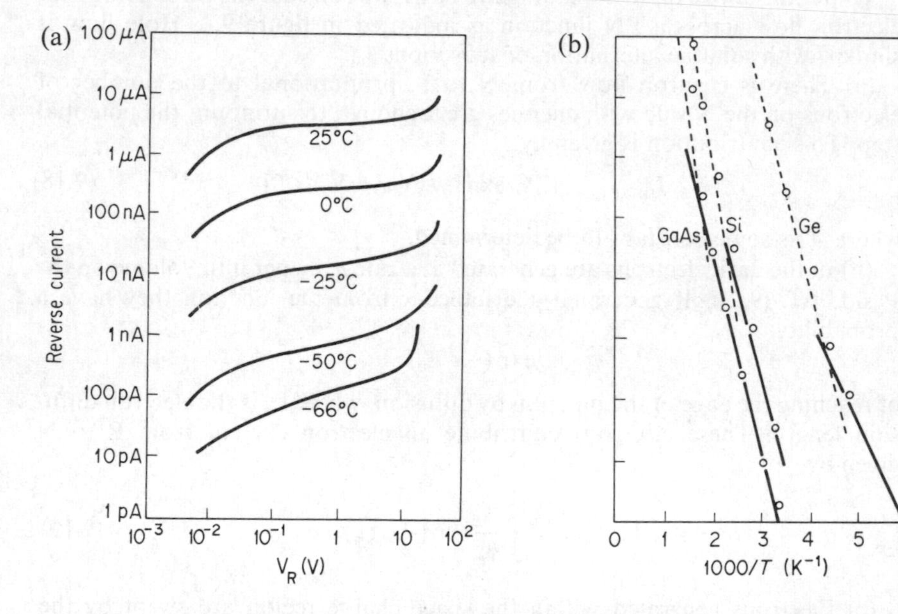

Figure 9.5 (a) The reverse current of a germanium PN junction as a function of voltage at various temperatures. The rise in reverse current at high voltages is due to the onset of avalanche breakdown (see § 9.2.4). Note that saturation (a substantially constant current over an extended voltage region) only occurs at the higher temperatures
(b) The temperature dependence of the reverse current of representative Ge, Si and GaAs PN junctions at a reverse voltage of 1 V. Dashed lines represent a temperature dependence proportional to n_i^2, solid lines proportional to n_i. Compare equation (9.22) and accompanying text. (After A. S. Grove, *Physics and Technology of Semiconductor Devices*, Wiley (1967).)

independent of voltage) when $eV_J \gg kT$. In fact J_0 increases slowly with voltage as ℓ increases with voltage (see table 9.1) and ideal saturation does not occur. At high temperatures, however, the term containing n_i will be the largest and J_0 very nearly independent of voltage. Furthermore, the temperature dependence of J_0 will change from being proportional to n_i at low temperatures ($n_i \propto \exp(-E_g/2kT)$, cf. (5.8)) to being proportional to n_i^2 at high temperatures, the changeover temperature being a function of E_g. These features are illustrated by some experimental results given in figure 9.5. Notice that the term in ℓ/τ_0 is quite negligible in germanium at room temperature because E_g is small but it cannot be neglected for silicon PN junctions.

The forward current of a PN junction follows by the same argument and is simply given by (9.21) with the signs of V_J and J reversed, namely

$$J(\text{forward}) = J_0[\exp(eV_J/kT) - 1] \qquad (9.23)$$

where the electronic contribution to J_0 is given by (9.22). In this polarity, ℓ decreases with V_J and, if this term is important, the forward current is less than would be predicted from (9.23) assuming a constant value of J_0. The current density at a forward biased PN junction can, however, be very substantial when $eV_J \gtrsim 10kT$. The assumption that the whole of the applied voltage appears across the junction may not then be valid since there will be some ohmic voltage drop across the P and N regions remote from the junction. Equation (9.23) is then seriously optimistic. The ratio of the forward and reverse currents at 0.6 V of a typical commercial silicon PN junction diode at room temperature is about 10^3.

Finally we note what happens if one side of the junction is much more heavily doped than the other, e.g. if $N_A \gg N_D$, a P^+N junction. Reference to (9.22) shows that the electron contribution to J_0 will then be smaller than the hole contribution, given by the equivalent expression

$$J_0(\text{holes}) = e\,n_i\left[\frac{n_i D_h}{N_D L_h} + \frac{\ell}{\tau_0}\right] \qquad (9.24)$$

which is very much greater than J_0 (electrons) if $N_A \gg N_D$. All terms in this expression (including the space charge thickness, ℓ, since $\ell_N \gg \ell_P$) are determined by the *purer side* (N) even though the current is primarily carried by holes. Similarly the properties at an N^+P junction are determined primarily by the P side. It is for this reason that the alloying process is capable of making good PN junction diodes (see § 5.5.4).

9.2.4 Avalanche breakdown at PN junctions (d)

The field in the space charge region of a reverse biased PN junction increases with voltage (see table 9.1). The reverse current increases slowly (9.22). This state of affairs cannot continue to indefinitely high fields. In fact the current starts to increase rapidly at a 'critical' voltage, V_c, as shown in figure 9.6. This is

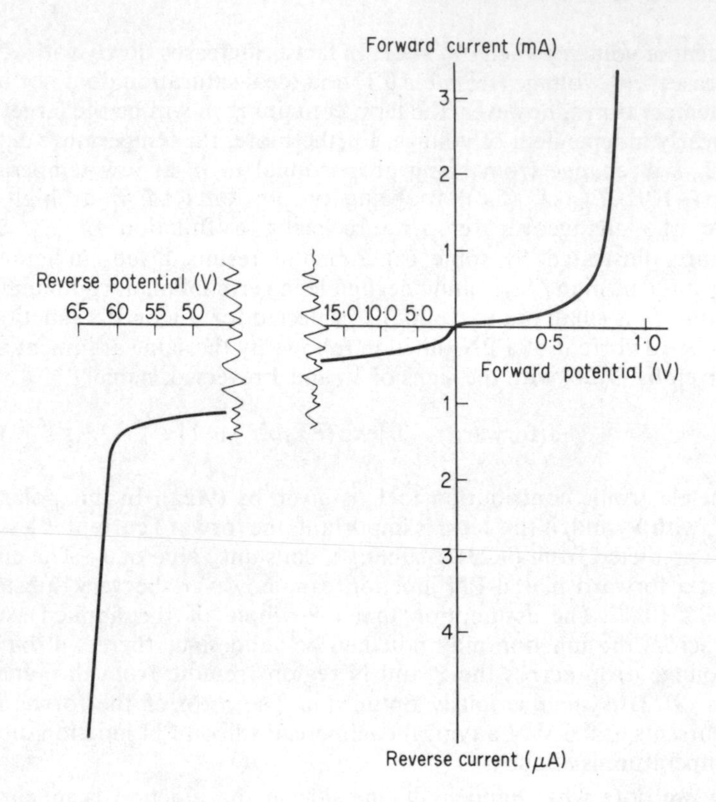

Figure 9.6 Illustration of avalanche breakdown characteristic of a PN junction

due either to impact ionisation and avalanching or tunnelling, depending on the conditions: avalanching occurs in lightly doped diodes and tunnelling is the dominant process in heavily doped diodes.

The avalanche mechanism is analogous to that occurring in a gas discharge. An electron (or hole) passing through the junction is accelerated by the field to an energy greater than E_g when it can generate a hole-electron pair by impact ionisation. We have seen that this requires an electric field of about 10^5 V cm^{-1} in germanium (§ 8.7.3). The generated hole travels back to produce a further electron-hole pair and the process is self-sustaining.

It is very difficult to analyse avalanche ionisation quantitatively, as will be clear from the discussion on hot electrons in Chapter 8. Order of magnitude agreement between theory and experiment can be obtained by assuming the chief energy loss mechanism for such hot electrons is optical phonon emission, which seems reasonable. The complicating features are (a) the complex band structure and (b) the tendency for an avalanche discharge to be laterally unstable. This is evidenced by the spots of light emission shown in figure 6.18, each spot corresponding to a single 'microplasma' i.e. a region of avalanche-sustained current flow. As the reverse voltage on a diode is steadily increased the number of microplasmas also increases, in the same sequence each time.

In view of the difficulties we content ourselves by noting that if we assume avalanching begins at some critical field F_A we can equate this to the maximum field in the junction to find the doping dependence of the critical reverse voltage. Using table 9.1 we find, for an abrupt junction,

$$V_c \approx \frac{\varepsilon F_A^2}{2e}\left(\frac{N_A + N_D}{N_A N_D}\right)$$

$$\propto 1/N_D \qquad \text{if} \qquad N_A \gg N_D$$

Experimentally, V_c is found to be proportional to $(1/N_D)^{0.75}$ for germanium and to $(1/N_D)^{0.7}$ for silicon.

9.2.5 Tunnelling at PN junctions (d)

Tunnelling is limited to junctions whose space charge regions have a width of the order of or less than a mean free path. Such thin regions are produced by heavy doping (see table 9.1). The tunnelling process, as it applies to a reverse biased PN junction, is illustrated in figure 9.7 (cf. figures 8.40 and 8.41).

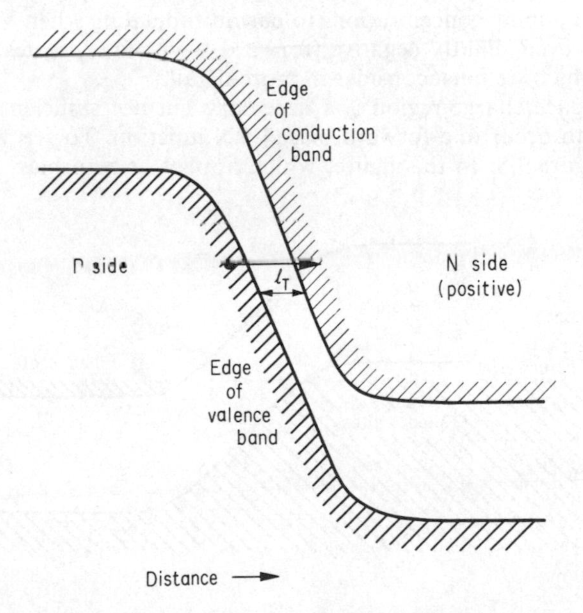

Figure 9.7 Illustration of the mechanism of tunnelling at a reverse biased PN junction. If ℓ_T is sufficiently small (of the order of an electron wavelength or less) electrons may tunnel at nearly constant energy from the valence band to empty states in the conduction band

Since the potential step is much greater than V_{BO} when the junction is reverse biased there are vacant states in the conduction band at the same energy as occupied states in the valence band on the P side and tunnelling can occur if ℓ_T is small enough. ℓ_T is about, but less than, the space charge thickness. The probability of tunnelling, $\mathscr{P}_T$, will be of the form

$$\mathscr{P}_T \sim \exp(-\kappa \ell_T)$$

since the electron wave vector, $k(=i\kappa)$, is imaginary in the energy gap (see §2.10.1). Assuming a simple band structure and noting from the geometry of figure 9.7 that ℓ_T is approximately equal to $E_g/e\mathrm{F}$ we find

$$\mathscr{P}_T \sim \exp\left[\frac{-E_g(2m^* E)^{1/2}}{e\,\mathrm{F}\hbar}\right] \tag{9.25}$$

where the field F is some average field in the junction but is not very different from F_{max} (see table 9.1). This expression gives a reasonable description of the tunnel current and accounts for its very rapid onset at a 'critical' reverse voltage. This rapid onset permits such diodes to be used as voltage reference sources. They are frequently called Zener diodes: Zener described the mechanism long before technology permitted their fabrication. Sets of Zener diodes covering a range of onset voltages are available commercially.

We now note, from the expressions in table 9.1, that ℓ_T may be small enough at very high doping concentrations to permit tunnelling when V_J is very small, or zero, or even slightly negative (forward bias). This creates a wholly new situation which we must consider in more detail.

A thin space charge region is a necessary but not sufficient condition for tunnelling to occur in a forward biased PN junction. To see why this is so, consider figure 9.8. In this figure, which applies at zero bias, both sides are

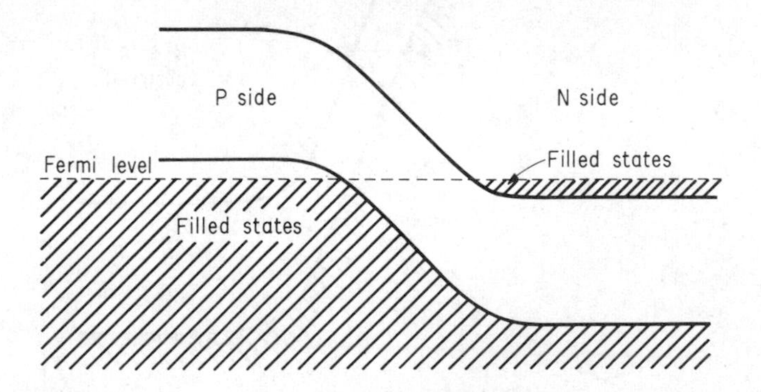

Figure 9.8 The electron potential near a PN junction which is heavily doped on both sides to achieve degeneracy. No external bias is applied. Due to the heavy doping the space charge layer is very thin and tunnelling is possible when the junction is biased to permit filled and empty states on each side to coincide in energy

assumed to be so heavily doped that the Fermi level lies within the valence band on the P side and the conduction band on the N side. Occupied states are shown shaded. Zero current requires that the Fermi level be at the same energy on each side. Forward current implies electron flow from N to P and forward bias reduces the height of the potential step, thus raising the potential on the N side relative to the P side. Tunnelling requires not only a thin junction but also empty states on the P side into which electrons from the N side can tunnel at constant energy. Figure 9.8 shows that this condition can be met for forward bias conditions if the doping is as heavy as indicated, at least up to a voltage equivalent to the sum of the two Fermi energies $[E_F(P) + E_F(N)]$. The tunnel current will be proportional to the degree of overlap of filled (N side) and empty (P side) states, i.e. a function of the form

$$\int \mathscr{P}_T \, \mathrm{S}(E_v)[1 - f_0(E_v)]\, \mathrm{S}(E_c)[f_0(E_c)]\, \mathrm{d}E$$

where f_0 is the Fermi function and $\mathrm{S}(E_v)$, $\mathrm{S}(E_c)$ are the densities of states in the valence and conduction bands. In spite of its rapid variation with field (9.25) $\mathscr{P}_T$ is a comparatively slowly varying term in this expression. The tunnel current passes through a sharp maximum at a forward voltage of about

$$[E_F(P) + E_F(N)]/2e$$

and should vanish at about twice this voltage. The normal forward current due to injection remains and this becomes dominant when tunnelling ceases.

PN junctions showing tunnelling in the forward direction were first made by Leo Esaki and are known as Esaki or simply tunnel diodes. The current–voltage characteristic of typical tunnel diodes are as shown in figure 9.9, the

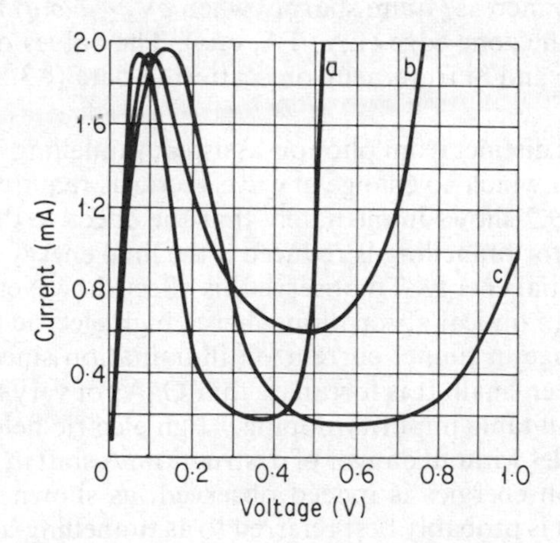

Figure 9.9 Current–voltage characteristics of experimental tunnel diodes; (a) germanium, (b) silicon (c) GaAs. (After N. Holonyak and I. A. Lesk, *Proc. I.R.E.*, **48**, 1405 (1960).)

main features being as described above. The fall in tunnel current with increasing voltage provides a region of negative resistance of the same type as the Gunn diode (§ 8.7.3) and capable of sustaining oscillations or for use as a switch. Switching times are usually in the nanosecond region (10^{-9} s) and limited by the very substantial capacitance of so thin a junction (cf. table 9.1). The resistance of the diode can be made very low by extremely heavy doping and the switching time, $|RC_B|$, where $|R|$ is the value of the negative resistance, reduced to as little as a picosecond (10^{-12} s) but R is then inconveniently small. At the other end of the scale, relatively lightly doped diodes may show no negative resistance region but pass more current (due to tunnelling) at small reverse voltages than forward voltages. Such 'backward diodes', as they are called, have found application where low impedances are no handicap and a very rapid change in resistance with bias is all important (e.g. microwave detection).

Apart from their device applications, work on tunnel diodes has shown up some interesting physical phenomena, two of which we shall now describe briefly.

Because of the band structure (§ 4.5) tunnelling in germanium or silicon (but not InSb or GaAs) involves the transfer of electrons from valleys near the extremities of the Brillouin zone to hole minima at the zone centre. A substantial change in wave vector is involved. Since electron wave vector must be conserved, we conclude that such transitions need the co-operation of a phonon. At low temperatures (in the liquid helium range) only phonon emission can be important, there being very few phonons to absorb, and phonon emission implies a loss of energy by the electrons of $\hbar\omega(\boldsymbol{q})$. Since $\boldsymbol{q}$ is of the order of a reciprocal lattice vector, $\hbar\omega(\boldsymbol{q})$ must be of order $k\Theta_D$ where Θ_D is the Debye temperature and hence much greater than kT at helium temperatures. Tunnelling is consequently not possible until the applied voltage is sufficient to heat the electrons to an energy $\sim\hbar\omega(\boldsymbol{q})$. The tunnel current under these conditions is indeed observed to increase quite sharply when $eV_J = \hbar\omega(\boldsymbol{q})$ for each of the phonon energies at the zone edge (LA, TA, etc.). The values of these are, of course, known in Ge and Si from neutron scattering data (§ 3.5) and infrared absorption (§ 6.6.2).

Photon-assisted, as distinct from phonon-assisted, tunnelling is also possible in semiconductors for which no change in wave vector is required (e.g. GaAs). Inspection of figure 9.7 shows immediately that the effective thickness of the space-charge region for tunnelling is reduced if the final energy of the electron is higher than the initial energy. To observe this effect, however, it is easier to measure the change in optical absorption when a high electric field is applied rather than the change in tunnel current on illumination since the latter is, proportionately, rather small. It is fortunate that GaAs of very high resistivity can be obtained by suitable impurity doping. High electric fields can then be applied to bulk samples without danger of destruction. A shift in the absorption edge to lower photon energies is indeed observed, as shown in figure 9.10. In this form the effect is probably best referred to as tunnelling-assisted photon absorption. As the absorption edge of GaAs is very steep quite a small shift produces a substantial change in absorption at a fixed photon energy and the effect can be exploited in a light modulator for use with lasers.

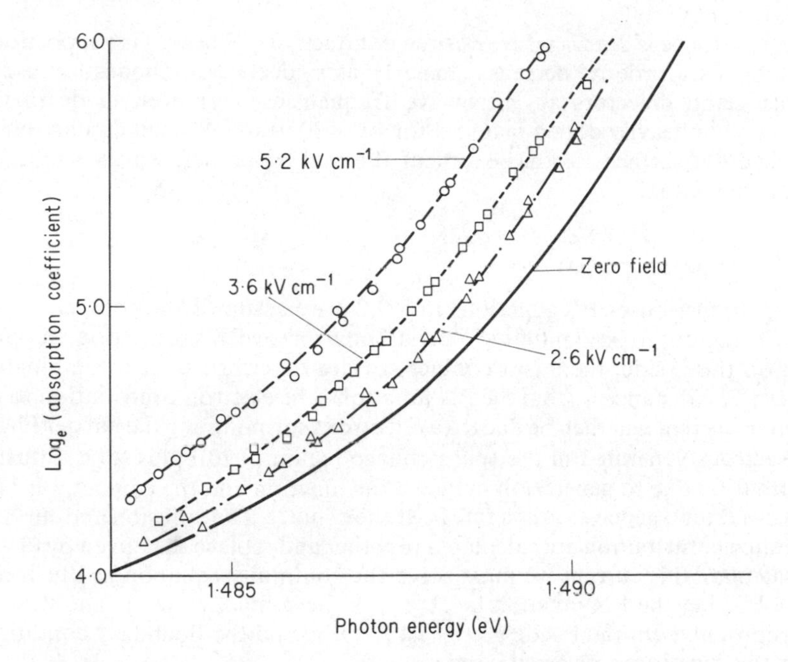

Figure 9.10 The absorption coefficient of GaAs at 77 K near its absorption edge at various electric fields. (After H. D. Rees, *Phys. Letters*, **21**, 629 (1966).)

9.2.6 High-frequency performance of PN junctions (D)

Two factors limit the a.c. performance of PN junctions: (a) the rate of change of carrier densities remote from the junction and (b) capacitative effects at the junction itself.

The rate of change of carrier densities must be found by solving the continuity equation (9.6), although it is clear by inspection that any disturbance will decay at least at a rate determined by the minority carrier lifetime and any diffusion in or out at junctions will assist. High-frequency performance, then, requires low τ values and thin structures, even if this implies some sacrifice in d.c. performance.

The degree to which the junction capacitance (C_B in Table 9.1) limits high-frequency performance depends on the resistance in series with it as it is the product of R and C_B that matters. This series resistance is the ohmic resistance of the N and P sections which can be reduced by the use of heavily doped material and by keeping dimensions small. Heavy doping, however, increases C_B and reduces the reverse voltage at which avalanching or tunnelling occurs, which is generally undesirable. The RC_B product may be reduced by the use of radial, as distinct from planar, geometry: in the limit the device becomes a small point contact on an extended flat surface. The contact area, and hence C_B, is then very small and the resistance determined by a small, approximately hemispherical, region near the contact. For a hemispherical contact of radius

r_0 the resistance is $2/\pi\sigma r_0$ where σ is the conductivity. Since C_B is proportional to r_0^2 the RC_B product decreases linearly as r_0 decreases. Diodes for use as rectifiers and detectors at microwave frequencies have been made in this geometry on heavily doped material for over 30 years. A small contact on an extended flat surface also has excellent thermal properties, which is an additional advantage.

9.2.7 LH junctions (D)

In our discussion of PN junctions in § 9.2.3 we assumed the contacts to the P and N regions to be at infinity. We shall now remove this restriction. Suppose that, on the P side, there is a contact surface described by a recombination velocity S_v a distance w from the PN junction. The electron contribution to the junction current will then be due to (a) electrons surmounting the barrier (9.18), (b) electrons generated in the space charge region (9.20), plus (c) a diffusion contribution due to generation between the junction and the contact, plus (d) the new feature, generation at a rate S_v at the contact. The diffusion and surface-generated contribution are calculated together and replace that given by (9.19). To calculate this current we must solve the continuity equation as illustrated in § 9.1.3. Let the PN junction be at $x = 0$, the contact at $x = w$ and flow by diffusion only. On the P side $g = n_0/\tau_e$, $r = n/\tau_e$ and the boundary conditions when the junction is reverse biased are

$$n = 0 \quad \text{at} \quad x = 0 \qquad \text{(reverse biased PN junction)}$$

$$S_v(n - n_0) = -D_e\left(\frac{dn}{dx}\right) \quad \text{at} \quad x = w \qquad \text{(by definition of } S_v\text{)}$$

The electron diffusion contribution to the reverse current, J_0, as defined by (9.19) is then given by $D_e(dn/dx)$ at $x = 0$. If, then, the contact to the P side is not at infinity (9.22) is replaced by

$$J_0 = e n_i \left\{ \frac{n_i D_e}{N_A L_e} \left[\frac{(S_v L_e/D_e)\cosh W + \sinh W}{(S_v L_e/D_e)\sinh W + \cosh W} \right] + \frac{\ell}{\tau_0} \right\} \tag{9.26}$$

where $W = w/L_e$. The term in rectangular brackets reduces to unity if $w = \infty$, in agreement with (9.22). A similar term, with $D_e/N_A L_e$ replaced by $D_h/N_D L_h$ and $S_v L_e/D_e$ by $S_v L_h/D_h$ will of course describe the hole current generated on the N side.

In practical PN junction diodes it is undesirable to place the contacts many diffusion lengths away from the junction since this results in a large ohmic resistance in series with the junction and reduces the forward current, as noted in § 9.2.3. On the other hand, (9.26) shows that a contact with large S_v close to the junction ($w < L_e$) increases the reverse current. It is clearly desirable to produce contacts with low effective values of S_v. This may be done using LH junctions, i.e. junctions between material of the same conductivity type (N or P type) but between sections of light (L) and heavy (H) doping. An N type LH junction is often written NN^+ and a P type, PP^+.

We may safely assume that minority carrier current in a very heavily doped region is entirely due to diffusion since the electric field in high conductivity material must be small. On the P^+ side of a P type LH junction (H side) the electron current density will be

$$e\, n_{0H}(D_e/\tau_e)^{1/2} \tag{9.27}$$

where n_{0H} is the equilibrium minority carrier density on the H side. (cf. (9.12)). Seen from the L side this is equivalent to a surface (the LH junction) at which electrons recombine or are generated at a rate $S_v n_{0L}$. As the electron current must be continuous across the junction we have

$$S_v = \frac{n_{0H}}{n_{0L}}\left(\frac{D_e}{\tau_e}\right)^{1/2} = \frac{P}{P^+}\left(\frac{D_e}{\tau_e}\right)^{1/2} \tag{9.28}$$

and S_v can be very small if $P^+ \gg P$, notwithstanding that τ_e is likely to be reduced in the heavily doped material. In practice it is comparatively easy to make $(S_v L_e/D_e) \ll$ unity at an LH junction. If, then, $w \ll L_e$ (9.26) simplifies very considerably and the electron contribution to the diode current is given by (9.21) where

$$J_0 = e n_i \left[\frac{n_i w D_e}{N_A L_e^2} + \frac{\ell}{\tau_0}\right] \tag{9.29}$$

which may be compared with (9.22). We conclude that improved performance will be obtained from PN junction diodes terminated at LH junctions relatively close to the main junction. The reverse current will, in general, be smaller and the forward current not seriously limited by ohmic resistance. Under these conditions the density of minority carriers injected under forward bias can be quite large and is given by an expression like (9.14). Physically this high injected carrier density arises because the LH junction is relatively impervious to minority carriers (low S_v value) which are being pumped in at the PN junction. Structures of this type have been used as infrared and microwave modulators as a large density of injected carriers implies high free-carrier absorption (Chapter 7).

Thus LH junctions can be considered as complementary to PN junctions. While the latter offer essentially no impedance to the flow of minority carriers they are impervious to majority carriers from either side. LH junctions of low S_v value show exactly the opposite behaviour and are relatively impervious to minority carriers.

9.2.8 Avalanche devices (D)

Consider a bar of N type material with an N^+ region at each end, as shown in figure 9.11. A large voltage is applied so that a substantial field in the N region drives holes (minority carriers) to one end. Since the negative NN^+ junction is impervious to holes they will accumulate at this junction. Furthermore, holes

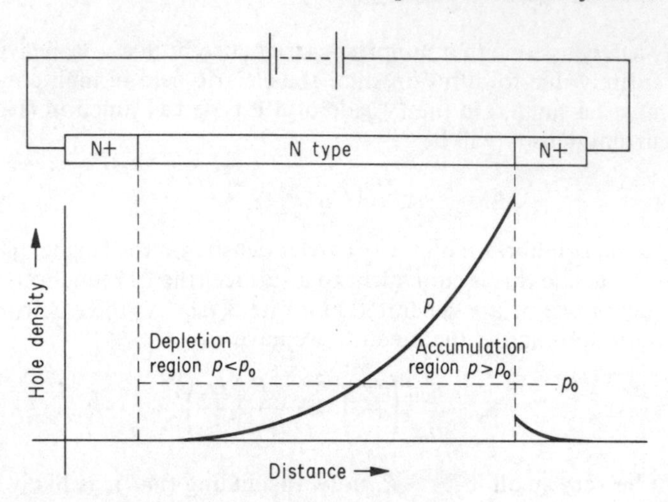

Figure 9.11 Structure in which avalanche injection and minority carrier accumulation can occur

cannot enter at the positive NN^+ junction to replace those swept along the bar: there will be a deficiency at this end. The resultant hole density distribution will be as shown in figure 9.11. As a consequence of this carrier distribution and the need for current continuity the electric field in the N type bar will be a maximum at the positive end. Suppose now the electric field is steadily increased until avalanche ionisation begins. Large numbers of electrons and holes are now generated at the positive (higher field) end so that even more holes are swept to, and accumulate at, the negative end. This in turn reduces the field (at constant current) at the accumulation end and consequently increases it where avalanching is occurring. The result is clearly unstable, a negative resistance appears and in the limiting situation the bar has a very high carrier density everywhere except in a small region near the positive LH junction where avalanching occurs. The avalanching junction is a source of minority carriers, much as a forward biased PN junction, and the process may be termed avalanche injection. The bar structure illustrated in figure 9.11 is not a practical one from the device viewpoint, since a high power dissipation is required. Using the more favourable thermal geometry of the small hemispherical contact on an extended surface (§ 9.2.6), current–voltage characteristics such as those shown in figure 9.12 can be obtained. Note that the negative resistance is of the current-controlled type and opposite to that of the Gunn diode (§ 8.7.3) or the tunnel diode (§ 9.2.5). We have already remarked that avalanching is laterally unstable and the current tends to flow in restricted filaments. This may be considered as the current analogue of the (longitudinal) electric field domains of the Gunn effect.

Avalanche instabilities of the above type can be observed in a number of junction structures, the essential feature being redistribution of the electric

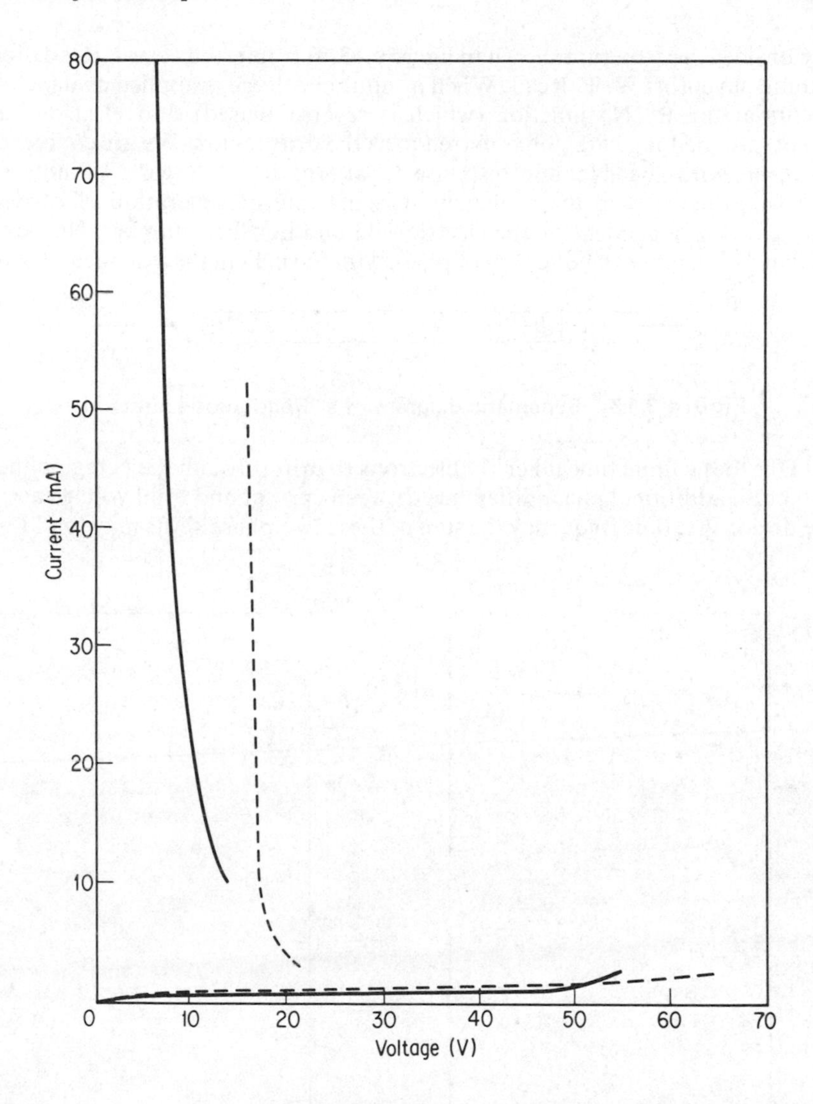

Figure 9.12 Current–voltage characteristics of avalanche injection (NN^+) diodes. Germanium (solid line), silicon (dashed line). (After A. F. Gibson and J. R. Morgan, *Solid State Electronics*, **1**, 54 (1960).)

field due to the space charge of the avalanche-generated carriers. Switching speeds in the nanosecond region are possible. The curve of figure 9.12 is, however, a 'static' characteristic; i.e. it could be observed at zero frequency. We now consider a new feature, namely the transit time of avalanche injected carriers, which can lead to a dynamic negative resistance at very high frequencies but not at zero frequency.

Consider the structure shown in figure 9.13. It is referred to as a Read diode after its inventor, W. T. Read. When a suitable voltage is applied avalanching occurs at the $P^{++}N^{+}$ junction (which is reverse biased) and electrons are swept into the long N region, referred to as the drift region. We are concerned, however, with the dynamic response to alternating voltages. We note the following: (a) that in the avalanche it is the rate of generation of carriers, dn/dt, that is dependent on the electric field and not their number. Hence the avalanche current can be 90° out of phase with the field in the avalanche region.

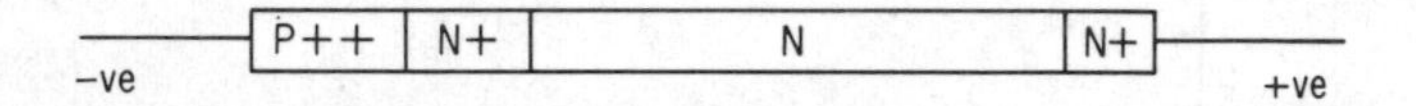

Figure 9.13 Schematic diagram of a 'Read' diode structure

(b) Due to the finite time taken for electrons to drift through the N region there will be an additional phase difference between current and total voltage across the device. At some frequency the sum of these two phase shifts may total 180°.

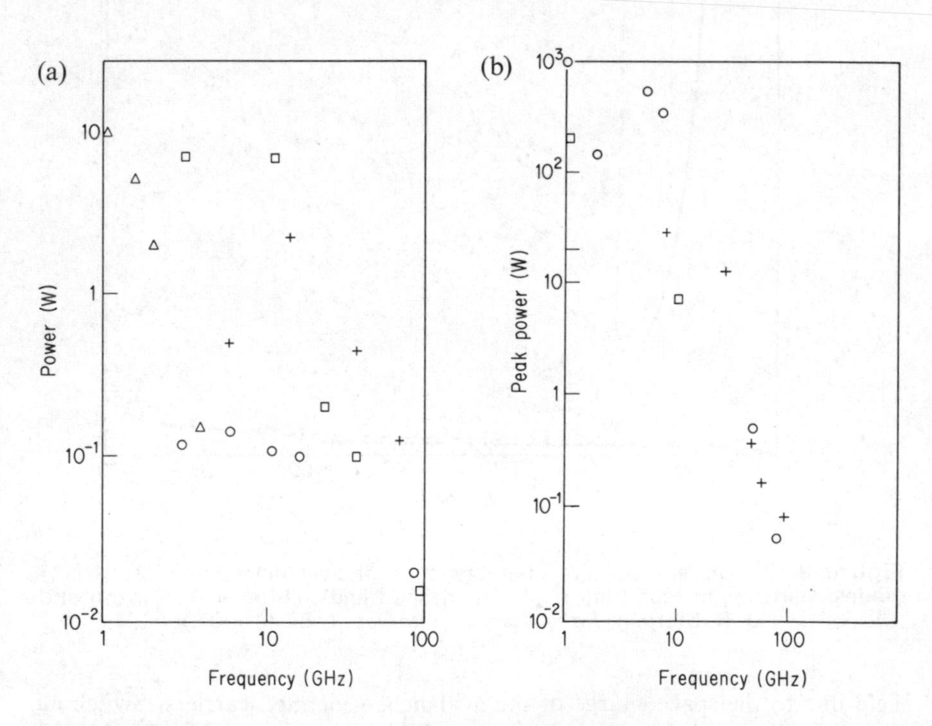

Figure 9.14 Comparison of performance of avalanche diodes, transferred electron (Gunn) diodes, variable capacitance (Varactor) parametric diodes and bipolar transistor oscillators as microwave sources. (a) c.w. operation, (b) pulsed operation. +, Read avalanche diode; ○, Gunn effect; □, variable capacitance parametric diode; △, transistor. (After C. Hilsum, *Brit. J. appl. Phys. Ser. 2*, **1**, 265 (1968).)

A 180° phase shift between a.c. current and voltage, superimposed on a direct current, represents a negative resistance since the current decreases when the voltage increases. The device will maintain oscillations at or near this critical frequency.

A number of variants on the Read diode have been examined. In general, theoretical analysis has to be done numerically on a computer since the continuity equation cannot be simplified by neglecting carrier space-charge effects. The physical principles, however, are as outlined above. Experimentally, a number of avalanche diode structures can achieve useful efficiencies as oscillators in the frequency range 1–50 gigahertz. We have now considered three types of semiconductor microwave source: the Gunn transferred electron amplifier (§ 8.7.3), parametric frequency converters (§ 9.2.2) and avalanche diodes. The relative merits of each device depends on the application. Their relative merits have been compared in a review article by Hilsum from which figure 9.14 is taken.

9.2.9 Photoconductive detectors and photodiodes (d)

We have seen in Chapter 6 that absorption of photons can lead to photoconductivity if either the photon energy exceeds the valence to conduction band energy gap or excites electrons (holes) from impurity centres. Commercial photodetectors are made either as thin bars of homogeneous material or in the form of PN junctions. In the absence of space charge effects (§ 9.1.2) the incremental change in conductivity, $\Delta\sigma$, of a uniformly illuminated homogeneous bar can be deduced by simply equating the carrier generation rate due to the illumination to the recombination rate. In turn the recombination rate determines the response time of the detector. It is, however, the fractional change in conductivity, $\Delta\sigma/\sigma$, which, together with the geometry, determines the photoresponse in a practical circuit. Since for a homogeneous bar of semiconductor in the dark

$$\sigma = e[n_0 \mu_e + P_0 \mu_h] \qquad \text{where} \qquad n_0 P_0 = n_i^2$$

σ passes through a minimum and the photoresponse is a maximum at a doping concentration which depends on μ_e/μ_h. For most semiconductors $(\mu_e/\mu_h) > 1$ so that optimal performance is obtained with slightly P type material.

A very substantial reduction in effective dark conductivity, and hence improved photoresponse, can be obtained by using a PN junction. We have seen that the current in a reverse biased PN junction is due to the thermal generation of minority carriers near the junction. Illumination in the fundamental absorption region simply adds to this generation rate. It will be evident from the previous sections how to calculate the increase in reverse current on illumination of a reverse biased diode. Suppose, for example, the light to be totally absorbed on the P side in a thin layer (the absorption coefficient will be $\sim 10^5$ cm^{-1}. see § 6.3) at a surface parallel to the junction and a distance w

from it. We may use the argument which led to (9.26) to obtain

$$J_G = \frac{eG}{\cosh W + (S_v L_e/D_e)\sinh W} \tag{9.30}$$

as the increment in reverse current density due to G photons per second being absorbed per unit area.

Thus the use of a PN junction can greatly enhance the photo response by reducing the dark current. The photo sensitivity, determined by the ratio of the signal to the electrical 'noise' generated in the detector (see § 9.5.3), is not necessarily improved. Indeed, the signal-to-noise ratio is frequently lower in a photodiode than a homogeneous photoconductive crystal. Since any photodetector may be followed by an amplifier photodiodes are generally unsuitable for the detection of very low light levels, but are convenient when signal-to-noise ratio is not a limiting factor, for example in infrared burglar alarms.

Illumination of an unbiased PN junction produces an e.m.f. across the junction and a current in an external circuit and this is sometimes called the photovoltaic effect. It occurs because the potential step and consequent electric field at the junction drives photo-generated holes into the P region and electrons into the N region (see figure 9.4). If no current is drawn the junction becomes forward biased and the maximum open circuit photovoltage is V_{BO}, the height of the unbiased potential step. The maximum possible short circuit photocurrent is one unit of charge around the external circuit per absorbed photon.

The photovoltaic effect represents the direct conversion of light into electricity and permits the provision of power to inaccessible equipment (e.g. earth satellites) by the direct conversion of sunlight. Solar radiation totals about 1.2 kW m^{-2} at high altitudes and can be nearly 1 kW m^{-2} at the earth's surface under good conditions. A large value of V_{BO} and consequently heavy doping is obviously desirable in this application, subject to the limitation imposed by free-carrier absorption. Since eV_{BO} cannot appreciably exceed E_g it is desirable to use a semiconductor with a relatively large energy gap. On the other hand, the diode is insensitive to photons with energies less than E_g so increasing E_g reduces the spectral range absorbed. For a given blackbody temperature, therefore, there is an optimum value for E_g. Fortunately silicon is not far from optimal for use with the sun and the impurity diffusion techniques required to prepare large area diodes near the surface are well established. Practical efficiencies of solar cells are about 10%.

9.3 Unipolar transistors

The term 'unipolar' implies that only one type of carrier is involved: no injection of minority carriers takes place in the structures we are about to discuss. We shall describe two main types of structure, using PN junctions and certain surface properties of silicon, respectively. A number of hybrid structures with similar properties can also be made. The performance of a unipolar transistor is similar to that of a triode or pentode valve, with the added feature that currents of either polarity can be controlled in a single device.

9.3.1 Junction field effect transistors (D)

Consider the structure shown in figure 9.15. It consists of a bar of N type material with contacts at either end referred to as the *source* and the *drain*. They are non-injecting, N^+ contacts on to N material. P^+ regions, prepared by diffusion, penetrate each side of the bar and are separated by a distance *d*. These two regions are electrically connected together and called the *gate*. We assume these regions are flat, parallel and extend entirely across the bar in one dimension. This simplifies the geometry of the problem. One practical version of this structure has circular symmetry around an axis down the centre of the rod; in another an N type epitaxial layer is grown on a P^+ substrate and a further P^+ junction diffused into the exposed epitaxial surface.

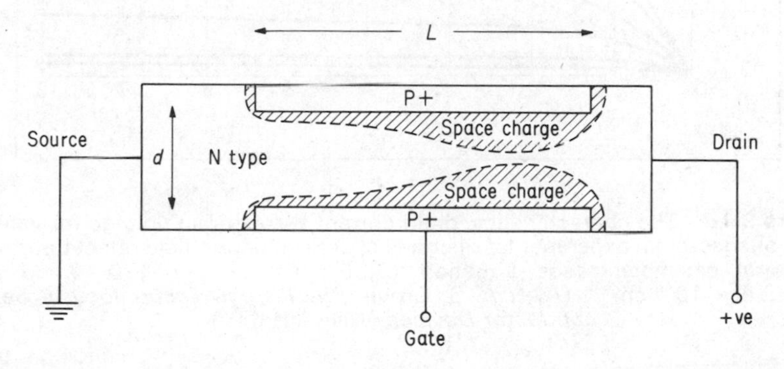

Figure 9.15 Schematic sectional diagram to illustrate operation of a junction field effect unipolar transistor

To understand the operation of this device, suppose that the source and gate are earthed while a positive voltage V_D is applied to the drain contact. The gate junction is then reverse biased due to the ohmic voltage drop between source and gate region and completely surrounded by a space charge region into which electrons cannot penetrate. An electron current flows from source to drain between the space charge regions of the upper and lower PN junctions. As V_D increases the reverse bias increases and the space charge penetrates further into the body of the bar. At some critical value of V_D, known as $V_D(\text{sat})$, the two space charge regions touch and the channel between source and drain is said to be 'pinched off'. Ignoring any resistance in series with the drain contact this occurs when the space charge penetration into N material, ℓ_N, is given by $\ell_N = d/2$ or

$$V_D(\text{sat}) + V_{BO} = \frac{e N_D d^2}{8\varepsilon} \tag{9.31}$$

where use had been made of table 9.1 (§ 9.2.1) and we have assumed $N_A \gg N_D$. The drain current does not, however, actually go to zero at or above pinch off but simply saturates, since it is the source-drain current that is producing the

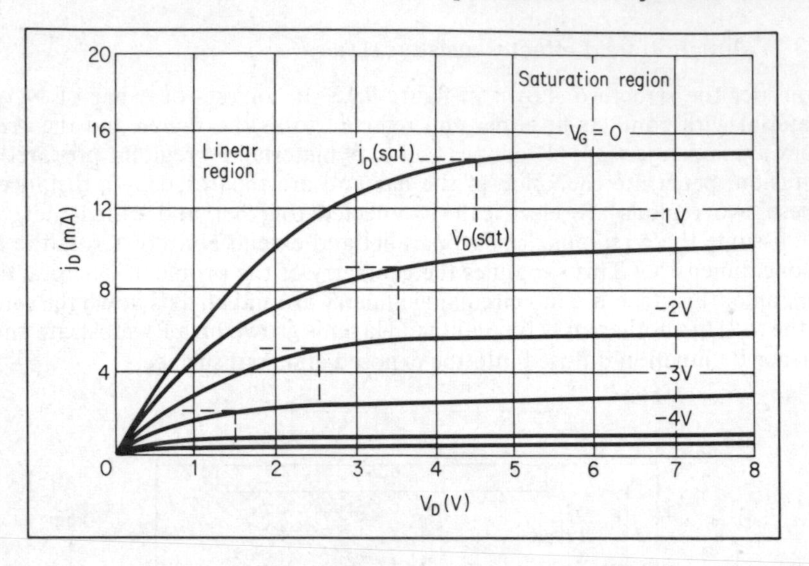

Figure 9.16 The characteristics, drain current *versus* drain voltage for various gate voltages, of an experimental silicon N channel junction field effect transistor. Structural parameters (see equation (9.35)) are: $z/L_G = 170$, $d = 3$ μm, $N_D = 2.5 \times 10^{15}$ cm^{-3}. (After A. S. Grove, *Physics and Technology of Semiconductor Devices*, Wiley (1967).)

bias on the gate. This behaviour is illustrated by some experimental results shown in figure 9.16. The results also show the effect of bias, V_G, on the gate electrode relative to the source. If $V_G \neq 0$ pinch-off occurs at

$$V_D(\text{sat}) + V_{BO} - V_G = \frac{e N_D d^2}{8\varepsilon} \tag{9.32}$$

and application of V_G can really reduce the source-drain current to zero.

To derive the characteristic curves shown in figure 9.16 theoretically, we note that the voltage drop dV across an element of the channel dy long is

$$\mathrm{d}V = \frac{j_D \, dy}{e \mu_e N_D z(d - 2\ell(y))} \tag{9.33}$$

where j_D is the drain current, z is the lateral dimension of the structure and $\ell(y)$ is the space charge thickness at y. From (9.32) $\ell(y)$ is given by

$$\ell(y) = \left[\frac{2\varepsilon(V(y) + V_{BO} - V_G)}{e N_D}\right]^{1/2} \tag{9.34}$$

Substituting into (9.33) and integrating between $y = 0$, where $V(y) = 0$, and $y = L_G$ where $V(y) = V_D$ (ignoring the series resistance at source and drain), we obtain

$$j_D = \frac{e\mu_e N_D z d}{L_G}\left[V_D - \frac{2}{3}\left(\frac{8\varepsilon}{e N_D d^2}\right)^{1/2}\{(V_D + V_{BO} - V_G)^{3/2} - (V_{BO} - V_G)^{3/2}\}\right] \tag{9.35}$$

as the characteristic equation of a unipolar junction transistor. This equation describes the curves of figure 9.16 up to the saturation point given by (9.32). Thereafter the current is constant. The main parameter of interest is the amplification which, as for a triode or pentode valve, is measured by the transconductance, g_m, where

$$g_m = \left[\frac{\partial j_D}{\partial V_G}\right]_{V_D = \text{constant}}$$

and this will be largest in the saturation region. Differentiating (9.35) with respect to V_G and substituting the condition for saturation given by (9.32) gives g_m in the saturation region as

$$g_m(\text{sat}) = \frac{e\mu_e N_D z d}{L_G}\left[1 - \left(\frac{8\varepsilon(V_{BO} - V_G)}{e N_D d^2}\right)^{1/2}\right] \tag{9.36}$$

which can in principle be made indefinitely large by increasing z. In practice, as inspection of figure 9.16 shows, g_m may be several mA per volt and compares quite favourably with thermionic valves. The input impedance of a junction field effect transistor is determined by the impedance of a reverse biased PN junction and can be quite substantial (10–100 MΩ). At high frequencies the PN junction capacitance, C_B, becomes important and this, together with the transconductance, determines the upper frequency limit of the device.

9.3.2 Insulated gate transistors (D)

Consider a parallel plate capacitor fabricated from, say, a P type semiconductor crystal, an insulating layer and a metal contact. The application of a positive charge to the metal will induce an equal and opposite negative charge on the semiconductor surface. The space charge in the semiconductor will reside in the negatively charged acceptor centres and any mobile carriers (electrons or holes) near the surface. This condition is the same as that applicable to PN junctions so we conclude that conditions similar to those pertaining at a PN junction can be induced on the surface. The electron potential is shown in figure 9.17 which should be compared with figure 9.1(c).

Inspection of figure 9.17 shows that the negative charge on the semiconductor surface arises not simply from the deficit of holes (repelled by the positive charge on the outer surface) and uncompensated acceptor impurities but also from electrons attracted to the surface. The degree of band bending, the

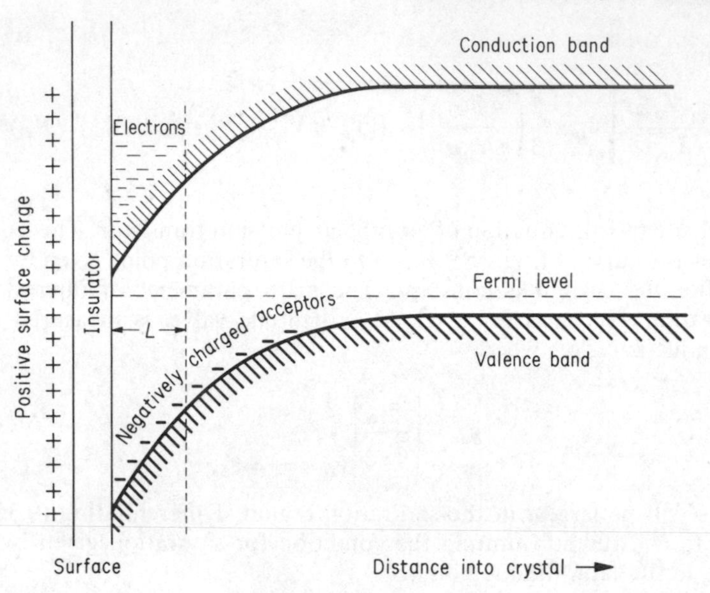

Figure 9.17 Electron potential energy near an insulated surface to which a positive potential is applied

electric field and the space charges are, of course, all related through Poisson's equation (cf. § 9.2.1). As the charge on the insulator is increased, band bending increases until there is a substantial electron density near the surface and a surface layer of thickness L (figure 9.17) is effectively N type. Any further increase in charge is then taken up primarily by an increase in free-electron density at the surface since quite a small change in band bending, once an N type surface has been achieved, results in a substantial change in electron density. We may write

$$\mathcal{Q}_s = \mathcal{Q}_e + \mathcal{Q}_A \tag{9.37}$$

where $\mathcal{Q}_s$ is the charge per unit area on the outer surface, and $\mathcal{Q}_e$ and $\mathcal{Q}_A$ are the free-electron and charged-acceptor contributions to the charge on the semiconductor. Once $\mathcal{Q}_e$ is appreciable, any further increase in $\mathcal{Q}_s$ is taken up by $\mathcal{Q}_e$ and $\mathcal{Q}_A$ is approximately constant at

$$\mathcal{Q}_{A(max)} \sim eN_A\ell_{max}$$

where ℓ_{max} is the maximum thickness of the space charge region. We conclude that it should be possible to induce a conducting N type layer on a P type crystal by the application of an electric field. If this layer connects two N type regions formed by impurity doping we have the elements of a unipolar transistor in which the channel conductance may be varied by varying the density of free electrons ($\mathcal{Q}_e$) rather than the channel thickness. A possible structure is shown in figure 9.18. Structures of this type were first discussed in the 1930s and have received sporadic attention ever since but it is only in the last few years that commercially successful devices have been fabricated. Success is primarily due to the special properties of Si/SiO_2 surfaces. To understand why previous attempts failed we must consider surface states.

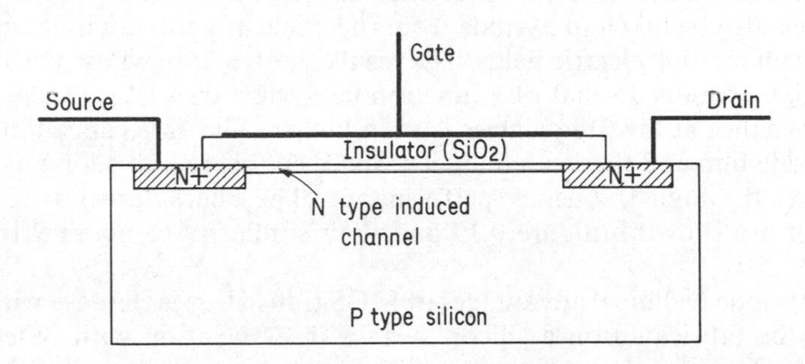

Figure 9.18 Schematic sectional diagram of a metal–oxide–silicon transistor

9.3.3 Surface states (D)

There is no reason to believe that the band structure of a solid is the same at the surface as in the bulk. The periodic boundary conditions are not satisfied at the surface and there will be unsaturated 'dangling' bonds, at least one per surface atom, or of order 10^{16} per cm^2. Such unsaturated bonds provide energy states for electrons immediately on the surface. This was first pointed out by Tamm and such surface states are sometimes known as Tamm states. In practice real surfaces are contaminated and the situation is confused, but Tamm states can be detected on clean surfaces, e.g. cleaved in an ultra-high vacuum.

The energies of surface states may lie between the valence and conduction bands. If such states do exist at a density anything like as high as 10^{16} cm^{-2} the space charge arising from a capacitatively applied field can be entirely accommodated by varying the occupation of the surface states and negligible charge penetrates the bulk of the semiconductor. This can be seen by a numerical example. If we write the charge on the insulator, $\mathcal{Q}$, as

$$\mathcal{Q} = CV \approx \frac{\varepsilon \mathcal{A} V}{x} \approx \varepsilon \mathcal{A} F$$

where V is the voltage across the insulator of thickness x, ε is the dielectric constant and $\mathcal{A}$ is the area. The electric field, F, cannot exceed the breakdown field (say 10^5 V cm^{-1}) and putting $\varepsilon \sim 10^{-10}$ F m^{-1} the electron density induced at breakdown is

$$\mathcal{Q}/e\mathcal{A} \sim 10^{12}\ cm^{-2}$$

Hence insulated gate unipolar transistor action requires a surface state density appreciably less than 10^{12} cm^{-2}.

9.3.4 Metal–oxide–silicon (MOS) transistors (D)

For reasons not clearly understood, thermally oxidised silicon surfaces carefully annealed (around 500°C) have surface state densities around 10^{10} cm^{-2} or less. It would appear that the Si/SiO_2 interface satisfies the periodic boundary conditions and annealing improves the 'match' between the two materials. Care must also be taken to exclude from the oxide any foreign ions capable of diffusion under high electric fields. The result is a structure whose performance is essentially similar to that of a junction field effect transistor except that the gate impedance at low frequencies is even higher. The gate capacitance—due to the oxide film and the semiconductor space charge region acting in series—still limits the high frequency performance. The characteristics of a MOS transistor are shown in figure 9.19 and their similarity to figure 9.16 is self-evident.

Amongst the technical advantages of MOS transistors is the ease with which they can be fabricated on a silicon surface in association with other device structures. They are thus well suited to microelectronic techniques. As circuit

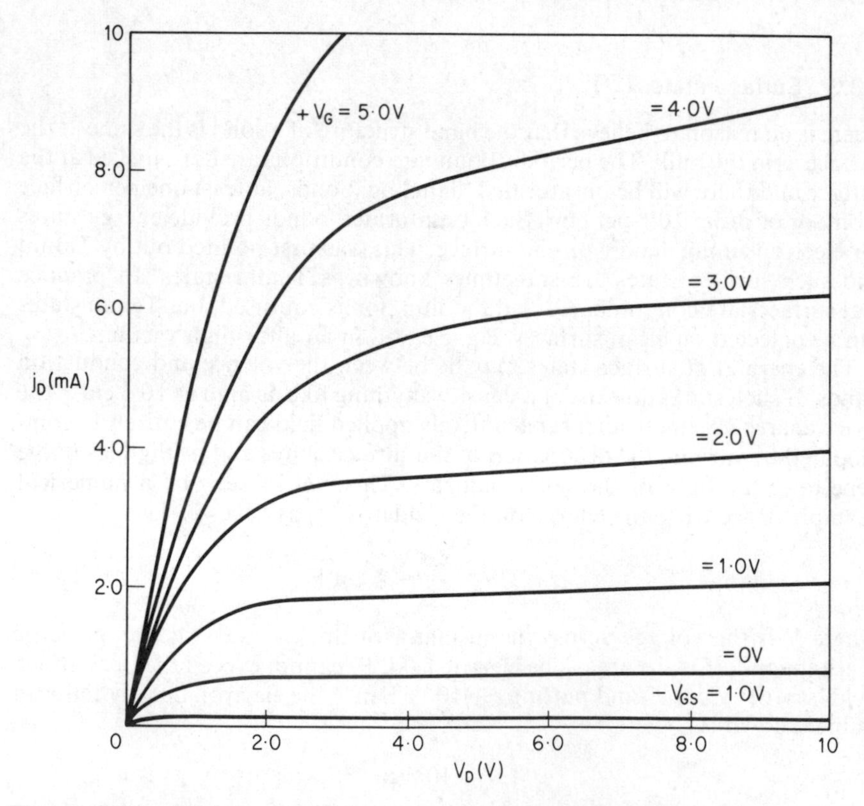

Figure 9.19 The characteristics, drain current *versus* drain voltage for various gate voltages, of a commercial silicon metal–oxide–silicon (MOS) transistor. (Courtesy of Mullard Ltd., England. Transistor type BFW 96.)

elements they have the great merit of near infinite input impedance (at least at moderate frequencies) and high output impedance, and so are fully specified by their transconductance, g_m. The interchangeability of source and drain is also valuable in some applications.

9.4 Bipolar transistors

9.4.1 The operation of bipolar transistors (D)

Bipolar transistors consist of a sequence of three sections which may be NPN or PNP. To be definite we consider a PNP structure of the type shown in figure 9.20. This structure may be fabricated by double impurity diffusion (see § 5.5.4) into a relativcly lightly doped P type substrate. Consistent with this technique is the assumption that the doping concentrations increase progressively from the P substrate to N type to the P^+ region, a state of affairs which we shall find electrically desirable.

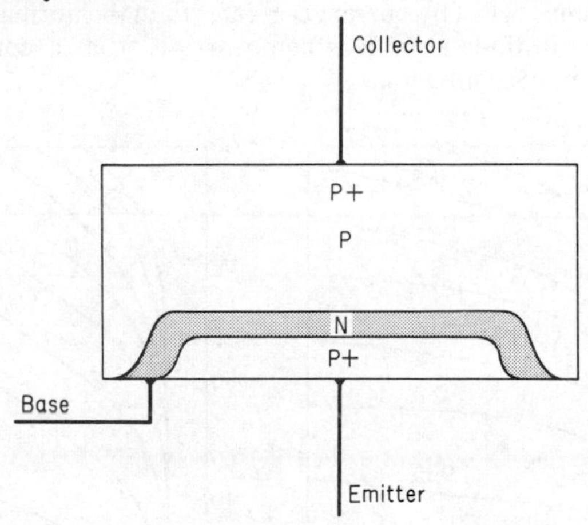

Figure 9.20 Schematic sectional diagram of a bipolar transistor structure. Structures of this type are normally fabricated by impurity diffusion through oxide masks (see Chapter 5). The series (ohmic) resistance of the collector region can be reduced by heavier doping remote from the base region. This is important, for example, in transistors to be used in switching applications

As shown in figure 9.20 the three regions and contacts of the transistor are referred to as the *emitter*, *base* and *collector*. The choice of the name 'base' is an historical accident,* but the other two names do describe the action performed. Imagine the emitter to be at zero potential, the base slightly negative (say –0.2 V) and the collector at, say –6 V. The emitter–base junction is then

* The name 'transistor' arose because it was thought that early transistors were best specified by a 'trans-resistancc' rathcr than a transconductance. The 'base' contact was used for mechanical mounting.

forward biased and a large fraction of the emitter current will be due to holes injected into the base region. These holes may reach the base–collector junction by diffusion. The latter junction is reverse biased and offers no impedance to hole flow (minority carriers) from the N region. Consequently a major part of the emitter current flows in the collector circuit and not the base, in spite of the fact that the collector is a high resistance (reverse biased diode). It is the nature of the emitter current (injected holes) which circumvents Kirchoff's Laws and thus makes amplification possible. Suppose unit current flows in the emitter and a fraction, α, of this reaches the collector. The base current, by subtraction, is then $(1 - \alpha)$. If the base current is used to control the collector current the current gain will be approximately

$$(\partial j_C/\partial j_B) \sim \frac{\alpha}{1-\alpha} \tag{9.38}$$

so that if α is near unity the gain can be quite large. This is illustrated by figure 9.21 which shows the collector characteristics of a commercial bipolar transistor for various values of base current. Notice that there is a significant collector current even when $j_B = 0$. This current is greater than the normal reverse current of a reverse biased diode since it is being drawn from a copious source of holes: the emitter–base junction.

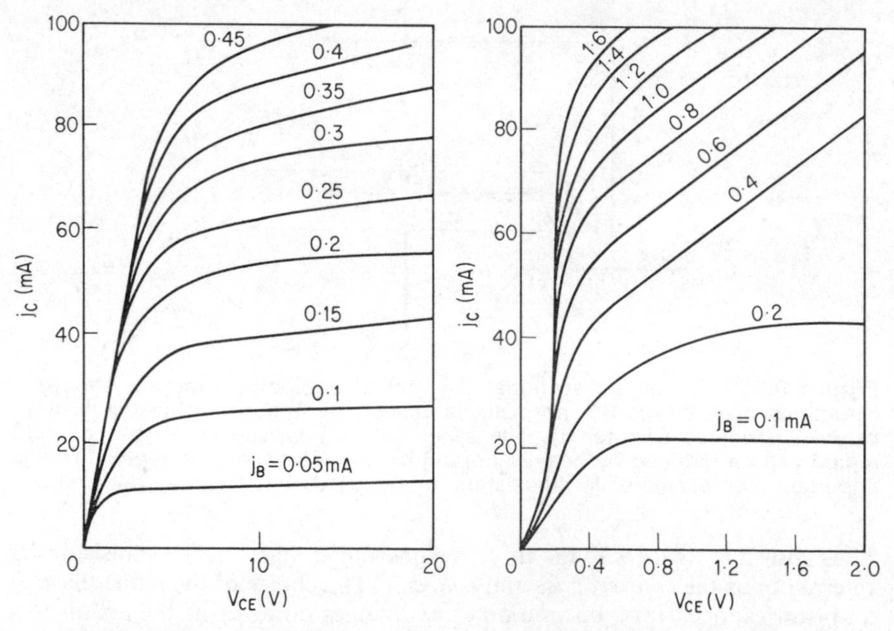

Figure 9.21 Output characteristics, collector current *versus* collector voltage at various values of base current, for a commercial silicon bipolar transistor. The h parameters (equation (9.39)) of this transistor are approximately $h_{11} \simeq 10$ kohms, $h_{12} \approx 2.5 \times 10^{-4}$, $h_{21} \approx 500$ and $h_{22} \approx 10^{-4}$ reciprocal ohms. (Courtesy of Standard Telephones and Cables Ltd., England. Transistor type BC 109.)

9.4.2 The specification of transistor parameters (D)

Since thermionic valves and unipolar transistors have essentially infinite input impedances the output and input are entirely decoupled (except, of course, at high frequencies). Furthermore, the output or drain impedance of a unipolar transistor, given by $(\partial V_D/\partial j_D)$ where V_D and j_D are the drain voltage and current, is high in the saturation region. Consequently the performance of a unipolar transistor is adequately described by one parameter, g_m, which immediately gives the incremental change in output current for a given change in gate voltage. If the transistor is operating into some load resistance, R_L, where R_L is much less than the output impedance, the voltage amplification is given by $|g_m R_L|$.

The same simple approach is not possible for bipolar transistors. Control of the output current, j_C, is by an input *current*, j_B, and the input impedance is low. Four parameters are required to specify any transistor fully. It is fortunate that only one of these is important in a unipolar transistor. In bipolar transistors two parameters (the current gain (9.38), and the emitter impedance) dominate the performance under most conditions but the remaining pair (base and collector impedance) cannot always be neglected. In this section, therefore, we discuss how transistors are specified, with particular reference to bipolar transistors. We shall then proceed to discuss how these parameters relate to semiconductor properties.

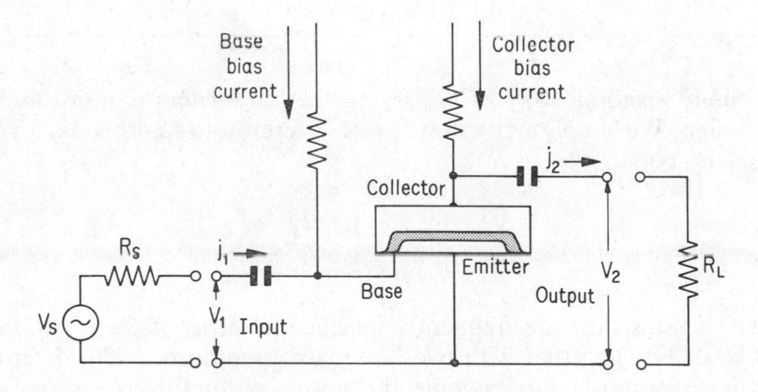

Figure 9.22 Bipolar transistor circuit to define V_1, V_2, j_1, j_2 of equation (9.39). (Though the above circuit would amplify, it is not recommended as an example of circuit design!)

Figure 9.22 shows a bipolar transistor in a possible a.c. signal amplifying circuit. Two constant current (infinite impedance) sources provide appropriate d.c. bias conditions. The incremental currents, j_1 and j_2 due to the signal are assumed small compared with the bias currents. From the point of view of the external circuit the transistor is then specified by the four quantities j_1, j_2, V_1 and V_2 which in turn imply four transistor parameters. There are two main ways in which these parameters may be specified: by an 'equivalent circuit' or purely algebraically. A possible equivalent circuit, which has the merit of being related to the physical action of the transistor, is shown in figure 9.23. It includes three impedances (base, emitter and collector impedances) and a current generator in parallel with the collector to indicate that a current αj_E flows in the collector but all currents in the equivalent circuit obey Kirchoff's Law. We shall relate Z_E, Z_B, Z_C and α to the physical operation of the transistor.

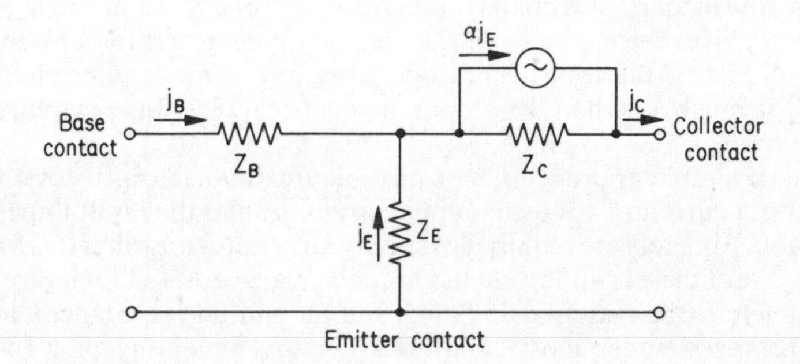

Figure 9.23 The equivalent 'T' network of a bipolar transistor. Other equivalent networks are possible, the minimal requirement being four components of which at least one must be a generator (current or voltage). The four components provide the same information as the four h parameters of equations (9.39)

The purely algebraic way of specifying the parameters is more useful in circuit design. We simply write down, with reference to figure 9.22, a pair of simultaneous equations

$$\begin{aligned} V_1 &= h_{11} j_1 + h_{12} V_2 \\ j_2 &= h_{21} j_1 + h_{22} V_2 \end{aligned} \tag{9.39}$$

Commercial transistors are frequently specified in terms of the h parameters defined by (9.39). Together with two boundary conditions, (9.39) determines the circuit performance. For example, the boundary conditions appropriate to figure 9.22 are

$$V_1 = V_s - j_1 R_s \qquad \text{and} \qquad V_2 = j_2 R_L$$

The voltage gain of the amplifier is then

$$\frac{V_2}{V_s} = \frac{h_{21} R_L}{h_{21} h_{12} R_L + (h_{11} + R_s)(1 - h_{22} R_L)} \tag{9.40}$$

The relationships between the h parameters and the equivalent circuit components can be found by equating similar conditions, e.g. the input impedance with the output open circuit, etc. With the approximation $Z_C(1-\alpha) \gg Z_E$ or Z_B the results are

$$h_{11} \approx Z_B + \frac{Z_E}{1-\alpha}, \qquad h_{12} \approx \frac{Z_E}{Z_C(1-\alpha)}$$

$$h_{21} \approx \frac{\alpha}{1-\alpha}, \qquad h_{22} \approx \frac{1}{Z_C(1-\alpha)}$$

Hence in circuit applications when $R_L \ll Z_C(1-\alpha)$, as is usual, the transistor parameters, α, Z_E, Z_B and Z_C are in approximately declining order of importance. In particular the voltage gain, given by (9.40) becomes

$$\frac{V_2}{V_s} \approx \left(\frac{\alpha}{1-\alpha}\right)\left[\frac{R_L}{R_s + Z_B + Z_E(1-\alpha)}\right]$$

9.4.3 Current gain in bipolar transistors (D)

It is clearly important that α be as near unity as possible since the gain is proportional to $\alpha/(1-\alpha)$. The value of α is determined by (a) the fraction of the holes injected into the base region from the emitter which reaches the collector, designated as β, and (b) the fraction of the emitter current carried by holes, γ. Then $\alpha = \beta\gamma$.

Since the collector junction is equivalent to a surface of large recombination velocity, S_v, we have already calculated (a) for zero field in the base. For a PNP transistor, it is given by (9.13) and is

$$\beta \approx 1 - \frac{w^2}{2L_h^2} \qquad (9.41)$$

where w is the base width and we have assumed $w \ll L_h$. As a result of the method of manufacture of diffused transistors, however, the donor impurity distribution in the base is rarely uniform but decreases towards the collector. This distribution produces an electric field—as at a PN junction, only much smaller—which assists hole diffusion from emitter to collector. β, then, is usually slightly larger than given by (9.41). Unfortunately, β is not usually the limiting factor: the emitter efficiency is more important.

The emitter efficiency, γ, is given by

$$\gamma = \frac{\text{diffusion current of holes across the emitter–base junction}}{\text{total current across the emitter–base junction}}$$

The factor $[\exp(eV_J/kT) - 1]$ is common to all terms and cancels. The numerator is then given by (9.26) with a suitable change in notation to apply to hole flow in a PNP transistor. We may again assume that $w \ll L_h$ and S_v is large at the collector surface. Hence the numerator is

$$\frac{e\,D_h\,n_i^2}{wN_D}$$

The denominator is the sum of the hole diffusion current, the electron diffusion current and the current due to generation in the space charge region. The latter terms are given by (9.22). We therefore obtain for γ

$$\gamma = \left(\frac{D_h}{wN_D}\right) \bigg/ \left(\frac{D_h}{wN_D} + \frac{D_e}{L_e N_A} + \frac{\ell}{n_i \tau_0}\right) \tag{9.42}$$

where we have assumed L_e to be much less than the thickness of the emitter region of the transistor. If the reverse is true the argument leading to (9.26) applies and L_e is replaced by the thickness of the emitter region.

The terms in the denominator of (9.42) are determined by the base region, the emitter region and the emitter–base space charge region respectively. For γ to be as near unity as possible we require

$$\frac{D_h}{wN_D} \gg \left[\frac{D_e}{L_e N_A} + \frac{\ell}{n_i \tau_0}\right]$$

It is clearly desirable to have N_A large (high doping in the emitter) and both N_D and w, the base doping and width, respectively, small. At room temperature the space charge region term is negligible for germanium transistors due to the high value of n_i but significant for silicon transistors (cf. figure 9.5). Since ℓ decreases with forward bias the emitter efficiency of silicon transistors is rather low at low currents but increases with emitter current (at least up to some maximum, see § 9.5.2).

Finally we note that α is usually observed to increase with collector voltage. This is due to the increased width of the collector junction space charge and a consequent decrease in the effective value of the base width, w. Decreasing w increases both β and γ (9.41) and (9.42).

9.4.4 The impedance parameters, Z_E, Z_B and Z_C (D)

At moderate frequencies the junction capacitances can be neglected and all the impedance parameters are real. Making this assumption we shall indicate the approximate values of each, beginning with the most important.

For small signals, as assumed, $Z_E = R_E$ is the slope resistance of a forward biased PN junction

$$\frac{1}{R_E} = \left[\frac{\partial j_E}{\partial V_{EB}}\right]_{V_{CE}=\text{constant}} = \frac{\partial}{\partial V_{EB}}\{j_0[\exp(eV_{EB}/kT) - 1]\} \tag{9.43}$$

using (9.23). j_0 is a slowly varying function of voltage through the space charge generation/recombination term but this may be neglected for large V_{EB} even in silicon transistors so that, to an adequate approximation

$$R_E \approx \frac{kT}{e}\left(\frac{1}{j_E}\right) \tag{9.44}$$

where j_E is the d.c. emitter bias current. Numerically kT/e is about 25 mV at room temperature so $R_E \approx 25$ ohms when $j_E = 1$ mA. Hence $R_E/(1-\alpha)$ is typically a few kilohms at this operating current.

R_B, which is purely resistive at all reasonable frequencies, arises from the ohmic resistance of the base material. The current carried by the base lead is due to the inefficiency of the emitter and recombination in the base region and, in the geometry of figure 9.20, is primarily a lateral current. The value of R_B can be reduced only by either increasing the base width or increasing the base doping concentration, either of which will have detrimental effects on the current gain. In practice, R_B ranges from a few hundred ohms to a few kilohms.

Finally, R_C is the slope resistance of a reverse biased diode. As the collector voltage is much larger than kT/e the only voltage-dependent part of the current is due to generation in the space charge region (9.22). Though this term is important in silicon devices it is offset by the reduced value of n_i so that even $R_C(1-\alpha)$ is large compared with any likely circuit load placed upon it. Since R_L always appears in parallel with $R_C(1-\alpha)$, the latter can be neglected.

9.5 The limitations of semiconductor junction devices

All electronic devices are limited in regard to frequency of operation, power handling capacity and by the extent to which they generate unwanted electrical signals in the form of 'noise'. To a large extent these are engineering problems and we will simply illustrate some of the problems encountered.

9.5.1 Frequency limitations (D)

The frequency response of diodes has been discussed earlier. Recovery after forward bias and minority carrier injection is determined by the carrier lifetime, or, if w is very much less than the appropriate diffusion length, by diffusion. The use of LH junctions increases the injected density at a given current and consequently increases the charge to be extracted. Thus any increase in speed, other than by decreasing geometrical factors, is at the expense of low-frequency performance.

The frequency limitations of transistors are naturally more complicated. Unipolar transistors, like triode valves, may be limited by either input capacitance or charge carrier transit time through the gate region. The effect of the former depends on the external circuit but the latter is a fundamental feature of the structure and the gate must be kept short. The input capacitance per unit area can be reduced only by reduced doping and, if the gate is more heavily doped than the bulk, as assumed in § 9.3.1, reduced doping also decreases the transconductance.

In bipolar transistors one or more of three factors may be important: collector junction capacitance, emitter junction capacitance or base transit time. The collector junction capacitance is reduced by lower doping in the collector region of the transistor and consequently thicker space charge region.

This has the added advantage of increasing the collector voltage at which avalanching occurs (§ 9.2.4) but adds a parasitic resistance in series with the collector which is particularly undesirable in transistor switching (i.e. computing) applications.

Little can be done about the emitter junction capacitance other than reducing the area. If the doping in either the base or the emitter region is reduced either R_B increases or γ decreases, respectively. Since the $R_B C_E$ product is often the limiting factor as far as frequency response is concerned a large R_B is the least tolerable.

The base transport factor, β, will decrease and become complex when $\omega t_r \sim$ unity, where ω is the angular frequency and t_r the transit time. Using the continuity equation (cf. 9.15)) and making the usual assumption that S_v is very large at the collector junction we get

$$t_r \approx w^2/2D$$

where w is the base width and D the appropriate diffusion constant. Clearly a small base width is vital and as reducing w improves the current gain there appear to be only technological problems of impurity diffusion control to solve. Reducing w, however, increases R_B which is obviously undesirable and a further limitation on w will be noted in the following section.

9.5.2 Power limitations (D)

We have seen that the reverse voltage that may be applied to diodes is usually set by the onset of avalanching. The avalanche voltage may be increased by lower doping but this increases the ohmic resistance in series with the diode which is important at large forward currents. The limitations on forward current are primarily thermal and the higher injection efficiency obtainable from germanium compared with silicon is an important consideration in high-power rectifiers. Note that the reverse current of a diode increases exponentially with temperature through the term in n_i^2. The reverse current of germanium diodes approximately doubles each 8°C around room temperature. An efficient heat sink, usually copper, is essential.

The maximum reverse voltage that may be applied to a bipolar transistor collector is also usually determined by avalanching. Since avalanche ionisation is initiated by a flow of current the voltage at which avalanche effects become significant decreases with collector current. If, however, the transistor base is very thin a different effect, called 'punch through', limits the maximum collector voltage. A transistor is said to be 'punched through' when the collector junction space charge region extends right through the base to meet the emitter junction. All control by the base contact is then lost.

The maximum current density at which a transistor can operate may be limited by heating but is frequently limited by one or both of two features of the emitter, as follows.

(a) At high emitter currents in a PNP transistor the hole density in the base, and consequently the excess electron density which must equal it (§ 9.1.2) can become equal to or exceed the donor density. This implies an increase in the effective value of N_D and, from (9.42), a fall in emitter efficiency. The current gain of most bipolar transistors decreases at high emitter currents for this reason.

(b) With the geometry of figure 9.20, a high emitter and consequently high base current produces a lateral field in the base. The result is to reduce the forward bias, relatively, on the emitter area remote from the base contact. This effect can be minimised by making base and emitter contacts as long parallel strips or as a set of interdigital lines, which is the common practice for high-power devices.

9.5.3 Noise limitations (D)

All electrical components generate random electrical signals or noise but some sources of noise are peculiar to, or more pronounced in, semiconductors. The main sources of noise are as follows:

(a) 'Johnson' or resistance noise, common to all conductors, is due to random *motion* of electrons in a conductor at finite temperatures. This motion constitutes a random current flowing in a resistor, which generates a voltage. For frequencies such that $\hbar\omega \ll kT$ its value (given in most textbooks of electricity) is

$$\overline{V_N^2} = \frac{2}{\pi} kTR(\omega_1 - \omega_2) \tag{9.45}$$

where $\overline{V_N^2}$ is the mean square noise voltage and $(\omega_1 - \omega_2)$ is the angular frequency bandwidth.

(b) Current noise is due to the *discrete* nature of the electronic charge. When a current flows, the random arrival of electrons (or holes) at, say, the collector of a transistor constitutes a fluctuation in the current given by

$$\overline{j_N^2} = \frac{e}{\pi} j(\omega_1 - \omega_2) \tag{9.46}$$

where j is the current flowing and $\overline{j_N^2}$ is the mean square fluctuation in the current. Current noise is usually the limiting factor in the performance of photoconductive radiation detectors.

(c) Generation-recombination noise, as its name implies, is due to random fluctuations in electron and hole *density* due to the random nature of the thermal generation and recombination processes. When a current flows it leads to a current fluctuation

$$\overline{j_N^2} = \frac{2j^2}{\pi \mathcal{N}} \int_{\omega_1}^{\omega_2} \frac{\tau \, d\omega}{1 + \omega^2 \tau^2} \tag{9.47}$$

where τ is the minority carrier lifetime and $\mathcal{N}$ is the total carrier number in the sample (not the density). Note that this type of noise, which is clearly important in bipolar transistors, increases as the physical dimensions of the device shrink and $\mathcal{N}$ decreases. Generation-recombination noise is small in unipolar transistors and this can give these devices a marked advantage over bipolar transistors.

(d) 'Flicker' noise, the nature of which is only partially understood, has a noise spectrum determined experimentally as

$$\overline{V_N^2} \propto j^{\eta} \int_{\omega_1}^{\omega_2} \frac{d\omega}{\omega^{\zeta}} \tag{9.48}$$

where $\eta \approx 2$ and $\zeta \approx 1$. The Flicker noise spectrum has been measured experimentally down to 10^{-4} Hz (roughly 1 cycle per 3 hours) and up to 10 MHz. Equation (9.48) is fairly well obeyed throughout this enormous range, which is difficult to explain theoretically. Flicker noise is reduced by careful attention to sample surfaces and dislocation density. In practice the value of the constant in, say, a bipolar transistor is such as to make Flicker noise dominant below about 1 kHz. MOS transistors sometimes show rather higher Flicker noise due to the importance of the surface in these devices but in general Field Effect Transistors develop considerably less noise above 1 kHz than bipolar transistors.

9.5.4 The development of transistors (D)

The importance of a low recombination rate for electrons and holes, and consequently a long diffusion length for minority carriers, will be clear from all that has gone before. Our treatment of PN junctions in § 9.2.3 is in fact valid only if the diffusion length of at least one carrier is large compared with the space charge thickness. It can be shown that no rectification occurs if the carrier diffusion lengths are both zero.

The need for a reasonably long diffusion length is, perhaps, the most fundamental limitation of all on junction devices for it severely limits the semiconductors that may be used. Stimulated by the success of germanium devices in the early 1950s considerable efforts were made to produce bipolar transistor action in other materials. Though the byproducts of this activity were very valuable—Gunn effect oscillators and injection lasers in GaAs are examples—germanium and silicon have continued to dominate the transistor field. Advances in transistors have been due primarily to the development of the fabrication techniques described in Chapter 5, notably impurity diffusion, epitaxial growth and oxide film masking. The state of development of transistors is probably best measured by the product of the maximum operating frequency and maximum power handling capacity available in a single transistor. Over the 20 years following the filing of Shockley's patent on junction transistors in 1948 this product increased by a factor of about 10^5.

The rate of development was a maximum in the late 1950s and early 1960s when the power-frequency product increased by about a factor of three per annum. Currently available transistors operate at power levels of 1 W at frequencies in excess of 1 GHz. The transistor, therefore, is now technically superior to the thermionic valve in all respects except in the microwave frequency range, when Gunn diodes, avalanche diodes, tunnel diodes and parametric devices take over. The most important application of unipolar, and particularly MOS transistors, is in the field of integrated circuitry since the structure lends itself to mass production of many identical units on one substrate. The acoustic amplifier, described in Chapter 8 must compete with this array of devices and its future is not yet certain.

PROBLEMS

(Answers, where appropriate, are given on page 482.)

9.1 Show that the minimum conductivity of a doped semiconductor is given by $\sigma = 2en_i(\mu_e\mu_h)^{1/2}$ and occurs when

$$N_A - N_D = n_i[(\mu_e/\mu_h)^{1/2} - (\mu_h/\mu_e)^{1/2}]$$

(Since μ_e is usually greater than μ_h, minimum conductivity material is usually P type.)

9.2 A bar of photoconductive material is illuminated transverse to the direction of current flow. Show that if the change in conductivity is small it is independent of the spatial distribution of the excess carriers in the bar.

9.3 Derive the Einstein relation

$$D = \mu kT/e$$

for electron diffusion (§ 9.1.2).

9.4 Show that the resistance of a small hemispherical contact of radius r to a large block of semiconductor of conductivity σ is $2/\pi\sigma r$ (§ 9.2.6).

9.5 Derive equation (9.30).

9.6 Suppose that a range of semiconductor alloys with energy gaps spanning that of silicon is available. How, in principle at least, could these be used to produce solar cells of improved efficiency?

9.7 A bar of germanium of rectangular cross section has transverse dimensions y and z where y is very much less than z and less the minority carrier diffusion length. The excess donor density, is 10^{12} cm^{-3} (i.e. less than n_i at room temperature). One of the broad faces, normal to y, is mechanically abrased to make the surface recombination velocity very large while the opposite face is etched to make S_v small. A magnetic field is applied parallel to these faces (i.e. in the z-direction) and a longitudinal current passed through the bar. In one current direction the resistance of the bar is observed to increase with increasing current while in the other the resistance decreases. Why does this happen? Ignoring contact and end effects, what is the minimum possible conductivity of the bar? Suggest a possible application of this effect.

9.8 Consider a bar of N type semiconductor of the same geometry as that of problem 9.7 except that all surfaces are etched. A P^+ region is formed by alloying over most of the area of one broad face though not extending as far as the end contacts. The structure is illuminated on the other broad face by a spot of light of frequency $\omega > E_g/\hbar$. Under open circuit conditions a voltage is observed across the contacts and as the light spot is traversed along the bar the polarity of voltage reverses. Zero voltage is observed when the spot is in the centre but otherwise the contact nearest the light spot is negative. Note that no connection is made to the P^+ region. Explain the origin of this effect and suggest a possible application.

9.9 A bar of N type semiconductor of the same geometry as that of problem 9.7 and with one broad face abrased as in problem 9.7 is illuminated uniformly on the other surface. The light ($\omega > E_g/\hbar$) is absorbed in a surface layer of thickness much less than the thin dimension of the bar, y, and maintains an excess density of electrons and holes, Δp_0, at the surface. Show that an electric field is generated in the y-direction whose magnitude is given by

$$F_y = \left(\frac{\Delta p_0}{y}\right)\left(\frac{kT}{e}\right)\left[\frac{b-1}{p_0 + bN_0 + \Delta p(1+b)}\right]$$

where y is the thinnest dimension of the bar, $b = \mu_e/\mu_h$ and the other symbols have their usual meaning.

9.10 A rod of N type germanium of length $\mathscr{L}$ has an ohmic contact at each end and a small P^+N junction at its centre. One end contact is earthed (zero potential) and a steady positive voltage V is applied to the other end contact. $\mathscr{L}$ and V are such that $2\mathscr{L}^2/V\mu_h$ is very much less than the minority carrier lifetime.

When the current–voltage characteristic of the P^+N junction is examined with the junction positive with respect to earth a region of current controlled negative resistance is observed. Why does this occur? At what potential on the junction would you expect the onset of negative resistance? (Structures of this type are called unijunction transistors and are available commercially. They are used, for example, in free running pulse generators.)

9.11 The P^+N junction on the bar of problem 9.10 is illuminated to generate excess electrons and holes at the junction at a rate $\mathscr{N}$ per second ($\mathscr{N}$ equals number, not density). Show that the onset voltage of the negative resistance is reduced by ΔV where

$$\Delta V = e\mathscr{N}(1+b)\,R/2$$

provided $\Delta V \ll V$ and where R is the resistance of the bar.

10

Magnetism

10.1 Introduction (a)

The magnetic properties of crystals are of great importance in elucidating the fundamental structure of the electron states in solids. It has already been pointed out in Chapters 4 and 7 that the behaviour of band electrons in magnetic fields allows the Fermi surface to be determined in metals and the effective masses in semiconductors. The magneto-optical effects discussed in Chapter 7 give additional methods of determining various aspects of band structure. In the isolated atoms of gases, magnetic properties like the Zeeman effect were historically one of the most powerful methods of determining the electronic structure. Collections of atoms display strong magnetism and those solids where some electrons are best described by atomic orbitals may have much larger magnetic effects than the simple metals described by band theory. It is these 'magnetic' crystals which will be discussed in detail in this chapter. The basic premise is that in these crystals the magnetism is fundamentally of atomic origin, but modified by the interactions between the atoms. This does not however mean that the effect is confined to insulators since in certain metals some of the electrons may show this type of behaviour while other electrons produce metallic behaviour. The electrons held in tight orbitals near the nucleus are most likely to contribute to magnetic properties and it is the transition metal elements with d electrons, the rare earths with 4f electrons, and the actinide elements with 5f electrons which display important magnetic properties.

The energy states of a single atom regarded as an impurity in a crystal were discussed in § 5.6. From this starting point we shall discuss the effect of magnetic fields on such states, and the magnetic interactions between different atoms which give co-operative effects. We shall then examine the nature of the co-operative states and the properties of ordered magnetic systems and, finally, discuss applications.

10.2 Atomic states in a magnetic field (a)

The energy of an electron in a crystal to which a magnetic field $\mathbf{B}$ with vector potential $\mathbf{A} = \frac{1}{2}(\mathbf{B} \times \boldsymbol{r})$ has been applied can be written in the form (called the Hamiltonian)

$$\frac{1}{2m}(\boldsymbol{p} + e\mathbf{A})^2 + 2\mu_B \boldsymbol{s}.\mathbf{B} + V \tag{10.1}$$

where $\boldsymbol{p}$ is the momentum, and $\boldsymbol{s}$ the spin. If the first term is expanded and the angular momentum written

$$\hbar \boldsymbol{l} = (\boldsymbol{r} \times \boldsymbol{p}) \tag{10.2}$$

(10.1) can be written

$$\frac{\boldsymbol{p}^2}{2m} + \mu_B(\boldsymbol{l} + 2\boldsymbol{s}).\mathbf{B} + \frac{e^2}{8m}(\mathbf{B} \times \boldsymbol{r})^2 + V \tag{10.3}$$

Here the first term is the kinetic energy and the last the potential energy. The second term represents the interaction between the effective magnetic dipole moment and the field. The dipole moment has two contributions, $\mu_B \boldsymbol{l}$ from the orbital motion and $2\mu_B \boldsymbol{s}$ from the intrinsic anomalous moment of the electron due to its spin. The Bohr magneton

$$\mu_B = e\hbar/2m = 0.9273 \times 10^{-23} \text{ Am}^2$$

gives a measure of this moment since $\boldsymbol{l}$ will have integral values and $s = \frac{1}{2}$. The third term in (10.3) is quadratic in the displacement $\boldsymbol{r}$ and is the dominant term in the treatment of free electrons (§§ 4.7 and 7.2), when $\boldsymbol{r}$ can become very large. In fact it gives rise to an effective potential well which restricts the orbits into circles. But in an isolated atom $\boldsymbol{r}$ is already restricted by the potential V into an orbit of size about 10^{-8} cm. In a field of 1 tesla (10^4 gauss) the energy associated with this term is then only 10^{-10} eV while the energy of the linear term is 10^{-4} eV. Thus the quadratic 'diamagnetic' term gives only a very small effect in atoms.

The linear 'paramagnetic' term leads to a splitting of any states with non-zero angular momentum. If the atom has several electrons and a total angular momentum L and spin momentum S this term becomes

$$\mu_{\mathrm{B}}(\boldsymbol{L}+2\boldsymbol{S}).\,\mathbf{B} \tag{10.4}$$

If these are coupled to give a total angular momentum $\boldsymbol{J}=\boldsymbol{L}+\boldsymbol{S}$, (10.4) may be written

$$\lambda_{\mathrm{g}}\mu_{\mathrm{B}}\boldsymbol{J}.\,\mathbf{B} \tag{10.5}$$

where λ_{g} is Landé's factor which gives the effective magnetic moment $\lambda_{\mathrm{g}}\mu_{\mathrm{B}}\boldsymbol{J}$. The value of λ_{g} may be obtained by projecting $(\boldsymbol{L}+2\boldsymbol{S})$ along $\boldsymbol{J}$

$$(\boldsymbol{L}+2\boldsymbol{S}).\boldsymbol{J}=\lambda_{\mathrm{g}}\boldsymbol{J}^2=\boldsymbol{J}^2+(\boldsymbol{S}.\boldsymbol{J}) \tag{10.6}$$

and since

$$(\boldsymbol{J}-\boldsymbol{S})^2=\boldsymbol{L}^2$$

i.e.

$$2\boldsymbol{S}.\boldsymbol{J}=\boldsymbol{J}^2+\boldsymbol{S}^2-\boldsymbol{L}^2$$

$$\lambda_{\mathrm{g}}\boldsymbol{J}^2=\tfrac{3}{2}\boldsymbol{J}^2+\tfrac{1}{2}\boldsymbol{S}^2-\tfrac{1}{2}\boldsymbol{L}^2 \tag{10.7}$$

Now the value of the square of an angular momentum operator $\boldsymbol{l}^2$ is $l(l+1)$ and hence

$$\lambda_{\mathrm{g}}=\tfrac{3}{2}+\frac{S(S+1)-L(L+1)}{2J(J+1)} \tag{10.8}$$

by substituting into (10.7) for a state with L, S and J defined.

If the crystal field effects discussed in § 5.6.2 are important the states with a given angular momentum will not be degenerate but electrons in different orbits will have different energies. As discussed there the orbital momentum is often 'quenched' and only the $(2S+1)$ spin degeneracy remains. This will be affected by a magnetic field through a term like

$$\mathrm{g}\mu_{\mathrm{B}}\boldsymbol{S}.\,\mathbf{B} \tag{10.9}$$

as if only the spin were present.

If the orbital momentum is completely quenched the atom appears to have a free spin and g in (10.9) is just 2 as in (10.4). But the spin-orbit coupling (4.18), which may be written $\zeta\boldsymbol{L}.\boldsymbol{S}$ where ζ has the dimensions of energy, will attempt to align some orbital moment parallel or antiparallel with the spin depending on the sign of ζ. If the shell of magnetic d or f electrons is less than half-full $\zeta>0$; if it is more than half-full $\zeta<0$. The crystal field opposes this alignment and if it causes a typical energy splitting of say Δ the actual alignment achieved will be about ζ/Δ. But this orbital moment also gives a magnetic moment and so a value of

$$\mathrm{g}\sim 2+\zeta/\Delta \tag{10.10}$$

is expected. Δg, the difference $g - 2$, gives a measure of the orbital contribution in this situation. When there is incomplete quenching small splittings of the S states may occur which reflect the crystal field. For example if there is a quadrupole field as in (5.19) it will be reflected in a term of the form

$$D_2(3S_z^2 - S(S+1)) \tag{10.11}$$

where D_2 is small. In some cases $D_2 \sim (\Delta g)^2 A_2^0$ reflecting the amount of the orbital contribution.

If the crystal field does not quench the orbital moment because of high symmetry or large ζ/Δ, it will reduce the degeneracy of the states. If the state so formed is a singlet it will have no magnetic moment. Doublets are always found if the atom has an odd number of electrons. Such a doublet may be ascribed a fictitious spin $\mathscr{S} = \frac{1}{2}$ and the two degenerate states labelled $\mathscr{S}_z = \pm\frac{1}{2}$. The doublet will be split by the applied magnetic field in a way described by

$$\mu_B \mathscr{S} . \mathbf{g} . \mathbf{B}. \tag{10.12}$$

Here in general $\mathbf{g}$ is a tensor and gives an energy $\pm\frac{1}{2} g\mu_B \mathrm{B}$ which depends on the direction of $\mathbf{B}$. The value of g will be determined by the actual orbit prescribed by the crystal field.

10.3 Observation of magnetic splittings (a)

The most direct way of measuring the splitting in the energy levels of atoms in crystals induced by a magnetic field is to observe transitions between them. The optical transitions between states coming from the same atomic configuration d^n or f^n are nearly forbidden to electric dipole transitions but can nevertheless be observed in solids. The Zeeman effect as demonstrated in figure 10.1 will

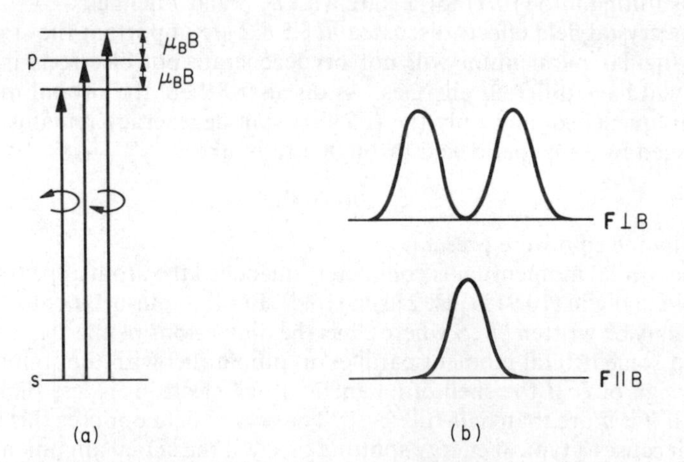

Figure 10.1 Transitions from an s to a p state split by a magnetic field B. (a) Transitions to states $l_z = -1, 0, +1$ occur respectively for radiation polarised left-handed, parallel and right-handed with respect to the direction of B. (b) The absorption for radiation polarised perpendicular to B shows a splitting if the lines are narrow

show the splitting, provided the lines are narrow enough. Since readily available fields are of order 1 tesla, and 10 tesla is the normal limit for solenoids with large generators or with superconducting solenoids, the Zeeman effect is not readily observable for situations where the electrons are strongly coupled to the lattice and the lines are broad—as for example in the F centre (figure 5.28). In these cases some information can be obtained from the Faraday effect (§ 7.2) as demonstrated in figure 10.2.

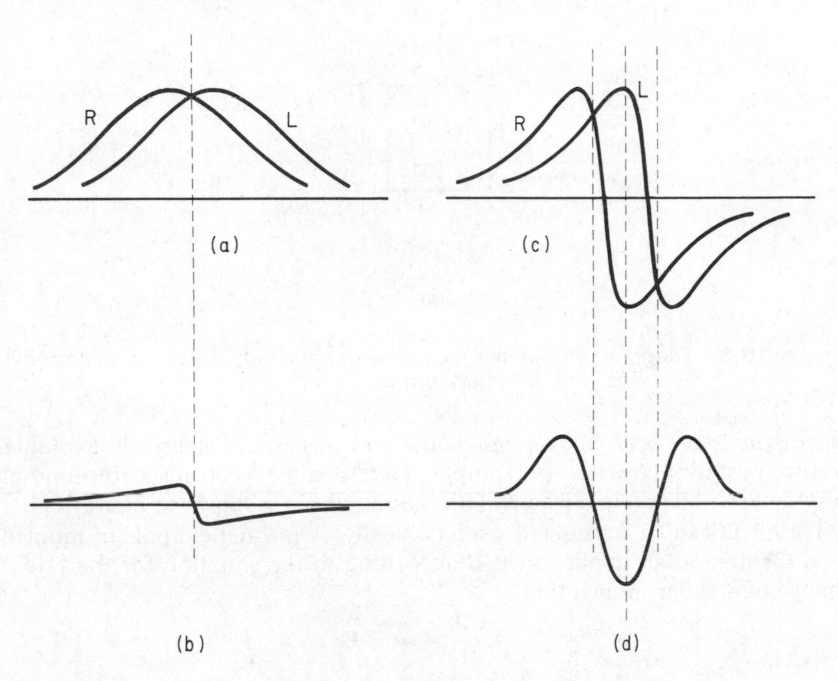

Figure 10.2 (a) Absorption as in figure 10.1 for left (L) and right (R) circularly polarised light in broad lines so that the splitting is not resolved. (b) Difference in absorption of right- and left-handed light (circular dichroism) (c) Variation of refractive indices of circularly polarised light due to absorption lines
(d) Difference in indices in (c) giving rotation of plane polarised light (Faraday effect)

Detailed information, particularly for the lowest multiplet, can be obtained by inducing transitions between the split states. For B ~ 1 tesla, $2\mu_B B \sim 10^{-4}$ eV so this requires electromagnetic radiation with a frequency of about 30 GHz and a wavelength of 1 cm—in the microwave region. The experiment must therefore be performed in a cavity or waveguide—a diagrammatic representation of the apparatus is shown in figure 10.3. Since it is relatively difficult to sweep the frequency in this range it is normal to work at fixed frequency and vary the field until resonance is achieved. This type of experiment is called

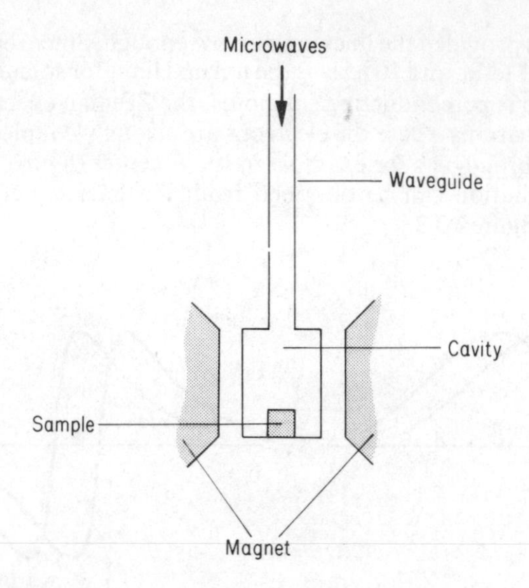

Figure 10.3 Diagram of paramagnetic resonance apparatus at microwave frequencies

paramagnetic or electron spin resonance and has been extensively exploited over the past twenty years. An example of a resonance spectrum corresponding to systems with energies given by (10.9) and (10.11) is shown in figure 10.4.

The effect can be considered semiclassically. A magnetic dipole of moment $\boldsymbol{m}$ will rotate in an applied field $\mathbf{B}$ according to the equation for the rate of change of angular momentum

$$\hbar \frac{\mathrm{d}\boldsymbol{J}}{\mathrm{d}t} = \boldsymbol{m} \times \mathbf{B}$$

or

$$\left[\frac{\mathrm{d}\boldsymbol{m}}{\mathrm{d}t}\right]_f = \left(\frac{\lambda_g e}{m}\right) \boldsymbol{m} \times \mathbf{B} \tag{10.13}$$

where $(\lambda_g e/m)$ is the gyromagnetic ratio. The suffix f indicates that this is the rate of change due to the field. If damping is included there is a further rate of change due to this

$$\left[\frac{\mathrm{d}\boldsymbol{m}}{\mathrm{d}t}\right]_\mathrm{d} = \frac{\boldsymbol{m}_0 - \boldsymbol{m}}{\tau} \tag{10.14}$$

where τ is the characteristic lifetime of $\boldsymbol{m}$. (10.13) and (10.14) give three scalar equations which, with $\mathbf{B}$ along the z-axis, take the form

$$\begin{aligned}
\mathrm{d}\boldsymbol{m}_z/\mathrm{d}t &= (\boldsymbol{m}_{0z} - \boldsymbol{m}_z)/\tau_1 \\
\mathrm{d}\boldsymbol{m}_x/\mathrm{d}t &= (\lambda_g e/m)\, \boldsymbol{m}_y \mathbf{B} - \boldsymbol{m}_x/\tau_2 \\
\mathrm{d}\boldsymbol{m}_y/\mathrm{d}t &= -(\lambda_g e/m)\, \boldsymbol{m}_x \mathbf{B} - \boldsymbol{m}_y/\tau_2.
\end{aligned} \tag{10.15}$$

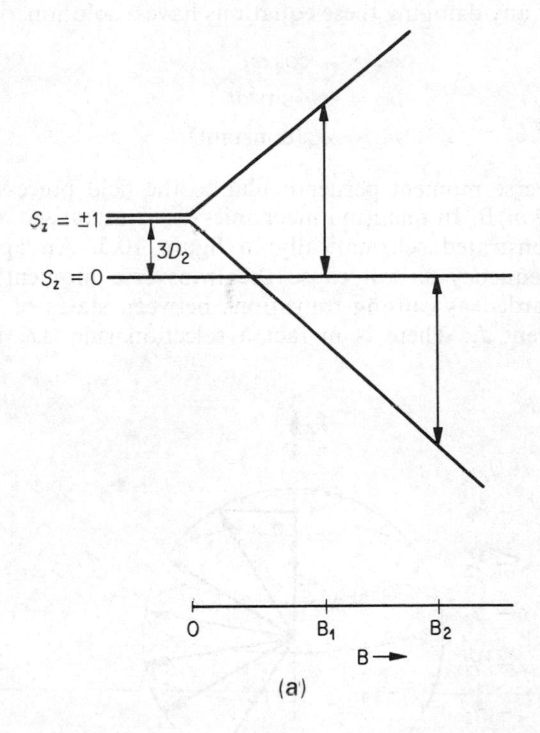

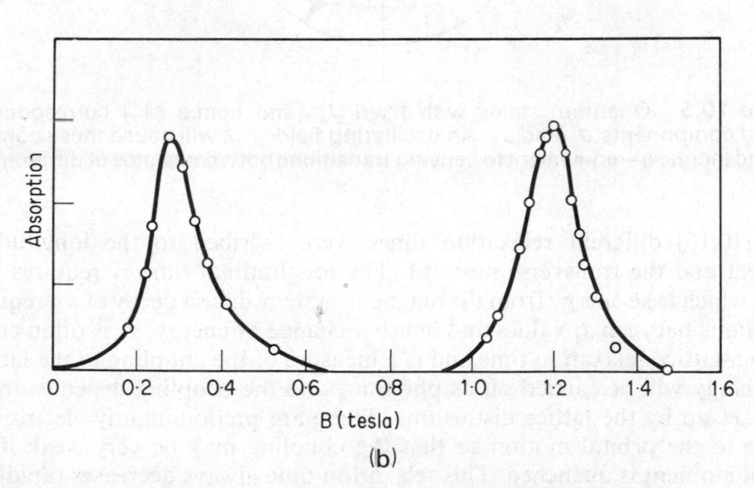

Figure 10.4 (a) Energy level variation of $S = 1$ states in a field **B** parallel to z-axis and axial anisotropy given by D_2 (10.11)
(b) Resonance spectrum of Ni^{2+} ions in nickel fluorosilicate corresponding to transitions marked in (a). (After A. N. Holden *et al.*, *Phys. Rev.*, **75**, 1443 (1949).) The resonances occur at $g\mu_B B_1 = \hbar\omega - 3D_2$, $g\mu_B B_2 = h\omega + 3D_2$. In this case $\omega = 2.44$ GHz, $g = 2.29$ and $3D_2 = -0.49$ cm^{-1}

In the absence of any damping these equations have a solution

$$m_x = m' \cos \omega t$$
$$m_y = -m' \sin \omega t$$
$$m_z = m_{z0}(\text{constant}) \qquad (10.15a)$$

where the transverse moment perpendicular to the field precesses with frequency $\omega_0 = (\lambda_g e/m)B$. In quantum mechanics m_z is quantised and equal to $\lambda_g \mu_B J_z$, as demonstrated schematically in figure 10.5. An applied transverse field of frequency ω will cause the transverse moment to increase and may be regarded as causing transitions between states of different m_z and hence different J_z. There is in fact a selection rule $\Delta J_z = 1$ for such transitions.

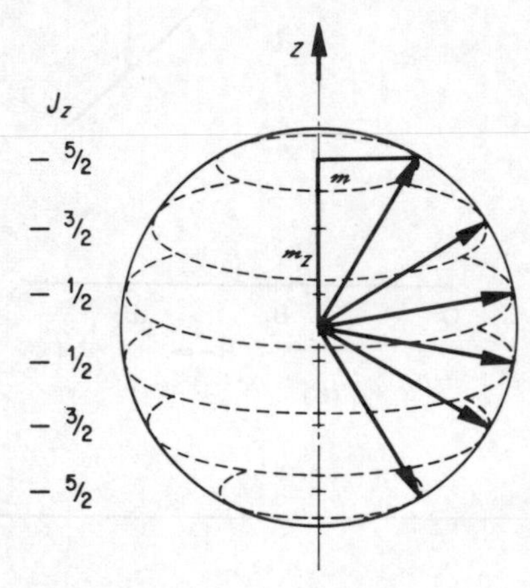

Figure 10.5 Quantum states with fixed J_z (and hence M_z) correspond to varying components J_x and J_y. An oscillating field ⊥ z will cause these components to increase—equivalent to causing transitions between states at different M_z

In (10.15) different relaxation times were ascribed to the longitudinal moment and the transverse moment. The longitudinal time τ_1 requires processes which take energy from the magnetic system since a decay of m_z requires transitions between J_z values and hence a change of energy. τ_1 is often called the spin-lattice relaxation time and is a measure of the coupling to the lattice. The energy will be carried off as phonons, and the coupling depends on the fields set up by the lattice distortions. These are predominantly electric and couple to the orbital motion so that the coupling may be very weak if the orbital moment is quenched. This relaxation time always decreases rapidly as T increases and the thermal agitation of the lattice grows (see figure 10.6).

The processes giving the transverse time τ_2 are not restricted to those which produce a decay of the energy and in general $\tau_2 < \tau_1$. τ_2 determines the width of the resonance and

$$\omega_0 \tau_2 > 1 \qquad (10.16)$$

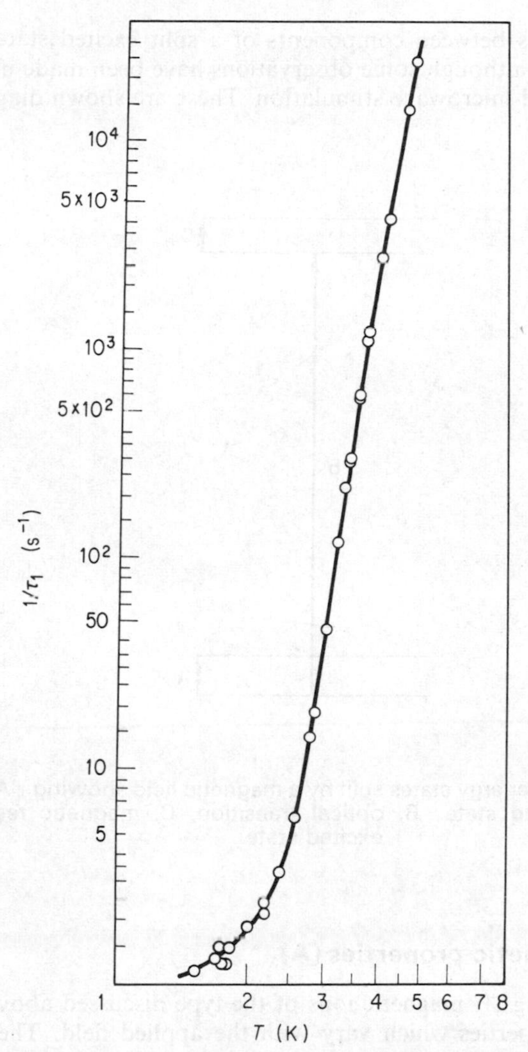

Figure 10.6 Variation of τ_1 with temperature for Sm^{3+} ions in yttrium ethyl sulphate. Note the rapid decrease in τ_1 (roughly proportional to T^{-9}) as T increases. (After G. H. Larson and C. D. Jeffries, *Phys. Rev.*, **141**, 461 (1966).)

is essential if resonance is to be observed. For $\omega_0 \sim 2\pi \times 30$ GHz this requires $\tau_2 > 5 \times 10^{-12}$ s. At high enough T the spin lattice component always eventually broadens the resonance away. For systems with large orbital contributions resonance may only be observable at $T < 4$ K but for spin only ions the resonance may be observable above room temperature. At low T random magnetic fields due to the surrounding magnetic dipoles—other atomic moments or nuclear moments—dominate the line width. This contribution to τ_2 is largely independent of temperature.

Direct transitions between components of a split excited state are more difficult to observe although some observations have been made using simultaneous optical and microwave stimulation. These are shown diagrammatically in figure 10.7

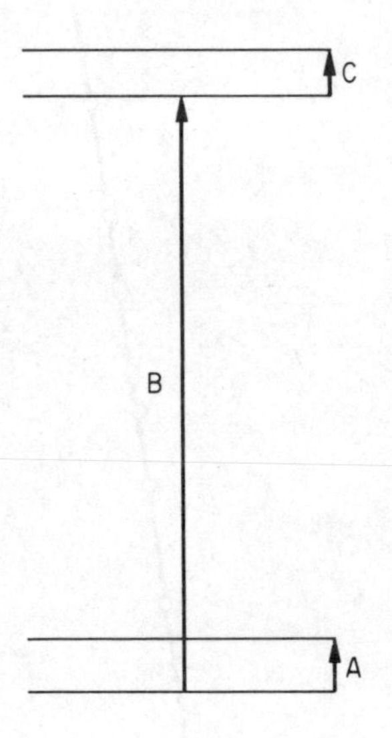

Figure 10.7 Two energy states split by a magnetic field showing : A, magnetic resonance in ground state; B, optical transition, C, magnetic resonance in excited state

10.4 Bulk magnetic properties (A)

A crystal containing $\mathcal{N}$ magnetic ions of the type discussed above will have bulk magnetic properties which vary with the applied field. These may be related to the behaviour of the energy levels in such a field *via* statistical mechanics. The variation in energy of each level i can be expanded in powers of the magnetic field

$$W^i = W_0^i + \mathrm{B}W_1^i + \mathrm{B}^2 W_2^i + \dots \qquad (10.17)$$

The term linear in B must arise from the linear term in (10.3). The quadratic term may arise from the quadratic term in (10.3) and also from the linear term used in second order of perturbation theory. The diamagnetic contribution always gives a positive contribution to W_2^i, while the second-order perturbation term will be negative on the lowest energy levels since these are repelled downwards. The relative occupation of each level is given by $\exp(-W^i/kT)$.

Since the magnetic energy is $-\mathcal{M}\,.\,\mathbf{B}$ the magnetic moment is related to $-\partial W^i/\partial \mathbf{B}$. In fact, averaging over the relative occupations

$$\mathcal{M} = -\mathcal{N}\,\frac{\sum_i \frac{\partial W^i}{\partial \mathbf{B}} \exp(-W^i/kT)}{\sum_i \exp(-W^i/kT)} \tag{10.18}$$

If we restrict considerations to a degenerate ground state where all W_0^i are equal and use only the linear term

$$W_1^i = \lambda_g \mu_B \, J_z \, (J_z = J, J-1 \ldots -J) \tag{10.19}$$

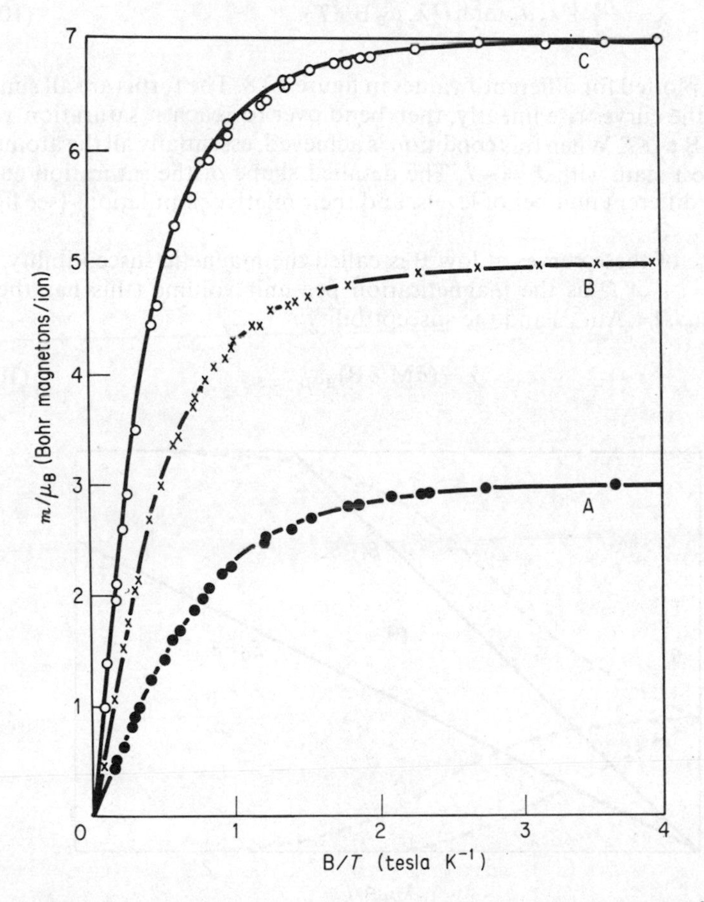

Figure 10.8 Theoretical variation of m as a function of $\lambda\mu_B/\mathrm{B}kT$. Experimental results are for Cr^{3+} ($S = \frac{3}{2}$) in potassium chrome alum (A), Mn^{2+} ($S = \frac{5}{2}$) in iron ammonium alum (B) and Gd^{3+} ($S = \frac{7}{2}$) in gadolinium sulphate octahydrate (C). (After W. Henry, *Phys. Rev.*, **88**, 559 (1952).)

in the splitting given by (10.5), and the algebraic sum in (10.18) can be performed to give

$$\mathscr{M} = -\mathscr{N}\lambda_g\mu_B[(J+\tfrac{1}{2})\coth[(J+\tfrac{1}{2})\lambda_g\mu_B \mathrm{B}/kT] - \tfrac{1}{2}\coth(\tfrac{1}{2}\lambda_g\mu_B\mathrm{B}/kT)]$$
$$= -\mathscr{N}\lambda_g\mu_B J\mathscr{B}_J(\lambda_g\mu_B\mathrm{B}/kT) \tag{10.20}$$

where $\mathscr{B}_J$ is called a Brillouin function. For $J(S$ or $\mathscr{S})$ equal to one-half $\mathscr{M}$ has a simple form

$$\mathscr{M} = \tfrac{1}{2}\mathscr{N}\lambda_g\mu_B \frac{\exp(-\frac{1}{2}\lambda_g\mu_B\mathrm{B}/kT) - \exp(\frac{1}{2}\lambda_g\mu_B\mathrm{B}/kT)}{\exp(-\frac{1}{2}\lambda_g\mu_B\mathrm{B}/kT) + \exp(\frac{1}{2}\lambda_g\mu_B\mathrm{B}/kT)}$$
$$= -\tfrac{1}{2}\mathscr{N}\lambda_g\mu_B\tanh(\tfrac{1}{2}\lambda_g\mu_B\mathrm{B}/kT) \tag{10.21}$$

$\mathscr{M}/\mathscr{N}\mu_B$ is plotted for different J values in figure 10.8. The forms are all similar. At low B, the curves rise linearly, then bend over to reach a saturation value when $\lambda_g\mu_B\mathrm{B} \gg kT$. When this condition is achieved, essentially all the atoms are in the lowest state with $J_z = -J$. The detailed shape of the saturation curves reflects the different number of levels, and their relative populations (see figure 10.9).

The slope of these curves at low B is called the magnetic susceptibility. We define $\mathrm{M} = \mathscr{M}/\mathscr{N}\Omega$ as the magnetisation per unit volume (this has the dimensions $\mu_B/\Omega = \mathrm{Am}^{-1}$) and the susceptibility

$$\chi = (\partial\mathrm{M}/\partial\mathrm{B})_{\mathrm{B}\to 0} \tag{10.22}$$

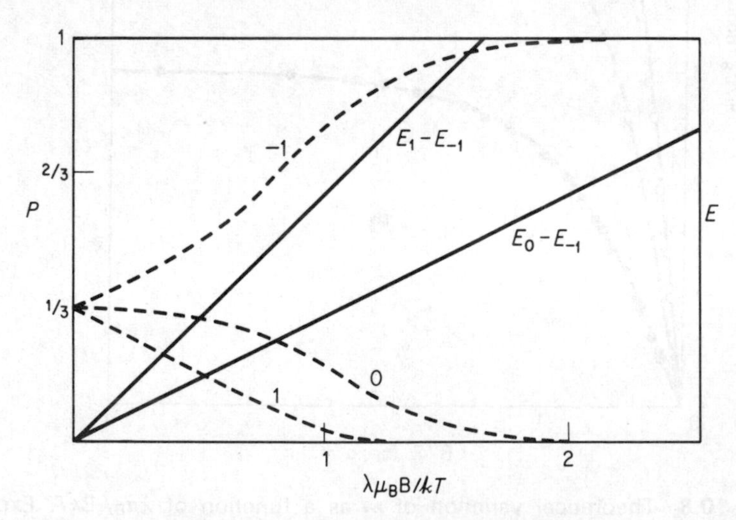

Figure 10.9 Variation of the energies E of excited states $J_z = \pm 1$, 0 of a triplet $J = 1$ in a magnetic field. Population variation P of the three J_z states is shown dashed

The general form is

$$\chi = \frac{\sum_i \left[\left(\frac{\partial W^i}{\partial \mathrm{B}}\right)^2 \Big/ kT - \frac{\partial^2 W^i}{\partial \mathrm{B}^2}\right] \exp(-W^i/kT)}{\Omega \sum_i \exp(-W^i/kT)} - \frac{\mathrm{M}^2 \Omega}{kT} \qquad (10.23)$$

differentiating (10.18). As $\mathrm{B} \to 0$

$$\chi = \sum_i \left[\frac{(W_1^i)^2}{kT} - 2W_2^i\right] \exp(-W_0^i/kT) \Big/ \Omega \sum_i \exp(-W_0^i/kT) \qquad (10.24)$$

For an assembly of effectively free atoms of angular momentum J

$$\chi = \frac{\lambda_g^2 \mu_B^2 \sum_{-J}^{J} J_z^2}{kT\Omega \sum_{-J}^{J} 1} = \frac{\lambda_g^2 \mu_B^2 J(J+1)}{3kT\Omega} = \frac{\mu_{\mathrm{eff}}^2}{3kT\Omega} \qquad (10.25)$$

This has the form of Curie's Law $\chi = C/T$ with an effective magnetic moment given by

$$\mu_{\mathrm{eff}} = \lambda_g \mu_B \sqrt{J(J+1)} \qquad (10.26)$$

The susceptibilities of most rare-earth compounds are in agreement with this formula around room temperature which indicates that at that temperature the ions behave as if they are free. Typical results are shown in the table 10.1.

	$4f^n$	L	S	J	λ_g	Calc. μ_{eff}	Exp. μ_{eff}
La	0	0	0	0	0	0	0
Ce	1	3	$\frac{1}{2}$	$\frac{5}{2}$	$\frac{6}{7}$	2.54	2.4–2.7
Pr	2	5	1	4	$\frac{4}{5}$	3.58	3.4–3.6
Nd	3	6	$\frac{3}{2}$	$\frac{9}{2}$	$\frac{8}{11}$	3.62	3.4–3.7
Pm	4	6	2	4	$\frac{3}{5}$	2.68	—
Sm	5	5	$\frac{5}{2}$	$\frac{5}{2}$	$\frac{2}{7}$	0.84	1.3–1.6
Eu	6	3	3	0	—	0	3.2–3.4
Gd	7	0	$\frac{7}{2}$	$\frac{7}{2}$	2	7.94	7.9–8.0
Tb	8	3	3	6	$\frac{3}{2}$	9.7	9.4–9.8
Dy	9	5	$\frac{5}{2}$	$\frac{15}{2}$	$\frac{4}{3}$	10.6	10.5–10.7
Ho	10	6	2	8	$\frac{5}{4}$	10.6	10.3–10.6
Er	11	6	$\frac{3}{2}$	$\frac{15}{2}$	$\frac{6}{5}$	9.6	9.4–9.6
Tm	12	5	1	6	$\frac{7}{6}$	7.6	7.4–7.6
Yb	13	3	$\frac{1}{2}$	$\frac{7}{2}$	$\frac{8}{7}$	4.5	4.4–4.6
Lu	14	0	0	0	0	0	0

Table 10.1 Atomic and Magnetic Properties of Trivalent Rare Earth Ions

	$3d^n$	S	L	Octahedral degeneracy	Spin only μ_{eff}	Exp.
Ti	1	$\frac{1}{2}$	2	3	1.7	1.6
V	2	1	3	3	2.8	2.5
Cr	3	$\frac{3}{2}$	3	1	3.9	3.9
Cr, Mn	4	2	2	2	4.9	7.8
Mn, Fe	5	$\frac{5}{2}$	0	1	5.9	5.9
Fe, Co	6	2	2	3	4.9	5.2
Co	7	$\frac{3}{2}$	3	3	3.9	4.9
Ni	8	1	3	1	2.8	3.1
Cu	9	$\frac{1}{2}$	2	2	1.7	1.9
Zn	10	0	0	1	0	0

Table 10.2 Atomic and Magnetic Properties of Transition Metal Ions

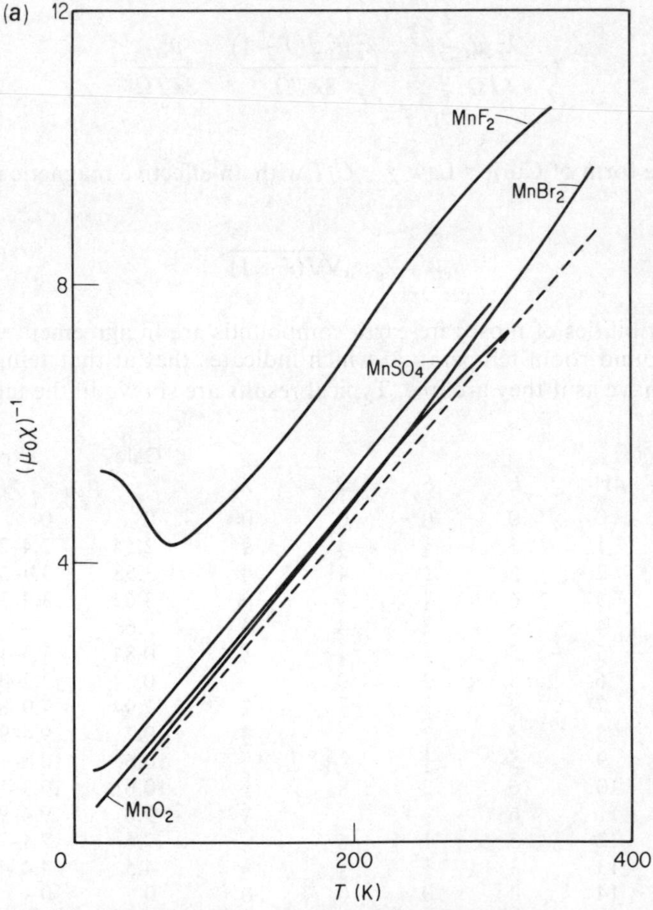

Figure 10.10 Experimental variation of paramagnetic susceptibility $1/\chi$ *versus* T in (a) manganese salts (susceptibility per molar volume, after W. J. de Haas *et al.*, *Physica*, **7**, 57 (1940))

At lower T deviations from Curie's Law are observed due to the crystal field splitting and in most cases near $T=0$, the magnetisation curves are typical of those of a doublet.

Some transition metal compounds have quenched orbital angular momentum and behave as free spins. In these cases Curie's Law is obeyed down to low T with $\lambda_g = 2$ and $J = S$. This is especially true of ions which have half-filled shells of electrons when $L=0$ (e.g. $Gd^{3+}4f^7$, Mn^{2+} and $Fe^{3+}3d^5$) but also occurs in many other cases. Some experimental results are shown in figure 10.10 and typical values of μ_{eff} in iron group compounds are given in the table 10.2.

The paramagnetic susceptibility given by Curie's Law arises from the term in $(W_1^i)^2/kT$ in (10.24). A further contribution, independent of temperature at low T arises from W_2^i. This is usually unimportant at low T unless the first term is absent. One may then have temperature independent paramagnetism from the negative part of W_2^i arising in second order from the paramagnetic term in (10.3) which is linear in B. This has a typical size

$$\mu_B^2/\Delta\Omega \qquad (10.27)$$

where Δ is the separation in energy between the ground and some excited level.

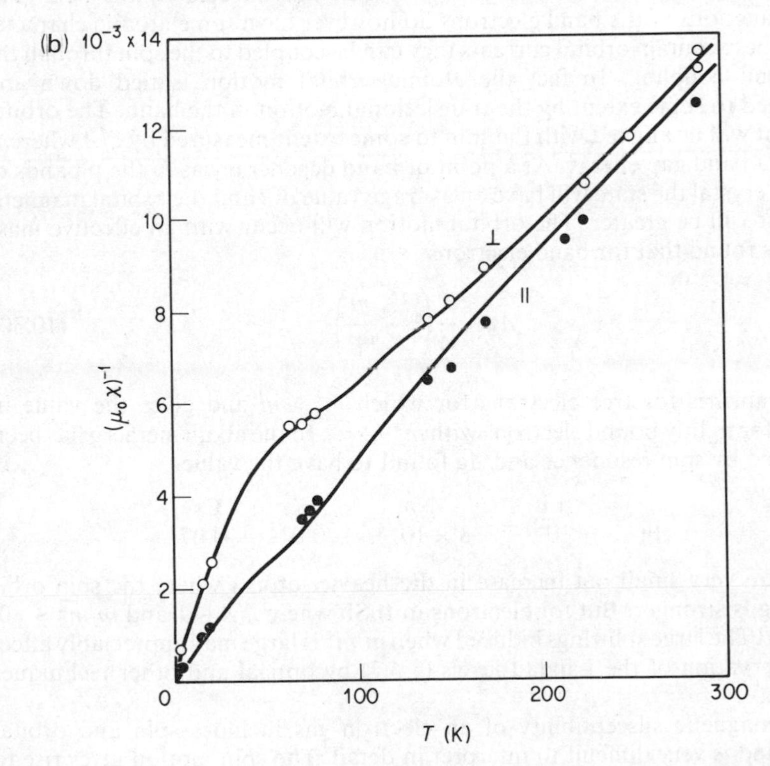

Figure 10.10 (b) neodymium ethyl sulphate (susceptibility per m³) (after R. J. Elliott and K. W. H. Stevens, *Proc. Roy. Soc.*, **A219**, 387 (1953))

For truly diamagnetic ions with $L = S = 0$ only the quadratic term in (10.3) contributes giving a positive W_2^i and a negative diamagnetic susceptibility. Its magnitude is

$$\chi = -\frac{e^2\langle r^2\rangle}{\Omega 6m} \tag{10.28}$$

where $\langle r^2\rangle$ is the average mean-square radius of all the atomic electrons. Note that the typical paramagnetic Curie Law susceptibility is always larger unless

$$kT \sim \hbar^2/m\langle r^2\rangle \tag{10.29}$$

which typically corresponds to $T \sim 10^3$ K.

10.5 Paramagnetism and diamagnetism of band electrons (A)

In §§ 4.7 and 7.2 the large-scale circular motion of free electrons in a magnetic field was considered in some detail. It was also pointed out that the electrons had an intrinsic spin motion which was affected by the field. The wave functions of the band electrons do however have some atomic character and if these contain orbital currents they can be coupled to the spin through the spin-orbit coupling. In fact the atomic orbital motion is tied down and quenched to some extent by the translational motion in the band. The orbital moment will be aligned with the spin to some extent measured by ζ/Δ where Δ is now a band gap energy. At a point of band degeneracy as in the p bands of a cubic crystal the state will have an average value of l and the orbital magnetic moment will be greater. The orbital motion will occur with an effective mass and it is found that for band electrons

$$\Delta g = \frac{\zeta}{\Delta}\left(1 - \frac{m}{m^*}\right) \tag{10.30}$$

i.e. it vanishes for free electrons for which $m^* = m$ and gives the value in (10.10) for tightly bound electrons with $m^* \to \infty$. In the alkali metals g has been measured by spin resonance and Δg found to have the values.

	Li	Na	K	Cs
Δg	$<10^{-4}$	8×10^{-4}	0.01	0.07

These are very small but increase in the heavier atoms where the spin orbit coupling is stronger. But for electrons in InSb where $\zeta/\Delta \sim 1$ and $m/m^* \sim 80$, $g \sim -50$! The large splittings induced when m/m^* is large may appreciably affect the observation of the Landau levels (§ 4.7) by optical and other techniques (§ 7.2.8).

The magnetic susceptibility of an electron gas includes spin and orbital effects and is very difficult to interpret in detail. The spin motion gives rise to two sub-bands separated by $g\mu_B B$ as shown in figure 10.11. The bands are filled to the same Fermi energy and the magnetisation is given by the surplus of

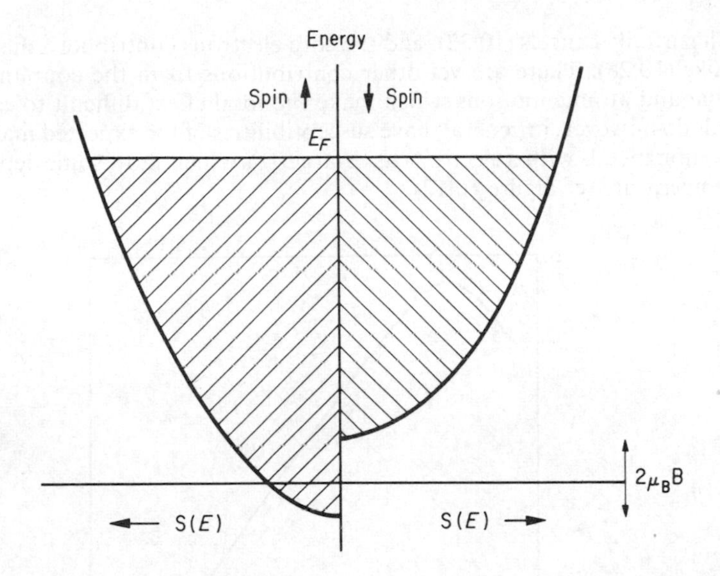

Figure 10.11 Density of states showing splitting of energy bands in a magnetic field

up spins over down spins. The difference in the number of electrons in the two bands is related to the density of states S(E) at the Fermi level (there is a density $\frac{1}{2}$S(E) in each sub-band). In fact

$$M = \tfrac{1}{2}g^2 \mu_B^2 \tfrac{1}{2}S(E_F)\,B \tag{10.31}$$

so that in a simple free electron band

$$\chi = 3g^2 \mu_B^2 / 8E_F \Omega \tag{10.32}$$

using (4.20). The susceptibility is independent of T. Comparing this formula with Curie's Law (10.25) and with (10.27) we see that the paramagnetic susceptibility of all these systems contains in the denominator the energy which opposes the alignment of the magnetic moments by the field; in the case of Curie's Law it is the thermal energy kT, but here the Fermi energy E_F. If there are few electrons in a band as in a semiconductor and Boltzmann statistics hold, the susceptibility reverts to the Curie Law

$$\chi = g^2 \mu_B^2 / 4kT\Omega \tag{10.33}$$

This spin susceptibility of band electrons can be measured separately on a few systems by means of paramagnetic resonance. But the bulk susceptibility measurements of real metals also include various other contributions. The diamagnetic susceptibility of free electrons is found to be

$$\chi = -\mu_B^2 / 2E_F \Omega \tag{10.34}$$

which partially cancels (10.32), and the core electrons contribute a diamagnetism like (10.28). There are yet other contributions from the coupling to the circular and atomic motions which make the total effect difficult to calculate. Metals do however, in general, have susceptibilities of the expected magnitude. Transition metals with d electrons have larger χ's which show some dependence on temperature (cf. figure 10.12).

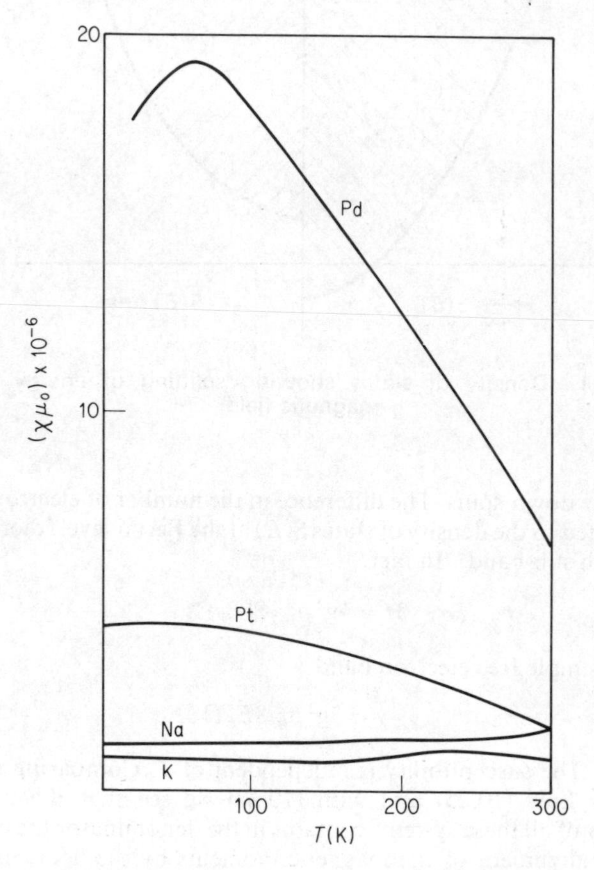

Figure 10.12 Total magnetic susceptibility of some real metals

10.6 Interactions between magnetic moments

10.6.1 Magnetic forces (A)

An ideal paramagnetic crystal as discussed in § 10.4 consists of a set of magnetic atoms or ions, each carrying a small magnetic moment, with no interaction between them. If a magnetic field is applied the moments align preferentially along the field—the alignment achieved is reflected in the magnetisation of the system. For a given field a larger magnetisation is achieved at low temperatures since the thermal disorder in the magnetic moments is more easily overcome there. For a large enough value of the ratio μ_B/kT the maximum saturation moment is achieved (cf. figure 10.8). But some substances appear to have a large magnetisation even in the absence of a field. These permanent magnets have attracted the attention of physicists for centuries. The commonest such material is of course metallic iron and this has given rise to the name of this phenomenon—ferromagnetism. The ordering of the constituent atomic moments suggests that there is some magnetic field within the system and ferromagnetism was 'explained' on this basis by Weiss in 1907.

The origin of the field remained a mystery until 1925. The normal interaction between two magnetic moments $\boldsymbol{m}_1, \boldsymbol{m}_2$ is *via* the dipole–dipole interaction which if the moments are a distance $\boldsymbol{R}$ apart, is given by

$$\frac{\mu_0}{4\pi}\left[\frac{\boldsymbol{m}_1 . \boldsymbol{m}_2}{R^3} - \frac{3(\boldsymbol{m}_1 . \boldsymbol{R})(\boldsymbol{m}_2 . \boldsymbol{R})}{R^5}\right] \tag{10.35}$$

where μ_0 is the permeability of free space. The energy of this interaction between atomic moments about 10^{-4} eV corresponding to a temperature of only 1 K when $m \sim \mu_B$ and $R \sim 10^{-8}$ cm. It is difficult to calculate the sum over all other magnets in a crystal to obtain the field at a given dipole except in certain simple situations. For a cubic crystal it is possible to do the sum over all neighbours in a sphere surrounding a given dipole and show that it is zero.

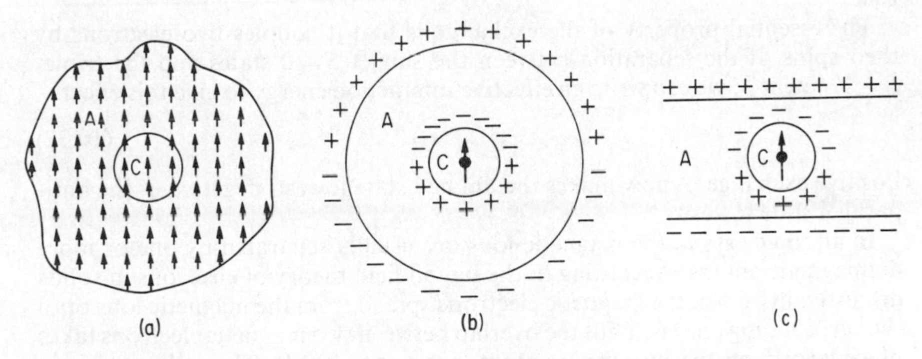

Figure 10.13 Origin of dipole field in a ferromagnet. The region C should be chosen to be large compared to an interatomic spacing. The dipoles in region A of (a) give an effect equivalent to a layer of poles on the surface as shown. For a spherical specimen (b) the total effect is zero, while for a plate (c) there is a demagnetising field

The effect of the distant dipoles (in the region A of figure 10.13) can be summed by assuming a continuous distribution of magnetisation. The field then arises from poles on the surface and depends on the shape of the specimen

$$\mathbf{B}_i = (\tfrac{1}{3} - \delta_m)\mathbf{M}\mu_0 \tag{10.36}$$

where δ_m is zero for a needle-shaped specimen, $\frac{1}{3}$ for a sphere and 1 for a flat plate. The first factor of $\frac{1}{3}$ comes from the inner surface of the spherical cavity in A. Thus this *demagnetising* field $\mathbf{B}_i$ is of order $\mathbf{M}\mu_0$. In Fe it is some 2 tesla and the energy of a dipole in this field corresponds to a temperature $\mu_B B_i/k \sim 2$K. But Fe maintains its permanent moment up to 1043 K, fields of order 2 tesla have only a small effect on it and the magnetisation is not noticeably shape-dependent. Thus this demagnetising field, while it does have some measurable effects (cf. § 10.11), cannot be the origin of the ferromagnetism.

10.6.2 Exchange forces (A)

The origin of the large interactions lies in what are called exchange forces—these are essentially electrostatic and influence the electron spins because of quantum mechanical restrictions on the antisymmetry of the wave function. The alignment of electron spins by Coulomb forces is familiar in atomic theory—the lowest energy level in atoms has the maximum value of total spin S available in the electronic configuration (cf. § 5.6.2). In simple molecules the effect is opposite in sign. Two electrons in the lowest state of the H_2 molecule have their spins antiparallel and a total $S = 0$. The triplet $S = 1$ state has a much higher energy (cf. figure 1.11) and does not correspond to a bound molecule. We are concerned with the exchange forces between two electrons largely confined to two separate atoms. It is difficult to predict *a priori* which sign the exchange coupling will have, if anything it appears to be nearer to the molecular case.

The essential property of the exchange is that it couples two electrons by their spins. If the separation between the singlet $S = 0$ states and the triplet $S = 1$ state is $\mathcal{J}$ we can write an effective interaction energy to give this effect

$$-\mathcal{J}(\mathbf{s}_1 . \mathbf{s}_2) \tag{10.37}$$

Positive exchange $\mathcal{J}$ now makes the aligned state lowest, negative $\mathcal{J}$ the anti-parallel state lowest.

In an ionic crystal the magnetic ions are usually separated by one or more diamagnetic anions. According to the ligand field theory of electrons in solids discussed in § 5.6.3, the magnetic electrons spread from the magnetic ions onto the surrounding anions. Thus the overlap between two magnetic electrons takes place largely on the intervening atom (see figure 10.14). There the magnetic electron has a wave function like that of an atomic electron but spends a fraction α^2 of its time on this atom. We may therefore expect an exchange derived from atomic exchange $\mathcal{J}_{at}$ favouring alignment of

$$\mathcal{J} = \alpha^4 \mathcal{J}_{at} \tag{10.38}$$

In an insulator there is, however, another mechanism which favours the other sign. We saw in § 4.10 that in an insulator the repulsive energy U between two magnetic electrons would greatly exceed the 'kinetic' energy E_δ which is gained from transfer of the electrons. However, some transfer will take place by including in the state some ionic character where one electron is transferred

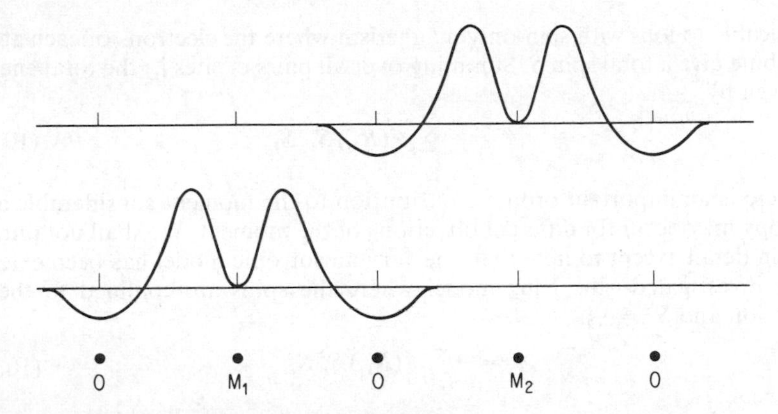

Figure 10.14 Variation of wave functions for two neighbouring magnetic ions M with an intervening anion O. The wave functions are d-like on the magnetic ions and s-like on the anions

to an adjacent site. This will lower the energy by E^2/U, but the transfer can only take place if the two atomic spins are antiparallel since otherwise they would be forbidden by the exclusion principle to reside on the same atom. Thus the antiparallel spin state is lowered by an effective exchange energy

$$\mathscr{J} = -E_{\delta}^2/U \tag{10.39}$$

As we shall indicate below, it is found that this latter form of exchange dominates in insulators.

In a metal like Fe the d electrons which give rise to the magnetism are in narrow bands and it is more difficult to devise a simple description of the origin of the exchange. Experimentally of course, it is positive and large but no satisfactory microscopic theory of these transition metals exists.

In the rare-earth metals the magnetic f electrons are well localised in atomic-like wave functions, and the electrical conductivity is provided by 6s, 6p and 5d electrons which form bands. The overlap between f electrons on adjoining atoms is very small and the exchange is thought to arise through the intermediary of the conduction electrons. An atomic exchange between a 4f and a 5d electron on one atom $\mathscr{J}_{\mathrm{fd}}$ will introduce a polarisation of $\mathscr{J}_{\mathrm{fd}}/E_F$ in the conduction electrons, and this will produce an energy

$$\mathscr{J} = \mathscr{J}_{\mathrm{fd}}^2/E_F \tag{10.40}$$

at a neighbouring site. The spatial dependence of this exchange is complicated because of the velocity distribution and the fact that electrons obey Fermi–Dirac statistics. In fact it varies like (5.34) as shown in figure 5.38.

Because of the difficulty of the microscopic theory of the exchange it is usual to work back from the experimental results to show that they can be explained by a given set of exchange constants which are regarded as parameters. In an insulator the Heisenberg model, based on (10.37), is most used. This is directly applicable to ions with spin-only magnetism where the electrons of each atom combine give a total spin S. Summing over all pairs of sites i,j the total energy is given by

$$\mathscr{H} = -\sum_{i,j} \mathscr{J}(\boldsymbol{R}_{ij})\, \boldsymbol{S}_i \cdot \boldsymbol{S}_j \qquad (10.41)$$

If there is an important orbital contribution to the moment considerable anisotropy may occur for different directions of the moment. We shall not pursue this in detail except to note that one very anisotropic model has been extensively investigated—the Ising model where the spins are confined to the z-direction and $S^z = \pm\frac{1}{2}$

$$\mathscr{H} = -\sum_{i,j} \mathscr{J}(\boldsymbol{R}_{ij})\, S_i^z\, S_j^z \qquad (10.42)$$

This is not directly applicable to many systems but it presents a simple mathematical model to investigate the nature of the magnetic ordering.

10.7 Molecular field theory

10.7.1 The Weiss field (a)

Weiss' basic idea was that a large internal magnetic field, called the molecular field, created by the magnets in the specimen would account for ferromagnetism. This is borne out by a simple calculation. In a small field B the magnetisation per unit volume defined by (10.22) is given by

$$\mathrm{M} = \chi \mathrm{B} \qquad (10.43)$$

Now if the effective field has a component proportional to M

$$\mathbf{B}_{\mathrm{eff}} = \mathbf{B} + \Gamma \mathbf{M} \qquad (10.44)$$

and this field is regarded as acting on an assembly of non-interacting ions so that

$$\mathrm{M} = \chi_0\, \mathrm{B}_{\mathrm{eff}} \qquad (10.45)$$

Comparing this with (10.43) we have

$$\chi = \chi_0(1 + \Gamma\chi)$$

or

$$\chi = \chi_0/(1 - \Gamma\chi_0) \qquad (10.46)$$

For systems obeying Curie's Law $\chi_0 = C/T$ this can be written

$$\chi = C/(T - T_c) \qquad (10.47)$$

which diverges at a critical temperature

$$T_c = \Gamma C \qquad (10.48)$$

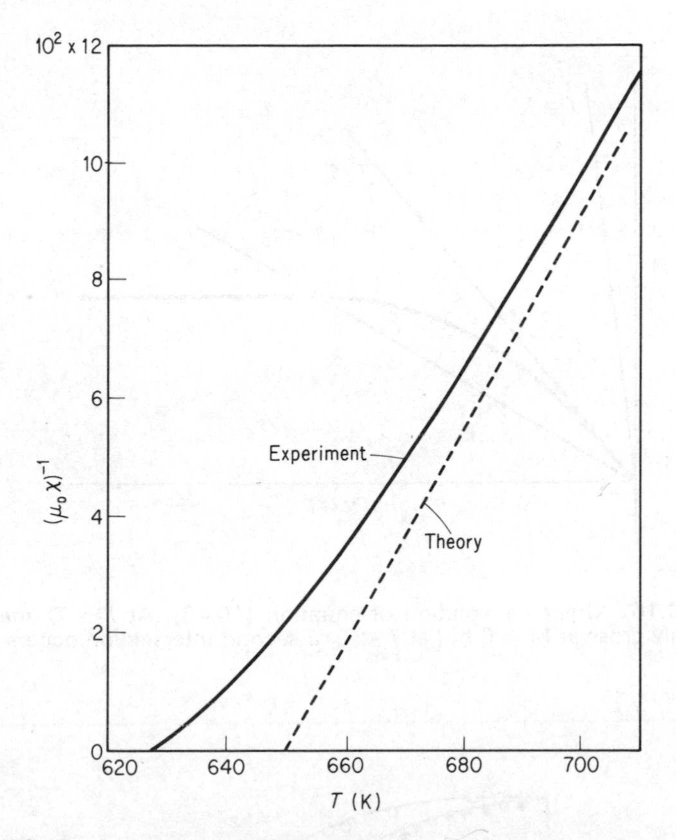

Figure 10.15 Inverse susceptibility of a ferromagnet on the molecular field theory compared with typical results on Ni

In fact the susceptibility of ferromagnets does show this kind of behaviour (cf. figure 10.15).

Below the critical temperature T_c a spontaneous non-zero value of M occurs even when $B = 0$. The effective field is now large and so the general form for M (10.20) must be used:

$$\mathrm{M} = (\mu_B \lambda_g J/\Omega)\, \mathscr{B}_J(\mu_B \lambda_g \Gamma \mathrm{M}/kT) \tag{10.49}$$

This transcendental equation may be solved graphically as in figure 10.16 to give a curve for $\mathrm{M}(T)$ which falls from a saturated value at $T = 0$ to zero at T_c, figure 10.17. The energy in the system per unit volume

$$E/\mathcal{N}\Omega = -\mathrm{M}\mathrm{B}_{\mathrm{eff}} = -\Gamma \mathrm{M}^2 \tag{10.50}$$

so the specific heat

$$C_V = -2\Gamma \mathrm{M}(\partial \mathrm{M}/\partial T) \tag{10.51}$$

This rises to a maximum at T_c and then falls abruptly to zero (figure 10.18).

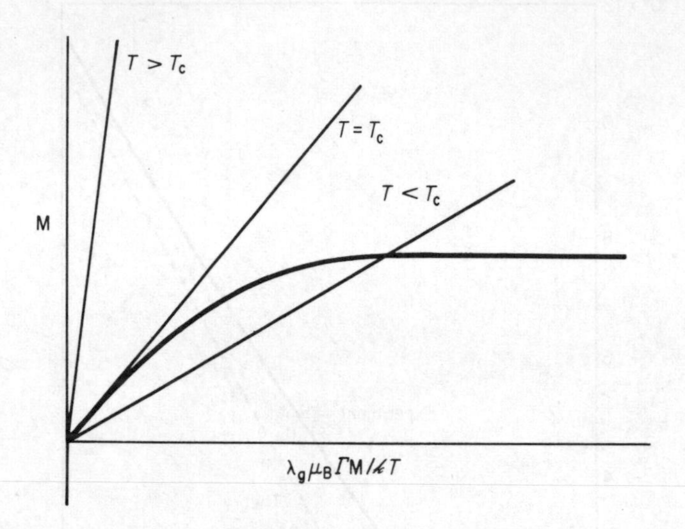

Figure 10.16 Graphical solution of equation (10.49). At $T > T_c$ the curves only cross at M = 0 but at $T < T_c$ a second intersection occurs

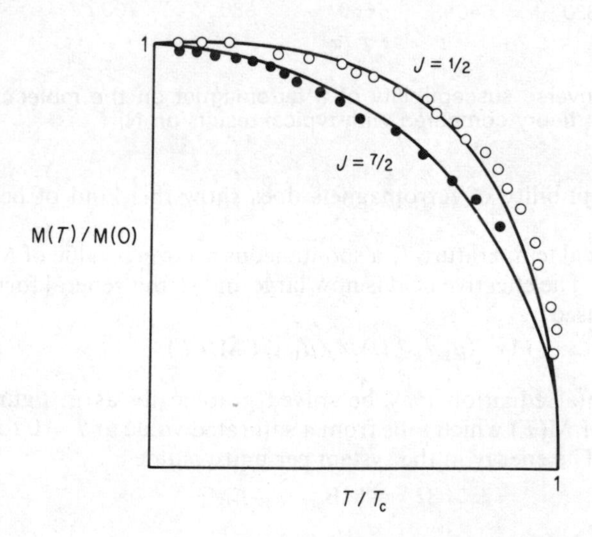

Figure 10.17 Spontaneous magnetisation as a function of T on the molecular field theory compared with experimental results for Ni (open circles) and Gd (full circles)

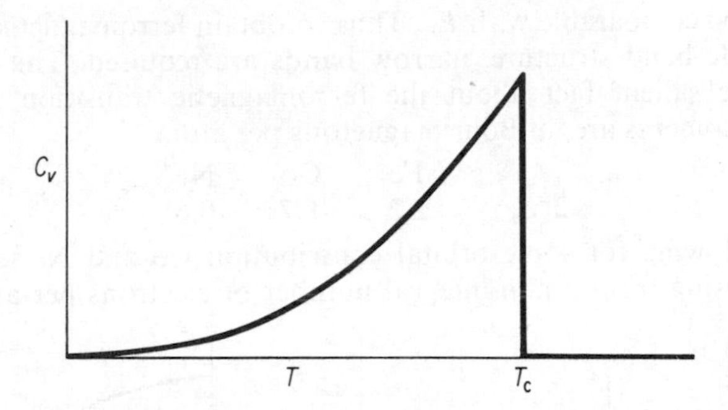

Figure 10.18 Specific heat anomaly predicted on the molecular field model

We see that a typical second-order phase transition occurs at T_c of the type described in § 3.8, equation (3.39).

The parameter Γ may be related to the exchange in the Heisenberg model by assuming the exchange energy acting on a single spin to be caused by $\mathbf{B}_{eff}$, i.e.

$$\sum_j \mathscr{J}(\boldsymbol{R}_{ij})\langle \boldsymbol{S}_j \rangle . \boldsymbol{S}_i = g\mu_B \mathbf{B}_{eff} . \boldsymbol{S}_i \tag{10.52}$$

and since

$$\mathbf{M} = g\mu_B \langle \boldsymbol{S}_j \rangle / \Omega \tag{10.53}$$

we find, by relating $\mathbf{B}_{eff}$ to $\mathbf{M}$ using (10.48) and (10.25) that

$$kT_c = \frac{S(S+1)}{3} \sum_j \mathscr{J}(\boldsymbol{R}_{ij}) \tag{10.54}$$

is directly proportional to the sum of the exchange integrals to all neighbours. Since the exchange only has this simple form for spin only systems we have put $\lambda_g = g \simeq 2$.

10.7.2 Molecular fields on band electrons (a model of ferromagnetic metals) (A)

The molecular field theory may be applied to other systems as well as those where $\chi_0 = C/T$. For example in a band one may assume an exchange splitting proportional to M. The system will assume a spontaneous magnetisation if

$$\Gamma \chi_0 > 1 \tag{10.55}$$

so that (10.46) cannot be satisfied. At $T=0$ with χ_0 given by (10.32) and $g = 2 = \lambda_g$ this requires the effective exchange splitting of the bands

$$\mathscr{J} = 4E_F/3 \tag{10.56}$$

i.e. it must be comparable with E_F. Thus to obtain ferromagnetism in a solid with a simple band structure, narrow bands are required. This simple idea explains one salient fact about the ferromagnetic transition metals. The saturation moments are, in Bohr magnetons per atom

	Fe	Co	Ni
$2\langle S_z \rangle$	2.2	1.7	0.6

and even allowing for some orbital contribution Co and Ni seem to have moments arising from a non-integral number of electrons per atom. This is

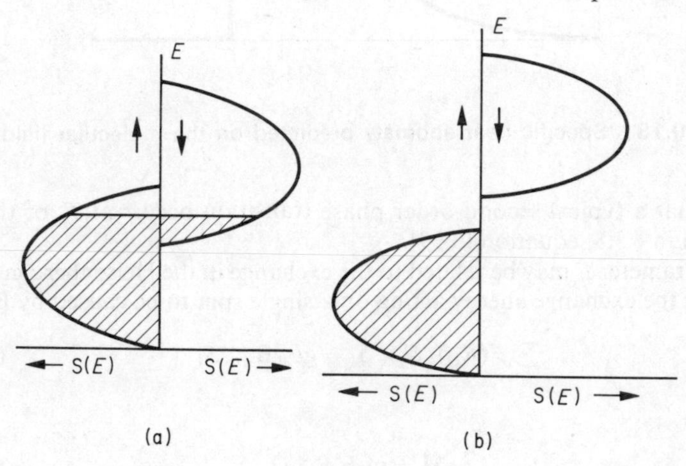

Figure 10.19 Distribution of electrons at $T = 0$ in the band model with an internal field showing a saturation moment (a) with a non-integral number of Bohr magnetons, (b) with an integral number

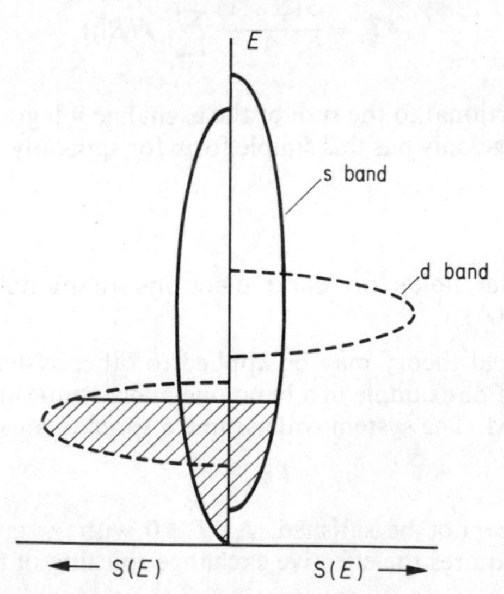

Figure 10.20 Simple band model of Ni giving a non-integral saturation number

possible in band theory with a correct choice of exchange integral as is shown schematically in figure 10.19. Actually in Ni there is a broad s band as well as narrow d bands (cf. figure 4.15) and some of the electrons can be regarded as not contributing to the moment. The best of these simple models for Ni is shown in figure 10.20. The saturation moment varies smoothly across an alloy series of the elements as if the extra electrons were varying the Fermi energy. Cu–Ni alloys lose all moment at a 40–60 concentration when the d band would be full (cf. figure 10.21). In fact this band model, due primarily to Stoner and Slater gives a good account of many aspects of the behaviour of the ferromagnetic metals. It has also been refined to give a much more detailed treatment of the difficult problem of electron–electron interactions in these materials.

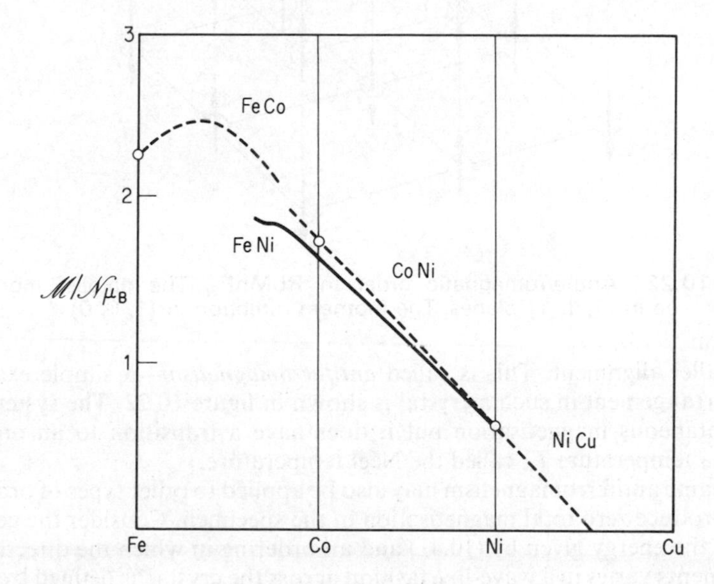

Figure 10.21 Variation of saturation moment with electron number in transition metal alloys (sometimes called the Slater–Pauling curve). Dashed curve gives adjacent mixtures, Fe–Co, Co–Ni, Ni–Cu. Full curve represents Fe–Ni alloys

10.8 Magnetic order

10.8.1 Types of magnetic order (a)

If all the exchange integrals in a system are positive the lowest energy state is clearly the ferromagnetic one in which all the spins line up parallel. But if some $\mathscr{J}$'s are negative there will be a tendency to produce an order where some spin pairs line up antiparallel. There are clearly several possibilities depending on the details of the interactions. The simplest situation occurs in those systems where $\mathscr{J}$ is only appreciable between nearest neighbours and is negative. In certain simple structures the spins can be arranged so that all neighbours have

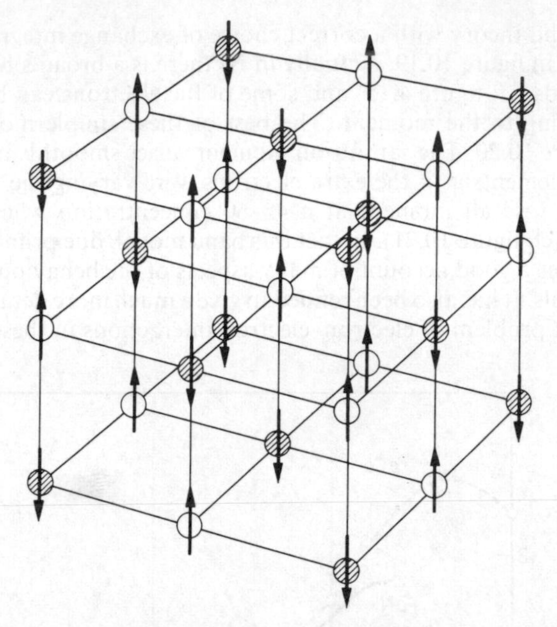

Figure 10.22 Antiferromagnetic order in $RbMnF_3$. The parallel moments lie in (1, 1, 1) planes. The moment direction in (1, 0, 0)

antiparallel alignment. This is called *antiferromagnetism*—a simple example of the arrangement in such a crystal is shown in figure 10.22. The system has no spontaneous magnetisation but it does have a transition to an ordered state at a temperature T_N called the Néel temperature.

The name antiferromagnetism may also be applied to other types of ordering which produce zero total magnetisation in the specimen. Consider the general form of the energy given by (10.41) and an ordering in which the direction of the moments varies in a wave-like fashion across the crystal as defined by some wave vector $\boldsymbol{w}$ such that

$$\left.\begin{aligned} g\mu_B\langle S_i^x\rangle &= m\cos(\boldsymbol{w}.\boldsymbol{R}_i) \\ g\mu_B\langle S_i^y\rangle &= m\sin(\boldsymbol{w}.\boldsymbol{R}_i) \\ g\mu_B\langle S_i^z\rangle &= 0 \end{aligned}\right\} \tag{10.57}$$

The energy is then

$$-\left(\frac{m}{g\mu_B}\right)^2\sum_j \mathscr{J}(\boldsymbol{R}_{ij})\cos\boldsymbol{w}.\boldsymbol{R}_{ij} = -\left(\frac{m}{g\mu_B}\right)^2\mathscr{J}(\boldsymbol{w})$$

where $\mathscr{J}(\boldsymbol{w})$ is the Fourier transform of the exchange $\mathscr{J}(\boldsymbol{R})$, as defined by the equation. The lowest energy is obtained when $\mathscr{J}(\boldsymbol{w})$ is a maximum. If the $\mathscr{J}(\boldsymbol{R})$ are all positive the maximum clearly occurs at $\boldsymbol{w} = 0$ and ferromagnetic ordering has the lowest energy. For $\mathscr{J}(\boldsymbol{R}_{ij})$ negative at the nearest neighbour position and zero elsewhere the maximum occurs if $\cos \boldsymbol{w} . \boldsymbol{R}_{ij} = -1$. In this case

$$g\mu_B\langle S_i^x\rangle = m, \quad g\mu_B\langle S_j^x\rangle = -m, \quad \langle S_i^y\rangle = \langle S_j^y\rangle = 0 \tag{10.58}$$

i.e. the near neighbours are aligned in opposite directions along x. This therefore corresponds to a simple antiferromagnetic arrangement as in figure 10.22. But if $\mathscr{J}(\boldsymbol{R})$ varies in sign with distance the maximum in $\mathscr{J}(\boldsymbol{w})$ may occur at a general point in which case helical order will occur as shown in figure 10.23

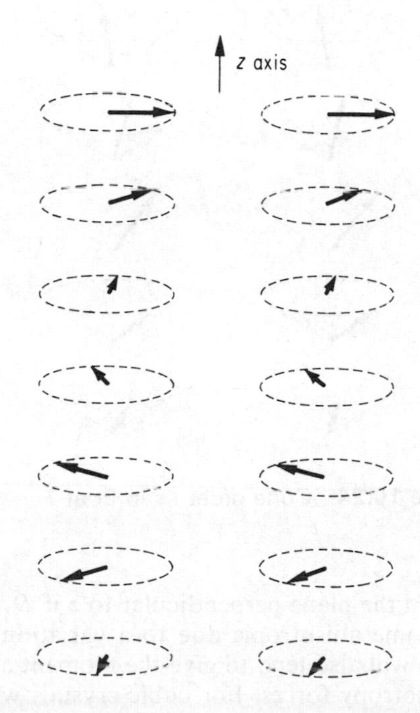

Figure 10.23 Helical order as in Dy at 180 K > T > 90 K. All atoms in the same layer have the same moment

for $\boldsymbol{w}$ parallel to z. Such an ordering occurs in the rare-earth metal Dy. In another rare earth Er ordering occurs with $\langle S_i^x\rangle$ and $\langle S_i^y\rangle$ given by (10.57) while $\langle S_i^z\rangle$ is constant so that the moments point along the generators of a cone as shown in figure 10.24. By analogy with the calculation in § 10.7 a molecular field theory may be developed and the transition occurs when

$$kT_N = \tfrac{1}{3}S(S+1)\mathscr{J}(\boldsymbol{w}) \tag{10.59}$$

The isotropic Heisenberg interaction (10.41) does not give any preferential direction for the magnetic moments. If, however, there is some orbital contribution to the moment this will lead to a slightly anisotropic charge cloud. The crystal field will then give a preferential axis for the magnetic moment. For example, a contribution like (10.11) will tend to make **M** point in the z-direction

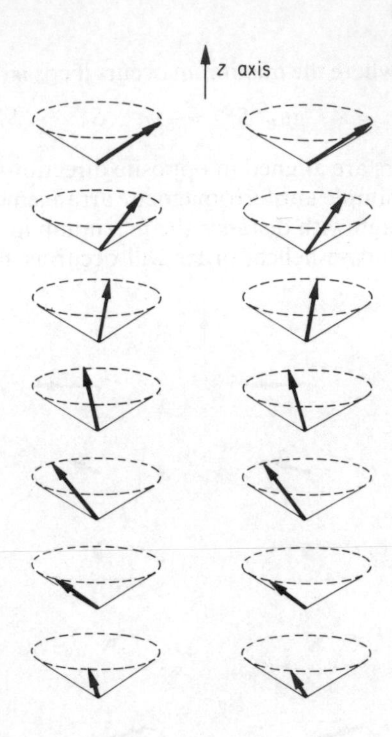

Figure 10.24 Cone order as in Er at $T < 50$ K

if D is positive; but in the plane perpendicular to z if D_2 is negative. The exchange will also become anisotropic due to a variation of the overlap (see figure 10.25) and this will also tend to give the moment a preferred direction. These are called anisotropy forces. For cubic crystals with nearly free spins they are small and the moment can be turned by a field of a 10^{-1} tesla as in Fe and Ni (figure 10.26). But in rare earths with large orbital moments the anisotropy fields may be 10^2–10^4 times larger and it is essentially impossible to force the moment into certain directions with laboratory fields.

In some crystals, atoms of different species with unequal moments occur together. Mixed oxides of the iron group with the spinel structure—the ferrites—are typical of this type of material. The garnets, mixed oxides with iron group and trivalent ions like rare earths, are a group which have been much studied. In these materials the coupling is negative and the moments point antiparallel but since they are unequal a net magnetisation results. This phenomenon is known as *ferrimagnetism* (figure 10.27). Magnetite Fe_3O_4, the lodestone of antiquity, is a ferrimagnet.

In a rare earth garnet there are eight magnetic atoms in a formula unit $M_3Fe_5O_{12}$. In the ordered phase three Fe^{3+}(S = 5/2) ions point in one direction and two in the opposite direction. The rare earth ions M point in the direction of the latter. They are coupled to the Fe spins rather weakly and lose a large

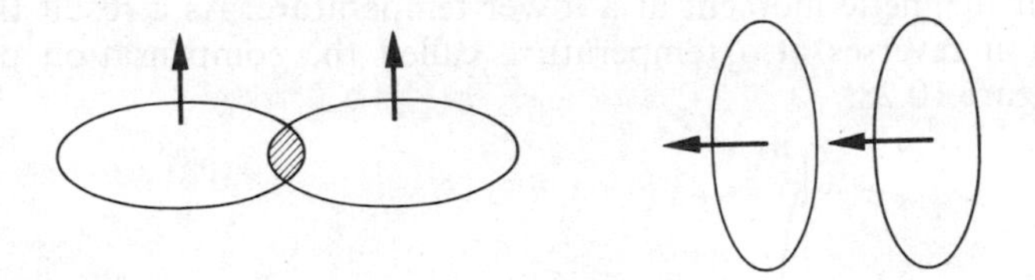

Figure 10.25 Schematic representation of variation of overlap between assymetric charge clouds in different orientations. The magnetic moment due to orbital effects has a definite direction relative to the charge cloud, hence the moments are coupled more strongly in some directions than in others

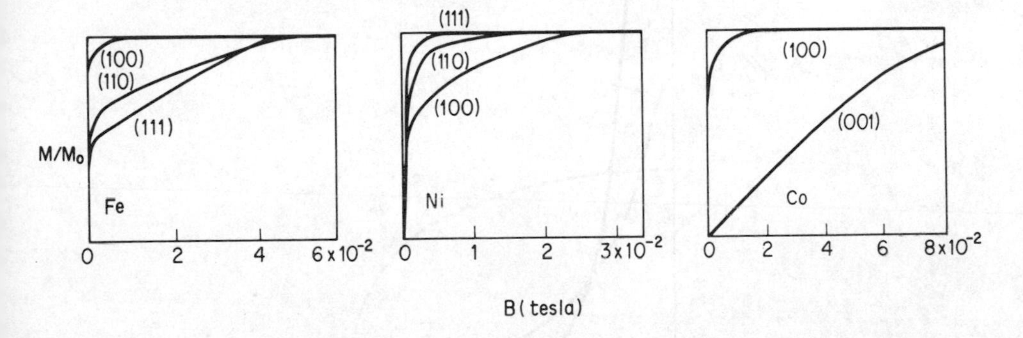

Figure 10.26 Magnetisation in Fe, Co and Ni for applied fields in different directions showing anisotropy fields

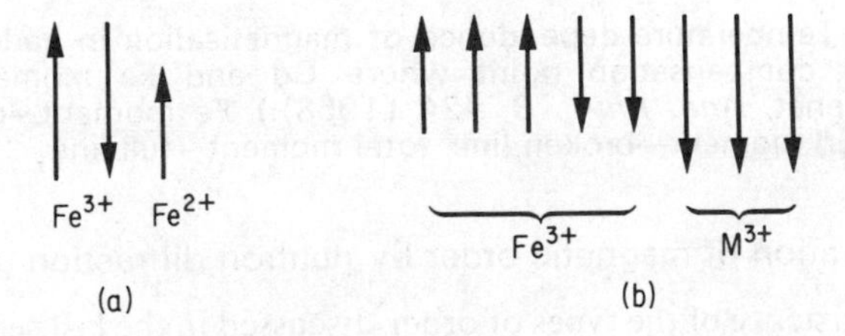

Figure 10.27 Schematic arrangement of spins in ferrimagnets (a) magnetite Fe_3O_4, (b) rare earth iron garnet $M_3Fe_5O_{12}$

part of their magnetic moment at a lower temperature. As a result the total magnetisation reverses at a temperature called the compensation point as shown in figure 10.28.

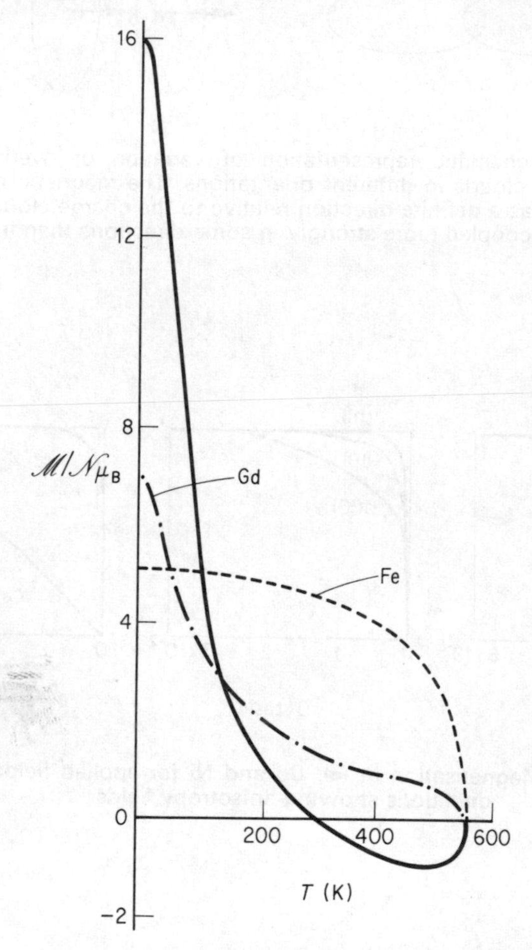

Figure 10.28 Temperature dependence of magnetisation in gadolinium iron garnet showing compensation point where Gd and Fe moments cancel. (After R. Pauthenet, *Ann. Phys.,* **3**, 424 (1958).) Fe moment—dashed line, Gd moment—broken line, total moment—full line

10.8.2 Observation of magnetic order by neutron diffraction (a)

Although the existence of the types of order discussed in the last section had all been inferred theoretically beforehand, their existence has only been directly verified in recent years with the advent of neutron diffraction techniques. The magnetic interaction between the magnetic moments of the electron and neutron is comparable in size to the neutron–nuclear interaction which allows the observation of Bragg scattering as discussed in Chapter 1. The amplitude of the scattered wave depends on the magnetic moment on the atoms. For a ferromagnetic order the crystal will have the same symmetry as the non-magnetic state and repeat itself after every lattice vector $\boldsymbol{R}_l$ as defined in (1.1). The magnetic scattering intensity will be proportional to m^2 and will appear for scattering from $\boldsymbol{K}$ to $\boldsymbol{K}'$ so that

$$\boldsymbol{K} - \boldsymbol{K}' = \boldsymbol{Q} \tag{10.60}$$

where $\boldsymbol{Q}$ is a reciprocal lattice vector as defined in (1.10). This scattering will be superimposed on the ordinary nuclear Bragg scattering. But in a magnet with more complex order defined with $\boldsymbol{w}$ as in (10.57), the magnetic scattering will appear when

$$\boldsymbol{K} - \boldsymbol{K}' = \boldsymbol{Q} + \boldsymbol{w} \tag{10.61}$$

at an entirely new set of reciprocal lattice vectors. The magnetic unit cell will be larger than the chemical unit cell since atoms which have different magnetic moments but are otherwise similar must be regarded as different scatterers. The crystal does not repeat after a displacement $\boldsymbol{R}_{ij}$ unless the moment is the same at the two sites. Experimental verification of this is shown for a simple antiferromagnet in figure 10.29.

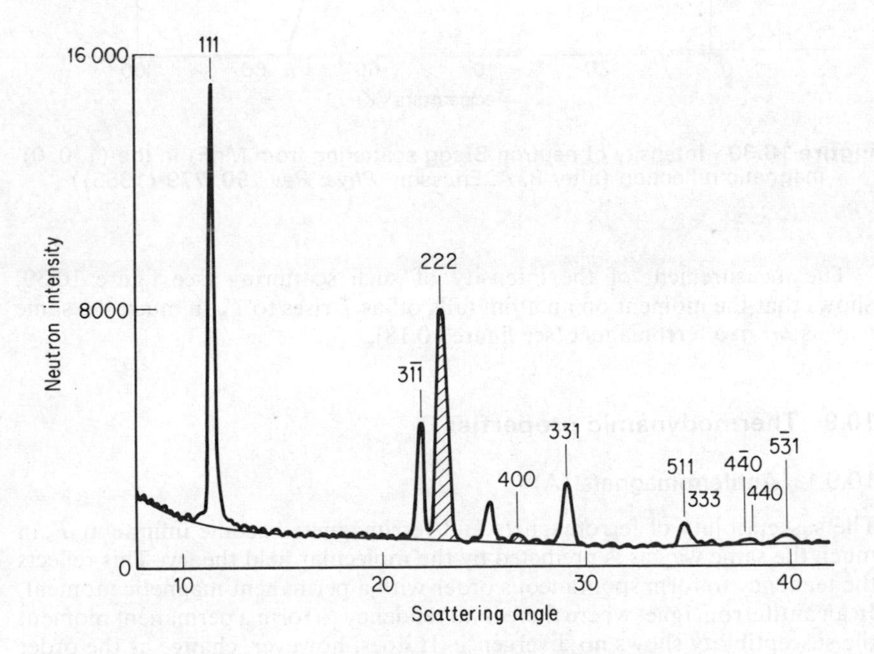

Figure 10.29 Neutron Bragg scattering from the antiferromagnet MnO. Nuclear peaks (shaded) persist at $T > T_N$ (after W. L. Roth, *Phys. Rev.*, **110**, 1333 (1958))

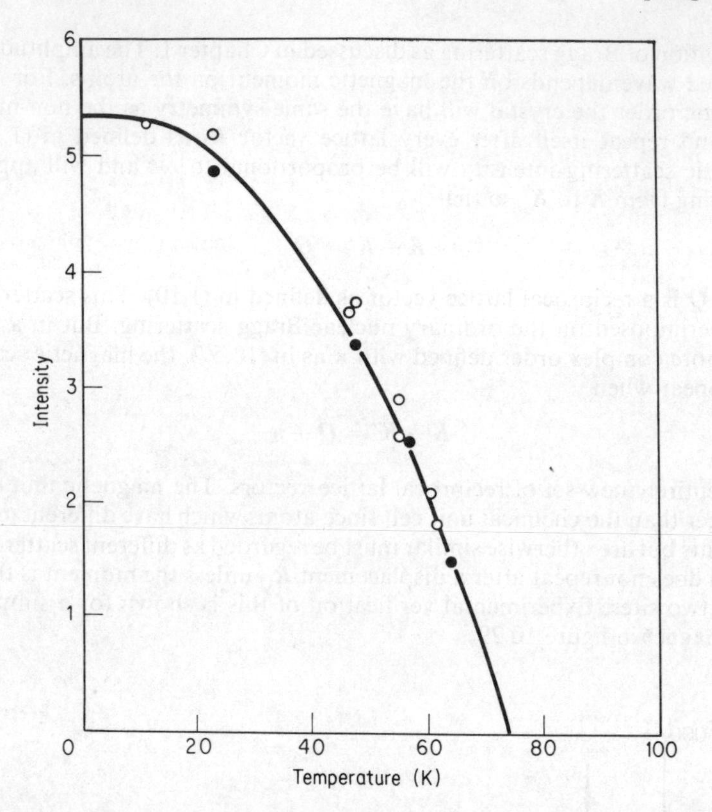

Figure 10.30 Intensity of neutron Bragg scattering from MnF_2 in the (1, 0, 0) magnetic reflection (after R. A. Ericsson, *Phys. Rev.*, **90**, 779 (1953))

The measurement of the intensity of such scattering (see figure 10.30) shows that the moment on an atom falls off as T rises to T_N in much the same way as $\mathscr{m}^2$ in a ferromagnet (see figure 10.18).

10.9 Thermodynamic properties

10.9.1 Antiferromagnets (A)

The susceptibility of ferromagnets and ferrimagnets become infinite at T_c in much the same way as is predicted by the molecular field theory. This reflects the tendency to form spontaneous order with a permanent magnetic moment. In an antiferromagnet where there is no tendency to form a permanent moment the susceptibility shows no divergence. It does, however, change as the order forms. Above T_N the susceptibility is still given by the analysis of § 10.7 as

$$\chi = C/(T + \Theta) \tag{10.62}$$

where now

$$\Theta = -\tfrac{1}{3}S(S+1)\sum_{j}\mathscr{J}(\boldsymbol{R}_{ij}) \tag{10.63}$$

For a simple antiferromagnet where all the $\mathscr{J}$ are negative $\Theta = T_N$ but in general this is not the case. Θ is sometimes called the Curie–Weiss constant.

In an isotropic antiferromagnet below T_N the moments will, in the presence of a field, set perpendicular to that field and turn slightly towards it through an

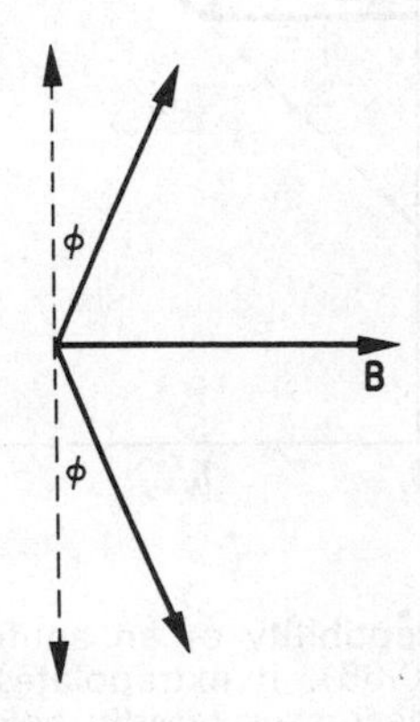

Figure 10.31 Moments of an antiferromagnet turned through an angle ϕ in an applied magnetic field

angle ϕ as shown in figure 10.31. The angle between the moments is then 2ϕ and the total energy

$$E = \mathscr{N}\left[\sum_{j}\mathscr{J}(\boldsymbol{R}_{ij})\left(\frac{m}{g\mu_B}\right)^2\cos 2\phi - Bm\sin\phi\right] \tag{10.64}$$

This is a maximum when $\partial E/\partial\phi = 0$ which for small ϕ gives

$$4\sum_{j}\mathscr{J}(\boldsymbol{R}_{ij})\left(\frac{m}{g\mu_B}\right)^2\phi = Bm$$

and

$$\chi = m\phi/B\Omega = g^2\mu_B^2/4\sum_{j}\mathscr{J}(\boldsymbol{R}_{ij})\,\Omega \tag{10.65}$$

This is constant and equal to $C/2T_N$, i.e. the high temperature value of χ given by (10.62) at $T = T_N$ since $\Theta = T_N$ in the simple case. This behaviour is plotted in figure 10.32. If there is anisotropy in the system this type of analysis only holds if the external field B is applied perpendicular to direction of the moment as determined by the anisotropy. It is more difficult to induce a moment parallel to this direction and $\chi_{\parallel}$ falls as T falls to zero (cf. figure 10.32). The susceptibility of a typical antiferromagnet shows a peak somewhere near T_N.

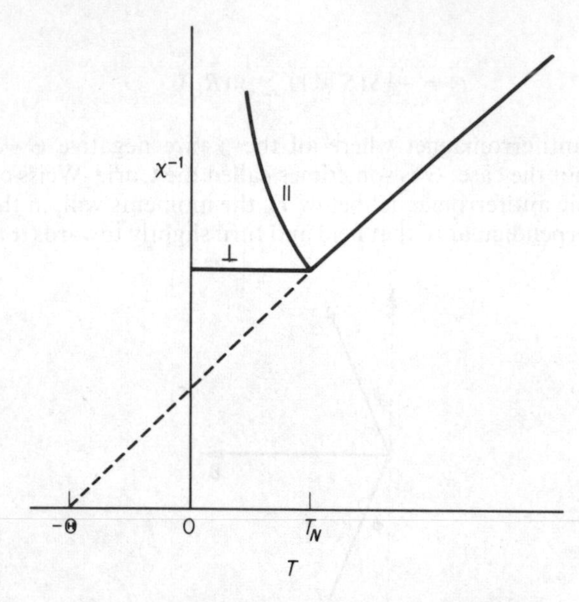

Figure 10.32 Inverse susceptibility of an antiferromagnet, using molecular field theory (10.62) and (10.65). It extrapolates to zero at $T = -\Theta$. In the presence of anisotropy χ^{-1} decreases rapidly below T_N (cf. MnF_2 results in Fig. 10.10(a))

10.9.2 Critical effects (A)

The specific heat of all types of magnetic crystals shows a maximum at the transition temperature which is much sharper than the break predicted by molecular field theory, figure 10.19. An example is shown in figure 10.33. In the best experiments the variation appears to be of the form

$$C_V \propto \log|T - T_c| \tag{10.66}$$

which become infinite at $T = T_c$. The nature of the infinity is such that $\int C_V \, dT$ remains finite and there is no latent heat, i.e. the transition remains second order.

This peak is one example of the 'critical phenomena' which occur just in the region of the critical temperature in a second-order transition. Just at this point where the system is about to order quite large but finite regions of order are built up, as shown in figure 10.34. Since these regions of short-range order are very effective scatterers of neutrons another critical effect is a peak in the scattering cross section around T_c. This effect is even more striking at the critical point of transition from gas to liquid when the scattering of light gives a remarkable critical opalescence although both gas and liquid are transparent. The theoretical problem of calculating the critical properties is currently of great interest in theoretical physics, much of the work being centred on the Ising model of (10.42).

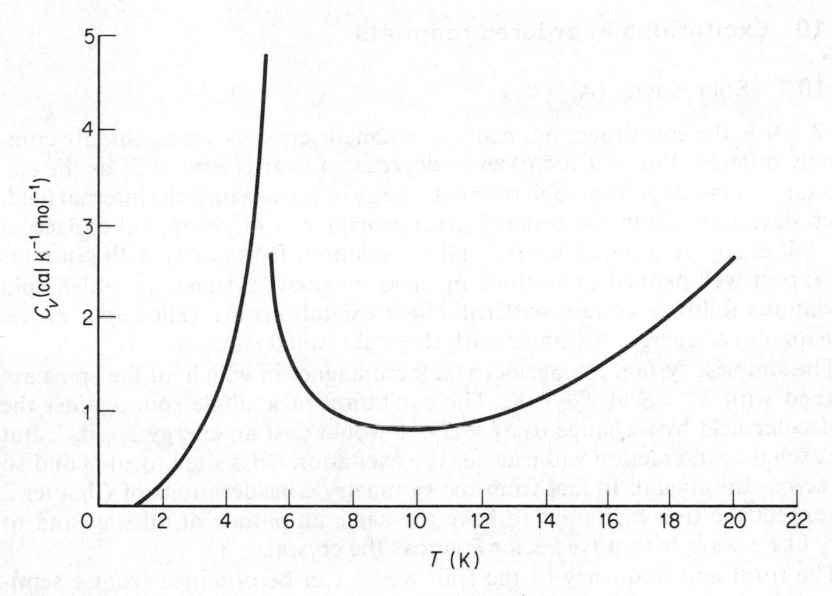

Figure 10.33 Specific heat anomaly in an antiferromagnet nickel salt $NiCl_2$, $6H_2O$ showing critical effects. (After W. K. Robinson and S. A. Friedberg, *Phys. Rev.*, **117**, 402 (1960).)

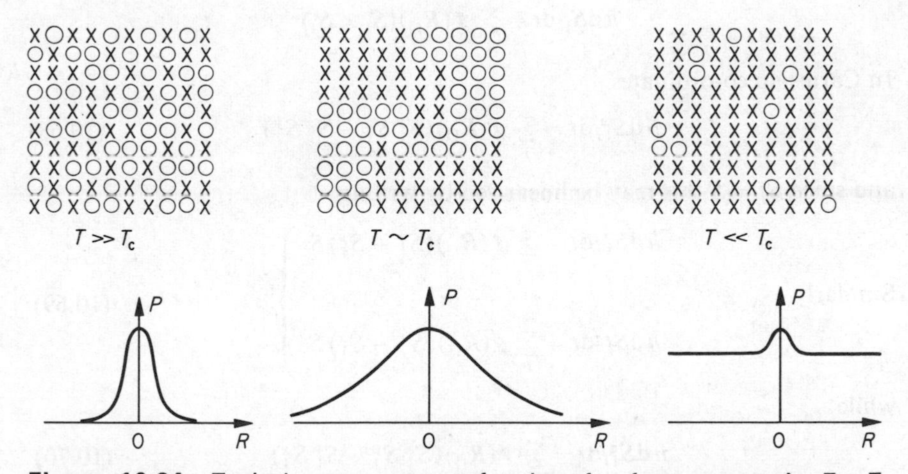

Figure 10.34 Typical arrangements of order of a ferromagnet. At $T \gg T_c$ the up spins (crosses) and the down spins (blanks) are arranged at random. Near T_c they congregate into large regions and cause critical effects. Below T_c there is a preponderence of up spins because of the long range order and the down spins are distributed at random. The probability of having two spins of the same kind at a distance R are also shown for the three cases. The width of the peak is a maximum at T_c. Below T_c it decreases but the wings give rise to long-range order

10.10 Excitations in ordered magnets

10.10.1 Spin waves (A)

At $T = 0$ K the constituent moments of magnetic crystals are essentially completely ordered. But as T increases m decreases towards zero at T_c as the elementary moments gain enough thermal energy to turn against the internal field. Such deviations from the ordered arrangement can of course take place at $T = 0$ if energy is supplied by external stimulation. By analogy with phonons we expect well defined excitations in these magnetic systems, in which spin deviations follow a certain pattern. These excitations are called *spin waves*. The quanta of energy associated with them are called *magnons*.

The simplest system to consider is a ferromagnet in which all the spins are aligned with $S_i^z = S$ at $T = 0$ K. The excitation of a single spin against the molecular field by a change to $S_i^z = S - 1$ would cost an energy $2\mu_B B_{eff}$. But the exchange interaction will transfer the excitation from site i to site j and so on across the crystal. In fact from the symmetry considerations of Chapter 2 we expect the true excitation to have the same amplitude at all sites and to vary like a wave with wave vector $\boldsymbol{k}$ across the crystal.

The form and frequency of the spin waves can be obtained from a semi-classical calculation. The equation of motion of a single spin is similar to (10.13)

$$\hbar\, \mathrm{d}\boldsymbol{S}_i/\mathrm{d}t = g\mu_B \boldsymbol{S}_i \times \mathbf{B}_{eff} \tag{10.67}$$

which from (10.52) may be written

$$\hbar\, \mathrm{d}\boldsymbol{S}_i/\mathrm{d}t = \sum_j \mathscr{J}(\boldsymbol{R}_{ij})(\boldsymbol{S}_i \times \boldsymbol{S}_j)$$

In Cartesian co-ordinates

$$\hbar\, \mathrm{d}S_i^x/\mathrm{d}t = \sum_j \mathscr{J}(\boldsymbol{R}_{ij})(S_i^y S_j^z - S_i^z S_j^y) \tag{10.68}$$

and since $S_j^z = S$ this may be linearised by writing

$$\hbar\, \mathrm{d}S_i^x/\mathrm{d}t = \sum_j \mathscr{J}(\boldsymbol{R}_{ij})(S_i^y - S_j^y)\, S$$

Similarly

$$\hbar\, \mathrm{d}S_i^y/\mathrm{d}t = \sum_j \mathscr{J}(\boldsymbol{R}_{ij})(S_j^x - S_i^x)\, S \tag{10.69}$$

while

$$\hbar\, \mathrm{d}S_i^z/\mathrm{d}t = \sum_j \mathscr{J}(\boldsymbol{R}_{ij})(S_i^x S_j^y - S_i^y S_j^x) \tag{10.70}$$

which is zero to first order since S_i^x and S_i^y are both small. By analogy with the vibrational problem of § 3.2 we look for wave-like solutions

$$\left.\begin{aligned} S_i^x &= \alpha \exp i(\boldsymbol{k}.\boldsymbol{R}_i - \omega t) \\ S_i^y &= \beta \exp i(\boldsymbol{k}.\boldsymbol{R}_i - \omega t) \end{aligned}\right\} \tag{10.71}$$

These are indeed solutions if

$$\left.\begin{aligned}-\mathrm{i}\omega\alpha\hbar &= \sum_j \mathscr{J}(\boldsymbol{R}_{ij})[1-\exp(\mathrm{i}\boldsymbol{k}.\boldsymbol{R}_{ij})]S\beta \\ -\mathrm{i}\omega\beta\hbar &= -\sum_j \mathscr{J}(\boldsymbol{R}_{ij})[1-\exp(\mathrm{i}\boldsymbol{k}.\boldsymbol{R}_{ij})]S\alpha\end{aligned}\right\} \tag{10.72}$$

i.e.

$$\hbar\omega = \sum_j \mathscr{J}(\boldsymbol{R}_{ij})[1-\exp(\mathrm{i}\boldsymbol{k}.\boldsymbol{R}_{ij})]S = S[\mathscr{J}(\boldsymbol{k}=0)-\mathscr{J}(\boldsymbol{k})] \tag{10.73}$$

and

$$\alpha = -\mathrm{i}\beta$$

This represents a circular motion of each spin with a well defined phase variation across the crystal as shown in figure 10.35. A typical variation of ω as $\boldsymbol{k}$ varies across a Brillouin zone is shown in figure (10.36) and reflects the Fourier transform $\mathscr{J}(\boldsymbol{k}) = \sum_j \mathscr{J}(\boldsymbol{R}_{ij})e^{i\boldsymbol{k}.\boldsymbol{R}_{ij}}$. For small k

$$\hbar\omega(\boldsymbol{k}) = Pk^2 \tag{10.74}$$

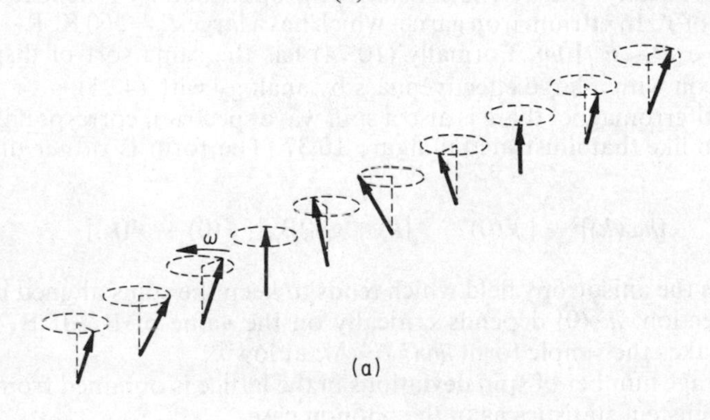

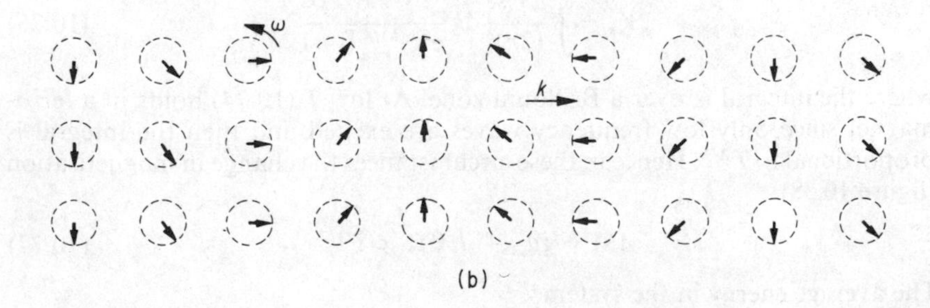

Figure 10.35 Schematic representation of spin motion in a ferromagnetic spin wave: (a) general view of a row of spins precessing with a phase lag; (b) plan view of an array with a spin wave

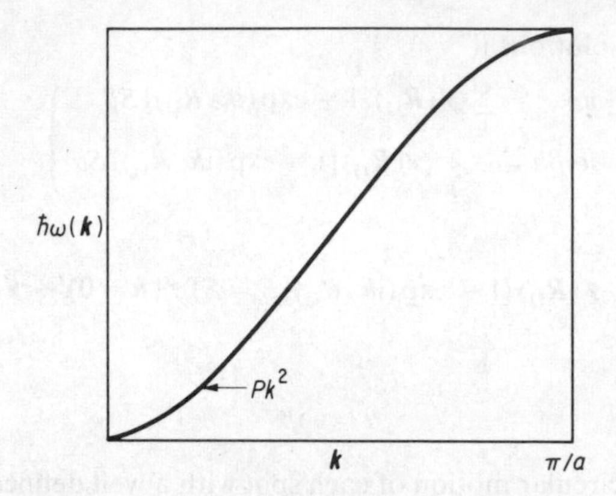

Figure 10.36 Spin wave spectrum in a simple ferromagnet from (10.73). The frequency is proportional to k^2 at low k

varies quadratically with k. The constant of proportionality P depends on the magnitude of $\mathscr{J}$. In yttrium iron garnet which has a large $T_c = 560$ K, $P \sim \hbar^2/20m$ while for Fe, $P \sim \hbar^2/10m$. Formally (10.74) has the same sort of dispersion as an electron with a large effective mass by analogy with (4.25).

In an antiferromagnet there is also a spin wave spectrum corresponding to a spin motion like that illustrated in figure 10.37. The form is rather different, given by

$$[\hbar\omega(\boldsymbol{k})]^2 = [\mathscr{J}(0) - \mathscr{J}(\boldsymbol{k}) + g\mu_B B_A][\mathscr{J}(0) + \mathscr{J}(\boldsymbol{k})] \qquad (10.75)$$

where B_A is the anisotropy field which tends to keep the spins aligned in a particular direction. $\hbar\omega(0)$ depends critically on the value of B_A. If $B_A = 0$ the spectrum takes the simple form $\hbar\omega(\boldsymbol{k}) = Ak$ at low k.

The average number of spin deviations in the lattice is obtained from applying Bose–Einstein statistics as in the phonon case

$$\mathscr{N}_s = \int \frac{\mathscr{N}\Omega}{(2\pi)^3} \mathrm{d}\boldsymbol{k} \frac{1}{e^{\hbar\omega(\mathbf{k})/kT} - 1} \qquad (10.76)$$

where the integral is over a Brillouin zone. At low T (10.74) holds in a ferromagnet since only low frequency waves are excited and then the integral is proportional to $T^{3/2}$. Hence in these circumstances the change in magnetisation (figure 10.38)

$$\Delta M = g\mu_B \mathscr{N}_s / \mathscr{N}\Omega \propto T^{3/2} \qquad (10.77)$$

The average energy in the system

$$\langle E \rangle = \int \frac{\mathscr{N}\Omega}{(2\pi)^3} \frac{\mathrm{d}\boldsymbol{k}\hbar\omega(\boldsymbol{k})}{e^{\hbar\omega(\mathbf{k})/kT} - 1} \propto T^{5/2} \qquad (10.78)$$

under the same conditions and $C_V \propto T^{3/2}$ (figure 10.38).

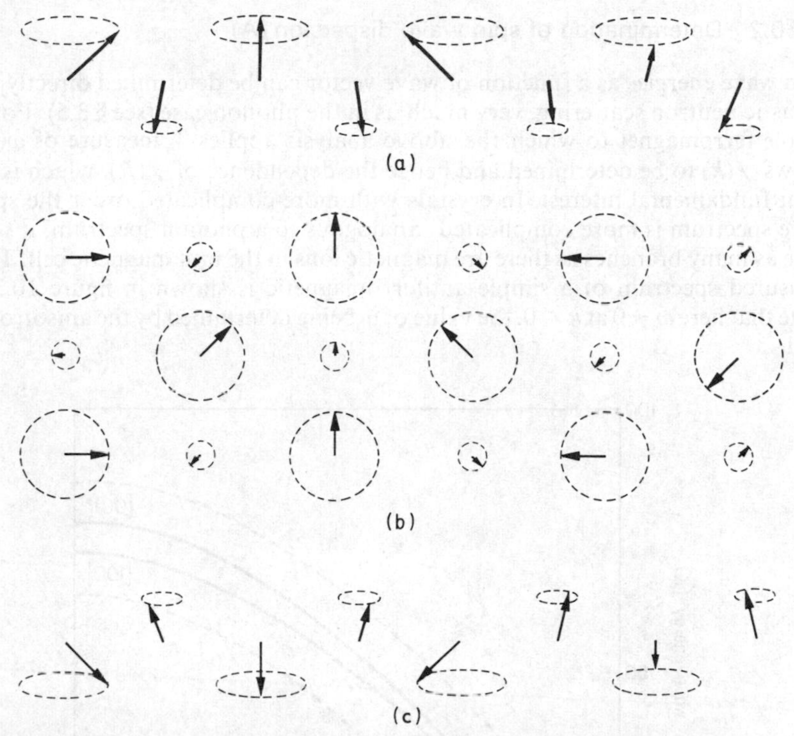

Figure 10.37 Schematic representation of spin waves in an antiferromagnet. (a) and (c) two spin waves with the same frequency propagating along a line of spins; (b) plan of an array containing a spin wave. The large circles are on spins which point up in equilibrium. The small circles on spins which point down

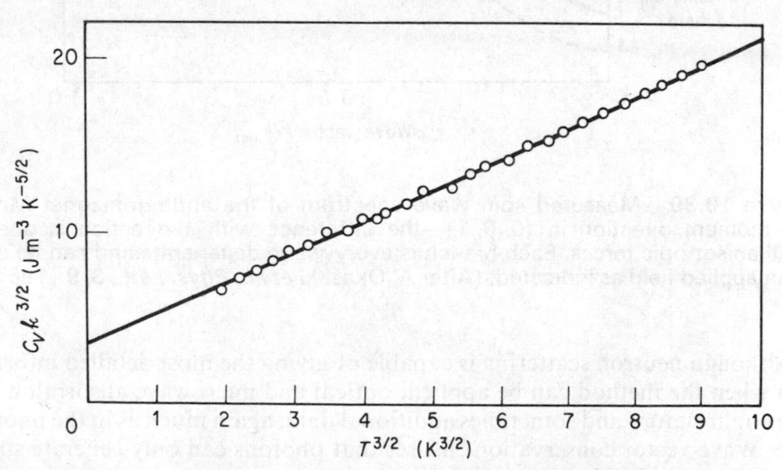

Figure 10.38 Variation of C_V at low T in a ferrimagnet YIG, showing variation with $T^{3/2}$ (after S. S. Shinozaki, *Phys. Rev.*, **122**, 383 (1961))

10.10.2 Determination of spin wave dispersion (A)

Spin wave energies as a function of wave vector can be determined directly by inelastic neutron scattering, very much as in the phonon case (see § 3.5). For a simple ferromagnet to which the above analysis applies, a measure of $\omega(\boldsymbol{k})$ allows $\mathcal{J}(\boldsymbol{k})$ to be determined and hence the dependence of $\mathcal{J}(\boldsymbol{R})$, which is of great fundamental interest. In crystals with more complicated order the spin wave spectrum is more complicated: analogous to a phonon spectrum, it will have as many branches as there are magnetic ions in the unit magnetic cell. The measured spectrum of a simple antiferromagnetic is shown in figure 10.39. Note that here $\omega \neq 0$ at $k = 0$, the value of ω being determined by the anisotropy field.

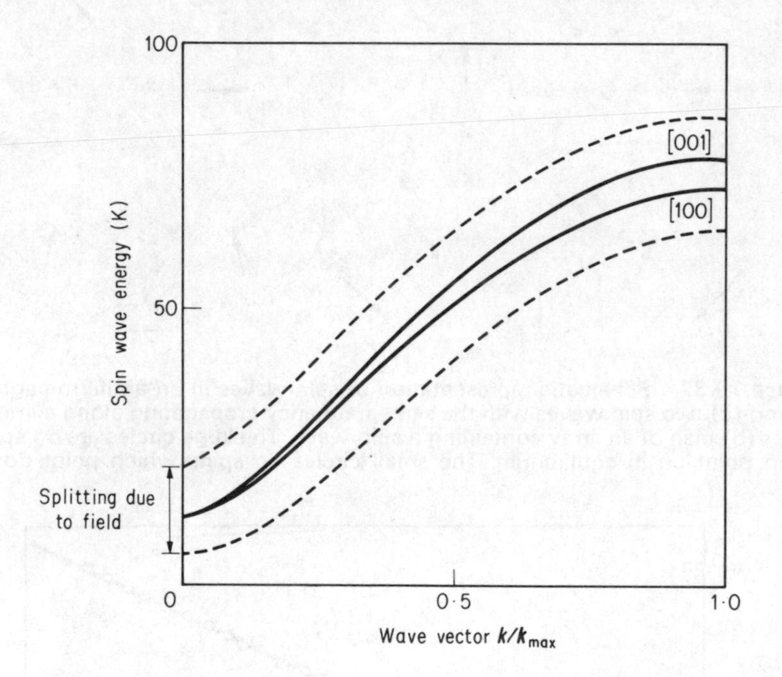

Figure 10.39 Measured spin wave spectrum of the antiferromagnet MnF_2. The moment direction in (0, 0, 1)—the difference with k direction is due to small anisotropic forces. Each branch is everywhere degenerate and can be split by an applied field as indicated. (After A. Okasaki *et al., Phys. Lett.,* **8**, 9 (1964).)

Although neutron scattering is capable of giving the most detailed information when the method can be applied, optical and microwave absorption can give confirmatory and sometimes additional data, again much as in the phonon case. Wave vector conservation ensures that photons can only generate single magnons with $\boldsymbol{k} \sim 0$. If pairs are generated, the magnons have oppositely directly wave vector (cf. two phonon absorption). Absorption due to single magnon absorption is usually called resonance by analogy with the paramagnetic resonance discussed in § 10.3. When g = 2 ferromagnetic resonance occurs at the same frequency as paramagnetic resonance namely

$$\hbar\omega(0) = g\mu_B B \tag{10.79}$$

but B contains a contribution from the demagnetising field which depends on the shape of the specimen. Geometric resonances, at discrete $\boldsymbol{k}$ values, can be observed in thin magnetic films (figure 2.2). With experimentally achievable fields, ferromagnetic resonance (like electron spin resonance) is normally observed at microwave frequencies. Antiferromagnetic resonance, on the other hand, does not require an externally applied field since each sublattice effectively experiences the field (the anisotropy field) due to the other. As B_A may be high and, from (10.75), $\hbar\omega(0)$ also depends on $\mathcal{J}(0)$, antiferromagnetic resonance occurs in the far infrared (cf. figure 6.3). On the application of an external magnetic field the degeneracy due to two sublattices is lifted and the resonance split by an amount $g\mu_B B_0$, as in figure 10.39.

Two magnon absorption can be observed in anti-ferromagnets, also in the infrared. For example, in FeF_2 the single magnon resonance occurs at 0.0065 eV and the two-magnon line at 0.019 eV, when observations are made at 4 K. The two-magnon absorption is surprisingly strong, comparable in strength to the single magnon process. It is also invariant with magnetic field. This suggests that the photon creates two spin deviations of opposite sign so the net change in crystal spin is zero and the external field raises the energy of one magnon by as much ($\frac{1}{2}g\mu_B B$) as it lowers the other.

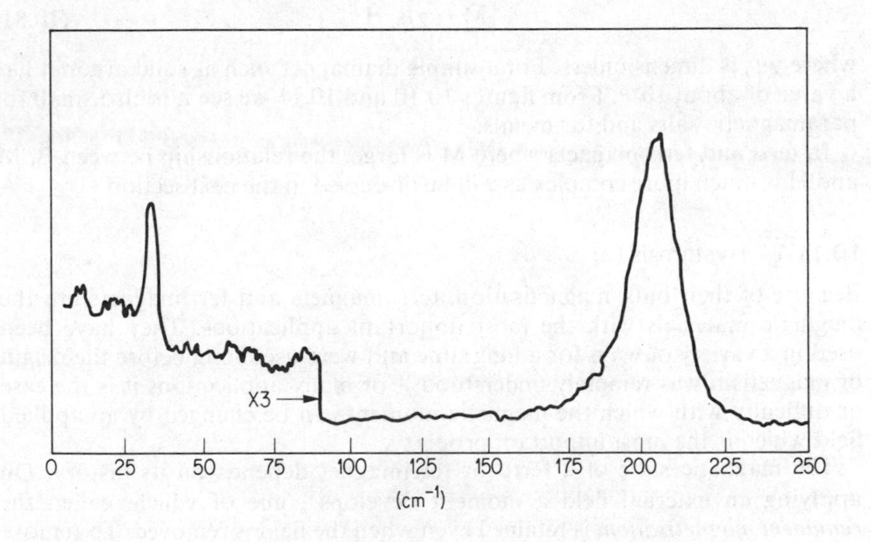

Figure 10.40 Raman scattering due to magnons in NiF_2 (after P. A. Fleury *Inter. J. Mag.*, **1**, 75, (1970)). One-magnon absorption occurs at 30 cm^{-1}, two-magnon processes at ~100 cm^{-1}

The experimental difficulties associated with far infrared spectroscopy makes Raman scattering from Laser sources (see § 6.8.5) an attractive method of measuring magnon spectra. Raman scattering from single and two-magnon processes has been observed in, for example, NiF_2 (figure 10.40).

10.11 The magnetisation of ferro- and ferrimagnets (a)

The effect of a magnetic field on an electron in a solid has up to this point been described entirely in terms of the magnetic field or flux **B**. In magnetostatics it is found convenient to define a second field* **H** related to **B** by

$$\left.\begin{aligned} \mathbf{B} &= \mu \mathbf{H} \\ &= \mu_0(\mathbf{H}+\mathbf{M}) \end{aligned}\right\} \tag{10.80}$$

where μ is the permeability. In free space where $\mathbf{M} = 0$

$$\mathbf{B} = \mu_0 \mathbf{H}$$

External measurements are readily made in terms of H. For a long solenoid of n turns per metre carrying a current of I amperes $H = nI$ and has the dimensions A/metre. This is the same as the dimensions of the magnetisation. μ_0 has the value $4\pi \times 10^{-7}$ henry per metre.

In all solids without a permanent moment (except a paramagnet near saturation) **M** is small and $\mu \simeq \mu_0$. The susceptibility χ defined in (10.22) can be conveniently written

$$\mathrm{M} = \chi\mu_0 \mathrm{H} \tag{10.81}$$

where $\chi\mu_0$ is dimensionless. For a simple diamagnet such as solid argon it has a value of about 10^{-5}. From figures 10.10 and 10.11 we see it is also small for paramagnetic salts and for metals.

In ferri and ferromagnets where M is large, the relationship between B, M and H is much more complex as will be discussed in the next section.

10.11.1 Hysteresis (a)

Because of their bulk magnetisation, ferromagnets and ferrimagnets are the magnetic materials with the most important applications. They have been used in a variety of ways for a long time and were used long before the origin of magnetism was remotely understood. For many applications it is the ease or difficulty with which the magnetic moment can be changed by an applied field which is the most important property.

The magnetic state of a ferro or ferrimagnet depends on its history. On applying an external field a moment develops some of which, called the *remanent magnetisation*, is retained even when the field is removed. To remove

* See for example: Feynman, Leighton and Sands, *Lectures on Physics*, Vol. II. Panofsky and Phillips, *Classical Electricity and Magnetism*.

all magnetisation from a specimen then requires the application of a field in the opposite direction termed the *coercive field*. The system is then said to show hysteresis—a lagging behind the field. The relationship between magnetisation and applied field is now quite complex and a function of both time and the detailed conditions of the experiment.

It is usual to display the relation between field and magnetisation in terms of **B** and **H** (10.80) which are more amenable to measurement than is **M** itself. **B** and **H** can both be measured in a toroidal specimen wound with both a primary and secondary coil as in figure 10.41. The field **H** is now related to the current in the primary while **B** is related to the integral of the electromotive force over time which may be determined from the secondary circuit. Representative curves of B against H, frequently known as the B/H hysteresis loop are shown in figure 10.42.

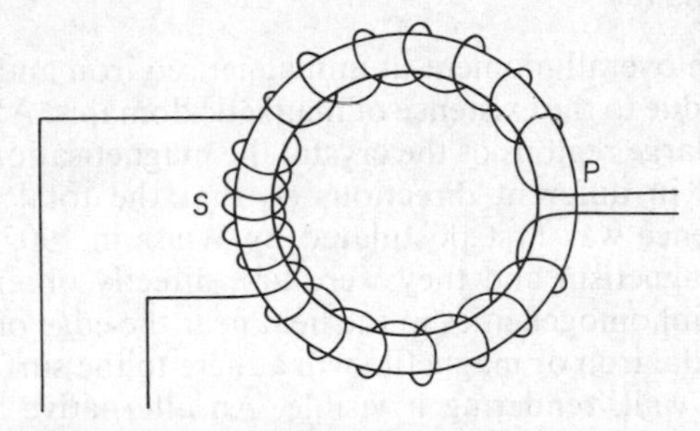

Figure 10.41 Apparatus for measuring B-H hysteresis in a ferromagnet. H is related to the current in the primary circuit P while B is related to the integral of the e.m.f. in the secondary circuit S

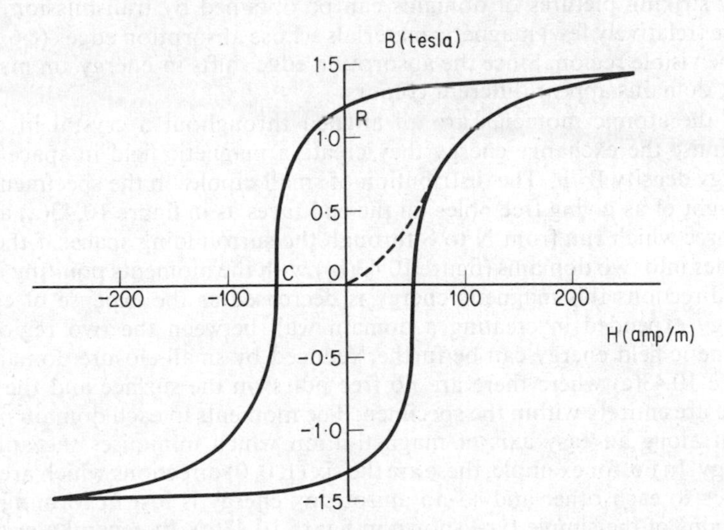

Figure 10.42 B-H loop for Fe. Initial magnetisation follows the dashed curve, subsequently the full hysteresis loop. *R* is the remanent induction and *C* the coercive field

The magnitude of the coercive field, both absolutely and in relation to the saturation field, is important in applications. Materials with small coercive fields in relation to their saturation field are termed *soft*, as this characteristic was early associated with soft (i.e. malleable) iron. Conversely a magnetically hard material has a high coercive field and is hard to magnetise, but once magnetised it retains a large magnetisation near the saturation value. This type of behaviour used to be associated with mechanically hard alloys. Both extremes of characteristics can, however, be obtained in ferrites of suitable composition. It should also be noted that in anisotropic media the shape of the B/H loop is a function of crystal orientation, so that 'hard' and 'soft' (or 'easy') directions may be observed in the same material.

10.11.2 Domains (a)

The absence of an overall moment in unmagnetised iron and also the form of the hysteresis are due to the existence of magnetic domains. Although magnetic order exists over large regions of the crystal the magnetisation in these regions (domains) points in different directions so that the total magnetisation is zero. Their existence was first postulated by Weiss in 1907 as a part of his theory of ferromagnetism and they were first directly observed by Bitter in 1931. Because of inhomogeneities in the field near the edge of a domain, small particles of colloidal iron or magnetite will adhere to the surface of the sample near the domain wall, rendering it visible. An alternative method is to use Faraday rotation (see § 7.2 and figure 10.2) either in reflection or (if the sample is transparent) in transmission, since the degree of rotation of the plane of polarisation of the light depends on the magnetisation of the sample. Particularly striking pictures of domains can be obtained by transmission through those (relatively few) magnetic materials whose absorption edges (§ 6.3) occur in the visible region. Since the absorption edge shifts in energy on magnetisation, domains appear different colours.

If the atomic moments are all aligned throughout a crystal in order to minimise the exchange energy they create a magnetic field in space with an energy density B^2/μ. The distribution of small dipoles in the specimen may be thought of as giving free poles on the end faces as in figure 10.43(a) and lines of force which run from N to S through the surrounding space. If the crystal divides into two domains (figure 10.43(b)) with the moments pointing in opposite directions this magnetic energy is decreased as the expense of exchange energy expended in creating a domain wall between the two regions. The magnetic field energy can be further reduced by small closure domains as in figure 10.43(c) where there are no free poles on the surface and the lines of force are entirely within the specimen. The moments in each domain normally point along an easy axis of magnetisation which minimises the anisotropy energy. In Fe, for example, these are the six (1, 0, 0) directions which are at right angles to each other and so no anisotropy energy is lost in forming closure domains of the simple type shown in figure 10.43(c). In general a crystal will contain many domains of this type as shown in figure 10.44 and the arrangement is always such as to minimise the energy.

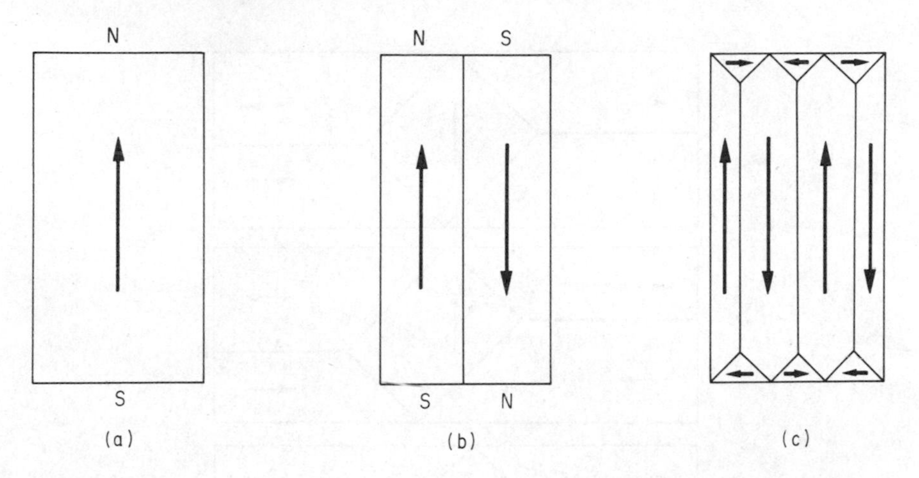

Figure 10.43 A sample which is (a) uniformly magnetised, (b) divided into two domains, (c) with simple closure domains (see text)

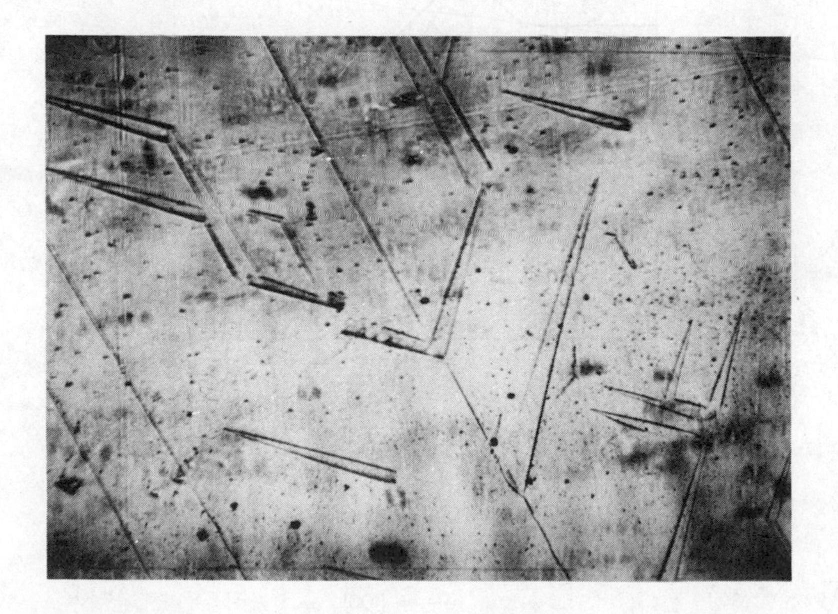

Figure 10.44 Actual domain pattern on (1, 0, 0) surface of a single crystal of silicon–iron. It shows 180° Bloch walls with 'fir tree' closure domains. (Courtesy of Prof. L. F. Bates and Dr. A. Hart.)

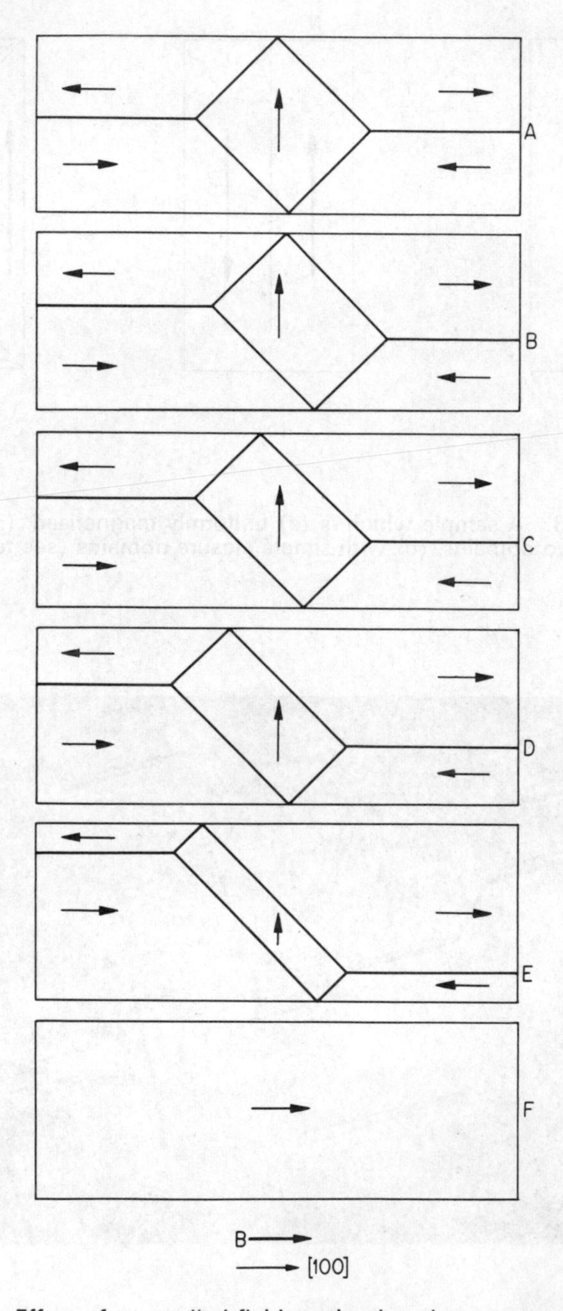

Figure 10.45 Effect of an applied field on the domain pattern on the surface of a single crystal iron whisker showing domain wall displacement, as H increases from 0 at A to 1.5 x 10^4 A/m at F. (After G. G. Scott and R. V. Coleman, *J. Appl. Phys.,* **28**, 1512 (1957).)

The crystal magnetises by moving the domain walls to create larger domains with favourable moments along the field and smaller domains with unfavourable moments as shown in figure 10.45. After this process, for which comparatively small fields are required, further magnetisation is produced by rotating the moments into the direction of the applied field against the anistropy field These processes give some explanation of the shape of the B/H loop but they appear reversible and hence would not at first sight give rise to hysteresis since the magnetisation could reorient and the walls re-appear when the field is removed. The hysteresis in fact arises from imperfections such as impurities, dislocations, and the boundaries of the grains, which tend to prevent domain wall motion and rotation. A sufficiently strong external field will push the wall past an obstacle but this is irreversible, and an equally strong field in the opposite direction is required to reverse it.

Thus, magnetically soft materials should be relatively free of impurities and have low anisotropy. Hard materials should be imperfect and have the easy axes aligned as far as possible. In fact it is found that the formation of domain walls becomes energetically unfavourable in very small particles. Thus the best hard magnets are precipitates of small magnetic particles which are aligned as far as possible (see figure 10.46).

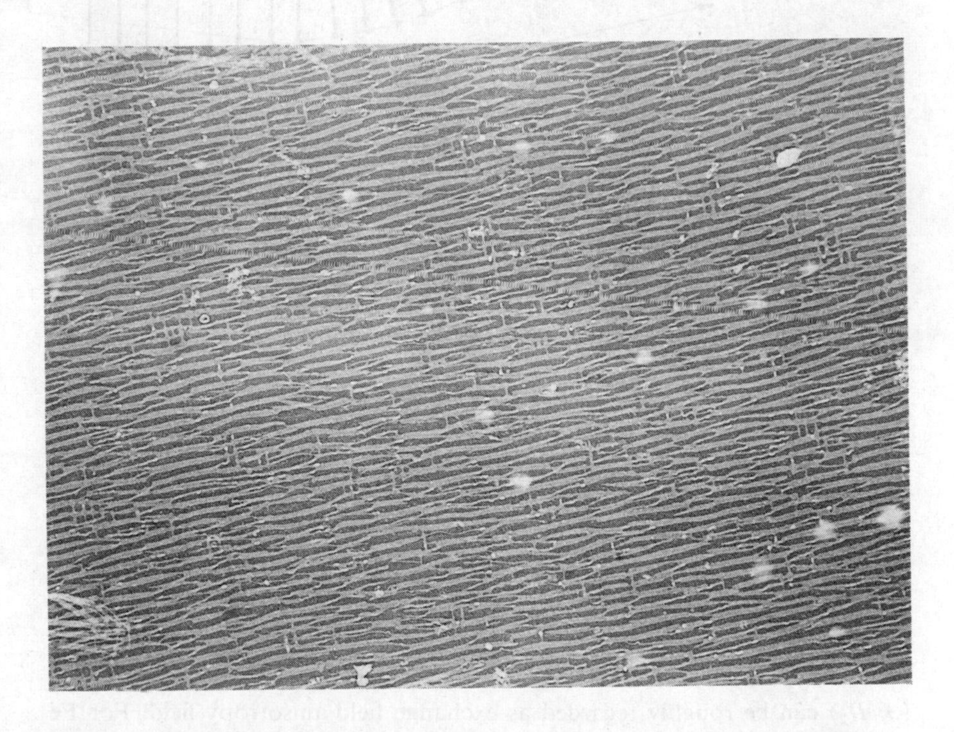

Figure 10.46 Microstructure of alnico steel showing oriented particles to produce a high coercive force. (Courtesy of Dr. K. J. de Vos.)

10.11.3 Domain walls (a)

The walls between domains of different orientation have a structure which minimises their energy. A rotation of the atomic moments through 180° between two planes would give an energy $\mathcal{J} S^2$ for each neighbouring pair of atoms. But if the rotation takes place gradually over n planes so that there is a small angle $\phi = \pi/\mathrm{n}$ between adjacent spins as shown in figure 10.47 the energy change is $\mathcal{J} S^2(1 - \cos\phi)$ for each pair and for $\cos\phi = 1 - \frac{1}{2}(\pi/\mathrm{n})^2$ the total energy is

$$\mathcal{J} S^2 \pi^2/2\mathrm{n} \tag{10.82}$$

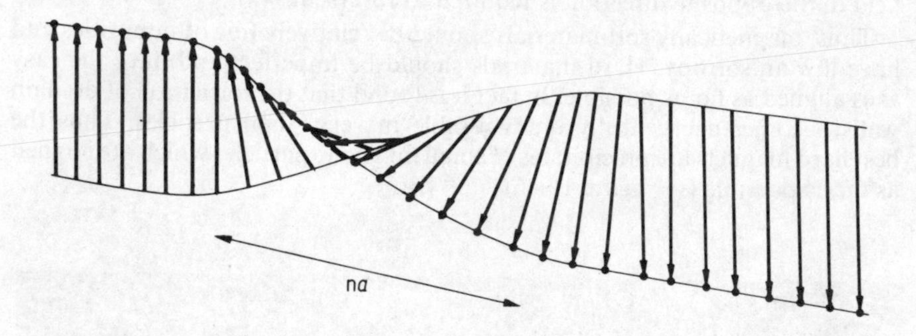

Figure 10.47 Diagram of a domain wall, thickness n*a*.

Thus the energy decreases as n increases and it would appear favourable to have thick walls. However, the anisotropy energy increases as the number of spins at an angle to the axis increases. If the anisotropy energy per atom is $3\, D_2 S^2 \cos^2\theta$ as from a term like (10.11) the sum for the set of n spins is

$$\sum_{r=0}^{\mathrm{n}} 3 D_2 S^2 [1 - \cos^2(r\pi/\mathrm{n})] \tag{10.83}$$

The total energy is the sum of (10.82) and (10.83) and has a minimum if $\partial E/\partial \mathrm{n} = 0$, i.e. if

$$\mathcal{J} S^2 \pi^2/2\mathrm{n}^2 = \tfrac{3}{2} D_2 S^2$$

or

$$\mathrm{n} = (\mathcal{J} \pi^2/3 D_2)^{1/2} \tag{10.84}$$

$(\mathcal{J}/D_2)$ can be roughly regarded as exchange field/anisotropy field. For Fe and Ni n is typically of the order of 1000 though it varies with the direction of the domain walls and the relative angle of the magnetisation.

10.12 Applications of magnetic materials (D)

Magnetic materials have found a variety of applications in microwave devices, servo-control systems and data storage, quite apart from the more familiar applications in transformers and as permanent magnets. We discuss a number of these below.

10.12.1 Applications of paramagnetic resonance (D)

Most of the applications of paramagnetic or electron spin resonance are in the scientific study of materials and E.S.R. equipment, usually working in the microwave region at 10 and 35 GHz, is commercially available. Paramagnetic impurities deliberately introduced into a material can 'probe' the crystal field and its symmetry since the field splittings can be measured (§§ 10.2 and 10.3). The invention of the MASER (acronym for Microwave Amplification by Stimulated Emission of Radiation) which preceded the Laser by several years (see Chapter 6, §§ 6.2.2 and 6.7.6), is of great historic interest. Population inversion in masers is normally obtained by pumping with a local microwave source across the outer pair of a three-level system, as explained in § 6.7.6. One of the first, and still the most popular material, was ruby. We have seen (§ 5.6.2) that the ground state of the paramagnetic impurity (Cr^{3+}) in ruby is split by the crystal field. Both levels are further split on the application of a magnetic field, the splitting being a function of orientation. Four levels result, any three of which may be used, and the level separations tuned to the desired frequency by varying **B**. The pump power required to achieve population inversion naturally depends on τ_1 and τ_2 (cf. §§ 6.7 and 10.3) and refrigeration to 4 K normally used. In fact the maser has been largely superseded as a low-noise microwave amplifier by parametric amplifiers using variable capacitance diodes (see § 9.2.2) except when extremely low-noise performance is needed. A long length of waveguide on the amplifier input (e.g. between maser and receiving aerial) can, however, introduce so much thermal noise that the inherently low-noise performance of the maser is lost.

It is of interest to note that amplification by the inversion of population in spin split states of paramagnetic impurities is not limited to photons. We have seen that the states can be coupled to lattice phonons through the crystal field. The amplification of 'microwave' phonons (i.e. phonons with frequencies in the GHz range) can and has been obtained in maser type systems. This development may prove important in the growing field of microwave acoustics (see § 8.4).

10.12.2 The applications of Faraday rotation (D)

Faraday rotation is not, of course, limited to magnetic materials, as we have seen in Chapter 7. The similarity between a spiralling electron and a precessing spin will be evident on comparing equations (10.15) with equations (7.15) and (7.16) and noting that the interaction is with the magnetic, rather than the electric, component of the radiation field. Due, however, to their high magnetic permeability ferrites show much larger Faraday rotations per unit field than semiconductors. Being insulators they are relatively transparent at microwave and some infrared frequencies, though they show phonon absorption over an extended range in the infrared. Thus, though the devices discussed in this section can be made from any material showing Faraday rotation, ferrites are, when applicable, the most effective.

In both optics and microwave engineering it is valuable to have non-reciprocal devices. Faraday rotation provides this property. Non-reciprocity, perhaps surprisingly, arises from the fact that the direction of rotation, clockwise (+ve) or anticlockwise (−ve) of the plane of polarisation is *independent* of the direction of propagation of the radiation, provided only that it is parallel to the magnetic field (as it must be in the Faraday orientation, by definition, see § 7.2.1). The simplest device to illustrate this property is a microwave isolator.

Isolators. Consider two sections of rectangular waveguide rotated with respect to each other by 45° around the propagation axis and separated by a length of circular guide containing a ferrite rod. The arrangement is illustrated in figure 10.48. A longitudinal field applied to the ferrite provides 45° of Faraday rotation in, say, the positive clockwise direction. If the rectangular guide is being used in its lowest order mode (as is usual) the electric vector must be perpendicular to the broad face of the guide. The circular section, by symmetry, has no preferred polarisation direction.

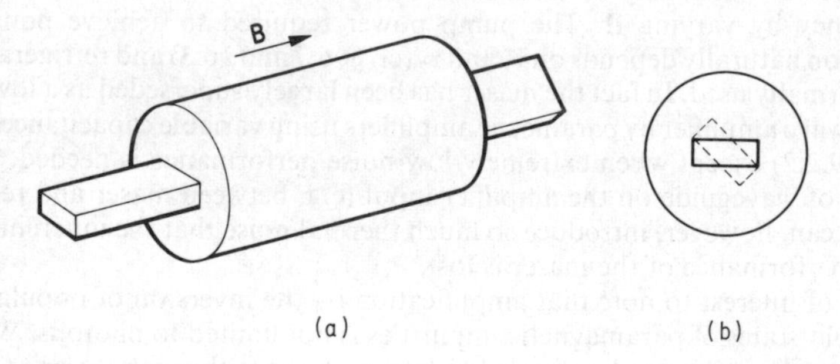

Figure 10.48 An isolator (a) in perspective, (b) end view

Operation of the device is then as follows. Radiation propagating in one direction (into the paper of figure 10.48(b)) is rotated clockwise by 45° and emerges with the correct polarisation to propagate freely in the second section of rectangular guide. Radiation propagating in the opposite direction, however, starts with its electron vector at +45° and the Faraday rotator adds a further rotation of +45° in the *same* direction. It emerges orthogonally polarised with respect to the original input guide and cannot propagate. At microwave frequencies an isolation of 20 dB (i.e. a forward to reverse transmission ratio of 100) can be achieved quite readily. The circuit symbol for an isolator is shown in figure 10.50(a).

The same result can be obtained in an optical system without guides by simply using two polarisers set at 45° with respect to one another and separated by a 45° Faraday rotator. It is, however, usual to angle one polariser with respect to the propagation axis so that rejected radiation, reflected by a polariser, is not simply reflected back down the propagation axis. The arrangement is illustrated in figure 10.49.

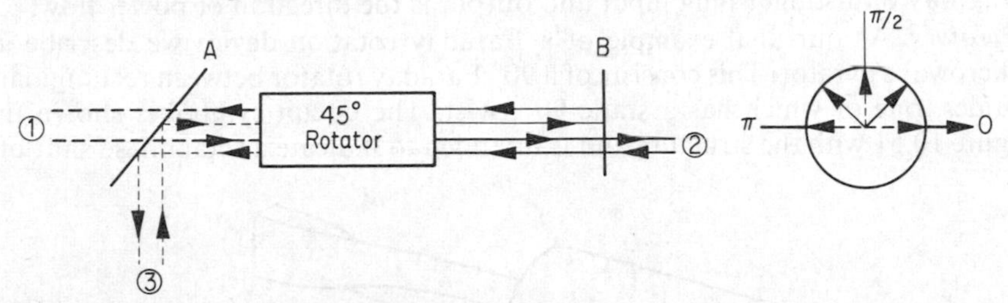

Figure 10.49 A circulator (see text). The directions of F for the various rays along the 1–2 axis are also shown

Circulators. Figure 10.49 also describes the optical equivalent of another important microwave device, namely a three-port circulator. Its operation and properties can be deduced by simply following a beam of given polarisation through the structure. We define, for convenience, a beam at port 1 to have a polarisation angle of 0°. Light with this polarisation entering at port 1 is transmitted by polariser A, rotated by 45°, transmitted by polariser B and emerges at port 2 with a polarisation angle of +45°. Light entering port 2 with an initial polarisation of +45° is transmitted by polariser B, rotated by a further 45° to +90° and hence reflected by polariser A to emerge at port 3. Finally light entering at port 3 with polarisation +90° is reflected at A, rotated to 135° and therefore reflected at B, rotated a further 45° to return to polariser A with a total polarisation angle of 180°. To polariser A this is indistinguishable from 0° so the radiation is transmitted to emerge at port 1. These properties are neatly summarised by the circuit symbol for a circulator shown in figure 10.50(b) which indicates that power entering at port 1 leaves at 2, power in at 2 leaves at 3 and power in at 3 leaves at 1.

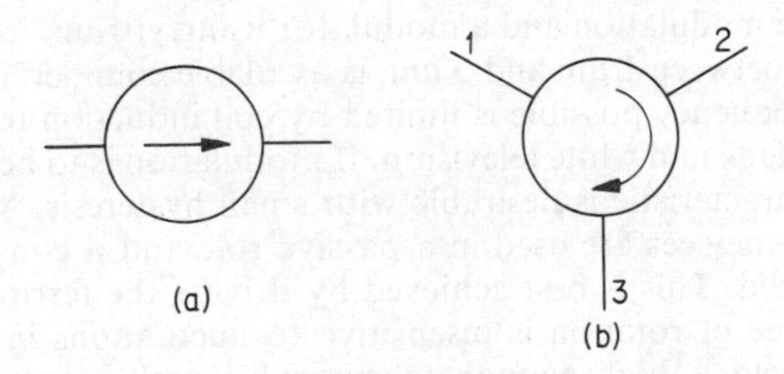

Figure 10.50 Circuit symbol for circulator

Circulators are used, for example, when a single aerial is to be employed for transmission and reception. The transmitter can be connected to port 2, the aerial to port 3 and the receiver to port 1. They are used in E.S.R. equipment when the sample under study replaces the aerial at the end of a waveguide run connected to port 3. A circulator (or gyrator, see below) is essential when a two-contact device, such as a parametric diode, is used as an amplifier since the only way of distinguishing input and output is the direction of power flow.

Gyrators. As our final example of a Faraday rotation device we describe a microwave gyrator. This consists of a 90° Faraday rotator between rectangular guides, one of which has a static 90° twist. The circuit symbol is shown in figure 10.51 with the structure and is intended to indicate a total phase shift of

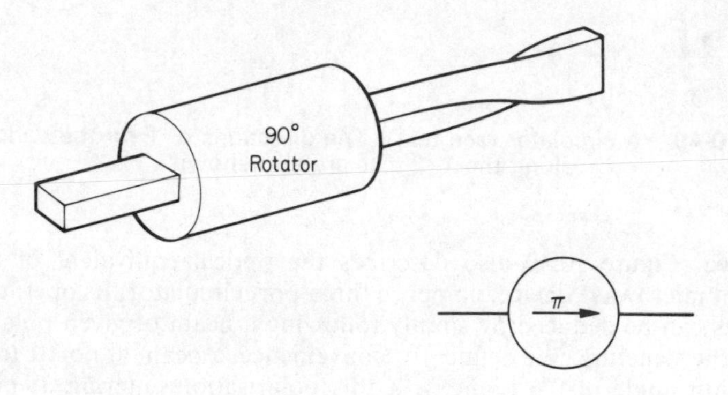

Figure 10.51 Diagram of microwave gyrator and circuit symbol

ϕ in one direction $(\phi + \pi)$ in the other. The mode of operation should be self-evident. The twist is reciprocal and produces a rotation of ±90° depending on the propagation direction. To this is added +90° in both directions due to the Faraday rotator. The result is either 0° or 180°. The optical equivalent of the microwave gyrator requires a birefringent crystal to provide the static 90° rotation.

Microwave versions of all the above devices are available commercially and optical equivalents have been made. It is, of course, possible to modulate the applied magnetic field and hence the transmission. Thus the isolator can provide amplitude modulation and a modulator using yttrium iron garnet, which is transparent between 1 μm and 5 μm, is available commercially. The highest modulation frequency possible is limited by coil induction to a few MHz but sufficient for black and white television. If modulation is to be applied a nearly linear B/H characteristic is desirable with small hysteresis. More commonly, however, these devices are used in a passive role and a constant rotation of 45 or 90° desired. This is best achieved by driving the ferrite into saturation when the degree of rotation is insensitive to fluctuations in applied field. If the coercive field is high enough (comparable with saturation) permanent magnetisation can be used.

10.12.3 Permanent magnets and transformers (D)

Permanent magnets are made from materials with high coercive fields to prevent changes in the field produced. We have seen that such magnetically hard materials should be imperfect to pin the domains. The most widely used materials are precipitates of small magnetic particles which are oriented so that the easy axis of magnetisation in the various particles is more or less aligned. Alnico steels which have a magnetic precipitate of Fe and Co with a non-magnetic Ni–Al component have coercive fields up to 10^{-1} tesla and materials with larger coercive fields are known. The development of new materials, which proceeds along semi-empirical lines, is indicated by figure 10.52.

Magnetically soft materials with small coercive fields are essential for use in transformers, motors and generators. Delay in the magnetisation due to hysteresis causes power loss and the total cost of such loss in the electric power

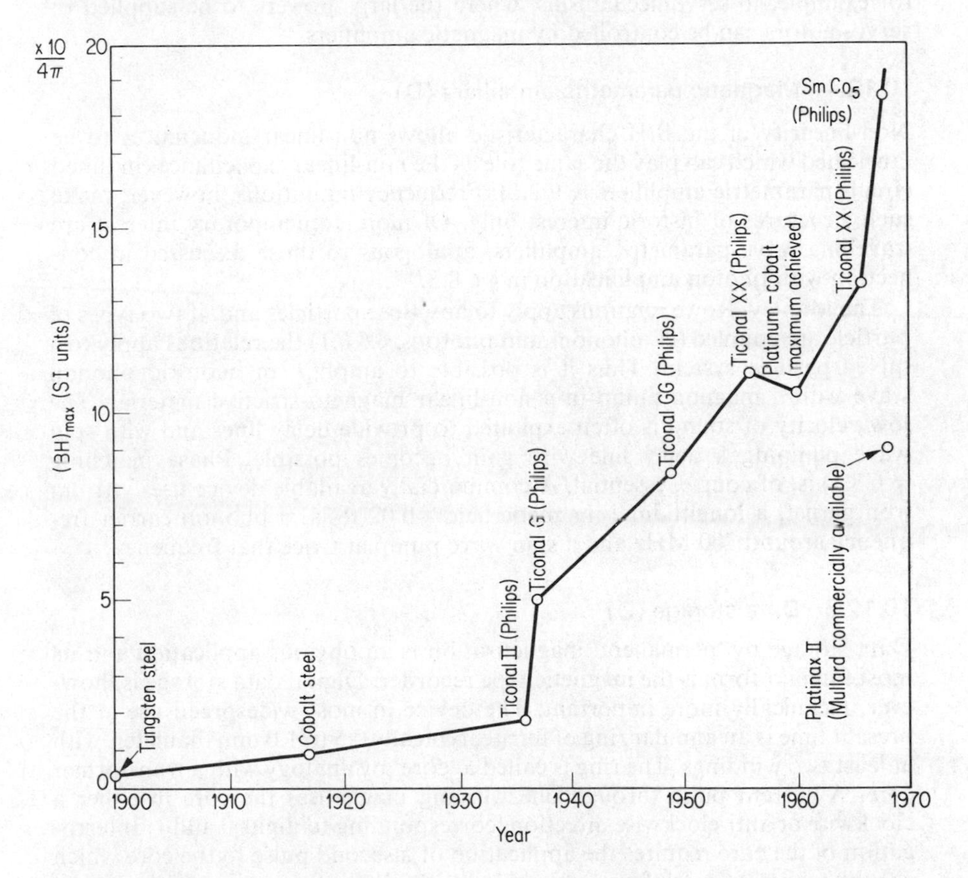

Figure 10.52 Chronological development of permanent magnet materials. The ordinate is BH, effectively the product of saturation magnetisation and coercive field, which is a convenient figure of merit. (After C. L. Boltz, *Phys. Bull.*, **20**, 189 (1969).)

generating industry is very large indeed. Again, there is steady development of new materials. Fe with 3% Si impurity, suitably treated, can show a coercive field as small as 10^{-4} tesla. In supermalloy, an alloy consisting of 79% Ni, 15.7% Fe, 5% Mo and 0.3% Mn the coercive field is a hundred times smaller. The more widespread use of the latter material is limited by cost.

If the output of a transformer is to faithfully reproduce the input without distortion the B/H characteristic must be used over a linear portion only and a high saturation field is desirable. Non-linearity of B/H characteristics is, however, exploited in magnetic amplifiers which are essentially transformers with an additional (3rd) winding to bias the transformer core into various regions of the B/H curve. If the additional winding has sufficient turns a comparatively small current can control a substantial a.c. power transfer between the remaining windings. Hence the name 'magnetic amplifiers'. Applications occur, for example, in servomechanisms, where the large powers to be supplied to servo-motors can be controlled by magnetic amplifiers.

10.12.4 Magnetic parametric amplifiers (D)

Non-linearity of the B/H characteristic allows non-linear inductances to be fabricated which can play the same role as the non-linear capacitances in tuned circuit parametric amplifiers (§ 9.2.2). Frequency limitations, however, make such structures of historic interest only. Of more contemporary interest are travelling wave parametric amplifiers, analogous to those discussed in connection with photon amplification in § 6.8.3.

The Manley–Rowe relations apply to any Bose particles and, if two types of particle are coupled (cf. phonons and photons, § 6.6.1) the relations apply to a mixed particle system. Thus it is possible to amplify an acoustic phonon wave with a magnon pump in a non-linear magneto-strictive material. The low velocity of sound is often exploited to provide delay lines and with spin wave pumping a delay line with gain becomes possible. Phase matching (§ 6.8.1) is, of course, essential. A commercially available device uses yttrium iron garnet, a longitudinal magnetic field ~0.02 tesla, a phonon carrier frequency around 700 MHz and a spin wave pump at twice that frequency.

10.12.5 Data storage (D)

Data storage by 'permanent' magnetisation is an obvious application and its most familiar form is the magnetic tape recorder. Digital data storage is, however, technically more important. The device in most widespread use at the present time is an annular ring of ferrite, typically 0.5 to 1.0 mm diameter, with at least two windings. The ring is called a 'core' by analogy with a transformer core. A current pulse through one winding magnetises the core in either a clockwise or anti-clockwise direction, corresponding to digits 0 and 1. Interrogation of the core requires the application of a second pulse to the core which either changes its state of magnetisation or not, depending on its previous state. If there is a change of magnetisation a voltage proportional to dB/dt appears across the second winding. Read out of data by this method destroys the information stored which, if required again, must be re-cycled and restored.

In practice ferrite cores are used in two-dimensional arrays (e.g. 100 × 100) and planes stacked in the third dimension for use in a computer store. Multi-turn windings are excluded by cost considerations. Writing in of data to a selected core in the two-dimensional matrix can be achieved by coincident current pulses in two 'write' wires which thread the core (see figure 10.53). It is clearly important that a current 2I be sufficient to change the state of magnetisation of the core but a current I be insufficient. The coercive field, therefore, must be well defined. Read out similarly requires a two-dimensional matrix of wires.

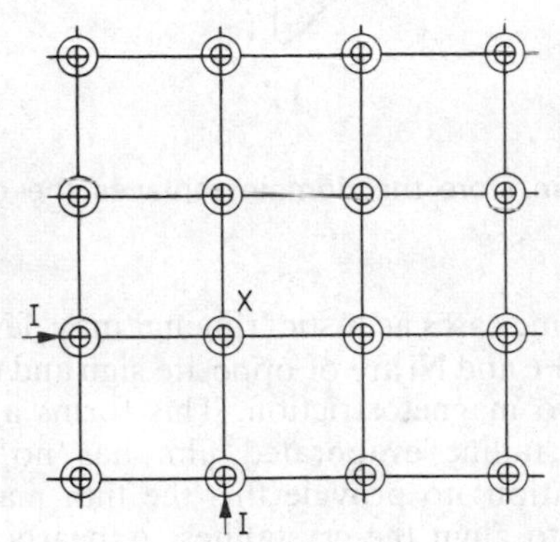

Figure 10.53 Diagram of a planar array of ferrite cores to make a store. Simultaneous current along the wires indicated will magnetise the core at their intersection X

Sintered ferrite rings with a nearly square B/H hysteresis loop are used in this application. H_c and B_c are then sharply defined. A large coercive field, B_c, is obviously desirable to give a large output voltage on switching the state of the core during interrogation. The high permeability of ferrites ensures that H_c is small so that the drive currents required (typically a few hundred milliamps) are within the capability of transistors. The time taken to reverse the magnetisation is approximately inversely proportional to drive current.

Macroscopic cores and conventional wiring is inconsistent with planar microelectronic circuitry produced by evaporation and diffusion (§ 5.5) and during recent years considerable effort has been devoted to the development of thin magnetic film stores. Some hybrid film/ferrite systems have been examined. The structure of the simplest film store follows that of a core store. A thin magnetic film is produced by evaporation on to a suitable substrate (which may be conducting) and a matrix of wiring evaporated on top. Selection of a particular *area* of the film, within which a domain is formed, may be by the coincident current method (see figure 10.54). Read out is by domain reversal, as with a core store.

The properties required of a magnetic film are somewhat different from those of a ferrite core. Firstly any magnetostriction (change in sample dimensions with magnetisation) is undesirable as this generates strain between film

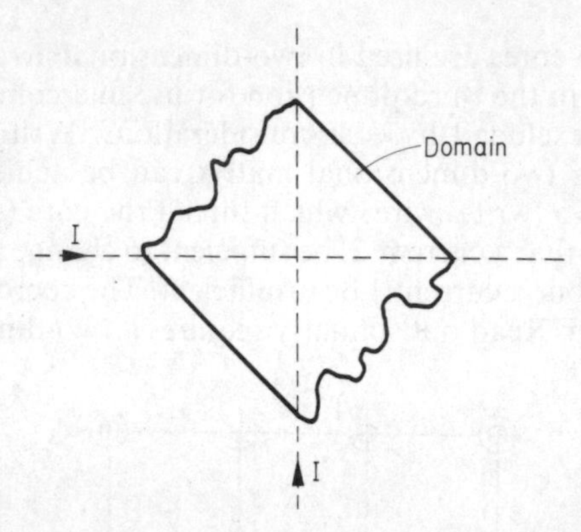

Figure 10.54 In a film store the domain replaces the core in figure 10.53

and substrate; in extreme cases acoustic 'ringing' may develop. The magnetostrictive coefficients of Fe and Ni are of opposite sign and the alloy 'permalloy' (20:80 Ni Fe) has zero magnetostriction. This forms a suitable evaporant. Secondly, a microcrystalline evaporated film has no naturally preferred direction of magnetisation: to provide this the film may be deposited in a strong magnetic field to align the crystallites. A nearly square B/H loop is obviously still desirable. Since the magnetised area under a cross over point is bounded by a domain wall (which tends to have the shape shown in figure 10.54 to minimise the energy) reversal of the magnetisation may take place by both wall movement and domain rotation. For most geometries the speed of both processes (at a given drive current) are comparable. Typical switching times are in the range 10–100 ns.

PROBLEMS

(Some answers, where appropriate, are given on page 483.)

10.1 A paramagnetic crystal contains $\mathscr{N}$ atoms of spin $\frac{1}{2}$ and $g = 2$. Show that

(i) the magnetisation $\mathscr{M}/\mathscr{N} = \mu_B \tanh(\Delta/2kT)$;

(ii) the specific heat $C/\mathscr{N} = (\Delta/2kT)^2 \operatorname{sech}^2(\Delta/2kT)$;

(iii) the entropy $S/\mathscr{N} = \dfrac{\Delta}{2T}\tanh(\Delta/2kT) + k\ln(2)\cosh(\Delta/2kT)$.

where $\Delta = 2B\mu_B$. How do these functions vary with high and low values of T?

Note that under adiabatic conditions ($S =$ constant) a change in B produces a change in T such that

$$(B_1/B_2) = (T_1/T_2)$$

10.2 Using the results of problem 10.1 above, or by counting states, show that the entropy can be written

$$S/\mathscr{N}k = (1+\nu)\ln(1+\nu) + (1-\nu)\ln(1-\nu)$$

where

$$\nu = \mathscr{M}/\mathscr{N}\mu_B$$

10.3 If the interaction energy between spins is written as $\frac{1}{2}\mathscr{J}v^2$ (using 10.2) show that the condition that the Gibbs free energy, $U - TS$, be a minimum leads to the results of the molecular field theory.

10.4 Estimate the diamagnetic susceptibility of a gas of hydrogen atoms in their lowest state.

10.5 In an antiferromagnet there are two sub-lattices of spins 1 and 2. If the effective molecular field on the lattices is written

$$\mathrm{B}_{\mathrm{eff}}^{(1)} = \mathrm{B} - \Gamma \mathrm{M}^{(1)} - \Gamma' \mathrm{M}^{(2)}$$

and

$$\mathrm{B}_{\mathrm{eff}}^{(2)} = \mathrm{B} - \Gamma \mathrm{M}^{(2)} - \Gamma' \mathrm{M}^{(1)}$$

show that the Néel temperature is given by

$$T_N = (\Gamma + \Gamma')\mathrm{C}$$

and the Curie Wiess temperature by

$$\Theta = (\Gamma - \Gamma')\mathrm{C}$$

where C is Curie's constant.

10.6 A semiconductor has 3×10^{16} electrons per cm^3 in a single conduction band characterised by an effective mass $m^* = m/10$. Estimate the temperature below which the susceptibility is independent of temperature. How does the susceptibility vary with T above this temperature?

10.7 In a uniaxial material, domains of width a and length l form as in Figure 10.47(b). The energy of the walls is given by (10.82) and (10.84) and, in a fixed volume of sample, varies as l/a. The magnetic energy is approximately $M^2 a/l$. Estimate the average wall spacing in equilibrium in cobalt if $\mathscr{J} \sim 10^{-2}$ eV, $D \sim 10^{-4}$ eV and $M \sim 1$ tesla.

10.8 When a small spherical magnetic particle of radius r is saturated it has a magnetic energy of about $\mathrm{M}^2 r^3$. This is effectively reduced to zero if a domain wall forms in an equatorial plane. Find the radius below which cobalt particles will be single domains.

10.9 Derive equation (4.57) using the uncertainty principle.

Answers to problems

Chapter 1

1.3 Reciprocal lattice vectors have lengths $8\pi\sqrt{3}/9$ and $2\pi/\sqrt{3}$, perpendicular to AB and AC.
1.4 Direct lattice has vectors $a\sqrt{3}$ at $\pi/3$ to each other. Reciprocal lattice is triangular with basic vectors $4\pi/3a$.
1.6 If all ν_i are odd, only Cu reflections; two odd and one even, only O reflections; two even and one odd, no reflection; all even, mixed.
1.7 (i) $\sin^{-1}(0.3) - \sin^{-1}(0.2) \approx 6°$; (ii) 90°.
1.8 Back scattering from (2,0,0) reflection and four symmetrical (1,1,1) reflections of neutrons with $\lambda = 0.4$ nm and 0.267 nm, respectively.
1.9 So that (1,1,1) reflection is allowed for this energy, i.e. a (1,1,1) crystal direction makes an angle of about 50° with the initial beam.
1.10 (i) $V(\partial P/\partial V) = nU_0/9V$ where U_0 is the minimum value of U. (ii) Add a term $-FR$ to $U(R)$ and show that U has no minimum if F is large enough.

Chapter 2

2.1 $$\frac{\mathscr{L}}{2\pi}\left[\frac{\mathrm{d}E}{nc^{1/n}E^{1-1/n}}\right], \qquad \frac{\mathscr{A}}{2\pi}\left[\frac{\mathrm{d}E}{nc^{2/n}E^{1-2/n}}\right]$$

2.4 −3A, −A, A, 3A.

Chapter 3

3.1 Use $\mathrm{d}\omega/\mathrm{d}q$ and the argument leading to (3.9).
3.3 $\omega(q)$ is practically independent of q_z. Graphite therefore behaves as if it were two-dimensional.

3.5 $$\omega^2(q) = \frac{\Lambda_1 + \Lambda_2}{M} \pm \frac{1}{M}[(\Lambda_1 + \Lambda_2)^2 - 4\Lambda_1\Lambda_2 \sin^2 qa]^{1/2}.$$

There are two branches because there are two atoms per unit cell.
3.6 $9\mathcal{N}\hbar\omega_m/8$.

Chapter 4

4.1 $\omega_p = 1.4 \times 10^{14}$ radians per second.
4.2 (i) $\frac{2\sqrt{2}}{3}\left(\frac{\pi}{a^3}\right)$; (ii) $\sqrt{3}(\pi/a^3)$.

4.4 $m^* = \hbar^2/8E_1a^2$.

Chapter 5

5.2 1.6×10^{11} cm^{-2}.
5.3 Both situations can be obtained with pairs of edge dislocations.
5.4 2×10^{-4} eV at B = 0; about 1×10^{-3} eV at B = 1 tesla.
5.5 (i) 83 K (use (5.12)). (ii) At impurity level, since these half occupied. (iii) About 2×10^{16} cm^{-3} (exhaustion range). (iv) About 487 K (use (5.11) and (5.8)).
5.7 Use equations (5.35) to (5.44) at low temperatures (i.e. $\rho^2 = \frac{1}{2}\hbar/M\omega$). Then: $\hbar\omega_1 = 0.022$ eV; $\hbar\omega_2 = 0.018$ eV; $\frac{1}{2}MQ^2\omega_1^2 = 0.617$ eV; $\frac{1}{2}MQ^2\omega_2^2 = 0.418$ eV; $n_2 \sim 23$.

Chapter 6

6.2 30 MW.
6.3 Decrease.
6.4 $E_g = 2.16$ eV, $C = 0.096$ eV. Two photon absorption or Raman scattering.
6.7 A trap controlled decay, if the traps are distributed in energy.
6.8 $|E_1 - E_2|$ = energy of phonon at zone edge. High density of phonon states and wave vector conservation requirement relaxed as localised state.
6.10 Phase matching condition implies equal refractive indices and hence lossless condition gives $F_0^2(\omega) = F^2(2\omega) + F^2(\omega)$. Substitute and integrate.
6.11 (1,0,0) GaP conduction band minima below central (0,0,0) GaAs conduction band minimum for GaP concentration greater than 0.6. Therefore indirect gap alloy with lowered transition probability.

Chapter 7

7.4 Electron density ~2.6×10^{22} cm^{-3}.

7.6 $$N = \frac{\varepsilon(\infty)m^*}{e^2}\left[\frac{\omega^2(\omega_L^2 - \omega_T^2)}{\omega_T^2 - \omega^2}\right].$$

7.7 The absence of a shear force in a free electron plasma ensures $\omega_T = 0$.

Chapter 8

8.1 3.7×10^{-4} m^3 C^{-1}.
8.2 $(\mu_e/\mu_h) \sim 80$ (the mobility ratio in InSb is unusually high).
8.3 About 110 cm^{-1}. $\omega\tau \sim 3.5$. 'Semiconductor' and 'metal' approximations for K not valid. Use equations (7.3) and (7.4).
8.6 None.
8.7 $(A - A_0)/A_0 = F^2/3F_0^2$. $F_0 \sim 9 \times 10^2$ V cm^{-1}.
8.9 Absorption by electrons in a simple band must be phonon assisted (see § 7.1). Phonons then take up some of photon momentum. Hence assumption not justified.
8.10 From Wiedemann–Franz Law, 4.4 W cm^{-1} deg^{-1}.
8.11 From figure 8.3. Debye temperature = 333 K. As ρ proportional to T at and above 273 K, ρ (333 K) = 1.86×10^{-8} ohm-m. $\omega_c\tau = eB\tau/m^* = R_H B/\rho$. Condition $\omega_c\tau = 1$ requires $\rho = 2.94 \times 10^{-2} \times \rho$ (333 K) and figure 8.3 then gives $T \sim 50$ K.
8.12 About 0.45×10^8 cm^{-1}.
8.13 $k_F \approx 1.1 \times 10^8$. $(k_F/d) \sim 0.85$ (cf. equation (4.21)).
8.15 Possible techniques include:
(a) improved detectors, e.g. the electronic bolometer of § 8.7.2;
(b) the use of interferometers and Fourier transform spectroscopy; or
(c) the use of microwave harmonic generators and variable T to sweep energy gap through fixed frequencies.

Chapter 9

9.3 At zero current electron density N related to energy reference (e.g. Fermi level) by N = constant exp $(-eV/kT)$ where V is electric potential so that grad V = F. Hence

$$\text{grad } N = -(eN/kT)\,\text{grad V} = -(eN/kT)\,\text{F}$$

and using (9.3) Einstein relation follows.
9.6 Place series of PN junctions of decreasing E_g optically in series but electrically in parallel. All junctions to be anti-reflection coated. Junctions of lower E_g utilise radiation transmitted through earlier layers.
9.7 Due to Lorentz force, electrons and holes either forced towards or out of high recombination velocity/generation rate surface. Since material is near intrinsic, carrier density decreased or increased when effective recombination rate enhanced or generation rate enhanced. Minimum conductivity when all minority carriers (holes) and equal number of electrons vanish. Conductivity then limited by excess donor density to $e\mu_c(10^{12}) = 6.25 \times 10^{-4}$ ohm cm. Possible application: magnetically controlled rectifier—polarity of rectification reverses on field reversal. Could replace slip rings or commutator in electrical machines.

9.8 Holes (minority carriers) generated by light can diffuse to and be captured by field at junction. On entering P^+ region holes become majority carriers, P^+ region becomes positively charged and junction forward biased. Holes must therefore be injected back into N region at an equal rate. But holes may be injected back anywhere along junction with equal probability: on average at the centre. Net result is motion of holes toward centre of structure and current flow leads to voltage observed.

Possible application: light position sensor for use with, say, star tracking servo-mechanism.

9.9 Use equations (9.3) with condition that total current is zero.

9.10 Effect of potential gradient along bar is to reverse bias P^+N junction. When potential applied to junction exceeds V/2 junction becomes forward biased and holes are injected which flow towards zero potential terminal. Injection of holes increases conductivity in injected half of bar, reduces potential difference across it and consequently increases effective forward bias on P^+N junction so that hole injection rate enhanced. Negative resistance results. Condition on $2L^2/V\mu_h$ ensures injected hole density uniform in injected half of bar—not strictly necessary condition for observation of negative resistance but simplifies description. Condition is, however, necessary for solution of problem 9.11 to be valid.

Chapter 10

10.4 Use equation (10.28). Susceptibility 3×10^{-3} SI units.

10.6 Temperature about 50 K. Above this temperature varies as $1/T$ (use equation (10.33)).

Subject index

Substance index

The properties of various solids are quoted throughout the book, primarily as examples. An index of the more important of such references follows.